Unmanned Aerial Vehicles Swarm for Protecting Smart Cities

Future Trends and Challenges

Oroos Arshi
Inam Ullah Khan
Keshav Kaushik
Nadeem Iqbal
Inam Ullah
Khadija Slimani

Apress®

Unmanned Aerial Vehicles Swarm for Protecting Smart Cities: Future Trends and Challenges

Oroos Arshi
Member IEEE,
Dehradun, India

Inam Ullah Khan
(Senior Member IEEE, Advisor Cisco Community Pakistan) Multimedia University,
Cyberjaya, Malaysia

Keshav Kaushik
Sharda University,
Greater Noida, India

Nadeem Iqbal
Abdul Wali Khan University,
Mardan, Pakistan

Inam Ullah
Department of Computer Engineering,
Gachon University,
Seongnam 13120, Republic of Korea

Khadija Slimani
Higher School of Computer Science,
Electronics and Automation (ESIEA),
Paris, France

ISBN-13 (pbk): 979-8-8688-1046-6
https://doi.org/10.1007/979-8-8688-1047-3

ISBN-13 (electronic): 979-8-8688-1047-3

Copyright © 2025 by The Editor(s) (if applicable) and The Author(s), under exclusive license to APress Media, LLC, part of Springer Nature

This work is subject to copyright. All rights are reserved by the Publisher, whether the whole or part of the material is concerned, specifically the rights of translation, reprinting, reuse of illustrations, recitation, broadcasting, reproduction on microfilms or in any other physical way, and transmission or information storage and retrieval, electronic adaptation, computer software, or by similar or dissimilar methodology now known or hereafter developed.

Trademarked names, logos, and images may appear in this book. Rather than use a trademark symbol with every occurrence of a trademarked name, logo, or image we use the names, logos, and images only in an editorial fashion and to the benefit of the trademark owner, with no intention of infringement of the trademark.

The use in this publication of trade names, trademarks, service marks, and similar terms, even if they are not identified as such, is not to be taken as an expression of opinion as to whether or not they are subject to proprietary rights.

While the advice and information in this book are believed to be true and accurate at the date of publication, neither the authors nor the editors nor the publisher can accept any legal responsibility for any errors or omissions that may be made. The publisher makes no warranty, express or implied, with respect to the material contained herein.

    Managing Director, Apress Media LLC: Welmoed Spahr
    Acquisitions Editor: Spandana Chatterjee
    Development Editor: Laura Berendson
    Editorial Assistant: Jessica Vakili

Cover designed by eStudioCalamar

Distributed to the book trade worldwide by Springer Science+Business Media New York, 1 New York Plaza, Suite 4600, New York, NY 10004-1562, USA. Phone 1-800-SPRINGER, fax (201) 348-4505, e-mail orders-ny@springer-sbm.com, or visit www.springeronline.com. Apress Media, LLC is a California LLC and the sole member (owner) is Springer Science + Business Media Finance Inc (SSBM Finance Inc). SSBM Finance Inc is a **Delaware** corporation.

For information on translations, please e-mail booktranslations@springernature.com; for reprint, paperback, or audio rights, please e-mail bookpermissions@springernature.com.

Apress titles may be purchased in bulk for academic, corporate, or promotional use. eBook versions and licenses are also available for most titles. For more information, reference our Print and eBook Bulk Sales web page at http://www.apress.com/bulk-sales.

Any source code or other supplementary material referenced by the author in this book is available to readers on GitHub. For more detailed information, please visit https://www.apress.com/gp/services/source-code.

If disposing of this product, please recycle the paper

I dedicate this book to my beloved parents, Mr. Anzar Ahmad Tauseef and Mrs. Mumtaz Arshi, my brother Mr. Aadil Bari, and my sister Dr. Zarrin Orooj. Without their unwavering support and encouragement, I could not have achieved this. I also dedicate it to my grandparents in their honor and to my friends who have helped me throughout this journey.

—Oroos Arshi

I dedicate this book to my parents and sister.

—Dr. Inam Ullah Khan

Table of Contents

About the Editors xxvii

About the Technical Reviewers xxxvii

Acknowledgments xliii

Introduction xlv

Chapter 1: Introduction to Unmanned Aerial Vehicles Swarm and Smart Cities 1

Introduction 2
    Application of Drones in UAVs 3
    UAV Applications for Smart Cities and ITS 4
    Interconnection Between UAV Swarms and Smart Cities 5
Fundamentals of UAV Swarms 6
Definition and Characteristics 7
    Key Characteristics: Autonomy, Cooperation, and Scalability 7
    Technological Components 9
    Software: Control Algorithms, Coordination Protocols, and AI Integration 10
    Key Technologies Enabling UAV Swarms 11
    Artificial Intelligence and Machine Learning 12
    Machine Learning Applications in Swarm Behavior and Optimization 14
Applications of UAV Swarms in Smart Cities 16
    Surveillance and Security 16
    Monitoring Public Spaces 17
    Rapid Response to Emergencies 17

TABLE OF CONTENTS

- Infrastructure Inspection and Maintenance 18
- Inspection of Bridges, Roads, and Buildings 18
- Urban Mobility 19
- Challenges and Solutions 20
  - Technical Challenges 20
  - Regulatory and Ethical Issues 21
  - Proposed Solutions 22
- Case Studies and Real-World Examples 23
  - Global Perspectives 25
- Future Trends and Directions 26
  - Technological Advancements 27
  - Emerging Applications 28
  - Predicted Impact 31
- Conclusion 32
  - Summary of Key Points 33
- References 34

Chapter 2: Machine Learning Applications in Unmanned Aerial Vehicle Swarms 45

- Introduction 46
- History of Swarm 47
- Swarms Machine Learning and Deep Learning 48
  - Advantages of Using Multiple Drones in Coordinated Missions 49
  - Mention of the Potential of Machine Learning in Enhancing UAV Swarm Capabilities 50
- Fundamentals of UAV Swarms 53
  - Definition and Key Characteristics of UAV Swarms 53
  - Benefits of Using UAV Swarms for Various Applications 55
  - Challenges and Limitations of Conventional UAV Swarm Technologies 56

TABLE OF CONTENTS

An Overview of Machine Learning 60
    Explanation of Machine Learning Concepts and Techniques 61
Introduction to Neural Networks and Deep Learning for UAV Swarm Applications 64
    Deep Learning for UAV Swarm Applications 64
    Deep Learning Challenges in UAV Swarm Applications 65
Data Collection and Analysis 66
Autonomous Navigation and Control 66
Swarm Intelligence and Coordination 67
Adaptive Mission Planning 67
    Predictive Analytics and Optimization Algorithms 68
Communication and Networking 68
    Importance of Communication and Networking in UAV Swarm Operations 69
    Optimization of Communication Protocols and Network Efficiency with Machine Learning 69
    Machine Learning Algorithms for Distributed Computing in UAV Swarms 69
Security and Threat Detection 70
    Security Measures in UAV Swarms 70
    Threat Detection Techniques 72
    Integration and Response 73
    Anomaly Detection and Intrusion Detection Algorithms 74
Real-Time Response to Security Threats Using Machine Learning 75
Persistent Challenges and Prospective Advancements to Pursue 76
    Battery Recharging and Scheduling 76
    Swarm-Level Active Fault Detection 77
    Controlling and Supervising Swarms of MAVs 77
Conclusion 78
    Summary of Key Points Discussed in the Chapter 78
References 81

TABLE OF CONTENTS

Chapter 3: Real-World Deployments in Unmanned Aerial Vehicle Swarms 91

Introdumction..92
    The Significance of UAV Swarms in Modern Applications.............................93
    Real-World Deployments Advance Technology..94
    Aims and Contributions 96
Theoretical Background and Literature Review...97
    Fundamental Concepts of UAV Swarms 97
    Historical Milestones in UAV Swarm Development..99
    Theoretical Frameworks Guiding Real-World Deployments 100
Methodological Approach 100
    Research Design Rationale in Drones 100
    Data Sources and Collection Techniques for Drones...................................101
    Case Study Methodology for Drones in Buildings...101
    Basic Drone Swarm Error Analysis Tools and Metrics..................................103
Technical Foundations of UAV Swarms...103
    Communication Networks and Protocols 104
    Advanced Navigation and Sensing 106
    Swarm Intelligence and Coordination Mechanisms 106
Real-World Deployment Case Studies 107
    Urban Services and Infrastructure Management..108
    Security and Surveillance Operations 108
Challenges and Implications 109
    Ethical Dimensions and Public Safety 110
Economic, Environmental, and Social Considerations 111
    Cost Analysis and ROI of UAV Swarm Deployments 113

TABLE OF CONTENTS

Future Horizons and Technological Innovations 114
    Integration with Advanced Technologies (AI, IoT, and 5G) 115
Best Practices and Strategic Recommendations 117
    Guidelines for Effective UAV Swarm Deployment 118
Conclusion 118
References 120

Chapter 4: Machine Learning Applications in UAV Swarms 127
Introduction 128
Literature Review 128
    Supervised Learning for Object Detection and Classification 129
    Reinforcement Learning for Search and Rescue 131
    Advanced Machine Learning Techniques 132
Machine Learning Landscape for UAV Swarms 132
    Supervised Learning 133
    Unsupervised Learning 135
    Reinforcement Learning (RL) 137
Deep Learning for Enhanced Capabilities 139
Applications of Machine Learning in Smart City Protection 140
    Surveillance and Monitoring 140
    Search and Rescue 141
    Traffic Management 141
Challenges and Future Directions 142
Discussion 145
    Challenges and Future Directions 147
Conclusion 149
References 149

TABLE OF CONTENTS

Chapter 5: 5G Integration for Enhanced UAV Swarm Connectivity 171

Introduction 171
The Evolution of 5G Technology 173
    Historical Context 173
    Key Features of 5G 175
5G Deployment and Infrastructure: Key Components and Their Roles in UAV Swarm Operations 176
    Small Cell Networks 176
    Distributed Antenna Systems (DAS) 177
    Cloud and Edge Computing 177
Integration for UAV Swarm Operations 178
UAV Swarms: Current State and Challenges 179
    Coordination Algorithms 179
Coordinated Search Patterns 180
    Real-Time Communication and Coordination 181
    Real-Time Data Capturing and Processing 181
    Applications in Ecological Studies 181
Agriculture 182
Technical Challenges 183
    Connectivity and Communication 183
    Solutions and Approaches 183
    Coordination and Control 184
    Solutions and Approaches 184
Autonomy 184
    Addressing the Challenges 185
5G Enhancements for UAV Swarm Connectivity 186
    Increased Data Transmission Speed 186

TABLE OF CONTENTS

    Improved Reliability 186
    Low Latency 186
    Massive Device Connectivity 187
    Network Slicing 187
    Edge Computing Integration 187
Enhanced Data Transmission 188
    Massive IoT Integration for UAV Swarms 189
Applications of 5G-Enhanced UAV Swarms 193
Technological Challenges and Solutions 195
Network Interference and Congestion 197
    Cybersecurity 197
    Energy Consumption 198
Future Prospects 198
    Advancements in AI Capabilities 198
    Improvements in Swarm Intelligence 199
    Expansion of Applications 199
Conclusion 200
References 202

Chapter 6: Augmented Reality (AR) and Virtual Reality (VR) for UAV Swarm Visualization 207

Introduction 207
    Challenges in UAV Swarm Visualization 209
    The Role of AR and VR in UAV Swarm Visualization 210
Source of Data 211
    Telemetry and Sensor Data 212
    Environmental Information 212
    Real-Time Video Feeds 213

xi

TABLE OF CONTENTS

Communication Networks 214
Literature Review 214
Applications of AR and VR in UAV Swarm Visualization 214
Challenges and Limitations 216
Advancements and Future Directions 217
Experimental Setup 218
Hardware Infrastructure 219
Software Platforms 220
Experimental Methodologies 221
Validation and Evaluation 223
Legal Considerations 224
Regulatory Frameworks 225
Privacy Concerns 225
Liability Issues 225
Intellectual Property Rights 226
Ethical Considerations 226
Limitation and Conclusion 227
Limitations of AR and VR in UAV Swarm Visualization 227
Conclusion 228
Future Directions and Opportunities 229
Future Scope 230
Advancements in Hardware and Software 230
Emerging Applications and Use Cases 231
Ethical and Societal Implications 231
Research and Innovation Ecosystem 232
References 233

Chapter 7: Surveillance and Monitoring in Smart Cities with UAV and Machine Learning Integration 241

Introduction 242

    Importance of Surveillance and Monitoring 244

    Role of UAV Swarms in Smart Cities 245

Current Technologies in Smart City Surveillance 247

Unmanned Aerial Vehicles (UAVs) in Surveillance........................249

    Types of UAVs Used in Surveillance..250

    Technological Advancements in UAV Surveillance 252

UAV Swarm Technology...254

    Definition and Concept 254

    Communication and Coordination Among UAVs 255

    Algorithms and Protocols for Swarm Intelligence 256

Applications of UAV Swarms in Smart Cities 257

    Entertainment...258

    Delivery 259

    Remote Sensing 259

    Surveillance..260

    Traffic Management and Monitoring 261

    Agriculture Management and Environmental Monitoring..........262

    Public Safety and Security 262

    Disaster Response and Recovery 263

Technical Challenges and Solutions..263

    Speed and Security 264

    Protection of Communication Channels 264

    Inventory Resource Management System................................265

    Global Mediums for Connecting and Communicating 265

    Interservice Operational Capabilities...266

TABLE OF CONTENTS

    Limited Storage and Processing Power 266
    Self-Controller Security Governance System 266
    Securely Sending Data to End Users 267
    Line of Sight Blockade 267
    UAV Surveillance Network 267
    UAV Networks for Edge Computing 268
    Limited Battery Issue 268
Regulatory and Ethical Considerations 271
Future Trends in UAV Swarm Surveillance 273
    Advancements in AI and Machine Learning 274
    Integration with Other Emerging Technologies 274
    Potential Innovations in UAV Design and Capabilities 275
    Autonomous Behavior and Learning 275
    Improved Sensor Technologies 275
    Multidomain Surveillance 276
    Collaborative Decision-Making 276
Case Studies and Real-World Examples 276
Conclusion 278
References 279

Chapter 8: Artificial Intelligence-Enhanced IoT and UAV Integration in Ad Hoc Networks for Smart Cities 285

    Introduction 285
    Literature Study 288
    AI for Ad Hoc IoT Networks 293
        Deep Learning for IoT in Ad Hoc Networks 294
        Federated Learning for Distributed IoT Devices in Ad Hoc Topology 295
        Reinforcement Learning for Dynamic Resource Management in Ad Hoc IoT Networks 296

Blockchain Technologies for Secure Data Sharing in Ad Hoc IoT Systems 296
AI-Enabled Ad Hoc Networks 297
    Artificial Intelligence-Enabled Mobile Ad Hoc Networks 297
    Vehicular Ad Hoc Networks Using Artificial Intelligence............................300
    Artificial Intelligence for Flying Ad Hoc Networks 301
    Artificial Intelligence for Robot Ad Hoc Networks......................................303
    Under Water Sensor Networks Using Artificial Intelligence........................305
    Artificial Intelligence Assisted Wireless Powered Internet of Things (IoT)...... 307
    Artificial Intelligence-Enabled Ship Ad Hoc Networks...............................309
AI and Machine Learning Applications for Ad Hoc Networks 310
AI-Enabled Cyberattack Detection System for Ad Hoc Networks and IoT 314
Results from Literature Review on AI Applications in Ad Hoc Networks...........315
Future Directions...317
Conclusion 318
Major Contribution Points 319
References...320

Chapter 9: Blockchain Application in UAV Swarm Security 333

Introduction...333
    Unveiling the Significance of Security in UAV Swarms 334
    Harnessing Blockchain: A Revolutionary Approach to Security...................334
Navigating the Objectives and Scope of Exploration 335
Background..336
    Key Concepts..336
Need for Secure Communication 339
    The Collaborative Nature of UAV Swarms..339
    Vulnerabilities in Communication 339

TABLE OF CONTENTS

Challenges in Ensuring Security 340
- Scalability 340
- Resource Constraints 341
- Dynamic Network Topology 341

Security of UAV Swarms Using Blockchain System 342
- Overview of Blockchain Technology 342
- Enhancing Security with Blockchain 343
- Benefits of Blockchain in UAV Swarms 344

Applications of Blockchain in UAV Swarm Security 345
- Secure Communication Protocols 345
- Data Storage and Access Control 347
- Identity Management and Authentication 347
- Inconsistency: Consensus Mechanisms and Smart Contracts 348

Challenges and Future Directions 349
- Scalability and Performance 349
- Energy Efficiency 350
- Interoperability with Existing Systems 351
- Future Research Directions 352

Conclusion 353

References 354

Chapter 10: Security and Privacy in AG-IoT-Enabled Unmanned Aerial Vehicles for Smart Cities 361
- Introduction 361
- Literature Review 364
- Security Issues in AG-IoT-Enabled UAVs for Smart Cities 367
  - Physical Threats 369
  - Digital Threats in UAVs 373
  - AI-Enabled Cyber Threats for UAVs 377

IoT Attacks for Smart Cities 378
Privacy Concerns in AG-IoT-Enabled UAVs 381
    Ethical Concerns Using UAVs for Smart Cities 382
    Legal Concerns 387
Future Research Directions for Secure AG-IoT-Based UAVs 388
Conclusion 389
References 389

Chapter 11: The Essential Role of Cybersecurity in UAV Swarm Operations 403

Overview 404
    Importance of Cybersecurity in UAV Swarm Operations 408
Evolution of UAV Technology 411
    Early Developments 411
    Cold War Era 412
    Technological Advancements 413
    Modern UAVs 414
    Emergence of UAV Swarms 415
Cyber Threats and Attacks on UAV Swarms 418
    Data Transmission Link Attacks 418
    GPS Spoofing 419
    Authentication Attacks 421
    Malware and Software Exploits 422
    Physical Attacks on Hardware 423
Main Challenges and Issues in Cybersecurity for UAV Swarm Operations 425
    Communication Security 425
    Data Protection and Privacy 425
    Vulnerability Management 426
    Autonomy and Decision-Making 426

TABLE OF CONTENTS

    Integration and Interoperability 426

    Resilience Against Advanced Threats 427

    Regulatory and Ethical Considerations 427

    Human Factors 428

  Future Trends and Research Directions in UAV Swarm Cybersecurity 429

    Advanced Threat Detection and Response 430

    Secure Autonomy and Decision-Making 430

    Blockchain Technology for Secure Communication 430

    Resilient Edge Computing and Cyber-physical Systems 431

    Regulatory Frameworks and Ethical Guidelines 431

    Human-Centric Security Solutions 432

    Interdisciplinary Collaboration and Innovation 432

  Conclusion 434

  References 435

Chapter 12: Cloud Computing for UAV Swarms 443

  Introduction 443

  Atmosphere of Swarm Studies 447

    Several Factors Explain Why Simulation Is Prevalent in UAV Research 447

    Preliminary Development and Validation 450

    The strengths and weaknesses of simulation experiences in UAV swarm research are primarily attributed to aspects such as modeling, scope, and purpose 452

  The Importance of Combining Simulation and Hardware Validation 455

  Having Publications for Analyzing Research Trends 456

  Data Collection and Questions 458

    Steps in Refining the Dataset for Analysis 459

    Creating an Analysis-Ready Dataset 461

  Simulation Trends in Comparison to Hardware-Based Experiments 462

Regional Analysis of Research Approaches 464

Challenges in Bridging Simulation and Real-World Implementation 469

Conclusion 481

References 482

Chapter 13: Enhancing UAV Swarm Security with Blockchain: Solutions for Secure Communication, Data Integrity, and Autonomous Operations 489

Introduction 490

Role of UAV Swarm Security 491

Blockchain Application 494

    Secure Communication 494

    Data Integrity and Tamperproof Logs 495

    Decentralized Coordination 495

    Authentication and Authorization 496

    Resilience Against Cyberattacks 496

Importance of Blockchain in UAV Swarm Security 496

Evolution of UAV Technology 499

    Early Developments and Military Applications 499

    Technological Advancements in the Late 20th Century 501

    Emergence of Multirole UAVs 502

    Development of UAV Swarms 503

    Integration of Blockchain and Advance Technology 504

Main Challenges and Issues 505

    Technical Challenges 505

    Security Challenges 506

    Regulatory and Ethical Issues 507

    Environmental Impact 509

    Reliability and Safety 509

TABLE OF CONTENTS

Future Trend and Research Direction in UAV Swarm Security 510
Integration of Advanced AI and Machine Learning 511
Blockchain and Distributed Ledger Technologies 511
Quantum Computing in Cryptography 512
Improved Communication Channels 513
Robust Cybersecurity measures 513
Ethical and Regulatory Frameworks 514
Impacts on the Environment and Society Environmental 514
Conclusion 514
References 515

Chapter 14: Fundamentals and Applications of Unmanned Aerial Vehicle Swarms 523

Introduction 524
Basic Concepts of UAV Swarms 525
    Comprehending UAVs 526
    Natural Swarm Behavior: An Inspiration 527
    Fundamental Features of UAV Swarms: Scalability, Autonomy, and Cooperation 528
Classification of UAVs 529
    Based on Size and Weight 529
    Based on Range and Endurance 534
    Based on Application 537
    Based on Design and Configuration 541
Critical UAV Components 545
    Airframe 546
    Propulsion System 547
    Power Source 548
    Flight Control System 548

TABLE OF CONTENTS

    Navigation and Communication Systems 549
    Autonomous Systems 550
UAVs with Communication Systems 551
    Types of Communication Systems 551
    Communication Protocols 553
    Communication Security 554
    Swarm Communication 555
UAV-Based Sensor Technologies 556
    Camera Sensors 556
    Thermal Imaging Sensors 558
    Light Detection and Ranging (LiDAR) Sensors 560
    Radar Sensors 562
    Ultrasonic Sensors 563
    Acoustic Sensors 565
Control Algorithms and Swarm Intelligence 567
    Swarm Intelligence Fundamentals: Concepts and Methods 567
    Basic Algorithms: Genetic, Ant Colony, and Particle Swarm Optimization ... 568
    Mechanisms for Real-Time Control and Adaptation 569
Challenges and Opportunities 570
    Advanced Sensor Integration 570
    Autonomous Navigation and AI 570
    Swarm Technology 570
    Robust Communication Systems 571
    Regulatory Frameworks and Standards 571
    Application-Specific Innovations 571
Conclusion 572
References 573

TABLE OF CONTENTS

Chapter 15: Integration of IoT Devices with UAV Swarms 591
Introduction 591
Unmanned Aerial Vehicles (UAVS) 594
    UAV Swarm 596
Integration of IoT with UAV Swarms 598
    Technologies Enabling UAV Swarms 599
Technological Integration 601
    IoT Devices Used in UAV Swarms 601
    Communication Protocols 603
    Data Collection and Management 604
Applications of UAV Swarms 606
    Security and Monitoring 607
    Emergency Response and Disaster Management 608
    Environmental Monitoring and Remote Sensing 609
    Rescue and Search 609
    Construction Monitoring and Infrastructure Inspection 610
    Traffic Monitoring and Management 611
    Power Line Inspection and Maintenance 612
    Urban Planning 613
    UAVs for Automated Forest Restoration 613
Challenges and Solutions 614
    Technical Challenges 614
    Cybersecurity Issues 615
    Regulatory and Ethical Issues 617
    Solutions 617
Future Trends and Innovations 618
Conclusion 621
References 622

Chapter 16: Navigating the Urban Landscape: A Comprehensive Review of IoT and UAV Integration in Smart Cities..........................633

Introduction..634
Related Works..635
Overview of Smart Cities...636
    Evolution of Smart Cities..637
    Components of Smart Cities..641
Importance of Applications of UAVs in Smart City...............................646
    Characteristics of UAVs..649
Internet of Things for Smart Cities...650
    IoT Architectures for Smart Cities...651
    Cloud Computing Model...654
    Fog Computing Model...654
    Edge Computing Model...655
Sensing Technologies for Smart Cities..656
Integration of IoT in Urban Infrastructure...659
    Smart Infrastructure..661
    Smart Parking...661
    Smart Waste Management System...662
    Smart Environment Monitoring...663
    Smart Grid...663
IoT in Healthcare Services..664
    Remote Patient Monitoring..664
    Smart Hospitals and Healthcare Facilities....................................669
    Public Healthcare Surveillance...670
IoT Challenges for Smart City...671
    Security and Privacy...672
    Smart Sensors...672

TABLE OF CONTENTS

Networking 673

Big Data Analytics 674

Conclusion 674

References 675

Chapter 17: Advancement in ML Techniques and Applications for UAV Swarm Management 691

Introduction 691

Literature Review 694

Classification of ML Techniques for UAVs 698

Supervised Learning 699

Unsupervised Learning 701

Reinforcement Learning 702

Semisupervised Learning 703

Role of Modern Machine Learning Techniques for UAV Swarms 703

Applications of Machine Learning for UAV Swarms 706

UAV Path Planning Based on Machine Learning 707

ML-Based Security and Surveillance in UAV Swarms 708

Resource and Network Optimization 709

Swarm UAV System 709

Fault Diagnosis in UAVs Using Machine Learning 710

Future Research Directions 711

Conclusion 712

References 713

Chapter 18: AI Integration in Drone Technology: Revolutionizing Applications in Agriculture, Security, and Beyond 721

Introduction 722

Historical Development of AI in Drone Technology 725

Applications of AI in Drone Technology 729

TABLE OF CONTENTS

Machine Learning and Drones: Enhancing Capabilities and Applications 734

Challenges in the Integration of AI and Drones 735

Future Prospects of AI in Drone Technology 736

Future Outlook for AI in Drone Technology 738

Conclusion 741

References 745

Chapter 19: Infrastructure Resilience and Disaster Management 751

Introduction 751

Literature Study 753

Role of AI in Disaster Management 755

Cybersecurity in AI-Enhanced Network 756

Resilient Network Protocol 758

Integrated Sensor Networks 759

Future Direction 761

Conclusion 762

References 762

Chapter 20: Innovations and Future Directions in UAV Swarm for Protecting Smart Cities 767

Introduction 767

Source of Data 770

Literature Review 772

Experimental Setup 775

    Hardware Infrastructure 775

    Software Platforms 776

    Experimental Methodologies 778

Legal Considerations 779

    Regulatory Frameworks 779

xxv

TABLE OF CONTENTS

- Privacy Protection 780
- Liability Management 781
- Ethical Governance 781
- Public Engagement and Stakeholder Consultation 781
- Limitation and Conclusion 782
  - Limitations 782
  - Conclusion 784
- Future Scope 785
  - Advancements in Autonomy and Intelligence 785
  - Integration with Emerging Technologies 785
  - Multidomain Collaboration 785
  - Environmental Sensing and Monitoring 786
  - Human–Machine Interaction and Augmented Reality 786
  - Ethical and Societal Implications 786
  - Resilience and Fault Tolerance 787
  - Privacy-Preserving Technologies 787
  - Collaborative Governance Models 788
- References 788

Index 795

About the Editors

Oroos Arshi (Member IEEE) is a Research Scientist at AI-EYS, where she is architecting the future through innovation and academic brilliance. She was awarded the Young Researcher award at the International Conference on Emerging Trends and Innovations (ICETI) 2024. She is an active member of the IEEE Women in Engineering, IEEE Robotics & Automation Society member, and IEEE Young Professional. She is also a member of the International Association of Engineers (IAENG). She completed her postgraduate degree M. Tech in Cyber Security and Forensics from the School of Computer Science, University of Petroleum and Energy Studies, Dehradun, India. She completed her undergraduate degree in Bachelor of Technology in Computer Science and Engineering from Integral University, Lucknow, India. She has around 15 international publications in reputed journals, conferences, book and book chapters. Her 8 books have been accepted as an editor by prestigious publishers including Wiley, IGI Global, Springer, Apple Academic Press, and Taylor & Francis. She served as session chair for AIOT's Emerging Technologies and Future Applications at the 4th International Conference on Advances in Computation Technology, Computing, and Engineering in Morocco. Additionally, she was the chief organizer and publication chair at the 2nd International Conference on Emerging Trends & Innovation (ICETI) in July 2024. More interestingly, she has served as a speaker at numerous conferences. Moreover, she has earned a gold badge as well as a silver badge for problem-solving for Design Analysis and Algorithm (DAA) from Hacker Rank. Her expertise has

ABOUT THE EDITORS

been showcased on various occasions, demonstrating her commitment to sharing insights and contributing to advancements in her field on a global platform. Her research interests include Artificial Intelligence, Unmanned Aerial Vehicles, the Internet of Things, Computer Vision, and Natural language Processing.

Dr. Inam Ullah Khan is a distinguished academic and industry leader, celebrated for his extensive contributions to Artificial Intelligence, Artificial General Intelligence, Unmanned Aerial Vehicles, Routing Protocols, Intrusion Detection Systems, Machine Learning, Deep Learning, and Evolutionary Computing. He is the founder of AI-Explain Your Science (AI-EYS) and a Senior Member of IEEE, with affiliations that showcase his influence in the field. These include memberships in the International Association of Engineers (IAENG), IEEE Young Professionals, IEEE Systems Council, and various IAENG societies focused on Artificial Intelligence, Computer Science, Internet Computing and Web Services, Information System Engineering, Scientific Computing, Software Engineering, and Wireless Networks. Additionally, he is Advisor Cisco Community Pakistan. Dr. Khan currently serves as a Mentor in Artificial Intelligence at Corvit Systems in Rawalpindi, Pakistan. He is also a Global Mentor and Guest Lecturer at Impact Xcelerator, School of Science & Technology, IE University, Madrid, Spain, and a Trainer at the National Vocational and Technical Training Commission (NVTTC) in Pakistan. He is currently working as Postdoctoral Research Fellow at Multimedia University, Malaysia. Also, Dr. Khan is Adjunct Faculty at PSGR Krishnammal College for Women, College in Coimbatore, India. His previous roles include serving as a Visiting Researcher at King's College London, UK, and holding faculty appointments at several prestigious universities in Pakistan, including the National University of Technology, Islamabad; Center for Emerging Sciences, Engineering & Technology (CESET), Islamabad; Abdul Wali Khan University (Garden Campus and Timergara Campus); University of Swat; Shaheed Zulfikar Ali Bhutto Institute of Science and Technology (SZABIST), Islamabad Campus; and the Riphah Institute of Systems Engineering, Riphah University, Islamabad Campus.

Dr. Khan holds a Ph.D. in Electronics Engineering and an M.S. in Electronic Engineering from Isra University, Islamabad Campus, as well as a Bachelor's degree in Computer Science from Abdul Wali Khan University, Mardan, Pakistan. His M.S. thesis, titled "Route Optimization with Ant Colony Optimization (ACO)," was published as a book in Germany and is available on Amazon. He has also completed the Huawei Technologies Pakistan Train the Trainer (TTT) Program. With over 100 research articles published in reputable journals, conferences, and book chapters, Dr. Khan has made significant scholarly contributions. He has edited around 20 books on various topics and remains actively involved in organizing academic conferences. He has served as a Technical Program Committee member for several prestigious international conferences, including the EAI International Conference on Future Intelligent Vehicular Technologies in Islamabad, Pakistan; the 2nd International Conference on Future Networks and Distributed Systems in Amman, Jordan (2018); the International Workshop on Computational Intelligence and Cybersecurity in Emergent Networks (CICEN'21), held in conjunction with the 12th International Conference on Ambient Systems, Networks and Technologies (EUSPN2021) in Leuven, Belgium; and the International Workshop on Intelligent Systems for Sustainable Smart Cities in New Delhi, India (2022).

In addition, Dr. Khan has played key roles in international conferences, serving as the Special Session Chair at the International Conference on Advances in Communication Technology and Computer Engineering (ICACTCE'22), where he led discussions on "Fusion of Emerging Technologies: Communication Networks & Future Applications."

ABOUT THE EDITORS

He was also a Technical Program Committee Member at the EAI MTYMEX 2024 in Vancouver, Canada, and continues to serve as a Guest Editor for numerous international journals. In recognition of his expertise, Dr. Khan has frequently appeared on Pakistan National Television as a technology expert, further cementing his reputation as a thought leader. He has held several leadership positions in global conferences, including serving as the General Chair of the International Conference on Trends and Innovations in Smart Technologies (ICTIST'22), a role he currently holds for the 2nd ICTIST'24. Additionally, he was the President and General Chair of both the 1st and 2nd International Conferences on Emerging Trends & Innovation (ICETI). Dr. Khan's educational contributions extend beyond academia. He has designed two specialized short courses—AI for Banking and AI for AI for Financial Institutes—at the National Vocational and Technical Training Commission in Pakistan. His impact on AI education is further solidified by the fact that three of his authored books will be used as course material nationwide. These books are titled: Cognitive Machine Intelligence: Applications, Challenges, and Related Technologies, Future Tech Startups and Innovation in the Age of AI, and Artificial Intelligence for Intelligent Systems: Fundamentals, Challenges, and Applications. Dr. Inam Ullah Khan's ongoing contributions to academia, industry, and international research continue to establish him as a global thought leader and influencer in the realm of emerging technologies.

Keshav Kaushik is an accomplished academician, cybersecurity, and digital forensics expert currently serving as an Associate Professor in the Department of Computer Science & Engineering, Sharda School of Engineering & Technology, Sharda University, Greater Noida, India. As a key member of the Cybersecurity Centre of Excellence, he has been instrumental in advancing the field of cybersecurity through his dedicated teaching and innovative research. In addition to his academic role, he holds the prestigious position of Vice-Chairperson for the Meerut ACM Professional Chapter, highlighting his leadership and commitment to the professional community. In 2024, he is listed in the World's top 2% Scientist released by Stanford University and Elsevier. His academic journey includes a notable stint as a Faculty Intern during the Summer Faculty Research Fellow Programme 2016 at the Indian Institute of Technology (IIT) Ropar, reflecting his continuous pursuit of knowledge and professional development. His scholarly contributions are extensive and impactful, with over 150 publications to his credit. This includes 25 peer-reviewed articles in SCI/SCIE/Scopus-indexed journals and 50+ publications in Scopus-indexed conferences. He is also an inventor, holding one granted patent and six published patents, alongside five granted copyrights. His editorial expertise is showcased by publishing 30 books and 25 book chapters, further cementing his reputation as a thought leader in the field. His professional certifications are a testament to his expertise and commitment to excellence. He is a Certified Ethical Hacker (CEH v11) by EC-Council, a CQI and IRCA Certified ISO/IEC 27001:2013 Lead Auditor, a Quick Heal Academy Certified Cyber Security Professional (QCSP), and an IBM Cybersecurity Analyst. His recognition as a Bentham Ambassador by Bentham Science Publishers and his role as a Guest Editor for the IEEE Journal of Biomedical and Health Informatics underscore his influence and authority in cybersecurity. He is a dynamic speaker, having delivered over 50 national and international talks on cybersecurity and digital forensics topics. His mentorship was acknowledged during the Smart India Hackathon 2017, under the aegis of the Indian Space Research Organization (ISRO), with a certificate of appreciation from AICTE, MHRD, and i4c. A two-time GATE qualifier with an impressive 96.07 percentile (2012 and 2016), he has also received accolades from the Uttarakhand Police for his significant contributions to cybercrime investigation training.

ABOUT THE EDITORS

With a career marked by significant achievements and a profound impact on cybersecurity and digital forensics, he continues to inspire and lead in both academic and professional circles.

Nadeem Iqbal (Senior Member IEEE) received a Ph.D. degree in bio and brain engineering from the Korea Advanced Institute of Science and Technology (KAIST), Daejeon, South Korea, in 2013. He has completed a Fulbright Fellowship at St. John's University in New York, USA. Dr. Iqbal was a postdoctoral fellow at the School of Mechanical Engineering, University of Leeds, UK. He is currently working as an Associate Professor with the Department of Computer Science, Abdul Wali Khan University Mardan, Pakistan. His research interests include biological information processing mechanisms in the brain, Natural Language Processing, the Internet of Things, and Pattern recognition.

Inam Ullah (Member, IEEE) received a B.Sc. degree in Electrical Engineering (Telecommunication) from the Department of Electrical Engineering, University of Science and Technology Bannu (USTB), KPK, Pakistan, in 2016 and a Master's and Ph.D. degree in Information and Communication Engineering from the College of Internet of Things (IoT) Engineering, Hohai University (HHU), Changzhou Campus, 213022, China, in 2018 and 2022, respectively. He completed his postdoc with Brain Korea 2021 (BK21) at the Chungbuk Information Technology Education and Research Center, Chungbuk National University, Cheongju 28644, S Korea, in March

ABOUT THE EDITORS

2023. He is currently an Assistant Professor at the Department of Computer Engineering, Gachon University, S Korea. His research interests include Robotics, the Internet of Things (IoT), Wireless Sensor Networks (WSNs), Underwater Communication and Localization, Underwater Sensor Networks (USNs), Artificial Intelligence (AI), Big data, Deep learning, etc. He has authored more than 130 peer-reviewed articles on various research topics and five books as an editor. He is a TPC member of ACM RACS 2023, Poland, August 2023, and IEEE ICC'24-SAC-10, Denver, CO, USA), 2024, IIoTBDSC 2024), Wuhan, China & Birmingham, UK, ICTIST 2024. He serve as an Editor-in-Chief for IECE Transactions on Emerging Trends in Network Systems, as an Associate Editor of IECE Transactions on Intelligent Systematics, International Journal of Biomedical & Bioinformatics Technology, IECE Transactions on Data Augmentation, IECE Transactions on Emerging Topics in Artificial Intelligence, and Guest Editor for various journals such as Computers in Human Behavior, Sensors, Electronics, Journal of Marine Science and Engineering, Frontiers in Sensors, Artificial Intelligence and Applications, etc. Moreover, he serve as a Editorial board member for Computers, Materials & Continua (CMC), World Journal of Artificial Intelligence and Robotics Research, Computing and Artificial Intelligence, Journal of Artificial Intelligence and Robotics, Frontiers in Medical Engineering, Journal of Computer and Creative Technology, and so on. He is the reviewer of many prominent journals, including IEEE Transactions on Industrial Informatics KSII Transactions on Internet & Information Systems, IEEE Transactions on Vehicular Technology, IEEE Transactions on Intelligent Transportation Systems, Transactions on Sustainable Computing, IEEE ACCESS, Sustainable Energy Technologies and Assessments, Future Generation Computer Systems (FGCS), Computers and Electrical Engineering (Elsevier), Internet of Things (IoT) Journal, Digital Communications & Networks (Elsevier), Springer Nature, Wireless Communication & Mobile Computing (WCMC), Alexandria Engineering Journal Sensors, Electronics, Remote Sensing, Applied Sciences, Computational Intelligence and Neurosciences, etc.

ABOUT THE EDITORS

His awards and honors include the Best Student Award from the University of Science and Technology Bannu (USTB), KPK, Pakistan, in 2015 and the Prime Minister Laptop Scheme Award from the University of Science and Technology Bannu (USTB), KPK, Pakistan, in April 2015. Top-10 students award of the College of Internet of Things (IoT) Engineering, Hohai University, China in June 2019, Top-100 students award of Hohai University (HHU), China in June 2019, Jiangsu Province Distinguish International Students award (30,000 RMB) in 2019-2020, Certificate of Recognition from Hohai University (HHU), China in 2021 & 2022 both, Top-100 students award of Hohai University (HHU), China in May 2022, Top-10 Outstanding Students Award, Hohai University (HHU), China in June 2022, and Distinguished Alumni Award from University of Science and Technology Bannu (USTB), KPK, Pakistan in Oct. 2022, ranked in Top-2% World Influential Scientists Worldwide in 2023 by Stanford University & Elsevier, featured in the 2024 list, and so on.

Dr. Khadija Slimani completed her PhD, which was jointly supervised by the University of Ibn Tofail in Morocco and the University of Technology of Belfort Montbeliard (UTBM) in France. This collaborative international effort showcased her commitment to academic excellence and cross-cultural research. For her doctoral studies, Dr. Slimani delved into the realms of machine learning, deep learning, pattern recognition, and computer vision, with a specific focus on academic emotion recognition. This interdisciplinary approach, spanning both Moroccan and French academic environments, enriched her research experience and contributed to the global perspective of her work. Dr. Slimani's academic contributions extended to various engineering schools in Paris, France, where she took on teaching responsibilities across multiple modules, including data science, deep learning, machine learning, data bases, and computer vision. She currently holds the title of Associate Professor at the Graduate School of Automatic Electronic Computing in Paris, France. Through her diverse roles, Dr. Slimani consistently inspires and contributes to shaping the next generation of researchers.

About the Technical Reviewers

Muhammad Abdul Rafay, Shaheed Zulfiqar Ali Bhutto Institute of Sciences and Technology, Islamabad, Email: arafay880@gmail.com

Mansoor Khan, Shaheed Zulfikar Ali Bhutto Institute of Science and Technology, Email: mansoorkhan3799@gmail.com

Mubashir Ullah, SZABIST University, Islamabad, Pakistan, Email: 1.mubashirullah@gmail.com

Urwa Sajjad, Quaid-e-Azam University, Islamabad, Pakistan, Email: Urwa.sajjad1999@gmail.com

Hamed Taherdoost, University Canada West, Vancouver, Canada, Email: hamed.taherdoost@gmail.com

Hafiz Muhammad Attaullah, Lecturer of Cybersecurity at Mohammad Ali Jinnah University, Email: muhammad.attaullah@jinnah.edu

Mehak Mushtaq Malik, University of Cyprus, Email: mehakcomsat@gmail.com

Prof. Keshav Kaushik, University of Petroleum and Energy Studies, Dehradun, India, Email: officialkeshavkaushik@gmail.com

Dr. Shashi Kant Gupta, Post-Doctoral Fellow and Researcher, Eudoxia Research University, USA, Email: raj2008enator@gmail.com

Dr. Menaa Nawaz, Assistant Professor, National University of Technology, Islamabad, Email: menaanawaz@nutech.edu.pk

Bodhisattwa Baidya, Research Scholar, Ramakrishna Mission Vidyamandira, Belur Math, India, Email: bodhisattwabaidya@gmail.com

ABOUT THE TECHNICAL REVIEWERS

Abdullah Akbar, Department of Electrical Engineering, National University of Computer and Emerging Sciences, Peshawar, Pakistan, Email: abdullahakbar0209@pwr.nu.edu.pk

Dr. Faisal Rehman, NUST, Islamabad, Pakistan, Email: Faisalrehman0003@gmail.com

Dr. Muhammad Bilal, Sunway University, Malaysia, Email: muhammadb@sunway.edu.my

Dr. Tarandeep Kaur Bhatia, Deakin University, Australia, Email: drtarandeepkaurbhatia@gmail.com

Dr. Ahthasham Sajid, Capital University of Science & Technology (CUST), Islamabad, Pakistan, Email: ahthasham@cust.edu.pk

Dr. Sabitha Banu A, PSGR Krishnammal College for Women, India, Email: sabithabanu@psgrkcw.ac.in

Mahtab Khalid, Ripha International University, Islamabad, Pakistan, Email: mqazimahtab1162@gmail.com

Arslan Ali Khan, Ripha International University, Islamabad, Pakistan, Email: arsalan.ali@riphah.edu.pk

Hamza Razza, Ripha International University, Islamabad, Pakistan, Email: hamza.razzaq@riphah.edu.pk

Muhammad Farukh Sohail, Department of Cyber Security, Riphah Institute of Systems Engineering, Islamabad, Pakistan, Email: 44756@students.riphah.edu.pk

Malik Nadeem Sandeela, Department of Cyber Security, Riphah Institute of Systems Engineering, Islamabad, Pakistan, Email: a 44756@students.riphah.edu.pk

Arsalan Ali Khan, Department of Cyber Security, Riphah Institute of Systems Engineering, Islamabad, Pakistan, Email: arsalan.ali@riphah.edu.pk

ABOUT THE TECHNICAL REVIEWERS

Abdullah Haroon, Department of Cyber Security, Riphah Institute of Systems Engineering, Islamabad, Pakistan, Email: 44756@students.riphah.edu.pk

Shah Mohammad, Department of Computer Science, Faculty of ICT, Baluchistan University of Information Technology, Engineering and Management Sciences, Quetta, Baluchistan, Pakistan, Email: hamza.razzaq@riphah.edu.pk

Mirza Aamir Mehmood, Department of Cyber Security, Riphah Institute of Systems Engineering, Riphah International University, Islamabad, Pakistan, Email: ahthasham.sajid@riphah.edu.pk

Raja Asif, Department of Information Technology, Faculty of ICT, Baluchistan University of Information Technology, Engineering and Management Sciences, Quetta, Baluchistan, Pakistan, Email: hamza.razzaq@riphah.edu.pk

Hummayun Raza Shakoor Watoo, Department of Cyber Security, Riphah Institute of Systems Engineering, Riphah International University, Islamabad, Pakistan, Email: hummayun.raza@riphah.edu.pk

Muhammad Farukh Sohail, Department of Cyber Security, Riphah Institute of Systems Engineering, Islamabad, Pakistan, Email: 44756@students.riphah.edu.pk

Malik Nadeem Sandeela, Department of Cyber Security, Riphah Institute of Systems Engineering, Islamabad, Pakistan, Email: 44756@students.riphah.edu.pk

Muhammad Sajid Iqbal, Department of Cyber Security and Data Science, Riphah Institute of Systems Engineering, Riphah International University Islamabad, Pakistan, Email: sajid516@yahoo.com

Rabia Laraib, SEECS, National University of Science and Technology (NUST), Islamabad, Pakistan, Email: rabialaraib17@gmail.com

Ruhma Ahmad, SEECS, National University of Science and Technology (NUST), Islamabad, Pakistan, Email: ruhma.ahmed6386263@gmail.com

ABOUT THE TECHNICAL REVIEWERS

Huda Fatima, SEECS, National University of Science and Technology (NUST), Islamabad, Pakistan, Email: hudafatima8681@gmail.com

Umna Iftikhar, Faculty of Engineering Science and Technology, Iqra University, Karachi, Pakistan, Email: yamnaiftikhar@gmail.com

Mursal Furqan Kumbhar, Computer Science Department, Sapienza University of Rome, Rome, Italy, Email: kumbhar.2047419@studenti.uniroma1.it

Srinjan Ghosh, Computer Science Department, Sapienza University of Rome, Rome, Italy, Email: ghosh.2053796@studenti.uniroma1.it

Dr. Khadija Slimani, Higher School of Computer Science, Electronics and Automation (ESIEA), Paris, France, Email: Khadija.slimani@esiea.fr

Basheer Ahmad, Department of Electrical and Computer Engineering, International Islamic University, Islamabad, Pakistan, Email: basheerahmad389@gmail.com

Yasir Khan, Department of Science and Technology & Information Technology (ST&IT), Peshawar, Pakistan, Email: imyasir.308@gmail.com

Waqar Hussain, Department of Computer Science, Shaheed Zulfikar Ali Bhutto Institute of Science and Technology, Islamabad, Pakistan, Email: waqarabaaj@gmail.com

Touseef Sadiq, Centre for Artificial Intelligence Research (CAIR), Department of Information and Communication Technology, University of Agder, Jon Lilletuns vei 9, Grimstad, Norway, Email: touseef.sadiq@uia.no

Muhammad Ajmal Naz, Department of Electronics and Telecommunications (DET), Politecnico di Torino, Email: muhammadajmal.naz@studenti.polito.it

Muhammad Adnan Khan, Associate Professor, School of Computing, Skyline University College, Sharjah, United Arab Emirates, Email: muhammad.adnan@skylineuniversity.ac.ae

ABOUT THE TECHNICAL REVIEWERS

Tayyaba Basri, Department of Electrical Engineering, National University of Computer and Emerging Sciences (NUCES) Peshawar campus, Pakistan, Email: tayyababasri56@gmail.com

Nastaran Davudi Pahnehkolaee, University of Applied Science and Technology, Tehran, Iran, Email: nastaran.davudi.1988@gmail.com

Muhammad Fazal Ijaz, Assistant Professor, School of Engineering and Information Technology, Melbourne Institute of Technology, Melbourne, Australia, Email: mfazal@mit.edu.au

Sayed Zillay Nain Zukhraf, University of Tartu, Estinia, Email: syeda.zillay.nain.zukhraf@ut.ee

Muhammad Abul Hassan, Department of Information Engineering and Computer Science, University of Trento, Italy, Email: muhammadabul.hassan@unitn.it

Dr. Inam Ullah Khan, Department of Computer Science, SZABIST University, Islamabad Campus, Pakistan, Email: inamullahkhan05@gmail.com

Abdur Rehman Khan, Department of Creative Technologies, Air University, Islamabad, Pakistan, Email: abdurrehamn809@gmail.com

Maryum Munir, Department of Computer Science, Shaheed Zulfiqar Ali Bhutto Institute of Science and Technology, Islamabad, Pakistan, Email: 80ee32f9@gmail.com

Dr. Inam Ullah, Department of Computer Engineering, Gachon University, Seongnam, Sujeong-gu 13120, Republic of Korea, Email: inam@gachon.ac.kr

Farhood Nishat, Department of Information Security, Institute of System Engineering, Riphah University, Islamabad Campus, Pakistan, Email: farhoodnishat@hotmail.com

Muhammad Afaq, Intern, BS Computer Science, SZABIST University, Islamabad, Email: muhammadafaq1050@gmail.com

ABOUT THE TECHNICAL REVIEWERS

Oroos Arshi, University of Petroleum and Energy Studies, Dehradun, India, Email: oroosarshi523@gmail.com

Aprajita Kashyap, University of Petroleum and Energy Studies, Dehradun, India, Email: kashyapaprajita2@gmail.com

Awais Mahmood, Department of Computer Science, SZABIST University, Islamabad, Pakistan, Email: awais.bzu21@gmail.com

Acknowledgments

The completion of this book has been made possible through the collective efforts, expertise, and support of many individuals and organizations. As editors, we would like to express our deepest gratitude to all those who contributed to this work. First and foremost, we extend our sincere thanks to the authors who have shared their knowledge and insights. Their contributions have been invaluable in shaping this book into a comprehensive resource on the evolving role of UAV swarms in smart cities. We are also grateful to our reviewers for their meticulous evaluation and constructive feedback. Their expertise has significantly enhanced the quality and accuracy of the chapters included in this book.

Our appreciation goes out to the publishers for their unwavering support and guidance throughout the publication process. Their professionalism and commitment have ensured that this book meets the highest standards of academic excellence. We would like to acknowledge the institutions and organizations that provided the necessary resources and infrastructure, enabling us to bring this project to fruition. Special thanks to our academic institutions for fostering an environment that encourages research and innovation. Finally, we are indebted to our families and colleagues for their understanding and encouragement during the development of this book. Their patience and support have been a source of motivation throughout this journey. To all who have contributed, directly or indirectly, to the successful completion of this book, we offer our heartfelt thanks.

—Oroos Arshi
University of Petroleum and Energy Studies, Dehradun, India

Introduction

This book, titled *Unmanned Aerial Vehicles Swarm for Protecting Smart Cities: Future Trends and Challenges,* is a comprehensive exploration of the intersection between unmanned aerial vehicles (UAVs) and the evolving landscape of smart cities. With the increasing integration of technology into urban environments, there is a growing need to understand how UAV swarms can contribute to the safety, efficiency, and resilience of these complex urban ecosystems. The book aims to provide a technical understanding of UAV swarms and their applications within the context of smart cities. It begins by laying the groundwork with an introduction to UAV swarms and smart cities, establishing the foundational concepts and motivations behind their integration. This sets the stage for a detailed exploration of the fundamentals of UAV swarms, including their operation, coordination, and potential capabilities. As the book progresses, it delves into various aspects of smart cities, exploring concepts, technologies, and challenges inherent in their development and operation. This includes discussions on cloud computing, cybersecurity, machine learning applications, surveillance and monitoring systems, urban planning, and infrastructure management. Each chapter contributes to a holistic understanding of how UAV swarms can be leveraged to enhance the functionality and security of smart cities. Moreover, the book addresses critical considerations such as public perception and acceptance, infrastructure resilience, disaster management, and international perspectives on UAV swarm technologies. It also examines the integration of IoT devices with UAV swarms, highlighting the synergies between these emerging technologies and their potential impact on urban environments. Furthermore, the book explores cutting-edge topics such as

INTRODUCTION

edge computing, blockchain applications, 5G integration, and augmented reality/virtual reality (AR/VR) visualization techniques in the context of UAV swarm operations. It concludes with reflections on innovations and future directions, offering insights into the evolving landscape of UAV swarm technologies and their implications for the protection and advancement of smart cities. In summary, *Unmanned Aerial Vehicles Swarm for Protecting Smart Cities: Future Trends and Challenges* serves as a comprehensive guide for researchers, practitioners, and policymakers interested in understanding the technical, social, and economic dimensions of UAV swarm technology within the context of smart city development and management.

CHAPTER 1

Introduction to Unmanned Aerial Vehicles Swarm and Smart Cities

Sher Taj[1], Xuefei Ma[2*], Yonghao Liu[1], Rahim Khan[2], Xianhua Cheng[3], Hina Hassan[4], M. A. Al-Khasawneh[5,6,7], Inam Ullah[8]

[1]Daqing Normal University, Daqing, Heilongjiang, 163455, China; shertajkhan002@gmail.com, yonghaoliu1980@163.com

[2]College of Information and Communication Engineering, Harbin Engineering University, Harbin150001, China; maxuefei@hrbeu.edu.cn, rahim@hrbeu.edu.cn

[3]Shenzhen Qingda Xieli Technology Co., Ltd, Qingdao, China; cheng.xianhua@ictuniv.com

[4]College of Life Science and Technology, Harbin Normal University, Heilongjiang Province, Harbin, China; hinauoh999@gmail.com

[5]School of Computing, Skyline University College, University City Sharjah, 1797, Sharjah, UAE

[6]Applied Science Research Center. Applied Science Private University, Amman, Jordan

[7]Jadara University Research Center, Jadara University, Irbid, Jordan; mahmoudalkhasawneh@outlook.com

[8]Department of Computer Engineering, Gachon University, Seongnam 13120, Republic of Korea; inam.fragrance@gmail.com

Corresponding Author: Xuefei Ma (maxuefei@hrbeu.edu.cn)

© The Editor(s) (if applicable) and The Author(s), under exclusive license to APress Media, LLC, part of Springer Nature 2025
O. Arshi et al. (eds.), *Unmanned Aerial Vehicles Swarm for Protecting Smart Cities*, https://doi.org/10.1007/979-8-8688-1047-3_1

CHAPTER 1   INTRODUCTION TO UNMANNED AERIAL VEHICLES SWARM AND SMART CITIES

Introduction

Unmanned aerial vehicles (UAVs) are often called drones and have been widely adopted in the past few years, thanks to their versatility to execute numerous functions. A relatively recent adaptation of civil UAV use despite its already existing use in military practice is "UAV swarm," whereby several UAVs perform a unified task. This paradigm has the possibility of being applied in several fields to transform the way various industries work, not just in the construction of smart cities but also in the management of smart cities as well.

In the context of smart systems, UAVs can be used in various scenarios in cities: surveillance, monitoring, public utilities, disaster relief, delivery of goods and services, and repairing of infrastructures. These coordinated fleets of drones consist of multiple unmanned aerial vehicles that can work independently of one another, amass and process large amounts of information in actual time, and perform multifaceted duties that one UAV is incapable of doing.

This chapter discusses the potential of applying UAV swarms in smart cities, which ranges from increased productivity, safety, and reduced expenses to the effectiveness of the environment. Using technologies like artificial intelligence, machine learning, and networking for communication, the optimum functioning of UAV swarms can be effective for managing various operations in the urban setting that will, in turn, improve the quality of life of citizens.

To continue with our proposal of how UAV swarm technology could be implemented in smart cities, it may be useful to expand the current literature to supply the subject with adequate information. Shnayder and his team in their papers described the usage of UAV swarms in smart cities while Tully and his team focused on the UAV swarm coordination for urban centers in 2019. Jackman [1] highlighted the problem of drone swarm positioning in smart cities based on the maximum likelihood approach employing genetic algorithms. In line with the exploration of

CHAPTER 1   INTRODUCTION TO UNMANNED AERIAL VEHICLES SWARM AND SMART CITIES

various trends related to UAV swarm technology for the creation of smart cities, Siddiqui et al. [2] have paid particular attention to this subject. Authors such as Pham et al. [3] elaborated on collaborative UAV swarms in disaster response for smart cities. In the study conducted by Bibri et al. [4], specifically, unmanned aerial vehicle swarms were used in learning how to monitor traffic in the context of smart cities without supervision. Saunders et al. [5] explored a low energy consumption coordination mechanism for tangible UAV swarms to monitor the environment in smart cities. Khairy et al. [6] focused on the security and privacy concerns in the smart city–enabled UAV swarm applications. Akay and Yildirim [7] discussed traffic control using UAV swarms in smart cities, and Mahmood et al. [8] elaborated on the developments in UAV swarms in the background of smart cities.

Application of Drones in UAVs

UAVs have been used in many fields including product delivery, space exploration, military uses, search and rescue missions, agricultural uses, and as a communication tool as shown in Figure 1-1. When it comes to last-mile deliveries, drones have proved to be a viable solution for delivering products quickly to remote or difficult areas. In space exploration, unmanned aerial vehicles carry cameras and sensors to capture images and capture information or map changes in planetary or lunar surfaces. In the military area, drones act as scouts, performing reconnaissance, surveillance, and even carrying out strikes, which makes them valuable in battles. In search and rescue, drones help in aerial search in disaster areas and aerial delivery of goods to trapped persons and assist in looking for missing persons. In agriculture, drones fitted with specific sensors and vision systems are applied in crop surveillance, precise farming, and the application of pesticides increasing output and efficiency in farming. Also, drones have been used to transport Internet in areas that lack this essential service in a bid to close the digital divide. Drones

CHAPTER 1 INTRODUCTION TO UNMANNED AERIAL VEHICLES SWARM AND SMART CITIES

are incredibly useful in so many different types of operations and thus represent the promise of increased innovation and productivity in a great number of fields [9–11].

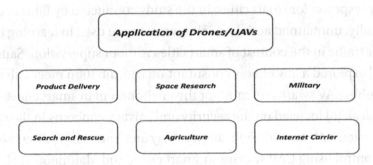

Figure 1-1. Application of drones in UAVs

UAV Applications for Smart Cities and ITS

In this chapter, we will focus on some of the chosen applications to understand the specific benefits of employing UAVs in ITS. Based on the evaluation of these applications, we list several prospects that can determine the further development of UAVs in the context of smart cities. Some examples of the variety of UAV-based ITS applications are represented in Figure 1-2. These include delivery services, traffic monitoring and surveillance operations, and media and entertainment dissemination, all of which impose different quality-of-service constraints and latency tolerances. The classification of these applications is predominantly centered on two key factors: it means the demand for bandwidth and the tolerance for delays. It is important to understand that different problems can occur in each application area depending on its goals and limitations.

CHAPTER 1   INTRODUCTION TO UNMANNED AERIAL VEHICLES SWARM AND SMART CITIES

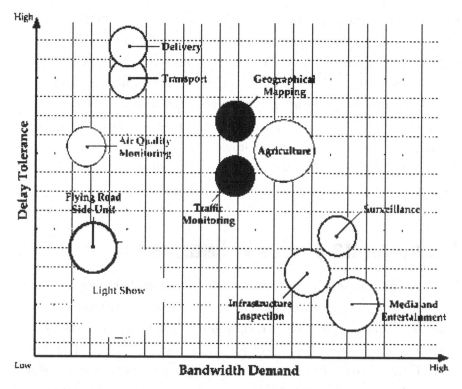

Figure 1-2. *Discussing the main and additional UAV uses for smart cities [12]. Movement up the x-axis means higher bandwidth demands, and movement up the y-axis means a higher level of tolerance to delay. The size of the circles shows the required number of UAVs in a swarm for providing reliable service to the population*

Interconnection Between UAV Swarms and Smart Cities

The interconnection between UAV swarms and smart cities is a crucial aspect of modern urban planning and management. UAV swarms, consisting of multiple unmanned aerial vehicles operating collaboratively, hold great potential to revolutionize various aspects of smart city infrastructure. For instance, in the realm of transportation and logistics, UAV swarms can be deployed for efficient traffic monitoring and management. By utilizing advanced sensors and communication technologies, these swarms can collect real-time data on traffic patterns, congestion points, and road conditions, enabling authorities to make timely and informed decisions to optimize traffic flow and enhance overall urban mobility. Moreover, UAV swarms can also be instrumental in emergency response scenarios, providing rapid aerial assessment of disaster-affected areas and delivering critical supplies to inaccessible locations in smart cities.

Fundamentals of UAV Swarms

UAV swarms refer to a group of drones that work together toward a shared objective. The concept behind these systems is to create a framework for multiple UAVs to communicate, navigate, and perform tasks simultaneously and in an organized manner. As per Kumar and Jaiswal [13], there are three primary components of a typical swarm which include the drones themselves; the communication network connects all individuals, enabling software to facilitate collective behaviors [14]. To exhibit such behaviors, each member must be endowed with sensors capable of detecting their surroundings' changes: a GPS receiver to locate its current position among others like radios or other means through which they can share information about what each one is doing or where it's going according to [15–17]. For instance, if we take a scenario where we are carrying out a search and rescue mission over a vast area, then it would be ideal if many UAVs could be used which will relay all found objects back and forth till none is left unattended.

To enhance the dependability as well as efficiency in the operation of these swarming robots, various algorithms and control techniques have been invented by different scholars [18]. The strategy proposed is a distributed control system that enables each group member to adapt to environmental changes, thereby fostering societal stability. Furthermore, they have found applications in fields like agriculture where large tracts need monitoring; environmental checking up against any possible degradation factors among others besides surveillance work has been enabled through this concept since [19]. However, before any such system takes off, there should be strict adherence to regulate them within specific airspaces only but also other international guidelines such as those stipulated by FAA 2020 [20].

Definition and Characteristics
Key Characteristics: Autonomy, Cooperation, and Scalability

1. **UAV Autonomy:** The term autonomy implies that each unmanned aerial vehicle can function on its own without any human input; this includes decision-making as well as task performance [14]. This means that a UAV with autonomy can adapt to different environments while responding instantly to unforeseen events. When used within a group or swarm setting, it also implies that these devices can communicate among themselves based on local knowledge and eventually reach an agreement toward achieving a common objective [21].

2. **UAV Cooperation:** The phrase cooperation means the capability for multiple unmanned aerial vehicles to operate in unison by sharing information and coordinating their activities toward attaining a shared goal [22]. With cooperation, drones can do complex tasks that would be hard or impossible for one drone alone such as surveillance and tracking among others. In a swarm setting, some drones may take up specific roles like sensing while others communicate to achieve an overall aim [23].

3. **UAV Scalability:** Scaling refers to how well systems tolerate increased workloads or potential for growing size. The term scalability when applied within context uncrewed aircraft systems (UAS) means the ability of a swarm of such vehicles to handle different scenarios as well as accomplish missions with varying numbers of individual units [24], for example, small-scale surveillance to large-scale environmental monitoring. More drones equal more capabilities hence greater effectiveness, thus making them suitable tools for a wide range of applications. Figure 1-3 shows overall autonomy, cooperation, and scalability in UAV swarm.

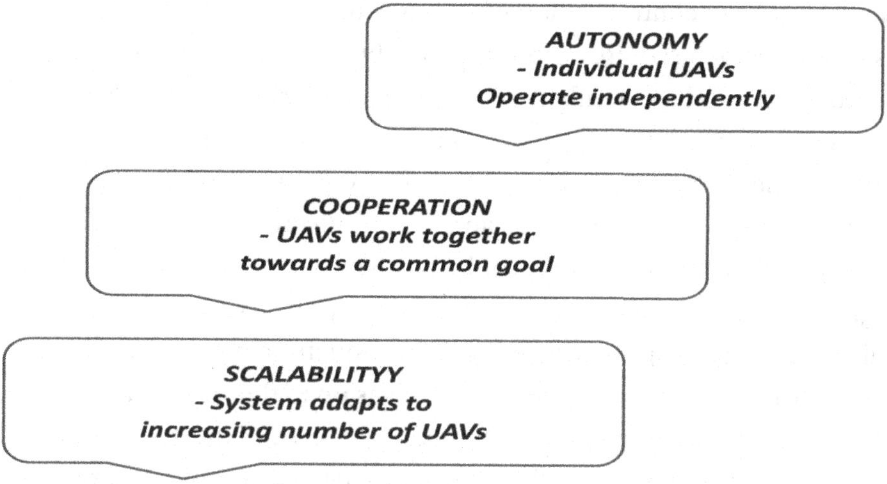

CHAPTER 1   INTRODUCTION TO UNMANNED AERIAL VEHICLES SWARM AND SMART CITIES

AUTONOMY
- Individual UAVs
Operate independently

COOPERATION
- UAVs work together
towards a common goal

SCALABILITYY
- System adapts to
increasing number of UAVs

Figure 1-3. *Autonomy, cooperation, and scalability in UAV swarm systems*

Technological Components

Hardware: Sensors, Communication Modules, and Power Systems

A UAV swarm's hardware components entail different sensors, communication modules, and power systems that enable efficient and effective operations by the UAV. GPS, accelerometers, and cameras are some of the sensors used for this purpose which provide necessary information on how to navigate through space, avoid obstacles, and perform tasks. Wi-Fi or Bluetooth among others is used as communication modules between drones themselves as well as with the ground control station while carrying out an activity such as this one which requires wireless capability for data exchange and coordination among members of the group. Batteries connected in series with converters alongside voltage regulators make up part of what supplies the energy needed for propulsion systems (among others), control systems (among others), and communication systems (among others) so that they may run reliably throughout the time taken by these flights. It is important to note here too that any mistake at any stage will affect everything else down the line so much, so it should be taken seriously right from the beginning as well as up to the very end. When these hardware components are carefully chosen and integrated, then such capabilities like artificial intelligence (AI) and cooperation among multiple agents or entities can be realized, hence making it possible for the successful swarming of drones during missions [25, 26].

Software: Control Algorithms, Coordination Protocols, and AI Integration

Advanced software incorporates control algorithms, coordination protocols, and AI to optimize processes for storing, retrieving, and shipping goods. For example, control algorithms, like model predictive control, do the optimization of robot motion planning and task assignment [27]. Coordination protocols, such as distributed constraint optimization, help to make robot-to-robot communication and task allocation efficient. In addition, AI integration through machine learning algorithms can be utilized to improve warehouse management through the prediction of demand and anomaly detection, as well as enhancement of inventory management. The use of multiagent systems in organizing robotic warehouse operations has been considered by researchers. Advanced software has also been proposed based on cloud-based software architectures to facilitate the real-time monitoring and control of robots in the warehouse. AI-powered warehouse management systems can also optimize the order fulfillment process and make it more productive by reducing delays.

Key Technologies Enabling UAV Swarms
Communication Systems

Communication systems are somewhat associated with the technique by which the data or information is sent or transmitted from one place to another through media; it can be wire, optical fiber, or any radiation signals such as radio waves. Figure 1-4 shows the complete overview of the communication system. These systems allow information exchange in formats like data, voice, and video on devices and networks as well as individuals. Effective communication system plays some roles in many areas of today's society, for instance, in business, learning institutions, hospitals, and in government. It refers to a class of technology that focuses on the aspects of transmitting information through telecommunication means, computer networks, and broadcasting over short and long distances. The transmitter and receiver are the two primary parts of the communication systems while the channel is an important component of the system that supports the flow of data with the help of communication protocols.

CHAPTER 1 INTRODUCTION TO UNMANNED AERIAL VEHICLES SWARM AND SMART CITIES

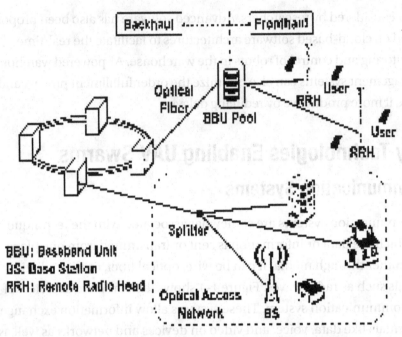

Figure 1-4. *RoF system-level solutions, for example, RoF-based backhaul for wireless base stations, RoF-based fronthaul for centralized small cells, and RoF-based fiber-wireless converged access networks such as passive optical networks (PONs) [26]*

Artificial Intelligence and Machine Learning

AI and ML in smart cities are the two innovative technologies that are revolutionizing urban planning and governance for better monitoring and faster decision-making [27]. A swarm intelligence that operates through a system domestically by an AI can potentially manage traffic flow, energy usage, and waste disposal in smart cities [28]. Cities can be made more human and resilient purposes through Internet of Things (IoT) sensors integrated with AI and ML [29, 30]. For instance, traffic management systems may cut down congestion and pollution levels, while AI-based predictive upkeep can avoid infrastructure breakdowns [31].

CHAPTER 1   INTRODUCTION TO UNMANNED AERIAL VEHICLES SWARM AND SMART CITIES

Role of AI in Decision-Making and Coordination

AI is reshaping decision-making and collaboration in different sectors such as smart cities, health, and finance domains [32]. AI algorithms refer to the capacity to handle massive amounts of data processing, recognize patterns, and offer decisions that can be used to guide actions [33]. Additionally, AI can coordinate and mutually interact with other agents, systems, or devices to constantly adjust optimization based on current trends [34]. For example, when it comes to traffic management, AI-driven traffic models can synchronize traffic lights in real time to intensify traffic jam and decrease travel time [35]. Likewise in the supply chain, AI-driven supply chain management systems could also make the right inventory, the right distribution channel, and the right delivery schedules more efficient and less costly [36]. Here, we can explain how AI is a support tool that improves decision-making and cooperation between different entities as shown in Figure 1-5.

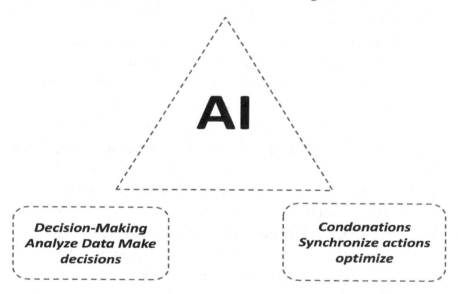

Figure 1-5. Role of AI in decision-making and coordination

CHAPTER 1 INTRODUCTION TO UNMANNED AERIAL VEHICLES SWARM AND SMART CITIES

Machine Learning Applications in Swarm Behavior and Optimization

Swarm intelligence is a subdivision of artificial intelligence that models the behavior of a group of individuals who exist in a system without obeying a central control authority, for example, bird flocks, shoal of fish, or bee colonies. Swarm techniques and optimization use machine learning in designing artificial algorithms in their quest to imitate the swarm systems to solve optimization difficulties. These algorithms are supposed to be scalable, flexible, and robust and known to solve efficiently those problems that may have a broad search space or multiple local optima [37–38].

Swarm intelligence optimization algorithms hold the characterized feature of adapting to an environment and learning form experience which is very important. This makes them appropriate for a broad list of applications in various fields such as image recognition, time-tabling, resource management, and controllers.

While Figure 1-6 shows different types of swarm behavior and optimization with their various applications in machine learning, the following diagram identifies the different algorithms of machine learning and their specific use. At the core of swarm intelligence optimization algorithms are several key techniques: a few of them are ant colony optimization, particle swarm optimization, artificial bee colony algorithm, artificial fish swarm algorithm, and firefly algorithm. They imitate the swarm behavior of ants, bees, sparrows, other animals, and birds to solve optimization problems systematically and effectively. Ant colony optimization (ACO) and PSO are popular metaheuristic techniques used to find optimal routes based on ants' foraging behavior and bird or fishes' social behavior. PSO is another metaheuristic technique that imitates the social behavior of a flock of birds or fishes in the given search space for optimization. The artificial bee colony (ABC) algorithm is one of the algorithms of colony with the behavior of honey bees being used in the optimization of numerical problems. The artificial fish swarm algorithm known as AFSA emulates fish swarms to serve best in different optimization activities. The firefly algorithm is an optimization metaheuristic that imitates the flashing patterns of fireflies and is used in addressing multimodal optimization issues.

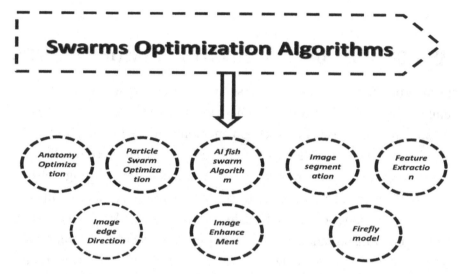

Figure 1-6. *Diagram of the swarm intelligence*

The swarm intelligence algorithms developed herein are incorporated in different image processing activities, as highlighted by the aforementioned diagram, because they are used in image segmentation, which is a process of dividing an image into multiple segments. Image matching, that is the task of aligning two images and finding the spatial correspondence, also utilizes these algorithms. Image classification is another application where it is used in categorizing images under some predetermined classes. In addition, these algorithms are employed on image feature extraction; this is the process whereby important characteristics about an image are highlighted and separated for further analysis. Two main killer applications are image edge detection, which is the ability to find the boundaries within images, and image enhancement, which is the general ability to enhance the visual quality of images. Lastly, image compression, which reclaims the size of the image files, is another area in which these swarm intelligence optimization algorithms are used effectively.

Applications of UAV Swarms in Smart Cities

The incorporation of UAV swarms in smart city systems can improve a number of processes in city development in different ways. For example, UAV swarms can be used in inspection, including bridge integrity and identification of possible damages on the surface [39]. Furthermore, they may be applied in security and combating criminal acts by relaying real-time videos to the relevant authorities. Moreover, UAV swarms enable air quality monitoring, tracking, and even detection of wildfires [40]. Within the area of intelligent transportation, UAV swarms can assist in the minimization of traffic density in areas that may have caused a jam by offering current traffic information [41]. Additionally, UAV swarms can contribute to the disaster response and recovery processes by effectively evaluating the extent of the damage and the places that require help [42, 43]. In light of global trends toward the growth of urban populations, the utilization of UAV swarms to address the needs of smart urban environments will be of significant importance in improving the outcome of urban systems.

Surveillance and Security

The implementation of surveillance systems has become common in recent years, and the main objective was for safety and to decrease crime. A study conducted by the Urban Institute reveals that surveillance cameras are effective in reducing crimes by bringing down property crimes by

20% to 30% and violent crimes by 10% to 20% [44]. Moreover, another research conducted in the Journal of Criminal Justice also concluded that surveillance cameras also help citizens in terms of reduction in fear of crime as they feel secure where security cameras are installed. However, the same systems have been criticized for invasion of privacy and infringement of civil liberties aspects with some scholars claiming that the benefits of surveillance outweigh the costs to individual freedom [45].

Monitoring Public Spaces

The monitoring of public spaces has become an important component of the modern development of cities, to ensure public safety and reduce crime rates. The use of closed-circuit television (CCTV) cameras has become a common feature in many cities, with cameras installed in public spaces such as streets, parks, and public transportation hubs. Research has shown that the presence of CCTV cameras can have a deterrent effect on criminal activity, with a study by the Scottish Government finding that CCTV cameras can lead to a reduction in crime rates of up to 30% [46]. Furthermore, a study published in the Journal of Urban Technology found that the use of CCTV cameras in public spaces can also improve the overall feeling of safety among citizens, leading to increased social cohesion and community trust [47]. However, the use of CCTV cameras in public spaces also raises concerns about privacy and surveillance, highlighting the need for effective regulation and governance of such systems.

Rapid Response to Emergencies

Rapid response to emergencies is critical in minimizing the impact of crises. Emergency response systems that integrate dispatch systems, emergency medical services (EMS), and other first responders can significantly reduce response times and improve outcomes. According to a study published in the Journal of Emergency Medical Services, the use of advanced dispatch systems can reduce response times by up to 30 seconds, which can be critical in life-threatening situations [48]. Furthermore, research has shown that the implementation of emergency response systems that integrate multiple agencies and services can improve coordination and communication, leading to faster and more effective response times [49]. Moreover, the application of sophisticated technological innovations like GPS tracking and real-time video transmission can also enhance emergency response efforts, enabling responders to arrive on the scene better prepared and equipped to handle the situation [50–52].

Infrastructure Inspection and Maintenance

Regular inspection and maintenance of critical infrastructure are essential to ensure public safety and prevent catastrophic failures. Advanced technologies such as drones, sensors, and AI can significantly improve the efficiency and effectiveness of infrastructure inspection and maintenance. For instance, drones equipped with high-resolution cameras and sensors can quickly and accurately inspect bridges, roads, and buildings, reducing the need for costly and time-consuming manual inspections [53]. Furthermore, AI-powered analytics can analyze sensor data to predict potential failures and optimize maintenance schedules, reducing downtime and improving overall infrastructure resilience [54]. Moreover, Studies have indicated that the utilization of advanced technologies can also improve the accuracy and reliability of infrastructure condition assessments, reducing the risk of failures and improving overall public safety [55].

Inspection of Bridges, Roads, and Buildings

The inspection of bridges, roads, and buildings is a critical task that ensures public safety and prevents catastrophic failures. Traditional visual inspection methods are often time-consuming, costly, and subjective, leading to the need for more efficient and accurate methods. Advanced technologies such as unmanned aerial vehicles (UAVs), sensors, and computer vision can significantly improve the efficiency and accuracy of infrastructure inspection. For example, UAVs with high-resolution cameras and sensors can quickly and accurately inspect bridge decks, roads, and building facades, detecting cracks, defects, and other anomalies [56]. Furthermore, computer vision algorithms can process images and sensor data to automatically detect defects, reducing the need for human interpretation and improving inspection accuracy [57]. Moreover, research has shown that using cutting-edge technology might also lessen the cost and duration of inspection while improving the overall quality of inspection data [57].

Urban Mobility

The integration of advanced technologies such as drones, sensors, and AI is transforming urban mobility, traffic monitoring and management, and the delivery of goods and medical supplies. For instance, AI systems can work on real-time traffic data to enhance traffic signal control, reducing congestion and lowering emissions [58–60]. Additionally, drones equipped with sensors and cameras can quickly gather traffic data, monitor traffic flow, and detect incidents, and AI-powered analytics can predict traffic patterns, enabling proactive traffic management [61]. Furthermore, drones are being used to deliver goods and medical supplies, such as blood and organs, increasing access to healthcare services and reducing delivery times [62]. Moreover, research has shown that the use of autonomous vehicles can also improve traffic efficiency, reduce congestion, and enhance mobility for the elderly and disabled [63]. Overall, the integration of advanced technologies is enhancing urban mobility, traffic management, and the delivery of essential goods and services.

CHAPTER 1   INTRODUCTION TO UNMANNED AERIAL VEHICLES SWARM AND SMART CITIES

Challenges and Solutions

A rapid rate of urbanization and growing population density of cities create numerous issues, such as traffic jams and pollutions and the load on infrastructure [64]. This paper discusses some major challenges related to rapid urbanization by using the United Nations (UN) as a reference. To deal with these challenges, it is possible to know about smart cities which is the promotion of using technologies like IoT, big data, and AI to improve planned city, ways of transport, and citizens [65]. However, the integration of swarm intelligence – an approach that focuses on the emergence of collective behavior through decentralized, self-organized networks such as ants – can push smart cities to the next level where traffic flows are optimized through traffic lights and smart energy consumption is coordinated through proper management of lightening and waste management systems [66]. These include intelligent transportation systems – new urban green structures – and participatory governance, which can be realized through interdisciplinary contortionism or through IT investments as acknowledged by the ITU [67]. Furthermore, it is also noted that defying the funding gap can also be done through new innovative sources of funding and participation of PPPs within smart cities [68].

Technical Challenges

1. **Communication Reliability and Latency:** Ensuring reliable and efficient communication between individual swarm components (e.g., sensors, drones, or robots) and the central hub or among themselves is crucial. However, maintaining consistent network connectivity, managing data transmission latency, and handling potential packet losses can be significant technical challenges.

2. **Battery Life and Power Management:** Swarm components often rely on batteries, which can quickly drain due to intense computational and communication activities. Effective power management strategies are necessary to extend battery life, optimize energy harvesting, and minimize maintenance needs, especially in large-scale deployments.

3. **Scalability and Interoperability:** As the number of swarm components increases, ensuring seamless integration, coordination, and communication between diverse devices and systems becomes increasingly complex. Technical challenges arise from managing heterogeneous networks, resolving compatibility issues, and ensuring seamless handovers between different communication protocols and technologies.

Regulatory and Ethical Issues

1. **Privacy Concerns:** The deployment of swarm intelligence in smart cities raises significant privacy concerns, as sensors and drones may collect sensitive information about citizens, including their location, behavior, and personal data. Ensuring that data is protected, anonymous, and used only for authorized purposes is crucial to maintaining trust and avoiding misuse.

2. **Regulatory Frameworks and Compliance:** The lack of clear regulatory frameworks for swarm intelligence in smart cities poses significant challenges. Existing laws and regulations may not be

sufficient to address the unique challenges of swarm technology, leaving cities to navigate a complex web of compliance issues, including data protection, safety, and liability.

3. **Ethical Considerations:** Swarm intelligence in smart cities also raises ethical concerns, such as unequal access to benefits, potential discrimination, and the impact on vulnerable populations. Cities must address these concerns by ensuring that swarm technology is designed and deployed in an inclusive, transparent, and accountable manner that prioritizes citizen welfare and well-being.

Proposed Solutions

1. **Advances in Battery Technology:** Breakthroughs in battery technology can help overcome power management challenges, enabling swarm components to operate for longer periods without recharging. This could include the development of more efficient batteries, energy harvesting technologies, or innovative power management strategies to minimize energy consumption.

2. **Development of Robust Communication Protocols:** The development of robust, reliable, and standardized communication protocols can help ensure seamless communication between swarm components and the central hub. This could include the creation of new protocols that prioritize low-latency, high-throughput, and fault-tolerant data transmission to support the complex interactions within the swarm.

3. **Implementation of Ethical Guidelines and Privacy Protections:** Establishing and implementing ethical guidelines and privacy protections can help address concerns around data privacy, security, and potential biases in swarm intelligence systems. This could include the development of transparent data management practices, anonymization techniques, and regular audits to ensure compliance with ethical standards and regulations.

Case Studies and Real-World Examples

- **Case Study 1: UAV Swarm for Disaster Response**

Background: Natural disasters such as hurricanes, wildfires, and earthquakes require rapid and effective response to minimize damage and loss of life. Unmanned aerial vehicles (UAVs) have the potential to revolutionize disaster response by providing critical information to emergency responders.

Deployment: A UAV swarm was deployed to Puerto Rico after Hurricane Maria in 2017 to assess damage and identify areas of need. The swarm consisted of 10 UAVs, each equipped with high-resolution cameras and sensors to capture detailed images and data on damage to infrastructure, homes, and the environment.

Results: The UAV swarm successfully captured over 10,000 images and gathered critical data on damage to critical infrastructure, including roads, bridges, and buildings. The data was transmitted in real time to emergency responders, enabling them to prioritize resources, allocate personnel, and plan effective response strategies.

CHAPTER 1 INTRODUCTION TO UNMANNED AERIAL VEHICLES SWARM AND SMART CITIES

Key Findings

- The UAV swarm reduced response times by up to 75%, enabling emergency responders to respond more quickly and effectively.
- The swarm's real-time data transmission enabled emergency responders to prioritize resources and allocate personnel more effectively.
- The UAVs' high-resolution cameras and sensors enabled detailed damage assessment, which informed decision-making and resource allocation.

Lessons Learned

- Prepositioning UAV assets in high-risk areas can significantly reduce response times and improve situational awareness.
- Seamless communication protocols are critical to ensure real-time data transmission and effective response.
- Integrating UAV data with existing emergency response systems can maximize effectiveness and improve decision-making.

Impact: The successful deployment of the UAV swarm for disaster response demonstrated the potential for swarm technology to revolutionize emergency response and improve outcomes in the face of natural disasters.

CHAPTER 1   INTRODUCTION TO UNMANNED AERIAL VEHICLES SWARM AND SMART CITIES

Global Perspectives

Swarm intelligence is being explored and implemented in various forms around the world. Here are some examples from different countries, followed by a comparative analysis of their implementations:

Examples from Various Countries

- **Singapore:** Singapore has launched the "Smart Nation" initiative, which includes a swarm intelligence-based traffic management system to minimize and enhance traffic flow congestion.

- **Barcelona, Spain:** The city has implemented a smart lighting system powered by swarm intelligence, which adjusts lighting levels based on pedestrian and vehicle traffic.

- **Tokyo, Japan:** Tokyo is using swarm intelligence to optimize its waste management system, with smart sensors and drones working together to detect and collect waste more efficiently.

Comparative Analysis of Different Implementations

While each country has its unique approach to implementing swarm intelligence in smart cities, some common trends and differences emerge:

- **Centralized vs. Decentralized Approaches:** Singapore's traffic management system is a centralized approach, where a central authority controls the system. In contrast, Barcelona's smart lighting system is more decentralized, with decision-making distributed among sensors and nodes.

- **Scale and Ambition:** Tokyo's waste management system is a large-scale implementation, covering a significant portion of the city. In contrast, Singapore's traffic management system is more focused on specific high-traffic areas.

- **Technology Integration:** Barcelona's smart lighting system integrates with existing infrastructure, while Tokyo's waste management system uses a combination of sensors, drones, and AI-powered algorithms.

Key Takeaways

- Swarm intelligence is being applied in diverse ways across the globe, with varying levels of centralization, scale, and technology integration.

- Comparative analysis highlights the importance of understanding local context, infrastructure, and needs when implementing swarm intelligence in smart cities.

- Global perspectives can inform and improve swarm intelligence implementations by sharing best practices and lessons learned across different regions.

Future Trends and Directions

As swarm intelligence continues to evolve, several technological advancements and innovations are expected to shape the future of this field.

Technological Advancements

Innovations in AI and Machine Learning

- **Explainability and Transparency:** Future swarm intelligence systems will prioritize explainability and transparency, enabling humans to understand and trust AI-driven decisions.

- **Edge AI:** With the proliferation of IoT devices, edge AI will play a crucial role in swarm intelligence, enabling real-time processing and decision-making at the edge.

- **Multimodal Learning**: Swarm intelligence will leverage multimodal learning, combining sensors, computer vision, and other modalities to create more accurate and robust models.

Next-Generation Communication Technologies

- **5G and 6G Networks:** Future swarm intelligence applications will leverage the low-latency, high-bandwidth, and massive machine-type communication capabilities of 5G and 6G networks.

- **Quantum Communication:** The integration of quantum communication will enable secure, high-speed data transmission, and advanced encryption methods for swarm intelligence applications.

- **Optical and Acoustic Communication:** Next-generation communication technologies will explore the use of optical and acoustic communication methods, enabling new forms of swarm intelligence applications.

CHAPTER 1   INTRODUCTION TO UNMANNED AERIAL VEHICLES SWARM AND SMART CITIES

Other Trends and Directions

- **Swarm Robotics:** The integration of swarm intelligence with robotics will enable autonomous systems to collaborate and adapt in complex environments.

- **Human-Swarm Interaction:** Future swarm intelligence systems will focus on intuitive human-swarm interaction, enabling humans to effectively communicate and collaborate with swarms.

- **Swarm Ethics and Governance:** As swarm intelligence becomes more pervasive, there will be a growing need for ethical frameworks and governance structures to ensure responsible development and deployment.

Key Takeaways

- The future of swarm intelligence is poised to be shaped by breakthroughs in AI, machine learning, and communication technologies.

- These advancements will unlock new possibilities for swarm intelligence applications, from autonomous systems to human-swarm collaboration.

- It is essential to address the ethical and governance implications of swarm intelligence to ensure its responsible development and deployment.

Emerging Applications

Swarm intelligence is poised to revolutionize various industries and domains, giving rise to innovative applications that transform the way we live and work.

Autonomous Urban Mobility Solutions

- **Smart Traffic Management:** Swarm intelligence can optimize traffic flow, reducing congestion and minimizing travel times. Autonomous vehicles can communicate with each other and the infrastructure to ensure smooth traffic flow.

- **Autonomous Public Transport:** Swarm intelligence can optimize public transport systems, such as buses and trains, to reduce waiting times, increase efficiency, and improve the overall passenger experience.

- **Intelligent Parking Systems:** Swarm intelligence can direct car owners to available parking spaces, reducing congestion and parking time and making urban mobility more efficient.

Integration with Other Smart City Technologies

- **IoT Integration:** Swarm intelligence can leverage IoT sensor data to optimize energy consumption, waste management, and public services in smart cities.

- **Big Data Analytics:** Swarm intelligence can be integrated with big data analytics to provide real-time insights, enabling cities to respond promptly to changing conditions and citizen needs.

- **Smart Buildings and Infrastructure:** Swarm intelligence can optimize energy consumption, maintenance, and security in smart buildings, making them more sustainable and efficient.

Other Emerging Applications

- **Swarm Robotics:** Swarm intelligence can be applied to robotics, enabling autonomous systems to collaborate and adapt in complex environments, such as environmental monitoring or search and rescue missions.

- **Healthcare and Biomedicine:** Swarm intelligence can be used to analyze medical data, identify patterns, and optimize treatment plans, leading to better healthcare outcomes.

- **Cybersecurity:** Swarm intelligence can be applied to cybersecurity, enabling systems to detect and respond to threats in real time, improving overall security and resilience.

Key Takeaways

- Swarm intelligence can transform multiple industries and domains, from urban mobility to healthcare and cybersecurity.

- Integration with other smart city technologies, such as IoT and big data analytics, will unlock new opportunities for innovation and improvement.

- Emerging applications of swarm intelligence will have a profound impact on the way we live, work, and interact with our environment.

Predicted Impact

The adoption of swarm intelligence in urban planning and management is expected to have a profound impact on the quality of life for citizens, the environment, and the economy.

Long-Term Benefits for Urban Living

- **Improved Efficiency:** Swarm intelligence can optimize urban systems, reducing waste and improving the overall efficiency of services, leading to cost savings and enhanced quality of life for citizens.

- **Enhanced Sustainability:** Swarm intelligence can help cities reduce their carbon footprint, optimize resource consumption, and promote sustainable development, ensuring a healthier environment for future generations.

- **Increased Safety and Security:** Swarm intelligence can enhance public safety and security by identifying patterns, predicting potential threats, and optimizing emergency response systems.

- **Better Decision-Making:** Swarm intelligence can provide city planners and administrators with real-time insights, enabling data-driven decision-making and more effective urban planning.

Potential for Global Adoption and Standardization

- **Global Cooperation:** The adoption of swarm intelligence in urban planning can facilitate global cooperation, knowledge sharing, and best practices, leading to more efficient and sustainable cities worldwide.

CHAPTER 1   INTRODUCTION TO UNMANNED AERIAL VEHICLES SWARM AND SMART CITIES

- **Standardization:** The development of standards and frameworks for swarm intelligence can facilitate its adoption, ensuring interoperability and enabling cities to learn from each other's experiences.

- **Scalability:** Swarm intelligence can be applied to cities of varying sizes, making it a scalable solution for urban planning and management, regardless of the city's size or complexity.

Conclusion

The Internet of Things (IoT) perspective is transforming real-world settings into intelligent, interactive platforms, offering a vast array of innovative services supported by the advent of 5G networks. Today, the popularity of unmanned aerial vehicles (UAVs) is high, and it is expected that their usage in everyday life will increase significantly in the coming years. These updated UAVs make use of new technologies encompassed by the term "Internet of Things." Currently, drones or unmanned aircraft systems can be autonomously programmed or remotely controlled, either by a remote control or a ground station, and are considered as an organized automated technology. Despite their potential, drones have yet to significantly impact agricultural practices. Recently, however, there has been a substantial amount of information regarding the applications of drone use in the field of agriculture and precision farming. Significant advancements have been made to automate agriculture, reducing costs and increasing benefits for farmers. The IoT can be utilized in agriculture to support the procedure for small- to large-scale design farming operations. In this chapter, we examined various areas of agriculture and the application of the IoT within these fields. Additionally, recent developments have used these UAVs for domestic use in a variety of fields.

CHAPTER 1   INTRODUCTION TO UNMANNED AERIAL VEHICLES SWARM AND SMART CITIES

Smart drones are a progressively more advanced concept, UAVs are connected to various devices via internet technologies, enabling communication and intelligence, making them indispensable for farmers in the vast market. Farmers cannot afford to remain outside this enormous market.

Summary of Key Points

- **Recap of the Importance of UAV Swarms in Smart Cities:** UAV swarms hold immense potential to revolutionize various aspects of smart cities, from improving infrastructure management to enhancing public safety. Their ability to perform coordinated and complex tasks with high efficiency makes them invaluable in urban environments.

- **Highlight Major Applications and Benefits:** Major applications include traffic monitoring, disaster response, infrastructure inspection, and delivery services. The benefits are vast, ranging from increased operational efficiency and reduced costs to enhanced data collection and real-time decision-making.

Final Thoughts

- **The Future Potential of UAV Swarms:** The future of UAV swarms in smart cities is promising. As technology advances, we can expect more sophisticated swarm behaviors and greater integration with other smart city technologies. This will lead to smarter, more resilient urban environments capable of addressing a wide range of challenges.

CHAPTER 1   INTRODUCTION TO UNMANNED AERIAL VEHICLES SWARM AND SMART CITIES

- **Call to Action for Further Research and Development:** There is a need for continued research and development in the field of UAV swarms. This includes improving the underlying technologies, developing new applications, and addressing challenges such as regulatory concerns and public acceptance. Collaboration between researchers, industry, and policymakers will be key to unlocking the full potential of UAV swarms in smart cities.

References

[1] Jackman, A., Domestic drone futures. Political Geography, 2022. **97**: p. 102653.

[2] Siddiqui, A.B., et al., Prioritized User Association for Sum-Rate Maximization in UAV-Assisted Emergency Communication: A Reinforcement Learning Approach. Drones, 2022. **6**(2): p. 45.

[3] Pham, Q.-V., et al., Swarm intelligence for next-generation networks: Recent advances and applications. Journal of Network and Computer Applications, 2021. **191**: p. 103141.

[4] Bibri, S.E., et al., Smarter eco-cities and their leading-edge artificial intelligence of things solutions for environmental sustainability: A comprehensive systematic review. Environmental Science and Ecotechnology, 2024. **19**: p. 100330.

[5] Saunders, J., S. Saeedi, and W. Li, Autonomous Aerial Delivery Vehicles, a Survey of Techniques on How Aerial Package Delivery Is Achieved. arXiv preprint arXiv:2110.02429, 2021.

[6] Khairy, S., et al., Constrained Deep Reinforcement Learning for Energy Sustainable Multi-UAV Based Random Access IoT Networks With NOMA. IEEE Journal on Selected Areas in Communications, 2020. **39**(4): p. 1101-1115.

[7] Akay, R. and M.Y. Yildirim, Multi-strategy and self-adaptive differential sine–cosine algorithm for multi-robot path planning. Expert Systems with Applications, 2023. **232**: p. 120849.

[8] Mahmood, S., N.Z. Bawany, and M.R. Tanweer, A comprehensive survey of whale optimization algorithm: modifications and classification. Indonesian Journal of Electrical Engineering and Computer Science, 2023. **29**(2): p. 899.

[9] Khalil H, Rahman SU, Ullah I, Khan I, Alghadhban AJ, Al-Adhaileh MH, Ali G, ElAffendi M. A UAV-Swarm-Communication Model Using a Machine-Learning Approach for Search-and-Rescue Applications. Drones. 2022 Nov 23;6(12):372.

[10] Khan HU, Sohail M, Ali F, Nazir S, Ghadi YY, Ullah I. Prioritizing the multi-criterial features based on comparative approaches for enhancing security of IoT devices. Physical Communication. 2023 Aug 1;59:102084.

[11] Khalil I, Khalil A, Ullah I, Tao Y, Khan I, Ashraf S, Ismael WM. Social Internet of Things (SIoT): Recent Trends and Its Applications. Future Communication Systems Using Artificial Intelligence, Internet of Things and Data Science.:159-92.

[12] Lucic, M.C., et al., Leveraging UAVs to Enable Dynamic and Smart Aerial Infrastructure for ITS and Smart Cities: An Overview. Drones, 2023. **7**(2): p. 79.

[13] Kumar, A. and A. Jaiswal, Swarm intelligence based optimal feature selection for enhanced predictive sentiment accuracy on twitter. Multimedia Tools and Applications, 2019. **78**: p. 29529-29553.

[14] Kumar, S.A., et al., Stable switched controllers for a swarm of UGVs for hierarchal landmark navigation. Swarm and Evolutionary Computation, 2021. **65**: p. 100926.

[15] Smith, R.N., et al., Planning and Implementing Trajectories for Autonomous Underwater Vehicles to Track Evolving Ocean Processes Based on Predictions from a Regional Ocean Model. The International Journal of Robotics Research, 2010. **29**(12): p. 1475-1497.

[16] Khan WU, Imtiaz N, Ullah I. Joint optimization of NOMA-enabled backscatter communications for beyond 5G IoT networks. Internet Technology Letters. 2021 Mar;4(2):e265.

[17] Khan HU, Hussain A, Nazir S, Ali F, Khan MZ, Ullah I. A Service-Efficient Proxy Mobile IPv6 Extension for IoT Domain. Information. 2023 Aug 14;14(8):459.

[18] Zhou, Y., B. Rao, and W. Wang, UAV Swarm Intelligence: Recent Advances and Future Trends. Ieee Access, 2020. **8**: p. 183856-183878.

[19] Zhang, H., et al., Optimum design of a multi-form energy hub by applying particle swarm optimization. Journal of cleaner production, 2020. **260**: p. 121079.

[20] Diller, J., et al. ICCSwarm: A Framework for Integrated Communication and Control in UAV Swarms. in Proceedings of the Eighth Workshop on Micro Aerial Vehicle Networks, Systems, and Applications. 2022.

[21] Shobeiry, P., et al. UAV Path Planning for Wildfire Tracking Using Partially Observable Markov Decision Process. in AIAA Scitech 2021 Forum. 2021.

[22] Yu, Y., et al., Distributed Multi-Agent Target Tracking: A Nash-Combined Adaptive Differential Evolution Method for UAV Systems. IEEE transactions on vehicular technology, 2021. **70**(8): p. 8122-8133.

[23] Wan, L., et al., Unmanned aerial vehicle-based field phenotyping of crop biomass using growth traits retrieved from PROSAIL model. Computers and Electronics in Agriculture, 2021. **187**: p. 106304.

[24] Matlekovic, L. and P. Schneider-Kamp, Constraint Programming Approach to Coverage-Path Planning for Autonomous Multi-UAV Infrastructure Inspection. Drones, 2023. **7**(9): p. 563.

[25] Mazhar T, Talpur DB, Shloul TA, Ghadi YY, Haq I, Ullah I, Ouahada K, Hamam H. Analysis of IoT Security Challenges and Its Solutions Using Artificial Intelligence. Brain Sciences. 2023 Apr 19;13(4):683.

[26] Asif M, Khan WU, Afzal HR, Nebhen J, Ullah I, Rehman AU, Kaabar MK. Reduced-Complexity LDPC Decoding for Next-Generation IoT Networks.

Wireless Communications and Mobile Computing. 2021;2021(1):2029560.

[27] Wang, Z., et al., Improved A* algorithm and model predictive control-based path planning and tracking framework for hexapod robots. Industrial Robot: the international journal of robotics research and application, 2023. **50**(1): p. 135-144.

[28] Kim, H. RoF-based Optical Fronthaul Technology for 5G and Beyond. in 2018 Optical Fiber Communications Conference and Exposition (OFC). 2018. IEEE.

[29] Ben Rjab, A. and S. Mellouli. Artificial Intelligence in Smart Cities: Systematic Literature Network Analysis. in Proceedings of the 12th International Conference on Theory and Practice of Electronic Governance. 2019.

[30] Chen, Z.-G., et al., Evolutionary Computation for Intelligent Transportation in Smart Cities: A Survey. IEEE Computational Intelligence Magazine, 2022. **17**(2): p. 83-102.

[31] Silva, B.N., M. Khan, and K. Han, Towards sustainable smart cities: A review of trends, architectures, components, and open challenges in smart cities. Sustainable cities and society, 2018. **38**: p. 697-713.

[32] Luo, J., A Bibliometric Review on Artificial Intelligence for Smart Buildings. Sustainability, 2022. **14**(16): p. 10230.

[33] Dear, K., Artificial Intelligence and Decision-Making. The RUSI Journal, 2019. **164**(5-6): p. 18-25.

CHAPTER 1   INTRODUCTION TO UNMANNED AERIAL VEHICLES SWARM AND SMART CITIES

[34] Atlam, H.F., R. Walters, and G. Wills, Internet of Things: state-of-the-art, challenges, applications, and open issues. International Journal of Intelligent Computing Research (IJICR), 2018. **9**(3): p. 928-938.

[35] Saboury, B., M. Morris, and E. Siegel, Future Directions in Artificial Intelligence. Radiologic Clinics, 2021. **59**(6): p. 1085-1095.

[36] Min, H., Artificial intelligence in supply chain management: theory and applications. International Journal of Logistics: Research and Applications, 2010. **13**(1): p. 13-39.

[37] Gupta D, Juneja S, Nauman A, Hamid Y, Ullah I, Kim T, Tag eldin EM, Ghamry NA. Energy Saving Implementation in Hydraulic Press Using Industrial Internet of Things (IIoT). Electronics. 2022 Dec 6;11(23):4061.

[38] Mazhar T, Irfan HM, Khan S, Haq I, Ullah I, Iqbal M, Hamam H. Analysis of Cyber Security Attacks and Its Solutions for the Smart grid Using Machine Learning and Blockchain Methods. Future Internet. 2023 Feb 19;15(2):83.

[39] Jacobsen, R.H., et al., Design of an Autonomous Cooperative Drone Swarm for Inspections of Safety Critical Infrastructure. Applied Sciences, 2023. **13**(3): p. 1256.

[40] Saffre, F., H. Karvonen, and H. Hildmann. Wild Swarms: Autonomous Drones for Environmental Monitoring and Protection. in International conference on FinDrones. 2023. Springer.

[41] Kumar, A., et al., A secure drone-to-drone communication and software defined drone network-enabled traffic monitoring system. Simulation Modelling Practice and Theory, 2022. **120**: p. 102621.

[42] Gladence, L.M., et al. Swarm Intelligence in Disaster Recovery. in 2021 5th International Conference on Intelligent Computing and Control Systems (ICICCS). 2021. IEEE.

[43] Rasheed Z, Ma YK, Ullah I, Tao Y, Khan I, Khan H, Shafiq M. Edge Computing in the Digital Era: The Nexus of 5G, IoT and a Seamless Digital Future. InFuture Communication Systems Using Artificial Intelligence, Internet of Things and Data Science (pp. 213-234). CRC Press.

[44] Phillips, C., A review of CCTV evaluations: Crime reduction effects and attitudes towards its use. Crime prevention studies, 1999. **10**(1): p. 123-155.

[45] Slobogin, C., Surveillance and the Constitution. Wayne L. Rev., 2009. **55**: p. 1105.

[46] McIvor, G. and H. Graham, Electronic Monitoring in Scotland. 2016.

[47] Tan, S.Y. and A. Taeihagh, Smart City Governance in Developing Countries: A Systematic Literature Review. sustainability, 2020. **12**(3): p. 899.

[48] Strnad, M., P. Jerot, and V. Borovnik Lesjak, Impact of dual dispatch system implementation on response times and survival outcomes in out-of-hospital cardiac arrest in rural areas. Signa Vitae, 2022. **18**(1).

[49]   McEntire, D.A., Disaster response and recovery: strategies and tactics for resilience. 2021: John Wiley & Sons.

[50]   Fan, C., et al., Disaster City Digital Twin: A vision for integrating artificial and human intelligence for disaster management. International journal of information management, 2021. **56**: p. 102049.

[51]   Pal R, Adhikari D, Heyat MB, Ullah I, You Z. Yoga Meets Intelligent Internet of Things: Recent Challenges and Future Directions. Bioengineering. 2023 Apr 9;10(4):459.

[52]   Adhikari D, Ullah I, Syed I, Choi C. Phishing Detection in the Internet of Things for Cybersecurity. InCybersecurity Management in Education Technologies 2023 Dec 6 (pp. 86-106). CRC Press.

[53]   Liang, H., et al., Towards UAVs in Construction: Advancements, Challenges, and Future Directions for Monitoring and Inspection. Drones, 2023. **7**(3): p. 202.

[54]   Collins, C., et al., Artificial intelligence in information systems research: A systematic literature review and research agenda. International Journal of Information Management, 2021. **60**: p. 102383.

[55]   Langåker, H.-A., et al., An autonomous drone-based system for inspection of electrical substations. International Journal of Advanced Robotic Systems, 2021. **18**(2): p. 17298814211002973.

[56]   Aliyari, M., E.L. Droguett, and Y.Z. Ayele, UAV-Based Bridge Inspection via Transfer Learning. Sustainability, 2021. **13**(20): p. 11359.

[57] Koch, C., et al., A review on computer vision based defect detection and condition assessment of concrete and asphalt civil infrastructure. Advanced engineering informatics, 2015. **29**(2): p. 196-210.

[58] Almatar, K.M., Implementing AI-Driven Traffic Signal Systems for Enhanced Traffic Management in Dammam. International Journal of Sustainable Development & Planning, 2024. **19**(2).

[59] Mazhar T, Irfan HM, Haq I, Ullah I, Ashraf M, Shloul TA, Ghadi YY, Imran, Elkamchouchi DH. Analysis of Challenges and Solutions of IoT in Smart Grids Using AI and Machine Learning Techniques: A Review. Electronics. 2023 Jan 3;12(1):242.

[60] Ullah I, Ahamd I, Anwar MS, Tao Y, Shafiq M. Data Science Meets Intelligent Internet of Things. InFuture Communication Systems Using Artificial Intelligence, Internet of Things and Data Science (pp. 73-91). CRC Press.

[61] Park, H., S. Byun, and H. Lee, Application of Deep Learning Method for Real-Time Traffic Analysis using UAV. Journal of the Korean Society of Surveying, Geodesy, Photogrammetry and Cartography, 2020. **38**(4): p. 353-361.

[62] Amukele, T., et al., Drone transportation of blood products. Transfusion, 2017. **57**(3): p. 582-588.

[63] Fagnant, D.J. and K. Kockelman, Preparing a nation for autonomous vehicles: opportunities, barriers and policy recommendations. Transportation Research Part A: Policy and Practice, 2015. **77**: p. 167-181.

[64] Ritchie, H. and M. Roser, Urbanization. Our world in data, 2018.

[65] Suresh, P. and S. Ramachandran, Development of Smart Cities in India–Dream to Reality. Scholedge International Journal of Business Policy & Governance, 2016. **3**(6).

[66] Bonabeau, E., Agent-based modeling: Methods and techniques for simulating human systems. Proceedings of the national academy of sciences, 2002. **99**(suppl_3): p. 7280-7287.

[67] FG-SSC, I., Smart sustainable cities: a guide for city leaders. FG-SSC: Geneva, Switzerland, 2015: p. 1-16.

[68] Băhnăreanu, C., World economic forum 2019: globalization 4.0–A better version. Strategic Impact, 2019(72+ 73): p. 79-82.

CHAPTER 2

Machine Learning Applications in Unmanned Aerial Vehicle Swarms

Sher Taj[1], Xuefei Ma[2*], Yonghao Liu[1], Rahim Khan[2], Xianhua Cheng[3], Hina Hassan[4], M. A. Al-Khasawneh[5,6,7], Inam Ullah[8]

[1]Daqing Normal University, Daqing, Heilongjiang, 163455, China; shertajkhan002@gmail.com, yonghaoliu1980@163.com

[2]College of Information and Communication Engineering, Harbin Engineering University, Harbin150001, China; maxuefei@hrbeu.edu.cn, rahim@hrbeu.edu.cn

[3]Shenzhen Qingda Xieli Technology Co., Ltd, Qingdao, China; cheng.xianhua@ictuniv.com

[4]College of Life Science and Technology, Harbin Normal University, Heilongjiang Province, Harbin, China; hinauoh999@gmail.com

[5]School of Computing, Skyline University College, University City Sharjah, 1797, Sharjah, UAE

[6]Applied Science Research Center. Applied Science Private University, Amman, Jordan

[7]Jadara University Research Center, Jadara University, Irbid, Jordan; mahmoudalkhasawneh@outlook.com

[8]Department of Computer Engineering, Gachon University, Seongnam 13120, Republic of Korea; inam.fragrance@gmail.com

CHAPTER 2   MACHINE LEARNING APPLICATIONS IN UNMANNED AERIAL VEHICLE SWARMS

Introduction

An unmanned aerial vehicle (UAV) group or collection of UAVs, also known as drones, works collaboratively to accomplish a specific objective. Each drone in the swarm or fleet is propelled by a specific number of rotors, enabling it to hover vertically, take off, and land (VTOL). The control of the drones' flight can be either manual, through remote control operations, or autonomous, utilizing processors installed on the drones themselves [1]. UAVs were initially developed for military purposes. However, in recent years, their civilian applications have gained significant attention. Affordable drones and their ability to operate in swarms offer promising prospects for innovative research projects and future commercial applications that could enhance people's work and daily lives [2, 3].

Drone swarms can be categorized in different ways. Figure 2-1 illustrates fully and partly (semi) autonomous swarms. For instance, they can be classified as fully autonomous or semiautonomous, where some drones have a certain level of autonomy while others are controlled externally. Another classification can be based on the hierarchy within the swarm. Some swarms may have a single-layered structure, where each drone operates independently as its leader, while others may have a multilayered structure with dedicated leader drones at each level, reporting to higher-level leaders. In such a hierarchy, a ground-based server station or a cloud-based system may serve as the highest level of control.

Within a drone swarm, each UAV can be assigned specific data collection and processing tasks and equipped with sufficient computing power to execute these tasks in real time. However, more complex processing may be offloaded to a more powerful server station or cloud infrastructure, leveraging their superior computational capabilities.

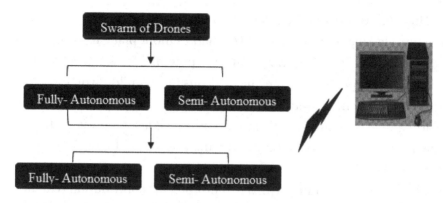

Figure 2-1. Classification of swarms

History of Swarm

The earliest documented instance of utilizing an unmanned aerial vehicle in warfare dates back to July 1849 [2], Austrian forces besieging Venice embarked on a groundbreaking mission using incendiary balloons. These balloons, launched from both land and the Austrian ship SMS Vulcano, marked the inception of air power in naval aviation, employing a balloon carrier concept that foreshadowed the development of aircraft carriers [4]. The outcome was mixed despite efforts to deploy approximately 200 of these balloons toward the besieged city. While a bomb did find its mark in the city, the shifting winds postlaunch caused the majority of the balloons to veer off course, with some inadvertently returning over Austrian territory and the launching vessel Vulcano [5].

The inception of significant drone development in the 1900s was primarily aimed at providing training targets for military personnel, marking the beginning of unmanned aerial vehicle (UAV) advancement in warfare applications [6]. A pivotal moment in this progress occurred in 1916 with A. M. Low's creation of the "Aerial Target," representing one of the earliest powered UAV attempts that laid the groundwork for further innovations in remote-controlled aviation systems [5].

CHAPTER 2   MACHINE LEARNING APPLICATIONS IN UNMANNED AERIAL VEHICLE SWARMS

This milestone was reinforced when Low successfully demonstrated the controlled flight of Geoffrey de Havilland's monoplane using his radio system on March 21, 1917, showcasing the potential for aerial vehicles to be remotely operated for specific missions [6, 7]. Building on the success of this demonstration, Low transitioned in the spring of 1917 to spearhead the development of aircraft-controlled fast motor launches, known as D.C.B.s, in collaboration with the Royal Navy for targeted strikes on shipping and port installations, highlighting the expanding role of drones in military operations [8]. Furthermore, his contributions extended to aiding Wing Commander Brock in the strategic preparations for the Zeebrugge Raid, underscoring the growing importance of unmanned technologies in tactical missions. This progression culminated in a series of British advancements in unmanned aerial systems, culminating in the introduction of the de Havilland 82 Queen Bee fleet of over 400 aerial targets in 1935, illustrating the continued evolution and utilization of UAVs in military settings [9].

Swarms Machine Learning and Deep Learning

Machine learning applications in UAV swarms involve utilizing algorithms to help drones work together in a coordinated and intelligent manner [10]. This can include navigation, coordination, decision-making, and communication between multiple drones. Some common machine learning techniques used in UAV swarms include reinforcement, deep, and distributed learning algorithms. These techniques can help drones learn from their environment, modify their actions to suit evolving circumstances, and adjust their conduct to accomplish particular goals.

CHAPTER 2 MACHINE LEARNING APPLICATIONS IN UNMANNED AERIAL VEHICLE SWARMS

Deep learning (DL) plays an important role in UAV swarm networks, as highlighted by [11]. DL offers various vital applications within UAV swarm networks typical application scenarios used in UAV swarms. Firstly, techniques like federated learning empower UAV swarms to develop a unified machine learning (ML) model globally without the need for data interchange, thereby enabling them to analyze locally gathered data and enhance the data confidentiality of individual UAVs. Secondly, DL algorithms such as multiagent reinforcement learning facilitate multiple UAVs to engage with the dynamic environment and determine the best scheduling strategy for complex UAV swarms. Thirdly, employing DL algorithms like distributed inference allows numerous UAVs to engage in swarm intelligence by exchanging limited inference details. Lastly, the implementation of DL frameworks like split learning enables multiple UAVs to efficiently train a substantial ML model by dividing it into smaller segments [12, 13].

Advantages of Using Multiple Drones in Coordinated Missions

- Employing several drones in coordinated missions allows for time-critical path following over time-varying networks, enhancing efficiency and adaptability in dynamic environments, as illustrated in Figure 2-2.

- Market-based distributed task assignment of multiple drones enables cooperative timing missions, optimizing resource allocation and task completion.

- Formulating network flow techniques for collaborative peer-to-peer refueling approaches showcases the collaborative nature of drone operations, emphasizing shared resources and strategic planning.

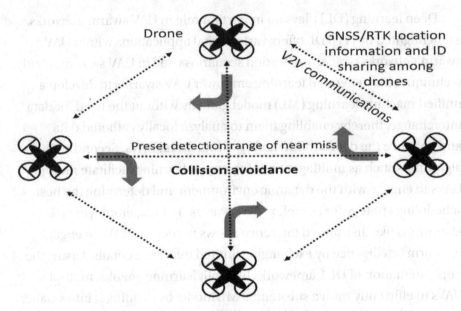

Figure 2-2. Drone V2V-based collision avoidance method [14]

Mention of the Potential of Machine Learning in Enhancing UAV Swarm Capabilities

The use of machine learning can greatly improve and expand the capabilities of UAV swarms by enabling them to operate more autonomously. For example, as shown in Figure 2-3, machine learning can facilitate adaptive and effective decision-making through various model types, utilization strategies, and algorithms. In this way, UAV swarms can operate more autonomously, making them more efficient and effective.

- Autonomous Navigation: Machine learning algorithms can be utilized to enhance the autonomy of UAV swarms in navigation tasks. Traditional UAV swarms often rely on preprogrammed routes or centralized control for navigation. However, with machine learning, individual UAVs can learn from their environment and make real-time decisions to navigate complex terrains, avoid obstacles, and optimize their trajectories. For example, reinforcement learning algorithms can enable UAVs to learn efficient flight paths by continuously evaluating the consequences of their actions and adjusting their behavior accordingly.

- Collaborative Decision-Making: Machine learning can facilitate collaborative decision-making among UAVs within a swarm. Through the analysis of data gathered from sensors, communication networks, and other UAVs, machine learning algorithms can facilitate the sharing of information among UAVs, coordinate their activities, and collaboratively achieve mission goals. For instance, clustering algorithms can group UAVs based on their proximity and task requirements, allowing them to work together more effectively in performing reconnaissance missions or surveillance tasks.

- Adaptive Behavior: Machine learning enables UAV swarms to adapt their behavior in response to changing environmental conditions, mission requirements, or unexpected events. For example, supervised learning algorithms can be trained on historical data to predict weather patterns or identify potential threats, allowing UAVs to adjust their strategies accordingly. Additionally, genetic algorithms or evolutionary algorithms can be employed to evolve optimal behaviors for specific tasks over time.

- Resource Optimization: Machine learning can optimize resource allocation and utilization within UAV swarms, improving their efficiency and effectiveness. For instance, predictive analytics algorithms can analyze data on fuel consumption, battery life, and payload capacities to optimize the distribution of resources among UAVs and minimize downtime. Moreover, machine learning can enable UAVs to learn from past experiences and make data-driven decisions to optimize their energy usage, extend their operational range, and maximize their endurance.

- Intelligent UAV swarms utilize a communication mechanism that entails utilizing mobile UAVs as transmitters or relays, necessitating multichannel access and networking amid potential interference. Cognitive radio systems with broad-spectrum sensing capacity enhance reinforcement learning processes, leading to improved policy sets, while machine learning algorithms, like deep learning models, enhance situational awareness by analyzing sensor data and detecting anomalies or patterns of interest in real time [15].

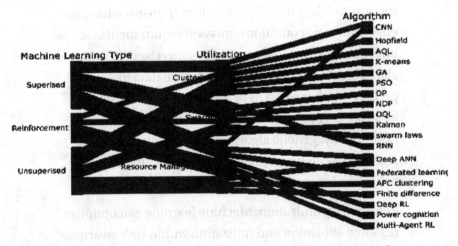

Figure 2-3. Machine learning model types, utilization, and algorithms

CHAPTER 2   MACHINE LEARNING APPLICATIONS IN UNMANNED AERIAL VEHICLE SWARMS

Fundamentals of UAV Swarms

Fundamentals of UAV swarms encompass grasping the intricacies of swarm intelligence and the synchronized efforts among unmanned aerial vehicles (UAVs) to attain shared objectives. For instance, akin to how a swarm of bees collaborates to forage for food, UAV swarms leverage principles inspired by the immune system, where antibodies collectively combat pathogens. In practical terms, this translates to coordinated path planning algorithms guiding multiple UAVs to intercept multiple targets simultaneously, much like a flock of birds maneuvering together in flight. Furthermore, the design and construction of swarm quadrotor UAVs, resembling the modular structure of ant colonies, enable seamless communication and task allocation within the swarm. Exploiting self-organization principles, exemplified by the coordinated movements of fish in a school, UAV swarms optimize their actions in dynamic combat environments, adapting autonomously to changing conditions for maximum efficiency and effectiveness [16].

Definition and Key Characteristics of UAV Swarms

UAV swarms represent a collection of autonomous or semiautonomous drones operating collaboratively to achieve a common goal, characterized by autonomy, scalability, robustness, adaptability, and collective intelligence. Figure 2-4 illustrates the capabilities and benefits of this approach, demonstrating the advantages of swarm intelligence.

1. Autonomy: Each UAV within the swarm possesses a degree of autonomy, capable of making decisions independently based on its programming and sensory inputs. This autonomy enables UAVs to operate without continuous human intervention, enhancing their versatility and effectiveness in various tasks.

CHAPTER 2   MACHINE LEARNING APPLICATIONS IN UNMANNED AERIAL VEHICLE SWARMS

2. Scalability: UAV swarms can vary in size from a few drones to hundreds or even thousands, depending on the specific application and requirements. This scalability allows for flexibility in accomplishing different missions, whether it involves surveillance of a small area or a large-scale search and rescue operation.

3. Robustness: UAV swarms are inherently robust against individual failures or disruptions. Even if one or several drones malfunction or are compromised, the overall functionality of the swarm remains largely unaffected. This robustness is crucial for mission-critical tasks where reliability is paramount.

4. Adaptability: UAV swarms exhibit adaptability in response to changing environmental conditions or mission objectives. Through communication and coordination among individual drones, the swarm can dynamically adjust its behavior and strategies to optimize performance and achieve desired outcomes.

5. Collective Intelligence: One of the most distinctive characteristics of UAV swarms is their ability to leverage collective intelligence. Similar to how social insects like ants or bees work together as a cohesive unit, UAVs in a swarm exchange information, share resources, and coordinate actions to accomplish complex tasks more efficiently than individual drones operating in isolation.

CHAPTER 2 MACHINE LEARNING APPLICATIONS IN UNMANNED AERIAL VEHICLE SWARMS

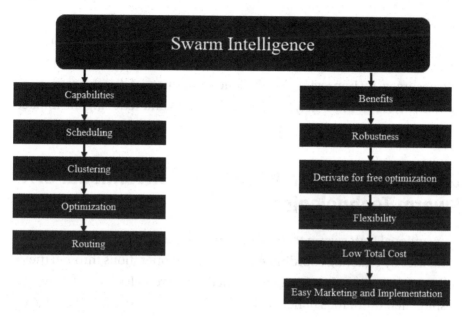

Figure 2-4. Key capabilities and benefits of swarm intelligence

Benefits of Using UAV Swarms for Various Applications

The utilization of UAV swarms presents numerous benefits across various applications, notably in disaster response scenarios. For instance, during natural disasters such as earthquakes or hurricanes, UAV swarms can rapidly survey affected areas, assess damage, and identify survivors in need of assistance. By leveraging their collective intelligence and scalability, these swarms can cover large areas more efficiently and comprehensively compared to individual drones or traditional search and rescue teams. Moreover, the robustness of UAV swarms ensures continued operation even if some drones encounter obstacles or malfunctions, thereby improving reliability and mission success rates. Additionally, the autonomy and adaptability of UAV swarms allow them to dynamically adjust their strategies and priorities based on real-time information, facilitating more effective coordination with ground teams and other response agencies. Overall, the use of UAV swarms enhances the speed and accuracy. By improving the efficiency and safety of disaster response operations, it can ultimately result in saving lives and reducing the impact of disasters on affected communities.

Challenges and Limitations of Conventional UAV Swarm Technologies

Conventional UAV swarm technologies have revolutionized various industries by enabling coordinated and efficient operations through the collective behavior of multiple unmanned aerial vehicles (UAVs). However, despite their numerous benefits, these technologies also face challenges and limitations that need to be addressed for further advancement and widespread adoption.

Challenges and Limitations of Conventional UAV Swarm Technologies

1. The difficulty lies in local sensing and control.

    When controlling multiple MAVs (micro air vehicles) simultaneously, the control architecture can be divided into two main types: (1) centralized and (2) decentralized. In the centralized approach, a single computer oversees the entire MAV swarm, possessing access to the relevant states of all MAVs, and can accordingly plan their actions, either in advance (a priori) or in real time (online). Conversely, in the decentralized approach, each MAV makes decisions autonomously. Additionally,

CHAPTER 2 MACHINE LEARNING APPLICATIONS IN UNMANNED AERIAL VEHICLE SWARMS

MAVs can sense their environment in two ways: (1) through external position sensing, typically utilizing a global navigation satellite system (GNSS) for outdoor operations or a motion capture system (MCS) for indoor environments, or (2) by relying solely on onboard sensors.

At present, the amalgamation of a centralized control framework and external positioning systems has reached the highest level of maturity, enabling flights involving multiple MAVs (micro air vehicles). Kushleyev et al. [17] demonstrated a swarm of 20 micro quadrotors capable of reorganizing into various formations. Northcott et al. [18], Augugliaro et al. [19], and Mirjan et al. [20] developed impressive collaborative construction schemes utilizing teams of [21] presented "Crazy Swarm," an indoor exhibition demonstrating 49 small quadrotors flying collectively in synchronization. These achievements highlight the advanced capabilities attained through the fusion of centralized control and external positioning systems, allowing for complex multi-MAV operations and formations.

2. A comprehensive overview across the design chain.

In this section, we will examine the latest advancements in MAV (micro air vehicle) technology within the realm of swarm robotics. To aid our examination, we will dissect the obstacles related to designing and controlling an MAV swarm into four distinct levels, spanning from "local" to "global."

CHAPTER 2   MACHINE LEARNING APPLICATIONS IN UNMANNED AERIAL VEHICLE SWARMS

1. MAV Design: MAV design pertains to the design elements of individual MAVs, encompassing factors such as processing capabilities, flight duration, dynamics, and functionalities. From a swarm engineering perspective, the design of individual MAVs is crucial because it defines the sensory information available in each unit. As shown in Figure 2-5, the design and modeling of UAVs ultimately shape the MAV's perception and understanding of its environment.

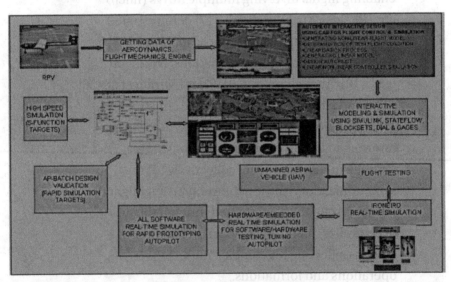

Figure 2-5. *Design and modeling of control laws for UAV [22]*

2. Enhancing Control: Local Ego-State Estimation and Control: At its core, an MAV must possess the fundamental capability to regulate its movement with precision. This core layer oversees the MAV's basic flight functions, encompassing attitude stabilization, altitude control, and velocity

estimation and regulation. Additionally, the MAV must be able to safely navigate its surroundings, with a minimum requirement to detect and circumvent potential obstacles.

3. Internal Swarm Sensing and Collision Avoidance: Two crucial technological pillars underpin the swarming capability of MAVs. The first is neighbor awareness, which is essential for MAV swarms as it enables advanced swarm behaviors and, more critically, prevents mid-air collisions. The second pillar is inter-MAV communication, which allows them to share data and collectively broaden their understanding of their environment.

4. Swarm behavior involves the overarching control strategy guiding the robots to produce collective swarm behavior. Advanced controllers in swarm operations utilize attraction and repulsion forces in flocking scenarios [23], as well as neural networks for aggregation, dispersion, or homing [24]. Other taxonomies have been established in the field. In [25], three levels of robotic cognition are outlined: sensory-motor autonomy, reactive autonomy, and cognitive autonomy. Additionally, Smeur et al. [26] categorize the control process for autonomous flight into four tiers: attitude control, altitude control, collision avoidance, and navigation. While these taxonomies share a similar conceptual structure (progressing from basic sensing and control to higher cognitive functions), the revised framework presented here is tailored to streamline our discourse within the realm of swarm robotics.

CHAPTER 2   MACHINE LEARNING APPLICATIONS IN UNMANNED AERIAL VEHICLE SWARMS

Furthermore, we incorporate the design of MAVs throughout the chain, as elucidated in this paper, which significantly influences the upper layers [27, 28].

The four identified phases exhibit a progressive level of abstraction. The lower stages establish the foundation for the resilience, adaptability, and scalability attributes essential at higher tiers. In contrast, the higher stages dictate, accommodate, and optimize the capabilities established at the lower levels. From a systemic viewpoint, the design of the MAV imposes limitations on the potential achievements of higher-level controllers, while these controllers, in turn, impose criteria that the MAV must meet. Figure 2-6 illustrates a simplified representation of the exchange of requirements and constraints.

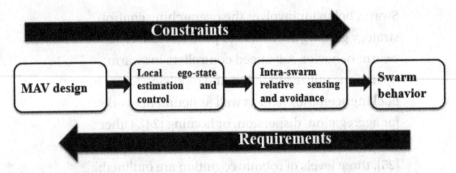

Figure 2-6. Generalized depiction of the flow of requirements and constraints for the design of MAV swarms

An Overview of Machine Learning

Machine learning, a branch of artificial intelligence, has revolutionized various aspects of our lives, and UAVs are no exception. By integrating machine learning algorithms into UAVs, also known as micro air vehicles (MAVs), we can enhance their autonomy, decision-making capabilities, and overall performance. In the context of MAV swarms, machine learning enables individual UAVs to learn from their environment, adapt to new situations, and make informed decisions in real time.

Explanation of Machine Learning Concepts and Techniques

Machine learning is a crucial component in UAV development, enabling these autonomous systems to learn from their environment, adapt to new situations, and make informed decisions in real time. Key machine learning concepts, such as supervised, unsupervised, and reinforcement learning, are employed in UAVs to improve their autonomy, navigation, and task execution. Techniques like neural networks, computer vision, and natural language processing are used in UAVs to analyze sensor data, recognize patterns, and detect anomalies. For instance, object detection algorithms, such as YOLO (You Only Look Once) and SSD (Single Shot Detector), are applied in UAVs for obstacle avoidance, tracking, and surveillance. Additionally, reinforcement learning enables UAVs to learn from trial and error, adjusting their behavior to optimize performance metrics, such as speed, accuracy, and fuel efficiency. By integrating machine learning capabilities, UAVs can effectively respond to dynamic environments, collaborate with other agents, and execute complex tasks with increased precision and reliability.

Overview of Supervised, Unsupervised, and Reinforcement Learning Algorithms

Supervised Learning

In supervised learning, the algorithm is trained using labeled data with known correct outputs. This process aims to establish a relationship between input data and output labels, enabling the algorithm to make accurate predictions on new, unseen data.

CHAPTER 2   MACHINE LEARNING APPLICATIONS IN UNMANNED AERIAL VEHICLE SWARMS

- Examples
  - Image classification (e.g., categorizing images as dogs or cats)
  - Sentiment analysis, such as categorizing text as positive, negative, or neutral
  - Regression (e.g., predicting house prices based on features like number of bedrooms and location)
- Algorithms
  - Linear regression
  - Logistic regression
  - Decision trees
  - Random forest
  - Support vector machines (SVMs)

Unsupervised Learning

In unsupervised learning, the algorithm is trained on unlabeled data, and the goal is to identify patterns, relationships, or structures in the data.

- Examples
  - Clustering (e.g., grouping customers based on buying behavior)
  - Dimensionality reduction (e.g., reducing the number of features in a dataset)
  - Anomaly detection (e.g., identifying outliers in a dataset)

- Algorithms
  - K-means clustering
  - Hierarchical clustering
  - Principal component analysis (PCA)
  - t-SNE (t-distributed stochastic neighbor embedding)

Reinforcement Learning

In reinforcement learning, the algorithm learns by engaging with an environment and receiving feedback through rewards or penalties. The objective is to develop a policy that maximizes the total cumulative reward.

- Examples
  - Robotics (e.g., teaching a robot to navigate a maze)
  - Game playing (e.g., playing chess or go)
  - Recommendation systems (e.g., personalizing product recommendations)
- Algorithms
  - Q-learning
  - Deep Q-networks (DQN)
  - Policy gradient methods
  - Actor-critic methods

These three categories of machine learning algorithms are not mutually exclusive, and many applications combine elements of supervised, unsupervised, and reinforcement learning to achieve complex tasks.

CHAPTER 2   MACHINE LEARNING APPLICATIONS IN UNMANNED AERIAL VEHICLE SWARMS

Introduction to Neural Networks and Deep Learning for UAV Swarm Applications

Deep learning is a specialized area of machine learning that emphasizes the use of neural networks with many layers, typically more than three. These networks are capable of learning complex patterns in data, such as images, speech, and text.

Types of Neural Networks

1. Feedforward Networks: Information flows only in one direction, from input to output.

2. Recurrent Neural Networks (RNNs): Information flows in a loop, allowing the network to keep track of previous inputs.

3. Convolutional Neural Networks (CNNs): Designed for image and video processing, using convolutional and pooling layers.

Deep Learning for UAV Swarm Applications

Deep learning techniques have numerous applications in UAV swarms, including the following:

1. Obstacle Detection: CNNs can be used to detect obstacles in the swarm's environment.

2. Object Tracking: RNNs can be used to track objects or teammates in the swarm.

3. Swarm Coordination: Deep reinforcement learning can be used to optimize swarm behavior and coordination.

4. Scene Understanding: CNNs can be used to analyze and understand the swarm's environment.

5. Autonomous Navigation: Deep learning can be used to enable autonomous navigation and decision-making in UAV swarms.

Deep Learning Challenges in UAV Swarm Applications

1. Computational Resources: UAVs have limited computational resources, making it challenging to run complex deep learning models.

2. Data Quality: Gathering high-quality, labeled data for training deep learning models can be challenging in UAV swarm scenarios.

3. Real-Time Processing: UAV swarms require real-time processing and decision-making, which can be challenging for deep learning models.

4. Safety and Security: Ensuring the safety and security of UAV swarms and their deep learning systems is crucial.

By leveraging deep learning techniques, UAV swarms can become more autonomous, efficient, and effective in various applications, like search and rescue, surveillance, and environmental monitoring.

CHAPTER 2   MACHINE LEARNING APPLICATIONS IN UNMANNED AERIAL VEHICLE SWARMS

Data Collection and Analysis

Data collection and analysis in UAV swarms involve collecting various types of data, including sensor data, visual data, communication data, and environmental data, and then processing and interpreting it using techniques such as machine learning, statistical analysis, and data mining [29]. The analyzed data can be used for swarm coordination, obstacle detection, environmental monitoring, search and rescue, and surveillance, among other applications. However, challenges arise from ensuring data quality, managing large data volumes, and ensuring real-time processing and security. Future directions include the use of edge computing, distributed machine learning, swarm intelligence, and autonomous decision-making, which can enhance the efficiency, effectiveness, and autonomy of UAV swarms in various applications.

Autonomous Navigation and Control

Autonomous navigation and control in UAV swarms enable the UAVs to operate independently without human intervention, making decisions in real time to achieve their goals. This is achieved through advanced algorithms and sensor fusion, combining data from sensors such as GPS, accelerometers, and cameras, to accurately estimate the UAV's state and surroundings. Autonomous navigation and control enable UAVs to adapt to changing environments, avoid obstacles, and maintain formation while in flight, as well as perform complex tasks such as search and rescue, surveillance, and package delivery. Machine learning and artificial intelligence techniques can be used to enhance autonomy, enabling UAVs to learn from experience, adapt to new situations, and make decisions based on incomplete or uncertain information.

Swarm Intelligence and Coordination

Swarm intelligence and coordination enable UAV swarms to operate collectively, making decentralized decisions and adapting to changing environments through self-organization and autonomous interactions. By leveraging swarm intelligence, UAVs can share information, delegate tasks, and allocate resources to achieve common goals, including search and rescue, surveillance, and environmental monitoring. This approach also enables UAVs to maintain formation, avoid collisions, and adapt to failures or losses through coordination algorithms inspired by natural swarming behaviors, ultimately ensuring the swarm's overall resilience and effectiveness. Thus, swarm intelligence and coordination empower UAV swarms to accomplish tasks that would be impossible for individual UAVs, while also improving their efficiency, flexibility, and scalability in various applications [30, 31].

Adaptive Mission Planning

Adaptive mission planning enables UAV swarms to dynamically adjust their plans and tasks in response to changing mission requirements, environmental conditions, and unexpected events. Through advanced algorithms and real-time data analysis, adaptive mission planning allows UAVs to reassess their objectives, reassign tasks, and replan their trajectories to ensure successful mission execution. This adaptability enables UAV swarms to respond to emerging threats, capitalize on new opportunities, and overcome unforeseen challenges, thereby enhancing their overall effectiveness, flexibility, and responsiveness in a wide range of applications, including search and rescue, surveillance, and environmental monitoring.

CHAPTER 2   MACHINE LEARNING APPLICATIONS IN UNMANNED AERIAL VEHICLE SWARMS

Predictive Analytics and Optimization Algorithms

Predictive analytics and optimization algorithms play an important role in UAV swarm operations, enabling the prediction of future states and optimization of mission outcomes. By leveraging machine learning, probabilistic modeling, and data-driven approaches, predictive analytics facilitate the forecast of environmental conditions, threat detection, and mission success. Meanwhile, optimization algorithms, including linear programming, dynamic programming, and genetic algorithms, are employed to optimize UAV trajectories, task assignments, and resource allocation, ensuring that mission objectives are achieved efficiently and effectively. By combining predictive analytics and optimization algorithms, UAV swarms can proactively adapt to changing conditions, minimize risks, and maximize mission success, leading to enhanced performance, safety, and decision-making in various applications.

Communication and Networking

Communication and networking are critical components of UAV swarm operations, enabling seamless interactions among the swarm members. This involves designing and implementing efficient communication protocols, network architectures, and data exchange formats to facilitate real-time data exchange, coordination, and control among the UAVs. The communication and networking infrastructure must ensure reliable, secure, and low-latency data transmission, while also addressing constraints such as limited bandwidth, energy consumption, and environmental interference [32, 33].

CHAPTER 2   MACHINE LEARNING APPLICATIONS IN UNMANNED AERIAL VEHICLE SWARMS

Importance of Communication and Networking in UAV Swarm Operations

Effective communication and networking are crucial in UAV swarm operations, as they enable seamless coordination, synchronization, and control among the UAVs, facilitating cooperative behaviors and ensuring successful mission execution. Reliable and efficient communication networks allow UAVs to share sensor data, exchange mission-critical information, and adapt to changing environments, thereby enhancing situational awareness, decision-making, and overall swarm performance.

Optimization of Communication Protocols and Network Efficiency with Machine Learning

Machine learning algorithms can be leveraged to optimize communication protocols and network efficiency in UAV swarms, enabling adaptive and dynamic management of communication resources. By analyzing network traffic patterns, signal strength, and environmental factors, machine learning models can predict and mitigate communication disruptions, optimize data routing, and allocate bandwidth efficiently. This leads to improved network reliability, reduced latency, and enhanced overall performance of the UAV swarm, ultimately enabling more efficient and effective mission execution [34, 35].

Machine Learning Algorithms for Distributed Computing in UAV Swarms

Machine learning algorithms for distributed computing in UAV swarms enhance their capabilities by enabling autonomous decision-making, optimizing resource allocation, and improving interdrone communication. Reinforcement learning algorithms like deep Q-learning are used for dynamic path planning, allowing UAVs to learn optimal navigation strategies through rewards and penalties. Federated learning, using methods such as federated averaging, facilitates decentralized model development enabling joint training while preserving data privacy, ensuring privacy and efficient use of local data across the swarm. Convolutional neural networks (CNNs) are employed for real-time target identification, enabling drones to process visual data onboard and detect objects or persons of interest. Consensus-based distributed task allocation algorithms help in evenly distributing search zones and minimizing redundancy, while multiagent deep deterministic policy gradient (MADDPG) algorithms manage adaptive communication protocols, optimizing information exchange and network resource usage. These integrated approaches ensure scalable, robust, and efficient operation of UAV swarms in complex scenarios like search and rescue missions [36].

Security and Threat Detection

Security and threat detection in UAV swarms are critical for ensuring the safety, integrity, and effectiveness of swarm operations. These tasks involve multiple layers of strategies and technologies to safeguard the swarm from both internal and external threats. Here's a detailed look into these aspects.

Security Measures in UAV Swarms

1. Encryption and secure communication
   - Algorithm: Secure data protection and authentication through sophisticated encryption and key management techniques.

CHAPTER 2   MACHINE LEARNING APPLICATIONS IN UNMANNED AERIAL VEHICLE SWARMS

- Description: UAVs use AES for encrypting data to prevent unauthorized access during communication. PKI enables secure key exchange and verification, ensuring that only authorized UAVs can participate in the swarm's communication network.

- Implementation: Each UAV encrypts its data packets before transmission. A central server or leader drone distributes public keys, and each UAV uses private keys to decrypt received data.

2. Authentication and authorization

    - Algorithm: Multifactor authentication (MFA) and role-based access control (RBAC).

    - Description: MFA ensures that only authenticated users and UAVs can join or control the swarm. RBAC restricts access to sensitive functions based on the role of the UAV or operator.

    - Implementation: UAVs and operators must provide multiple forms of identification, such as passwords and biometric verification. Access to various control functions is granted based on predefined roles.

3. Intrusion detection systems (IDS)

    - Algorithm: Anomaly detection using machine learning (e.g., isolation forest, autoencoders).

    - Description: IDS monitors the UAV network for unusual patterns that may indicate cyberattacks or intrusions. Machine learning models are trained to recognize normal behavior and flag deviations.

CHAPTER 2   MACHINE LEARNING APPLICATIONS IN UNMANNED AERIAL VEHICLE SWARMS

- Implementation: UAVs continuously send telemetry data to an IDS module, which analyzes it in real-time to detect anomalies such as unusual communication patterns or unauthorized access attempts.

Threat Detection Techniques

1. Physical threat detection

    - Algorithm: Convolutional neural networks (CNNs) and object detection models (e.g., YOLO, SSD).

    - Description: UAVs use onboard cameras and pretrained CNN models to detect physical threats such as hostile drones, ground threats, or obstacles.

    - Implementation: UAVs capture real-time video feeds, which are processed by CNN models to identify potential threats. Detected threats are flagged and communicated to the swarm for appropriate action.

2. Electronic threat detection

    - Algorithm: Spectrum analysis and signal classification using machine learning.

    - Description: UAVs monitor the radio frequency spectrum for jamming signals or unauthorized communications. Machine learning models classify these signals to identify potential electronic attacks.

CHAPTER 2   MACHINE LEARNING APPLICATIONS IN UNMANNED AERIAL VEHICLE SWARMS

- Implementation: UAVs are equipped with spectrum analyzers and signal processing algorithms that continuously scan for anomalies. Detected threats are analyzed and classified, triggering defensive measures.

3. Behavioral analysis for insider threats

    - Algorithm: Behavioral anomaly detection using machine learning (hidden Markov models and deep learning architectures like recurrent neural networks).

    - Description: UAVs monitor the behavior of other UAVs in the swarm to detect insider threats, such as compromised or malfunctioning drones.

    - Implementation: Each UAV logs its actions and communicates with a central system that uses behavioral models to detect deviations from expected behavior patterns. Suspected UAVs are isolated or taken offline.

Integration and Response

1. Real-time threat response

    - Algorithm: Rule-based systems and reinforcement learning.

    - Description: Upon detecting a threat, UAVs use predefined rules and learned strategies to respond, such as evasive maneuvers, returning to base, or neutralizing the threat.

CHAPTER 2   MACHINE LEARNING APPLICATIONS IN UNMANNED AERIAL VEHICLE SWARMS

- Implementation: UAVs follow a set of rules for immediate response to common threats. For complex situations, reinforcement learning helps in adapting and learning optimal responses over time.

2. Collaborative threat management

    - Algorithm: Distributed consensus algorithms (e.g., Byzantine fault tolerance)

    - Description: UAVs collaborate to validate and respond to detected threats, ensuring that decisions are robust against potential faults or compromised UAVs.

    - Implementation: UAVs use consensus algorithms to agree on the presence of a threat and the appropriate response, ensuring that even if some UAVs are compromised, the swarm can still function effectively.

By integrating these advanced security measures and threat detection techniques, UAV swarms can operate more securely and resiliently, capable of defending against a wide range of potential threats while maintaining mission effectiveness.

Anomaly Detection and Intrusion Detection Algorithms

Anomaly detection and intrusion detection algorithms are crucial for identifying unusual patterns and potential security breaches within a system. Anomaly detection typically employs machine learning techniques such as isolation forests and autoencoders, which learn the normal behavior of a system and flag deviations that may indicate anomalies.

These models analyze data for irregularities that deviate from established norms, helping to identify unexpected behaviors. Intrusion detection systems (IDS) leverage these techniques, along with others like hidden Markov models and recurrent neural networks, to monitor network traffic and system activities for signs of malicious activities. By continuously analyzing data in real time, these algorithms can detect unauthorized access attempts, unusual communication patterns, and other indicators of cyber threats, enabling timely responses to mitigate potential security risks.

Real-Time Response to Security Threats Using Machine Learning

Real-time response to security threats using machine learning involves continuous data collection and monitoring from various sources, where anomaly detection models like isolation forests, autoencoders, and recurrent neural networks (RNNs) identify unusual patterns indicative of threats. Intrusion detection systems (IDS) utilize both signature-based and behavior-based detection methods to flag potential intrusions. Stream processing frameworks enable real-time analysis, while automated response systems and dynamic incident response playbooks are triggered to mitigate threats immediately. Reinforcement learning enhances adaptive response strategies, and distributed consensus algorithms ensure robust decision-making. Continuous learning from past incidents and integrating external threat intelligence feeds to ensure that the system evolves and improves its detection and response capabilities over time.

CHAPTER 2 MACHINE LEARNING APPLICATIONS IN UNMANNED AERIAL VEHICLE SWARMS

Persistent Challenges and Prospective Advancements to Pursue

Battery Recharging and Scheduling

Flight time is a critical limitation for MAVs (micro aerial vehicles). Swarming can extend the overall flight time by allowing some MAVs to recharge while others continue operating. This involves two primary challenges: designing an integrated MAV and recharging system and managing distributed scheduling among the drones. Although there has been research progress, a fully automated and distributed recharging method for MAV swarms outside controlled environments is still under development. Leonard [37] created battery swapping stations for quick battery replacement in quadrotors, which, while efficient, require precise landing. Shen et al. [38] proposed a contact-based recharging station, which, although slower, simplifies the system. Leahy et al. [39] showcased a multiagent aerial system comprising three quadcopter drones, alternated between surveillance and recharging. Brommer et al. [40] developed formal strategies based on temporal logic to ensure proper queuing for recharging, though their experiments involved only one MAV operating at a time. Commercial charging stations, as noted by Bin Junaid et al. [41], are becoming available, potentially accelerating research. Wireless charging, discussed by Achtelik et al. [42] and Sá and Festa [43], offers a less precise landing requirement, although it is slower. Increasing MAV flight time can also be approached through design improvements, such as integrating solar cells, as seen in the "Skysailor" MAV [44], or hybrid designs like the one proposed by [45] which alternates between fixed-wing and quadrotor modes to store and use surplus energy. Fully solar-powered quadrotors, demonstrated by [46], require significant surface area and low altitudes. Combustion engines and fuel cells offer alternative power sources, with fuel cells showing promising results for long-endurance flights [47–49].

CHAPTER 2 MACHINE LEARNING APPLICATIONS IN UNMANNED AERIAL VEHICLE SWARMS

Swarm-Level Active Fault Detection

Distributed and proactive fault identification mechanisms is essential for MAV swarms to ensure the overall system's reliability. Highlighted the risk of one MAV's errors affecting the entire swarm. Pugh and Martinoli [50] used failure mode and effect analysis (FMEA) to assess swarm reliability by identifying potential failure points. Tarapore et al. [51] introduced the sensor analysis-based fault detection (SAFDetection) approach. The study utilizes clustering algorithms to simulate anticipated robot behavior and detect any anomalies that might indicate faults. A distributed version [52] allows each robot to learn its behavior model locally and share it, enhancing scalability. Lane et al. [53] proposed a strategy for classifying normal and abnormal behaviors using neighbors' behavior synthesis in a binary feature vector. Later, Freistetter [54] introduced a consensus algorithm for collective decision-making on team member behavior, tested on real robotic systems. Tsykunov et al. [55] reviewed this active research area, emphasizing the importance of robust fault detection for safe MAV swarm operations.

Controlling and Supervising Swarms of MAVs

Effective control interfaces for MAV swarms should allow operators to issue commands like takeoff, landing, mission objectives, and emergency procedures intuitively and with minimal effort [56]. Vincenzi et al. [57] explored using a gesture vocabulary for human operators to command MAV teams, detected by the MAVs' onboard cameras. Nam et al. [58] proposed a haptic glove interface that enables drone control emulating a spring-damper connection. Investigations have explored the use of gestural languages for interaction [23] and virtual reality applications for MAV competition. Additionally, strict MAV flight regulations, especially

CHAPTER 2   MACHINE LEARNING APPLICATIONS IN UNMANNED AERIAL VEHICLE SWARMS

outdoors, pose significant challenges, often requiring one pilot per drone. For a comprehensive overview of the challenges and current technologies in human control of aerial swarms, refer to Hocraffer and Nam.

Conclusion

The chapter on "Machine Learning Applications in UAV Swarms" highlights how integrating machine learning (ML) and deep learning (DL) technologies revolutionizes the capabilities of drone swarms by enabling autonomous navigation, collaborative decision-making, adaptive behavior, and resource optimization. Initially developed for military purposes, UAVs now serve numerous civilian applications due to their affordability and versatility. ML techniques like reinforcement learning and distributed learning empower drones to learn from their environment, optimize their behavior, and communicate effectively within the swarm. Despite these advancements, challenges such as local sensing, control, and interdrone communication remain, requiring further technological improvements to fully harness the potential of UAV swarms.

Summary of Key Points Discussed in the Chapter

The chapter on "Machine Learning Applications in UAV Swarms" covers several key points:

1. Introduction to UAV swarms

    - UAV swarms consist of multiple drones working collaboratively, either autonomously or semiautonomously, using internal processors or remote controls.

    - Initially developed for military purposes, UAVs now have significant civilian applications.

CHAPTER 2   MACHINE LEARNING APPLICATIONS IN UNMANNED AERIAL VEHICLE SWARMS

2. Historical context

- The development of UAVs began in the 19th century, evolving significantly in the early 20th century for military use, including target practice and carrying payloads for various missions.

3. Machine learning in UAV swarms

- ML techniques, such as reinforcement learning, deep learning, and distributed learning, enable drones to navigate, make decisions, and communicate within a swarm.

- Advanced distributed learning techniques, such as federated learning, multiagent reinforcement learning, decentralized inference, and split learning, are pivotal for enabling sophisticated collaborative UAV swarm systems.

4. Advantages of UAV swarms

- Coordinated missions involving multiple drones to enhance efficiency, adaptability, and resource optimization.

- ML improves UAV swarm capabilities by enabling autonomous navigation, collaborative decision-making, adaptive behavior, and efficient resource management.

5. Fundamentals and characteristics

- UAV swarms are characterized by autonomy, scalability, robustness, adaptability, and collective intelligence, allowing them to perform complex tasks more effectively than individual drones.

6. Challenges and limitations

    - Current UAV swarm technologies face challenges like local sensing, control, design complexities, and intraswarm communication and coordination.

    - Advancements are needed in decentralized control, local sensor integration, and robust interdrone communication systems to overcome these challenges.

7. Machine learning techniques

    - The chapter explains key ML concepts, including supervised, unsupervised, and reinforcement learning.

    - Techniques like neural networks, computer vision, and natural language processing are used for tasks such as obstacle avoidance, tracking, and real-time decision-making.

8. Prospects for the integration of machine learning with UAV swarms

    The integration of machine learning with UAV swarms holds promising prospects, significantly enhancing their operational capabilities through autonomous navigation, adaptive behavior, and collaborative decision-making. ML techniques like reinforcement learning and deep learning enable drones to learn from their environment, optimize performance, and improve coordination within the swarm. This integration facilitates efficient resource management and complex task execution, making UAV swarms more effective in diverse applications ranging from military missions to civilian uses such as disaster response and environmental monitoring. However, overcoming challenges like local sensing, control complexities, and robust interdrone communication is essential to fully realize the potential of this technology.

References

[1] Boriceanu, A.-M., THE USE OF UNMANNED AERIAL VEHICLES FOR MONITORING PURPOSES IN CIVILIAN APPLICATIONS. Review of the Air Force Academy, 2021(1): p. 17-26.

[2] Khalil H, Rahman SU, Ullah I, Khan I, Alghadhban AJ, Al-Adhaileh MH, Ali G, ElAffendi M. A UAV-Swarm-Communication Model Using a Machine-Learning Approach for Search-and-Rescue Applications. Drones. 2022 Nov 23;6(12):372.

[3] Adhikari D, Ullah I, Syed I, Choi C. Phishing Detection in the Internet of Things for Cybersecurity. InCybersecurity Management in Education Technologies 2023 Dec 6 (pp. 86-106). CRC Press.

[4] McKenna, A., The Future of Drone Use: Opportunities and Threats From Ethical and Legal Perspectives. The Future of Drone Use, 2016.

[5] Murphy, J.D., Military Aircraft, Origins to 1918: An Illustrated History of Their Impact. 2005: Bloomsbury Publishing USA.

[6] Bayliss, J., Unmanned Aerial Vehicle 25% Report. 2012.

[7] Mazhar T, Irfan HM, Haq I, Ullah I, Ashraf M, Shloul TA, Ghadi YY, Imran, Elkamchouchi DH. Analysis of Challenges and Solutions of IoT in Smart Grids Using AI and Machine Learning Techniques: A Review. Electronics. 2023 Jan 3;12(1):242.

[8] Adu-Gyamfi, S., R.M. Gyasi, and B.D. Darkwa, Historicizing medical drones in Africa: a focus on Ghana. History of science and technology, 2021. **11**(1): p. 103-125.

[9] Taylor, J.W.R. and K. Munson, Jane's pocket book of remotely piloted vehicles: robot aircraft today. 1977: Collier Books.

[10] Kurunathan, H., et al., Machine Learning-Aided Operations and Communications of Unmanned Aerial Vehicles: A Contemporary Survey. IEEE Communications Surveys & Tutorials, 2023.

[11] Ding, Y., et al., Distributed Machine Learning for UAV swarms: Computing, Sensing, and Semantics. IEEE Internet of Things Journal, 2023.

[12] Pal R, Adhikari D, Heyat MB, Ullah I, You Z. Yoga Meets Intelligent Internet of Things: Recent Challenges and Future Directions. Bioengineering. 2023 Apr 9;10(4):459.

[13] Khan HU, Sohail M, Ali F, Nazir S, Ghadi YY, Ullah I. Prioritizing the multi-criterial features based on comparative approaches for enhancing security of IoT devices. Physical Communication. 2023 Aug 1;59:102084.

[14] Shan, L., et al., Vehicle-to-Vehicle Based Autonomous Flight Coordination Control System for Safer Operation of Unmanned Aerial Vehicles. Drones, 2023. **7**(11): p. 669.

[15] Khalil, H., et al., A UAV-Swarm-Communication Model Using a Machine-Learning Approach for Search-and-Rescue Applications. Drones, 2022. **6**(12): p. 372.

[16] Weng, L., et al., Immune network-based swarm intelligence and its application to unmanned aerial vehicle (UAV) swarm coordination. Neurocomputing, 2014. **125**: p. 134-141.

[17] Kushleyev, A., et al., Towards a swarm of agile micro quadrotors. Autonomous Robots, 2013. **35**(4): p. 287-300.

[18] Northcott, P.A., et al., Subgroup-specific structural variation across 1,000 medulloblastoma genomes. Nature, 2012. **488**(7409): p. 49-56.

[19] Augugliaro, F., et al., The Flight Assembled Architecture installation: Cooperative construction with flying machines. IEEE Control Systems Magazine, 2014. **34**(4): p. 46-64.

[20] Mirjan, A., et al., Building a Bridge with Flying Robots. Robotic fabrication in architecture, art and design 2016, 2016: p. 34-47.

[21] Hausman, K., et al., Observability-Aware Trajectory Optimization for Self-Calibration With Application to UAVs. IEEE Robotics and Automation Letters, 2017. **2**(3): p. 1770-1777.

[22] Rachman, E. and R. Razali, A MATHEMATICAL MODELING FOR DESIGN AND DEVELOPMENT OF CONTROL LAWS FOR UNMANNED AERIAL VEHICLE (UAV). International Journal of Applied Science and Technology, 2011. **1**(4).

[23] Coppola, M., et al., A Survey on Swarming With Micro Air Vehicles: Fundamental Challenges and Constraints. Frontiers in Robotics and AI, 2020. **7**: p. 18.

[24] Krause-Jensen, D. and C.M. Duarte, Substantial role of macroalgae in marine carbon sequestration. Nature Geoscience, 2016. **9**(10): p. 737-742.

[25] Floreano, D. and R.J. Wood, Science, technology and the future of small autonomous drones. nature, 2015. **521**(7553): p. 460-466.

[26] Smeur, E.J., Q. Chu, and G.C. De Croon, Adaptive Incremental Nonlinear Dynamic Inversion for Attitude Control of Micro Air Vehicles. Journal of Guidance, Control, and Dynamics, 2016. **39**(3): p. 450-461.

[27] Ullah I, Ahamd I, Anwar MS, Tao Y, Shafiq M. Data Science Meets Intelligent Internet of Things. InFuture Communication Systems Using Artificial Intelligence, Internet of Things and Data Science (pp. 73-91). CRC Press

[28] Mazhar T, Talpur DB, Shloul TA, Ghadi YY, Haq I, Ullah I, Ouahada K, Hamam H. Analysis of IoT Security Challenges and Its Solutions Using Artificial Intelligence. Brain Sciences. 2023 Apr 19;13(4):683.

[29]   Asif M, Khan WU, Afzal HR, Nebhen J, Ullah I, Rehman AU, Kaabar MK. Reduced-Complexity LDPC Decoding for Next-Generation IoT Networks. Wireless Communications and Mobile Computing. 2021;2021(1):2029560.

[30]   Gupta D, Juneja S, Nauman A, Hamid Y, Ullah I, Kim T, Tag eldin EM, Ghamry NA. Energy Saving Implementation in Hydraulic Press Using Industrial Internet of Things (IIoT). Electronics. 2022 Dec 6;11(23):4061.

[31]   Khan WU, Imtiaz N, Ullah I. Joint optimization of NOMA-enabled backscatter communications for beyond 5G IoT networks. Internet Technology Letters. 2021 Mar;4(2):e265.

[32]   Li, J., et al., Microseismic joint location and anisotropic velocity inversion for hydraulic fracturing in a tight Bakken reservoir. Geophysics, 2014. **79**(5): p. C111-C122.

[33]   Mazhar T, Irfan HM, Khan S, Haq I, Ullah I, Iqbal M, Hamam H. Analysis of Cyber Security Attacks and Its Solutions for the Smart grid Using Machine Learning and Blockchain Methods. Future Internet. 2023 Feb 19;15(2):83.

[34]   Rasheed Z, Ma YK, Ullah I, Tao Y, Khan I, Khan H, Shafiq M. Edge Computing in the Digital Era: The Nexus of 5G, IoT and a Seamless Digital Future. InFuture Communication Systems Using Artificial Intelligence, Internet of Things and Data Science (pp. 213-234). CRC Press.

[35] Khan HU, Hussain A, Nazir S, Ali F, Khan MZ, Ullah I. A Service-Efficient Proxy Mobile IPv6 Extension for IoT Domain. Information. 2023 Aug 14;14(8):459.

[36] Khalil I, Khalil A, Ullah I, Tao Y, Khan I, Ashraf S, Ismael WM. Social Internet of Things (SIoT): Recent Trends and Its Applications. Future Communication Systems Using Artificial Intelligence, Internet of Things and Data Science.:159-92.

[37] Leonard, L.B., Children with Specific Language Impairment. 2017: MIT press.

[38] Shen, S., et al. Multi-sensor fusion for robust autonomous flight in indoor and outdoor environments with a rotorcraft MAV. in 2014 IEEE International Conference on Robotics and Automation (ICRA). 2014. IEEE.

[39] Leahy, K., et al., Persistent surveillance for unmanned aerial vehicles subject to charging and temporal logic constraints. Autonomous Robots, 2016. **40**: p. 1363-1378.

[40] Brommer, C., et al. Long-Duration Autonomy for Small Rotorcraft UAS Including Recharging. in 2018 IEEE/RSJ International Conference on Intelligent Robots and Systems (IROS). 2018. IEEE.

[41] Bin Junaid, A., et al., Autonomous Wireless Self-Charging for Multi-Rotor Unmanned Aerial Vehicles. energies, 2017. **10**(6): p. 803.

[42] Achtelik, M.C., et al. Design of a flexible high performance quadcopter platform breaking the MAV endurance record with laser power beaming. in 2011 IEEE/RSJ International Conference on Intelligent Robots and Systems. 2011. IEEE.

[43] Sá, D.C.d. and C. Festa, Inflammasomes and dermatology. Anais brasileiros de dermatologia, 2016. **91**(5): p. 566-578.

[44] Goh, G.D., et al., Additively manufactured continuous carbon fiber-reinforced thermoplastic for topology optimized unmanned aerial vehicle structures. Composites Part B: Engineering, 2021. **216**: p. 108840.

[45] Gong, A. and D. Verstraete. Design and Bench Test of a Fuel-Cell/Battery Hybrid UAV Propulsion System Using Metal Hydride Hydrogen Storage. in 53rd AIAA/SAE/ASEE Joint Propulsion Conference. 2017.

[46] De Wagter, C., et al., Design and Testing of a Vertical Take-Off and Landing UAV Optimized for Carrying a Hydrogen Fuel Cell with a Pressure Tank. Unmanned Systems, 2020. **8**(04): p. 279-285.

[47] Zhou, Y., et al., Secure Communications for UAV-Enabled Mobile Edge Computing Systems. IEEE Transactions on Communications, 2019. **68**(1): p. 376-388.

[48] Xu, B., et al., Failure analysis of unmanned autonomous swarm considering cascading effects. Journal of Systems Engineering and Electronics, 2022. **33**(3): p. 759-770.

[49] Li, X. and L.E. Parker. Sensor Analysis for Fault Detection in Tightly-Coupled Multi-Robot Team Tasks. in Proceedings 2007 IEEE International Conference on Robotics and Automation. 2007. IEEE.

[50] Pugh, J. and A. Martinoli, Distributed scalable multi-robot learning using particle swarm optimization. Swarm Intelligence, 2009. **3**: p. 203-222.

[51] Tarapore, D., A.L. Christensen, and J. Timmis. Abnormality Detection in Robots Exhibiting Composite Swarm Behaviours. in ECAL. 2015.

[52] Telli, K., et al., A Comprehensive Review of Recent Research Trends on Unmanned Aerial Vehicles (UAVs). Systems, 2023. **11**(8): p. 400.

[53] Lane, S.N., A. Gentile, and L. Goldenschue, Combining UAV-Based SfM-MVS Photogrammetry with Conventional Monitoring to Set Environmental Flows: Modifying Dam Flushing Flows to Improve Alpine Stream Habitat. Remote Sensing, 2020. **12**(23): p. 3868.

[54] Freistetter, A., Design and Implementation of Autonomous and Collaborative Drone Behaviour in a Smartphone App/submitted by Alexander Freistetter. 2020.

[55] Tsykunov, E., et al. SwarmTouch: Tactile Interaction of Human with Impedance Controlled Swarm of Nano-Quadrotors. in 2018 IEEE/RSJ International Conference on Intelligent Robots and Systems (IROS). 2018. IEEE.

[56] Serpiva, V., et al., DronePaint: Swarm Light Painting with DNN-based Gesture Recognition, in ACM SIGGRAPH 2021 Emerging Technologies. 2021. p. 1-4.

[57] Vincenzi, D.A., B.A. Terwilliger, and D.C. Ison, Unmanned Aerial System (UAS) Human-Machine Interfaces: New Paradigms in Command and Control. Procedia Manufacturing, 2015. **3**: p. 920-927.

[58] Nam, C., et al. Predicting trust in human control of swarms via inverse reinforcement learning. in 2017 26th ieee international symposium on robot and human interactive communication (ro-man). 2017. IEEE.

CHAPTER 3

Real-World Deployments in Unmanned Aerial Vehicle Swarms

Muhammad Danyal Javed[1], Sher Taj[2], Yonghao Liu[2], Rahim Khan[3*], Hina Hassan[4], Inam Ullah[8], M. A. Al-Khasawneh[5,6,7]

[1]Department of Bachelor Science Information Technology, Shaheed Benazir Bhutto University, Shaheed Benazirabad Nawab shah, 67450, Pakistan; dani.ai.practitioner@gmail.com
[2]Daqing Normal University, Daqing, Heilongjiang, 163455, China; shertajkhan002@gmail.com, yonghaoliu1980@163.com
[3]College of Information and Communication Engineering, Harbin Engineering University, Harbin150001, China; rahim@hrbeu.edu.cn
[4]College of Life Science and Technology, Harbin Normal University, Heilongjiang Province, Harbin, China; hinauoh999@gmail.com
[5]Department of Computer Engineering, Gachon University, Seongnam 13120, Republic of Korea; inam.fragrance@gmail.com
[6]School of Computing, Skyline University College, University City Sharjah, 1797, Sharjah, UAE
[7]Applied Science Research Center. Applied Science Private University, Amman, Jordan;
[8]Jadara University Research Center, Jadara University, Irbid, Jordan; mahmoudalkhasawneh@outlook.com
Corresponding Author: Rahim Khan (rahim@hrbeu.edu.cn)

CHAPTER 3 REAL-WORLD DEPLOYMENTS IN UNMANNED AERIAL VEHICLE SWARMS

Introdumction

A swarm of unmanned aerial vehicles consists of a group of drones and aerial robots that collaborate to accomplish specific objectives [1]. Each drone in a swarm has a specific number of rotors and can vertically hover, take off, and land [2]. The flight of the drones is controlled either manually, by remote control operations, or autonomously by using processors deployed on the drones. A common purpose for drones is a military one, but their civilian applications have attracted increased attention in recent times [3]. Indeed, low-cost drones and their swarms provide a promising platform for research projects and future commercial applications that will help people in their work and everyday lives [4].

Swarms of drones can be classified in different ways. For example, they can be categorized into fully and semiautonomous swarms [5]. From another point of view, the classification can be seen in single-layered swarms where every drone acts as its own leader and multilayered swarms that have dedicated leader drones at each layer, which report to their leader drones at a higher layer; the ground-based server station is the highest layer in this hierarchy, 2021. In each swarm, every drone can have dedicated data collection and processing tasks with sufficient computing capability to execute these tasks in real time. Central processing often takes place on a more performant server base station or even in the cloud.

The increasing use of drone swarms in civilian applications underscores their potential in diverse fields ranging from agriculture to disaster management and logistics [6]. As we move toward more sophisticated implementations of drone technology, understanding and optimizing the hierarchical structures of these swarms will be crucial for their effective deployment [7].

CHAPTER 3 REAL-WORLD DEPLOYMENTS IN UNMANNED AERIAL VEHICLE SWARMS

The Significance of UAV Swarms in Modern Applications

1. **Applications in Military and Defense Systems:** An example of swarming everyone knows is found in the military and defense applications. Large groups of drones can scout enemy territory, keep an eye on them, and be used as an attack tool. Working together provides a significant advantage in both offensive and defensive operations, leading to better target identification and data.

2. **Search and Rescue:** Swarms of drones in disaster areas can help search and rescue far more effectively. They can cover a lot of ground quickly, finding survivors and relaying that information to first responders for a fast and efficient rescue [8].

3. **Environmental Surveillance:** Swarm UAVs can help to keep track of and safeguard the environment. Wildlife populations, forest perimeters, and fire-prone areas can be monitored and secured through drones, which can relay live data to the concerned authorities to identify endangered life forms and assess fire risks. Figure 3-1 shows the application of drones in several fields like farming, filming, and rescuing.

CHAPTER 3 REAL-WORLD DEPLOYMENTS IN UNMANNED AERIAL VEHICLE SWARMS

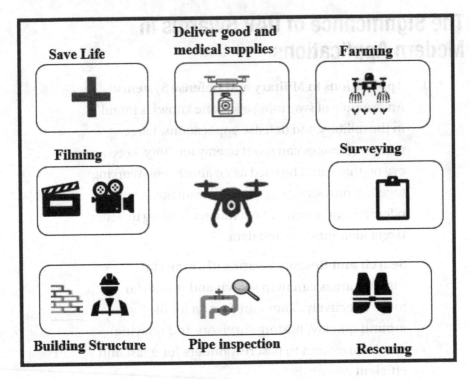

Figure 3-1. Applications of drones

Real-World Deployments Advance Technology

Technologies are becoming increasingly complicated and interconnecting together. Cars, airplanes, medical devices [9], financial transactions, and electricity systems all incorporate more software than ever before, making them seem both harder to understand and, in some cases, harder to control [10]. Government and corporate surveillance of individuals and information processing relies largely on digital technologies and artificial intelligence and therefore involves less human-to-human contact than ever before and more opportunities for biases to be embedded and codified in our technological systems in ways we may not even be able to identify or recognize [11]. Bioengineering advances are opening up challenging philosophical, political, and economic questions regarding human-natural relations. Additionally, the management of these large and small devices and systems is increasingly done through the cloud, so that control over them is both very remote and removed from direct human or social control. The study of how to make technologies like artificial intelligence or the Internet of Things "explainable" has become its own area of research because it is so difficult to understand how they work or what is at fault when something goes wrong [12].

This growing complexity makes it more difficult than ever and more imperative than ever for scholars to probe how technological advancements are altering life around the world in both positive and negative ways and what social, political, and legal tools are needed to help shape the development and design of technology in beneficial directions [13]. This can seem like an impossible task in light of the rapid pace of technological change and the sense that its continued advancement is inevitable, but many countries around the world are only just beginning to take significant steps toward regulating computer technologies and are still in the process of radically rethinking the rules governing global data flows and exchange of technology across borders.

These are exciting times not just for technological development but also for technology policy. Our technologies may be more advanced and complicated than ever but so, too, are our understandings of how they can best be leveraged, protected, and even constrained. The structures of technological systems are determined largely by government and institutional policies, and those structures have implications for social organization and agency, ranging from open source, open systems that are highly distributed and decentralized, to those that are tightly controlled and closed, structured according to stricter and more ranked hierarchical models [14].

CHAPTER 3   REAL-WORLD DEPLOYMENTS IN UNMANNED AERIAL VEHICLE SWARMS

Aims and Contributions

This chapter provides an analysis of real-world deployments of UAV swarms:

1. **Synthesize Existing Knowledge:** We will delve into the current state of the art by reviewing relevant research on UAV swarms, including their theoretical foundations and development history.

2. **Analyze Real-World Deployments:** We will explore case studies of real-world UAV swarm applications in diverse fields like disaster management, urban services, agriculture, and security. This will offer valuable insights into the practical applications and challenges of these deployments physically.

3. **Identify Research Gaps:** Through our analysis, we will identify gaps in existing knowledge and areas where further research is needed to advance the development and responsible use of UAV swarms.

4. **Critical Analysis:** We go beyond simply describing real-world deployments. We will critically analyze their impact on technological advancements, identifying both successes and limitations.

5. **Future Direction:** By highlighting research gaps and emerging trends, we will inform future research directions that can further optimize UAV swarm technology and its applications.

6. **Policy Considerations:** Our analysis will also contribute to discussions surrounding regulations and ethical considerations for the safe and responsible use of UAV swarms in various sectors.

Theoretical Background and Literature Review

This section delves into the theoretical foundations of UAV drones and reviews the current research in this field and its related areas. This chapter aims to comprehend the theoretical context of real-world deployments and their influence on technological advancements [15]. The fundamental concepts of UAV swarms revolve around swarm intelligence, which draws inspiration from biological behavior like bird flocking to create a collective intelligence from coordinated individual actions without a central controller. Decentralization and self-organization are key principles, as each UAV in the swarm makes decisions based on local information and communication with nearby members, allowing for adaptability to dynamic environments without centralized control. Various types of UAV swarms can be classified based on factors like size, similarity, dissimilarity, and communication ranges, enabling analysis of different swarm configurations for effective real-world deployment scenarios.

Fundamental Concepts of UAV Swarms

This subsection lays the foundation for understanding UAV swarms by delving into their core principles and characteristics.

Swarm Intelligence

UAV swarms are the concept of swarm intelligence, inspired by the collective behavior in nature. Imagine a flock of birds effortlessly or a swarm of ants collaborating to build structures. These behaviors emerge from the actions of individual entities like birds and ants operating with limited individual intelligence but achieving complex outcomes through coordinated interaction. Similarly, UAV swarms rely on decentralized decision-making and communication between individual UAVs to achieve a common goal without a central controller [16].

CHAPTER 3   REAL-WORLD DEPLOYMENTS IN UNMANNED AERIAL VEHICLE SWARMS

Here are some key aspects of swarm intelligence and smart in UAV swarms:

> **Decentralization:** Individual UAVs make decisions based on local information and communication with nearby swarm members. This allows for adaptation and flexibility in 3D environments.
>
> **Scalability:** UAV swarms can be easily scaled or down depending on the mission requirements. Adding or removing UAVs generally does not require significant changes to the overall control system.

Decentralization and Self-Organization

Unlike traditional robots controlled by a central computer, UAV swarms operate in a decentralized manner and in person. Each UAV possesses a degree of autonomy and makes decisions based on its sensors, communication with nearby persons, and preprogrammed rules of execution. This decentralized approach offers several advantages:

1. **Flexibility:** Decentralization allows for faster response times to changing environments as individual UAVs can react dynamically.

2. **Scalability:** Adding or removing UAVs is simpler in decentralized systems, making swarms adaptable to various mission needs.

Effective communication is vital for coordinated action within a UAV swarm, as individual drones exchange critical information regarding their location, environment, and mission objectives, allowing the swarm to operate cohesively toward a common goal. Different communication network topologies, such as mesh networks, can be utilized within UAV swarms to facilitate direct communication between all drones, ensuring seamless information flow within the network. The selection of communication protocols is influenced by considerations like bandwidth efficiency, aiming to optimize data transmission to conserve battery power on each UAV and enhance overall swarm performance and endurance.

Historical Milestones in UAV Swarm Development

The field of UAV swarm development has had significant advancements over the past few centuries. This section has some key historical milestones that have paved the way for real-world deployments.

The seeds of UAV swarm technology were sown in the realm of theoretical research. Pioneering works laid the groundwork for understanding and simulating collective behavior in UAV swarms:

> **1987: Boids Model by Craig Reynolds:** This seminal work introduced the Boids model, a computer simulation that mimicked the flocking behavior of birds. It demonstrated how simple rules governing individual movement could lead to complex collective patterns, inspiring research on swarm intelligence for UAV control.

Early demonstrations of UAV swarms served as proof and showcased their potential capabilities:

> **2000: US Military Research Projects:** The US Department of Defense funded research projects exploring the use of UAV swarms for military applications, such as reconnaissance and cooperative target engagement.
>
> **2016:** Defense Advanced Research Projects Agency

Experiment: DARPA's Locust program explored the concept of large scale, collaborative with UAV swarms for communications relay tasks, demonstrating the potential for robust and adaptable communication networks.

Theoretical Frameworks Guiding Real-World Deployments

The successful deployment of UAV swarms in real-world applications relies on robust theoretical frameworks. This section discusses key theoretical concepts that inform swarms' design, control, and operation.

Control theory principles provide a foundation for designing algorithms that govern the behavior of UAV swarms. These algorithms dictate how individual swarms move, communicate, and collaborate to achieve the desired mission objectives. Here are some related control theory concepts:

Network theory principles are valuable for analyzing communication patterns and information within a UAV swarm. Understanding these communication networks is for optimizing swarm performance and ensuring a robust system in swarms.

Methodological Approach

This section outlines the research methodology that would be employed to investigate a specific aspect of UAV swarms.

Research Design Rationale in Drones

The research design chosen for your study on UAV swarms will depend on your specific research question. Here are some common research designs used in this field, along with their rationales [17]:

CHAPTER 3   REAL-WORLD DEPLOYMENTS IN UNMANNED AERIAL VEHICLE SWARMS

- **Survey Research:** A survey approach is suitable for collecting data on current practices and experiences related to UAV swarms. This could be part of surveying industry professionals, drone operators, or the public depending on your research question rationale: surveys are efficient for collecting data from a large number of participants and can provide valuable insights into attitudes and trends.

- **Case Study Research:** This approach allows for in-depth exploration of a specific swarm deployment or project. You could analyze a successful real-world application of UAV swarms, like their use in disaster response.

Data Sources and Collection Techniques for Drones

The data collection techniques for your research on UAV swarms will depend on the chosen research design and the collection and storing of data monitored in real time using an IoT device as mentioned in [18] and efficient resource management of the interconnected networks (using the provided sensors) [46, 47].

Case Study Methodology for Drones in Buildings

If your research focuses on a specific application of UAV swarms in buildings, a case study methodology might be appropriate. Here's how you could structure it:

- **Case Selection:** Choose a specific building inspection or search and operation that utilized UAV swarms [19].

CHAPTER 3 REAL-WORLD DEPLOYMENTS IN UNMANNED AERIAL VEHICLE SWARMS

- **Data Collection:** Gather information through interviews with project managers, pilots, or data analysts involved in the deployment. Analyze relevant reports, technical experts, and sensor data from the UAVs.

- **Data Analysis:** Evaluate the effectiveness of the UAV drone deployment by examining factors like mission completion time, data quality, and challenges. Identify key success factors and potential areas for enhancement.

- **Case Discussion:** Present your finding point, and highlight the case studies' contribution to the understanding of swarms in building applications.

The methodology of UAV building is explained in Figure 3-2, as discussed in a case study on drone building.

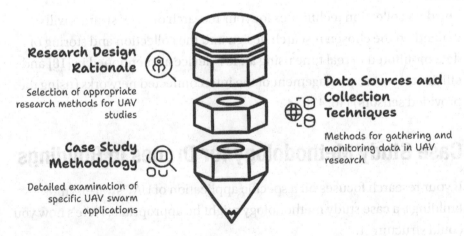

Figure 3-2. *Methodology of UAV [20]*

CHAPTER 3   REAL-WORLD DEPLOYMENTS IN UNMANNED AERIAL VEHICLE SWARMS

Basic Drone Swarm Error Analysis Tools and Metrics

Analyzing data from UAV swarms and detecting errors requires specialized tools and metrics as in [21]. The reliability analysis of a drone fleet can be implemented based on different mathematical approaches.

1. **Data Visualization Tools:** Software like MATLAB and Python (libraries) can be used to visualize sensor data collected by UAV swarms, helping to identify patterns and features.

2. **Machine Learning (Deep Learning) Techniques:** Machine learning algorithms can be trained more to detect errors in flight paths, sensor readings, or communication patterns within a swarm [22].

3. **Metrics:** Several metrics can be used to evaluate swarm performance and identify potential errors, like?

Technical Foundations of UAV Swarms

UAV swarms, consisting of multiple unmanned aerial vehicles working combined, hold potential for various applications. To navigate complex environments and achieve their goals, these swarms rely on complex several technical aspects. Here are some of the key values as seen in [23] that *systems that can act autonomously, learn, decide, and fulfil the mission given as a swarm.*

UAV within a swarm relies on robust flight control systems to ensure stability, efficiency, and obstacle avoidance. These systems incorporate sophisticated algorithms for motor control, sensor fusion, and autonomous navigation. Furthermore, reliable communication is essential for swarm coordination, with drones utilizing various methods such as

CHAPTER 3 REAL-WORLD DEPLOYMENTS IN UNMANNED AERIAL VEHICLE SWARMS

radio frequency, Wi-Fi, or even Li-Fi for data exchange. This facilitates the sharing of crucial information regarding position, environment, and tasks among the drones. Additionally, processing sensor data and making real-time decisions necessitate significant onboard computing power. Ongoing advancements in miniaturization have yielded powerful microprocessors capable of handling complex swarm algorithms on each drone [40].

Ensuring the security of the communication path and safeguarding the autonomy of the swarm against cyberattacks is paramount. Additionally, managing and coordinating swarms as their numbers continue to increase presents an ongoing scalability challenge. This necessitates effective systems to manage and coordinate larger groups of drones. Moreover, the establishment of regulations and safety protocols is crucial to guarantee the safe operation of swarm drones in populated areas, thereby addressing safety concerns and ensuring compliance with regulatory requirements.

Communication Networks and Protocols

UAV swarm networks have been used in many applications, such as disaster recovery, area surveillance, weather monitoring systems, and military communications networks. There are many challenging R&D issues in UAV swarm network designs, such as the hardware or software integration for large-scale UAV network management, long-distance data transmissions among UAVs, swarm shape or formation control, and intelligent UAV position prediction. This chapter will be the first one to cover the engineering designs for dynamic, large-scale UAV networks. It has the technical models and protocol specifications for practical UAV swarm network deployment.

Information and communication technology is essential for the growth of a nation [24], and effective communication is essential [25]. In UAV swarms, communication is the backbone for coordination and efficient mission execution. Here is a breakdown of the key aspects of communication networks and protocols:

CHAPTER 3   REAL-WORLD DEPLOYMENTS IN UNMANNED AERIAL VEHICLE SWARMS

Communication Network Architectures

- **Centralized Architecture:** A ground control station acts as a central hub, communicating directly with each swarm. This offers good control but suffers from a single point to another point of failure and limited scalability for large swarms.

Communication Protocols

- **Routing Protocols:** These protocols determine how data packets are routed between swarms to reach their destination within the swarm system. Some common protocols include on-demand distance vector and dynamic source routing protocols designed for mobile networks [26]. Figure 3-3 shows the protocols of communication between drones and person commands.

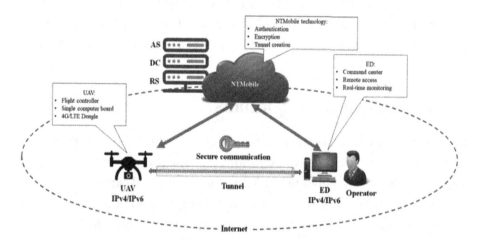

Figure 3-3. *Communication network protocols [27]*

CHAPTER 3   REAL-WORLD DEPLOYMENTS IN UNMANNED AERIAL VEHICLE SWARMS

Advanced Navigation and Sensing

In UAV swarms, effective navigation and robust sensing capabilities are essential for successful operation in complex environments. Precise localization and positioning of each drone within the swarm are crucial, which can be achieved through techniques such as global navigation satellite system or vision-based methods like landmark recognition. Additionally, path planning algorithms are necessary for optimal flight paths and obstacle avoidance, considering factors such as environmental conditions and mission objectives to ensure efficient and safe swarm operation in dynamic settings.

Swarm Intelligence and Coordination Mechanisms

The study explores the full potential of UAV swarms in their ability to coordinate actions and adapt collectively. Swarm intelligence algorithms, derived from nature, offer a robust framework for achieving goals, providing a detailed understanding of coordination mechanisms in swarms.

Swarm Intelligence Algorithms

Biological Inspiration of Swarms: Swarm intelligence algorithms draw inspiration from the collective behavior in natural swarms like birds flying. These algorithms enable individual drones with limited capabilities to achieve complex goals through local interactions and organization.

Swarm intelligence algorithms are grounded in key principles such as individual drones adhering to simple behavioral rules and engaging in local interactions with neighboring drones to enhance their understanding of the surrounding environment. However, challenges and considerations arise concerning scalability, with a focus on ensuring that swarm intelligence algorithms can efficiently manage large groups of drones. Moreover, emphasizing the importance of robustness, it is vital for the swarm's behavior to remain resilient in the face of failures or environmental changes. Additionally, addressing security concerns is crucial, as swarm coordination is susceptible to cyberattacks, underscoring the necessity for implementing secure communication protocols to safeguard against potential threats.

Real-World Deployment Case Studies

UAV swarms are moving beyond research labs to practical implementation, with case studies demonstrating their potential applications. In precision agriculture, drones forming swarms equipped with cameras and LiDAR sensors can effectively map fields, pinpoint crop health issues, and accurately administer pesticides or fertilizers. The adoption of UAV swarms in agriculture offers advantages such as heightened efficiency, enhanced performance, minimized waste, and improved crop yields, highlighting their valuable impact in optimizing agricultural practices [28].

Search and Rescue

> **Scenario:** Swarms of drones with thermal imaging cameras and loudspeakers can search vast areas for missing persons or survivors after natural disasters come.
>
> **Benefits:** Faster search times, covering larger areas compared to traditional methods, and potential for locating survivors in difficult-to-reach locations.

CHAPTER 3   REAL-WORLD DEPLOYMENTS IN UNMANNED AERIAL VEHICLE SWARMS

Urban Services and Infrastructure Management

Urban System and Infrastructure Management: Transforming Cities with UAV Swarms

UAV swarms hold the potential to revolutionize how cities manage services and infrastructure. Utilizing high-resolution cameras and LiDAR sensors, swarms can autonomously conduct inspections of crucial infrastructure such as bridges and buildings, detecting cracks, corrosion, or other damages effectively. This automation not only reduces risks for human inspectors but also allows for more frequent inspections. Furthermore, the data collected by these swarms can be analyzed using artificial intelligence algorithms to identify potential issues, facilitating preventive measures and cost-saving initiatives based on proactive maintenance strategies.

Urban Planning Side and Development

- **3D Mapping and Modeling:** Swarm-generated data can be used to build detailed 3D models of urban and rural areas, facilitating urban planning, construction monitoring, and disaster efforts [29–31].

- **Traffic Monitoring and Management:** Real-time data from swarms can be used to monitor traffic patterns and congestion, enabling dynamic adjustments to traffic lights or signalization for improved traffic flow.

Security and Surveillance Operations

Drones are used for security and surveillance by using UAV swarms to capture images and videos from high altitudes. Its main aim is to gather information about specific targets like individuals, groups, analysis, or the environment. Drones are small and can navigate challenging, making them excellent for surveying inaccessible. They offer a unique point

CHAPTER 3   REAL-WORLD DEPLOYMENTS IN UNMANNED AERIAL VEHICLE SWARMS

of view, providing a first-person perspective not usually accessible to traditional photographers. Security drones with livestreams video and sensors have become common in law enforcement power.

There are various types of drones, each with unique surveillance missions and innovative techniques [32]. Multirotor drones are highly effective in stationary positions, excelling in detailed asset inspection tasks but are restricted to smaller areas like mines, construction sites, and chemical plants. On the other hand, fixed-wing drones are specifically designed for patrolling vast regions such as coastlines, borders, and highways, enabling extended operational capabilities without the need for frequent recharging.

Understanding the key features of surveillance drones involves recognizing the enhanced capabilities they offer in surveillance operations. These technologies provide security agents with an elevated viewpoint as they can quickly ascend up to 100 m, granting a comprehensive overview of the entire area in a matter of seconds. Additionally, their rapid deployment speed of 20 to 40 km/h allows them to cover distances of 350 to 500 m per minute consistently. Surveillance drones are equipped with remote video controllers, enabling the transmission of live video feeds through a radio-controlled system. Furthermore, their onboard sensors facilitate data collection, enabling drones to gather significant information based on their model specifications for analysis and enhanced operational performance in surveillance tasks.

Challenges and Implications

The use of commercial UAVs relies on global positioning technology for navigation, providing controllers with precise vehicle location information even at long distances. However, a significant safety concern arises from the lack of awareness provided by GPS regarding surrounding airspace, potentially leading to unmanned aerial vehicles interfering with aircraft flight paths and posing safety risks. In terms of privacy, UAVs can gather extensive data through video surveillance and other sensors, raising concerns about personal privacy and prompting legislative action to address such issues. Additionally, the security of unmanned aerial vehicles is a critical consideration, as the data they collect and transmit may be vulnerable to interference or hacking attempts, leading to fears of sensitive information falling into unauthorized hands and impacting public security.

Ethical Dimensions and Public Safety

The increasing adoption of drones is a crucial ethical statement and public safety concerns that demand thoughtful consideration. Here is a breakdown of these issues and potential solutions and some more about it [33].

The growing use of drones for surveillance has brought forth various concerns, including privacy issues stemming from unrestricted data collection that can impede freedom of movement and assembly. Additionally, the potential weaponization of drones for military and law enforcement purposes raises ethical questions surrounding civilian casualties, proportionality, and the development of autonomous weapon systems. Moreover, the increasing presence of drones in airspace heightens the risk of mid-air collisions with manned aircraft, posing a significant threat of catastrophic accidents. Furthermore, the susceptibility of drones to hacking poses security risks, potentially enabling unauthorized individuals to misuse drones or manipulate their actions for malicious intents, posing threats to critical infrastructure and public safety.

The use of drones in public safety brings forth significant concerns regarding privacy violations, primarily due to their capability of capturing detailed images with high-resolution cameras, raising apprehensions around surveillance and potential data misuse. Efficient handling and storage of the collected information, especially in sensitive operations like disaster response, are essential considerations for public safety agencies. Questions regarding data access, retention periods, and secure databases must be addressed when deploying drone swarms. Furthermore, there is a potential risk of data misuse, emphasizing the importance of ensuring drones are employed strictly for their intended purposes and within ethical boundaries to prevent any misuse or unauthorized data access. We look at some more flow of ethical dimensions and public safety of drones in Figure 3-4.

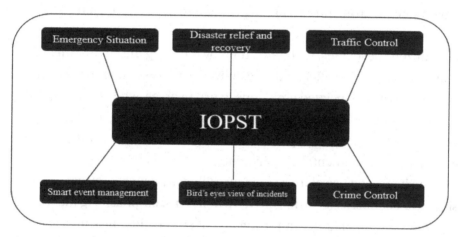

Figure 3-4. Ethical dimensions and public safety of drones

Economic, Environmental, and Social Considerations

Drone delivery is increasing globally and is estimated to reach a market value of USD 641 billion by 2032 [34]. The drone industry holds the promise of creating new job opportunities in areas such as manufacturing, operation, software development, and data analysis, thus contributing to job creation and economic growth. Drones can significantly enhance efficiency across diverse sectors like agriculture, logistics, and

CHAPTER 3 REAL-WORLD DEPLOYMENTS IN UNMANNED AERIAL VEHICLE SWARMS

infrastructure inspection, leading to cost savings through improved processes and resource management. By automating tasks that previously required manual labor, drones have the potential to boost productivity and economic output. However, challenges such as job displacement in certain sectors due to increased automation call for workforce retraining and adaptation. Additionally, integrating drones into existing infrastructure, such as air traffic management systems, requires substantial investment, impacting businesses and governmental budgets. On the environmental front, drones play a pivotal role in promoting precision agriculture by optimizing resource use and minimizing the application of fertilizers, pesticides, and water, thereby reducing the environmental footprint. Moreover, efficient drone delivery services can mitigate traffic congestion and associated emissions in urban areas, while drones utilized for environmental monitoring tasks contribute to tracking deforestation, pollution levels, wildlife populations, aiding conservation efforts, and enhancing environmental sustainability.

The increased use of drones for surveillance purposes raises significant concerns regarding privacy violations, emphasizing the urgent need for clear regulations and stringent data protection measures to ensure user privacy and data security. Safety concerns stemming from possible collisions between drones and manned aircraft or individuals highlight the necessity for robust regulations and responsible operational practices for safe integration into existing airspace. Negative public perception regarding noise pollution, privacy threats, and safety risks poses a challenge to the widespread acceptance of drone technology, underscoring the importance of addressing these issues to promote social acceptance and trust in drone-related applications. Figure 3-5 presents an analysis of drone usage and the percentage of industrial sectors involved.

CHAPTER 3 REAL-WORLD DEPLOYMENTS IN UNMANNED AERIAL VEHICLE SWARMS

Drone Delivery Impact and Considerations

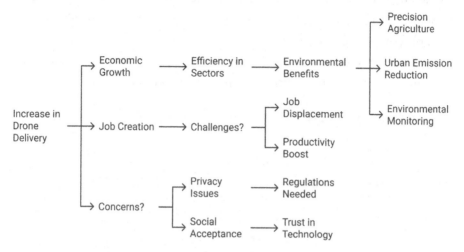

Figure 3-5. Drone analysis and usage

Cost Analysis and ROI of UAV Swarm Deployments

Deploying drone swarms entails substantial initial costs, including expenses for drone acquisition, infrastructure and support systems, and maintenance and operations. Factors such as payload capacity, range, and sensor capabilities influence the costs associated with purchasing multiple drones, while additional expenses arise from ground control stations, communication infrastructure, and software for swarm management. Ongoing costs encompass maintenance activities like battery charging, repairs, and personnel for mission planning and operation. Assessing the return on investment (ROI) potential of drone swarms' hinges on various factors, including reduced operational costs in sectors such as agriculture or infrastructure inspection, where drone swarms can streamline processes and potentially lower labor expenses compared to

CHAPTER 3 REAL-WORLD DEPLOYMENTS IN UNMANNED AERIAL VEHICLE SWARMS

traditional methods. Efficiency gains from faster data collection, enhanced monitoring capabilities, and quicker response times, particularly in tasks like search and rescue, can translate into economic benefits. Moreover, the deployment of drone swarms can open up new revenue streams by enabling services like drone-based delivery or aerial photography, thereby diversifying income sources and enhancing overall ROI.

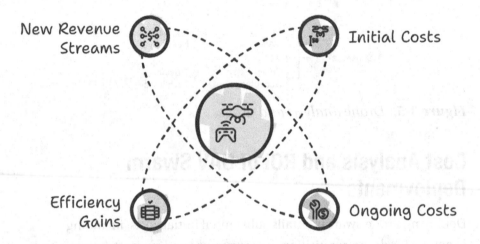

Figure 3-6. UAV swarm deployments [35]

Future Horizons and Technological Innovations

As technology progresses, the line between reality and virtuality may blur with the help of special glasses or headsets that immerse individuals in computer-generated worlds for entertainment or work purposes. Artificial intelligence will advance, enabling super smart machines to assist with various tasks and potentially solve complex problems that are currently beyond human comprehension. Furthermore, advancements in medical science may lead to remarkable breakthroughs in healthcare, allowing doctors to treat almost any illness and possibly extend human life expectancy while improving overall well-being. Additionally, the development of superfast computers capable of performing rapid calculations could pave the way for revolutionary inventions and innovations in various fields.

Integration with Advanced Technologies (AI, IoT, and 5G)

AI, IoT, and 5G are transforming technology, creating new industries, and revolution existence [41–43].

Enhanced Auto and Decision-Making: AI algorithms are taking central stage, automating tasks, and making real time with world data-driven decisions. This integration across various sectors improves accuracy and reduces human error more possible.

- **Seamless Connectivity and Data Exchange:** IoT devices are becoming ubiquitous, creating a vast network of interconnected objects collecting and sharing data.

- **5G High Speeds and Low Latency:** The device gap pertains to real-time communication and data exchange.

- **Advanced Analytics and Insights:** The massive amount of data generated by IoT devices presents valuable insights. AI acts as the muscle, analyzing this data to uncover patterns, predict outcomes, and optimize processes for the future.

Let's delve deeper into how each technology contributes to this powerful integration of AI [36, 37], IoT, and 5G.

CHAPTER 3   REAL-WORLD DEPLOYMENTS IN UNMANNED AERIAL VEHICLE SWARMS

IoT devices and applications utilize embedded sensors and software to collect various datasets from the physical world, enabling a wide range of applications [48, 49]. Environmental monitoring sensors track air quality, temperature, and humidity in smart cities and agricultural fields, while wearable devices such as fitness trackers gather personal health data for proactive health management. IoT connectivity protocols allow devices to communicate with each other, enabling the creation of smart homes where thermostats, lights, and appliances can be controlled remotely to optimize energy usage and enhance comfort. Additionally, connected cars can communicate with each other and infrastructure, improving safety through features like collision avoidance [44, 45]. After discussing IoT, Figure 3-7 shows how IoT and devices work and shows the most popular applications of IoT in many fields.

Technological Integration and Its Impact

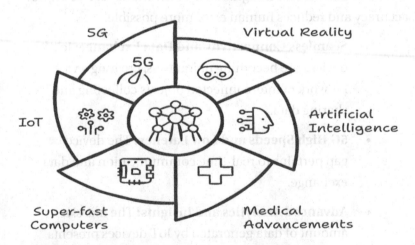

Figure 3-7. *Business segments for IoT implementation*

The ultrafast data speeds facilitated by 5G technology are vital for applications necessitating high bandwidth, such as virtual reality and augmented reality, which depend on rapid data transfer for immersive experiences, and remote surgery, where real-time data transmission with

CHAPTER 3 REAL-WORLD DEPLOYMENTS IN UNMANNED AERIAL VEHICLE SWARMS

minimal latency is crucial for surgeons conducting complex procedures remotely. Low latency in data transfer is essential for various applications, including autonomous vehicles, which require real-time data exchange between vehicles and infrastructure for safe and efficient autonomous driving, and industrial automation, where 5G's low-latency capabilities enable precise control of robots and machines in factories, enhancing production efficiency. The integration of 5G and drones through satellite connectivity is discussed in Figure 3-7, highlighting their relationship with each other.

Best Practices and Strategic Recommendations

A successful drone strategy involves more than just purchasing and flying drones; it involves successfully integrating these tools into operational systems within your business. This means that you are not achieving the results to enhance performance and enable high-quality data capture and recording videos. Maybe you are just starting to look at drones and have realized that it's a whole new technology which the company needs to be researched, formulated, and managed it. There are significant benefits to integrating drones into business operations and real scenarios; however, many program development and integration can be complex, time-consuming, and come with a unique set of risks during deploy [38].

Key Drone Strategy Considerations

- Who will run the program?
- Is there an expert who will champion the project?
- Who will be the Chief Remote Pilot (CRP)?
- Who will support the CRP in their absence or departure?
- Where will you keep the data?

Guidelines for Effective UAV Swarm Deployment

The successful deployment of UAV swarms hinges on careful planning, execution, and restriction to best practices. Before deploying drone swarms, thorough predeployment planning is essential, involving mission and risk assessment to define objectives, environmental conditions, and potential risks, conducting a comprehensive risk evaluation to gauge deployment feasibility. Swarm design and configuration require selecting suitable UAV platforms and aligning swarm configuration, including the number of drones and roles, with mission requirements, considering factors like payload capacity, endurance, and communication range. Regulatory compliance is crucial, necessitating adherence to airspace regulations and authorization procedures established by aviation authorities. Operational considerations involve developing detailed flight plans, including takeoff and landing points, waypoints, and designated operational areas, along with creating a map to restrict swarm movement within authorized airspace and normal weather conditions. Real-time monitoring and communication infrastructure are imperative for continuous surveillance of the swarm's status, health, and surrounding environment, ensuring a reliable communication link between the swarm and human operators for control and data exchange. Additionally, conducting a safety and security review postdeployment is vital to identify potential breaches, enabling continuous improvement and responsible swarm operation [39].

Conclusion

This chapter highlights the transformative potential of unmanned aerial vehicle (UAV) swarms in various sectors. By examining their use in both military and civilian contexts, we've shown how UAV swarms improve efficiency in tasks like urban management, environmental monitoring,

CHAPTER 3   REAL-WORLD DEPLOYMENTS IN UNMANNED AERIAL VEHICLE SWARMS

and emergency response. Our study categorizes UAV swarms into single-layered and multilayered systems, each offering unique organizational advantages. However, significant challenges remain. Current navigation technologies, including AI systems, struggle with object detection and avoidance, raising safety concerns. Ethical issues, such as privacy and the need for strong regulatory frameworks, are also critical. Economically, integrating UAV swarms with technologies such as artificial intelligence, Internet of Things, and 5G appears to be cost-effective, providing substantial returns on investment. Looking ahead, UAV swarms are poised to revolutionize industries, from everyday tasks like package delivery to critical roles in disaster relief. To utilize their full potential, we need strategic rules, guidelines, and best practices to ensure safe and efficient deployment. Unmanned aerial vehicle swarms are set to be game changers in the future of autonomous systems, reinforcement learning, and technological innovation, driving significant operational and societal advancements. Imagine a picture of a future where drone swarms easily deliver your packages and respond to action during natural disasters. The limitless capabilities of these entities are only scratching the surface of their potential.

References

[1] Floreano, D. and R.J. Wood, Science, technology and the future of small autonomous drones. nature, 2015. **521**(7553): p. 460-466.

[2] Kumar, P. and M.L. Clark, Kumar and Clark's Clinical Medicine E-Book. 2012: Elsevier health sciences.

[3] Brambilla, M., et al., Morality and intergroup relations: Threats to safety and group image predict the desire to interact with outgroup and ingroup members. Journal of Experimental Social Psychology, 2013. **49**(5): p. 811-821.

[4] Shakhatreh, H., et al., Unmanned Aerial Vehicles (UAVs): A Survey on Civil Applications and Key Research Challenges. Ieee Access, 2019. **7**: p. 48572-48634.

[5] Cai, G., J. Dias, and L. Seneviratne, A Survey of Small-Scale Unmanned Aerial Vehicles: Recent Advances and Future Development Trends. Unmanned Systems, 2014. **2**(02): p. 175-199.

[6] Mathew, D., et al., Deep immune profiling of COVID-19 patients reveals distinct immunotypes with therapeutic implications. Science, 2020. **369**(6508): p. eabc8511.

[7] Nyemba, W.R. and K.F. Carter, Royal Academy of Engineering Interventions, in Doctoral Training in Engineering: Developing Indigenous Capacities and Skills for Economic Growth in Industrialising Countries. 2024, Springer. p. 41-58.

[8] Erdelj, M., M. Król, and E. Natalizio, Wireless Sensor Networks and Multi-UAV systems for natural disaster management. Computer Networks, 2017. **124**: p. 72-86.

[9] Amin SU, Taj S, Hussain A, Seo S. An automated chest X-ray analysis for COVID-19, tuberculosis, and pneumonia employing ensemble learning approach. Biomedical Signal Processing and Control. 2024 Jan 1;87:105408.

[10] Clark, D.D. and K. Claffy, Anchoring policy development around stable points: An approach to regulating the co-evolving ICT ecosystem. Telecommunications Policy, 2015. **39**(10): p. 848-860.

[11] Demetris, A.J., et al., 2016 Comprehensive Update of the Banff Working Group on Liver Allograft Pathology: Introduction of Antibody-Mediated Rejection. American Journal of Transplantation, 2016. **16**(10): p. 2816-2835.

[12] Doshi-Velez, F. and B. Kim, Towards A Rigorous Science of Interpretable Machine Learning. arXiv preprint arXiv:1702.08608, 2017.

[13] Brynjolfsson, E. and A. McAfee, The Second Machine Age: Work, Progress, and Prosperity in a Time of Brilliant Technologies. 2014: WW Norton & Company.

[14] Van Est, Q., J. Gerritsen, and L. Kool, Human rights in the robot age: Challenges arising from the use of robotics, artificial intelligence, and virtual and augmented reality. 2017.

[15] Gorton, M., C. Hubbarda, and I. Fertő, Theoretical background and conceptual framework. 2013, Leibniz Institute of Agricultural Development in Central and Eastern Europe.

[16] Campion, M., P. Ranganathan, and S. Faruque, UAV swarm communication and control architectures: a review. Journal of Unmanned Vehicle Systems, 2018. **7**(2): p. 93-106.

[17] Rodriguez, S. and V. Hilaire, A methodological approach for the analysis and design of Human–Swarm interactions based upon feedback loops. Expert Systems with Applications, 2023. **217**: p. 119482.

[18] Petkovic, S., D. Petkovic, and A. Petkovic, IoT Devices vs. Drones for Data Collection in Agriculture. DAAAM International Scientific Book, 2017. **16**: p. 63-80.

[19] Sun, Z. and Y. Zhang, Using Drones and 3D Modeling to Survey Tibetan Architectural Heritage: A Case Study with the Multi-Door Stupa. Sustainability, 2018. **10**(7): p. 2259.

[20] Megahed, H.A., et al., Develop of a machine learning model to evaluate the hazards of sand dunes. Earth Science Informatics, 2024: p. 1-25.

[21] Zaitseva, E., et al., Review of Reliability Assessment Methods of Drone Swarm (Fleet) and a New Importance Evaluation Based Method of Drone Swarm Structure Analysis. Mathematics, 2023. **11**(11): p. 2551.

[22] Tufail AB, Ullah I, Rehman AU, Khan RA, Khan MA, Ma YK, Hussain Khokhar N, Sadiq MT, Khan R, Shafiq M, Eldin ET. On Disharmony in Batch Normalization and Dropout Methods for Early Categorization of Alzheimer's Disease. Sustainability. 2022 Nov 8;14(22):14695.

[23] Xiaoning, Z. Analysis of military application of UAV swarm technology. in 2020 3rd International Conference on Unmanned Systems (ICUS). 2020. IEEE.

[24] Khan R, Yang Q, Noor A, Altaf Khattak SB, Guo L, Tufail AB. An efficient adaptive modulation technique over realistic wireless communication channels based on distance and SINR. Frequenz. 2022 Jan 27;76(1-2):83-95.

[25] Khan, Rahim, Qiang Yang, Ahsan Bin Tufail, Alam Noor and Yong-Kui Ma. "Classification of Digital Modulated COVID-19 Images in the Presence of Channel Noise Using 2D Convolutional Neural Networks." Wirel. Commun. Mob. Comput. 2021 (2021): 5539907:1-5539907:15.

[26] Chen, X., J. Tang, and S. Lao, Review of Unmanned Aerial Vehicle Swarm Communication Architectures and Routing Protocols. Applied Sciences, 2020. **10**(10): p. 3661.

[27] Ko, Y., et al., Drone Secure Communication Protocol for Future Sensitive Applications in Military Zone. Sensors, 2021. **21**(6): p. 2057.

[28] Phadke, A. and F.A. Medrano, Examining application-specific resiliency implementations in UAV swarm scenarios. Intell. Robot, 2023. **3**(3): p. 436-461.

[29] Khan R, Yang Q, Ullah I, Rehman AU, Tufail AB, Noor A, Rehman A, Cengiz K. 3D convolutional neural networks based automatic modulation classification in the presence of channel noise. IET Communications. 2022 Mar;16(5):497-509.

[30] Khan R, Yang Q, Tufail AB, Ma YK, Noor A. Binary Classification of Modulation Formats in the Presence of Noise through Convolutional Neural Networks. In2020 15th IEEE International Conference on Signal Processing (ICSP) 2020 Dec 6 (Vol. 1, pp. 386-390). IEEE.

[31] Khan R, Qiang Y, Tufail AB, Noor A. Multiclass Classification of Modulation Formats in the presence of Rayleigh and Rician Channel Noise using Deep Learning Methods. In2020 3rd International Conference on Information and Communications Technology (ICOIACT) 2020 Nov 24 (pp. 297-301). IEEE.

[32] Kroener, I. and D. Neyland, New technologies, security and surveillance, in Routledge handbook of surveillance studies. 2012, Routledge. p. 141-148.

[33] Schneider, R.O., The ethical dimensions of emergency management. Southeastern Political Review, 1993. **21**(2): p. 251-267.

[34] Materna, R., R.E. Mansfield, and R.O. Walton, Aerospace industry report. 2015: Lulu. com.

[35] Boubin, J., et al. Adaptive Deployment for Autonomous Agricultural UAV Swarms. in Proceedings of the 20th ACM Conference on Embedded Networked Sensor Systems. 2022.

[36] Tufail AB, Ma YK, Zhang QN, Khan A, Zhao L, Yang Q, Adeel M, Khan R, Ullah I. 3D convolutional neural networks-based multiclass classification of Alzheimer's and Parkinson's diseases using PET and SPECT neuroimaging modalities. Brain Informatics. 2021 Dec;8:1-9.

[37] Khan R, Ma X, Taj S, Hassan H, Ullah I, Alwabli A, Tao Y, Ullah H. Bridging the Future: The Confluence of Internet of Things and Artificial Intelligence in Communication System. In Future Communication Systems Using Artificial Intelligence, Internet of Things and Data Science (pp. 30-54). CRC Press.

[38] Fowler, M., The Strategy of Drone Warfare. Journal of strategic Security, 2014. **7**(4): p. 108-119.

[39] Albani, D., et al. Dynamic UAV Swarm Deployment for Non-Uniform Coverage. in Proceedings of the 17th international conference on autonomous agents and multiagent systems. 2018.

[40] Khalil H, Rahman SU, Ullah I, Khan I, Alghadhban AJ, Al-Adhaileh MH, Ali G, ElAffendi M. A UAV-Swarm-Communication Model Using a Machine-Learning Approach for Search-and-Rescue Applications. Drones. 2022 Nov 23;6(12):372.

[41] Mazhar T, Irfan HM, Haq I, Ullah I, Ashraf M, Shloul TA, Ghadi YY, Imran, Elkamchouchi DH. Analysis of Challenges and Solutions of IoT in Smart Grids Using AI and Machine Learning Techniques: A Review. Electronics. 2023 Jan 3;12(1):242.

[42] Khan HU, Sohail M, Ali F, Nazir S, Ghadi YY, Ullah I. Prioritizing the multi-criterial features based on comparative approaches for enhancing security of IoT devices. Physical Communication. 2023 Aug 1;59:102084.

CHAPTER 3   REAL-WORLD DEPLOYMENTS IN UNMANNED AERIAL VEHICLE SWARMS

[43] Mazhar T, Talpur DB, Shloul TA, Ghadi YY, Haq I, Ullah I, Ouahada K, Hamam H. Analysis of IoT Security Challenges and Its Solutions Using Artificial Intelligence. Brain Sciences. 2023 Apr 19;13(4):683.

[44] Khan WU, Imtiaz N, Ullah I. Joint optimization of NOMA-enabled backscatter communications for beyond 5G IoT networks. Internet Technology Letters. 2021 Mar;4(2):e265.

[45] Asif M, Khan WU, Afzal HR, Nebhen J, Ullah I, Rehman AU, Kaabar MK. Reduced-Complexity LDPC Decoding for Next-Generation IoT Networks. Wireless Communications and Mobile Computing. 2021;2021(1):2029560.

[46] Gupta D, Juneja S, Nauman A, Hamid Y, Ullah I, Kim T, Tag eldin EM, Ghamry NA. Energy Saving Implementation in Hydraulic Press Using Industrial Internet of Things (IIoT). Electronics. 2022 Dec 6;11(23):4061.

[47] Mazhar T, Irfan HM, Khan S, Haq I, Ullah I, Iqbal M, Hamam H. Analysis of Cyber Security Attacks and Its Solutions for the Smart Grid Using Machine Learning and Blockchain Methods. Future Internet. 2023 Feb 19;15(2):83.

CHAPTER 4

Machine Learning Applications in UAV Swarms

Sadaf Hussain[1], Tanweer Sohail[2], Muhammad Adnan Khan[3,4,5*]

[1]Department of Computer Science, Lahore Garrison University, Lahore, Pakistan
Email: sadafhussain@lgu.edu.pk

[2]Department of Mathematics, University of Jhung, Jhung, Pakistan
Email: tanveersms@gmail.com

[3]School of Computing, Skyline University College, Sharjah, United Arab Emirates.

[4]RSCI, Riphah International University, Lahore Campus, Lahore, Pakistan.

[5]Department of Software, Faculty of Artificial Intelligence and Software, Gachon University, Seongnam-si, Republic of Korea.
adnan@gachon.ac.kr

CHAPTER 4 MACHINE LEARNING APPLICATIONS IN UAV SWARMS

Introduction

Swarms of UAVs, which are coordinated group of drones that are capable of operating on their own, can play a crucial role in increasing the protection of smart cities. Previously, the control for multiple UAVs can be done manually through a master control unit and communication "wire" which can be restricting because of distance, data capacity, and possible system failure [1]. On the other hand, the ML techniques provide much capable tools for decentralizing, providing more fault-tolerant and self-adapting control of UAVs [2, 3]. Current approaches in ML allow the developed swarm of drones to collectively gather information from various sensors, decide on possible threats, and apply the required action [2]. For instance, in a case of safety and security, computer vision and the application of deep learning algorithms can identify dangerous signs in videos. To achieve patrolling, searching, and responding to an incident, swarm uses reinforcement learning that locates a better strategy through testing.

However, applying ML in the UAV swarm also involves additional issues like safety, security, and reliability of UAV swarm system in presence of adversarial attacks or system failure. Current work is being conducted on methods such as federated learning, differential privacy, and blockchain to solve these problems. Altogether, it can be concluded that the smart city security enhancement, contributed by the machine learning aided UAV swarms, can benefit from using more autonomous, more flexible, and more scalable solutions. Thus, as the technology progresses, it will be important to address technical and socio-legal factors of this potentially revolutionary technology.

Literature Review

Artificial intelligence (AI), specifically ML, proves critical in improving the performance of drones for the protection of smart cities, especially in swarms' cooperation and adaption to the urban area's dynamics. Some operations using ML algorithms are image identification or object recognition, determining the best path in a territory with certain landmarks and avoiding obstacles [4]. Using ML and computer vision, monocular cameras used in UAVs can identify objects and also self-localize, useful in managing traffic and security in cities [5]. FL and SL are incorporated to enhance the efficiency of UAV swarm by controlling the communication load of 5G networks and improving the precision [6, 7]. Besides, hierarchical nested personalized federated learning (HN-PFL) and cooperative unmanned aerial vehicle (UAV) resource pooling mitigate the lack of raw computing resource and temporal and spatial variation in data in UAV swarms for enhancing the machine learning (ML) outcome and reducing the network resource consumption [8].

Supervised Learning for Object Detection and Classification

In the scope of UAV swarms, supervised learning approaches especially CNNs are studied within object detection and classification tasks as it can be seen from the works of Javaid et al. [5] and Bernardo et al. [9]. These papers demonstrate the effectiveness of CNNs in the processes like detecting suspicious activity, differentiating a pedestrian from a vehicle, and examining unauthorized drones from aerial images [10]. Such work illustrates that with the help of the CNNs, it is possible to identify the features inherent in the image data, thus improving the situational awareness of the UAV swarms [10]. Despite the significance of supervised learning for explicit applications such as object detection, additional research should be conducted on unsupervised learning approaches, which may be highly useful in providing predictions for threatening situations and resource allocation of UAV swarms.

Research by Guo, Zhang, Chen, and Yu (*Unsupervised Feature Learning with Emergent Data-Driven Prototypicality*) delves into leveraging unsupervised learning for threat forecasting in smart cities by proposing an innovative approach of unsupervised feature learning in hyperbolic space with sphere packing. This method is, in turn, aimed at mapping images to points in the feature space where the distance between the points constitutes the amount of visual similarity, while the position of the points in the space highlights the prototypicality within the dataset to help identify threats. Also, Kanevski [11] underlines the applicability of the unsupervised learning in the environmental data that is best illustrated by the feature selection, dimensionality reduction, and data clustering for the predictive modeling. Big data analytics is a vital component and consists of unsupervised learning like clustering and principal component analysis techniques, which assist the system in predicting threats in the smart city context [12].

Analyzing the historical crime data and sensor perception, it is possible to recognize the patterns of similar rates by employing the ML models, which helps to patrol the UAV and prevent crime more actively [13]. ML algorithms such as support vector machine, random forest, decision tree, and K-means can be utilized to analyze crime patterns and make accurate predictions based on the extracted features from demographic information and temporal crime data [13]. Additionally, deep learning techniques, including artificial neural networks with multiple layers, can further enhance predictive capabilities by identifying complex relationships in large datasets, aiding in the early detection of criminal behavior and potential crime hotspots (Watch From Sky: Machine learning and multi-UAV network for forensic surveillance prediction) [13]. The employment of UAVs in the predictive police surveillance, made by the means of employing ML models, provides an indication of how the law enforcement strategies can be advanced through the process of monitoring and patrolling the risky areas.

Ashush et al. [14], in their research, find an exciting approach regarding the application of unsupervised learning concerning UAV swarms concentrating on the elements of allocation of resources and their patrol formation in great urban territories. Stress that allocation of resources in multi-UAV should be efficient to keep vehicles safe and meet delays which can be solved by using a method based on Lagrange multipliers and reinforcement learning. Also provide a multi-UAV mission planning model that integrates energy consumption costs as well as impact costs and which demonstrates the impartation of the proposed lightning search algorithm in raising UAV execution coverage and reducing energy consumption [15]. Another automated swarming algorithm design approach for UAV is described in detail by based on the multiobjective reinforcement learning with coverage and stability remarkably better than in any conventional UAV swarm [16]. Moreover, the research by unnamed authors (2021) focuses on federated learning for UAV networks, in order to reduce the overall training energy consumption while achieving a global accuracy and latency among all UAVs, proven that distributed learning can enhance the UAV's intelligence and optimized energy saving [17]. Furthermore, the study by the unnamed authors (2021) delves into federated learning for UAV networks, aiming to minimize overall training energy consumption while ensuring global accuracy and latency constraints are met, showcasing the benefits of distributed learning in optimizing UAV intelligence and energy efficiency [18].

Reinforcement Learning for Search and Rescue

The use of reinforcement learning algorithms has demonstrated good potential for training UAV swarms to perform optimally when operating is unstructured and volatile conditions such as searching for victims in disaster-stricken areas [19]. Yet, the study by deals with the actualization of RL-learned behaviors within a leader–follower UAV swarm within the context of search and rescue operations and compares the efficiency based on the factors such as mission time, number of redundant moves, and time of stagnation, as well as goal achievement [20, 21]. Also, the paper by Khawaja et al. [22] highlighted that multiagent reinforcement learning can modify radar parameters and UAVs' positions for the improved capability of detection, tracking, and classification in swarms. Altogether, these studies convey the preliminary promise of RL in providing UAV swarms with responsive and effective behaviors for ambitious missions in hostile scenes. Due to this learning process, the UAVs are capable of reaching several locations and identifying the possibilities of victims, and they are also able to conduct communication among the swarm.

Advanced Machine Learning Techniques

New trends in the field of ML have proven to be quite useful for the UAV swarm application, especially concerning anomaly identification and real-time threat evaluation. ANNs have been noted to be critical in learning relations from data, making them suitable for such tasks regarding the UAV swarms [23, 24]. Such advances incorporate the use of unsupervised and supervised machine learning techniques such as stacked autoencoders federated learning LightGBM and GANs for efficient identification of the wanted anomaly or network attack in UAV swarms [23]. Moreover, HN-PFL and cooperative UAV resource pool have been introduced to overcome the computational resource scarcity and the data heterogeneity among geo-distributed device clusters within UAV swarms over time [8]. The application of these methodologies demonstrates that artificial neural networks have the capacity to boost the security as well as effectiveness for UAV swarm functionality.

Machine Learning Landscape for UAV Swarms

Current research is very active in the machine learning development for UAV swarms with topics in safety distances, resource scarcity, and communication issues. Studies focus on the application of machine learning models on multiple clusters of UAVs, and some of the discussed techniques include HN-PFL [8]. FL frameworks are necessary for training complex models that

CHAPTER 4   MACHINE LEARNING APPLICATIONS IN UAV SWARMS

can only be done on private data while not sharing the data especially in real-time applications such as obstacle avoidance during emergencies [6]. To enhance the training and transmission quality over a large number of UAV swarms, a clustered FL architecture has been introduced where UAV networks are divided into groups under the management of cluster head UAV for model update and intercluster consolidation [25]. These advancements are to improve usage of machine learning in UAV swarms and to maximize their performance in different civil and military implementations [26].

Supervised Learning

Supervised learning is among the basic categories of ML and is crucial in the training of models that are applicable in tasks that are important in the protection of a smart city through UAV swarms. This is how UAVs stands to benefit, for, in gist, it empowers them. To illustrate, in the case of a UAV swarm monitoring a city park, supervised learning is involved in training an object detection model [9, 27]. Using millions of images with highly detailed tags regarding the objects such as people, bicycles, and left-behind packages, the outline of each object is trained into the algorithm. This is an algorithm, which, once placed on a UAV, may analyze real-time video monitored at a certain territory, which makes it possible to recognize objects and alert on possible suspicious actions, for instance, on abandoned packages in a deserted area. Integration of machine learning techniques, computer vision, and deep convolutional neural networks improves the UAVs' monitoring capabilities which makes them useful in security, logistics, and live video surveillance in urban environments [5].

Supervised learning is of significant importance in improving the traffic management in smart cities using UAV swarms by endowing them with the feature of image classification [28, 29]. Supervised learning uses tagged images of urban traffic to recognize active traffic objects like pedestrians, vehicles, and bicycles. This "learning by example" approach reinforces

133

recognition of visual patterns in each object category, optimizing the classification process during traffic monitoring operations. This "learning by example" approach enhances object detection and classification. Also, combining supervised learning algorithms with UAV swarms helps them to improve the control of vehicle movement and traffic flow and thereby increase operational efficiency within cities [30].

For instance, with the help of high-end deep convolutional neural network (DCNN) algorithms, UAVs are capable of performing real-time detection as well as classification of objects within traffic scenes, including vehicles, pedestrians, and bicycles [9, 31]. This capability allows the UAVs to gather relevant data on determining the traffic flows since the various objects have been mapped in various regions. Thus, traffic authorities are in a position to decide, for instance, which signal lights to change, which road to change traffic direction on, or which road junctions may become a source for traffic jams in the near future [32]. In supervised learning, UAVs help create optimal traffic experiences in smart cities to make them as smooth and safe as possible by giving real-time data to the UAV for assisting in the improvement of traffic management and safety [2, 32].

To enhance UAVs' abilities as intelligent traffic observers in smart cities, supervised machine learning approaches have been underlined in numerous studies. By using the labeled image datasets containing such urban traffic scenes, the UAVs with supervised learning algorithms can analyze a real-time traffic condition and can improve the flow and safety of the traffic motion [30, 33]. By such image classification skills, these algorithms help UAVs to distinguish between pedestrians, vehicles, and bicycles to solve traffic issues. Convolutional neural networks make it possible to identify an object and detect anomalies enabling UAVs to positively contribute to urban traffic systems and their management [6, 34].

Incorporation of deep learning approaches with UAV-produced image classifications to analyze the live video streams aids in traffic supervision [35, 36]. By determining and categorizing object such as vehicles, pedestrians, and bikes in each frame, useful information on traffic flow is obtained to help traffic authorities to manage flow of traffic [37]. The collected data enables decisions to be made based on facts, such as adaptive traffic signal control for specified areas prone to traffic congestion, preemptively rerouting in cases of an event and the identification of new areas prone to congestion before this becomes a problem [38, 39]. It, therefore, helps boost overall traffic management approaches and the general capability for controlling traffic in cities.

To improve security for smart cities, it is suggested to use the supervised learning where such tools as anomaly detectors are trained to recognize some rather peculiar patterns [40]. In this case, the training data is constituted by video records with the authorized drones' patrolling trajectories, and the algorithm can then identify these drones' nonsuspicious behaviors [41]. When placed in an UAV flying over the area, the anomaly detection algorithm is capable of analyzing the live video streams and identify any anomalies that does not correspond to the expected flow, thereby allowing the UAV swarm to monitor for and identify potentially unsafe intruding UAVs [42, 43]. AIoT frameworks and CNN-SVM algorithms are adopted in this approach to achieve optimal and reliable real-time anomaly detection in surveillance systems, hence enhancing significantly the security aspects of smart cities.

Unsupervised Learning

Unsupervised learning is thus very instrumental in the strengthening of measures of security for UAV swarms within smart cities since the huge volumes of data generated can be analyzed to reveal characteristics that may be hitherto hidden from plain sight and yet crucial in the proactive taking of security measures that may prevent having to resort to reactive measures of mitigating harms relating to security. Despite the nature of the UAV swarm being mostly cooperative with predictable interaction patterns, using clustering, principal component analysis, or variational autoencoder, generative adversarial networks and other

CHAPTER 4   MACHINE LEARNING APPLICATIONS IN UAV SWARMS

related deep learning models give the ability to detect the flight anomalies, irregularities, and network attacks in UAV swarms [23, 44]. By applying the unsupervised techniques for characterizing the drone swarms via the RF signal analysis and the application of various machine learning techniques, the security systems can easily identify multiple UAVs, which will provide redundancy, survivability, scalability, and the enhanced self-organizing behavior in the complex environment; it will all be in favor of the protection of the smart city from possible threats and attacks [32].

One of the advantages of unsupervised learning algorithms is used in the analysis of the past crime records to find out regions that are most probable to experience the crime at a given time and day, weather conditions, and locations [44]. This predictive capability allows the enforcement agencies to mobilize UAV patrols in these reported hot areas to further the prevention of crime, thus safeguarding the public. Additionally, by using unsupervised learning to analyze the data of drone patrolling such as flight trajectories, response time to emergency occurrences, and covered area, the activity patterns can be determined concerning the utilization of UAV swarms and distributing the resources needed [45]. This optimization guarantees unique regions are adequately covered in order to attain the maximum utilization of the efficiency of the swarm to ensure that UAV patrols can increase public security and crime reduction effectivity.

The integration of algorithms in smart cities, such as the Postman Moving Voronoi Coverage (PMVC) algorithm [46] and the IMRFO-TS algorithm [47], allows for the detection of under-patrolled areas with faster reaction times and the discovery of duplicate patrol routes. Using these algorithms, authorities may efficiently change patrol patterns, resulting in a more balanced and efficient deployment of unmanned aerial vehicles (UAVs) for citywide security. Furthermore, using artificial intelligence and machine learning techniques to generate police patrol routes, as proposed in a study focusing on Quito City [48], improves the ability to optimize patrol routes based on spatial and temporal crime data, ultimately improving citizen security and resource allocation while lowering crime rates.

Unsupervised learning improves security measures for UAV swarms in smart cities by recognizing anomalies and potential threats [23, 44]. Unsupervised learning, using techniques such as stacked autoencoders, federated learning, and graph-oriented deep variational autoencoder (VAE), can identify flight anomalies, network attacks, and atypical incidents in UAV video data, allowing for the prediction of crime hotspots and proactive resource allocation [49]. Authorities can optimize security methods by analyzing large datasets and discovering hidden patterns, resulting in a more efficient and proactive approach to smart city security, which improves public safety and infrastructure protection [50].

Reinforcement Learning (RL)

Reinforcement learning (RL) is useful in training drones to achieve optimal behavior through repeated trials, which is suitable for dynamic environments such as smart cities [51]. In situations such as navigating to find victims in a dangerous area, where predicting the environment is difficult because of obstacles such as vegetation and uneven terrain, reinforcement learning is better because of its adaptation to unexpected challenges [21]. Reinforcement learning has been effective in various UAV operations missions, including search and rescue missions, where reinforcement learning behavior in UAV teams following leaders has shown positive results during the completion of the mission, the reduction of the work, and the achievement of the objectives. In addition, reinforcement learning has been used to adjust radar parameters and positions on drones to improve the detection, tracking, and classification of drones, demonstrating its effectiveness, to deal with bad behavior through psychological adjustments [22].

The remarkable thing about using reinforcement learning (RL) for UAVs is that it helps them learn by trial and error in places where there are no predefined instructions [52]. The RL algorithmically equipped drones get rewards for the right decisions they make and penalties for the poor ones. This way, they optimize their behavior without having to write code for specific activities [53]. In complex situations like disaster areas, this approach is very useful since these vehicles need to quickly locate victims. It has been proved that RL algorithms like multi-pass deep Q-network (MP-DQN) when combined with hindsight experience replay (HER) perform better than other approaches in UAV navigation especially relay operation in tough scenarios [53]. Moreover, deep reinforcement learning (DRL) has recently shown promise as an effective means for enabling drones to fly efficiently within 3D spaces, thereby demonstrating the power of RL models in the context of obstacle avoidance and energy saving during missions [54].

Smart city safety measure improvement, particularly in the search and rescue operations of unmanned aerial vehicles (UAVs), greatly depends on reinforcement learning (RL). By taking advantage of RL algorithms, UAVs can move efficiently in complex and rapidly changing environments as demonstrated by disaster management [51]. Learning from those experiences using RL makes it possible for UAVs to achieve maximum efficiency when locating victims or helping them during an emergency. Additionally, multiagent RL allows real-time adjustment of radar parameters and locations, thus improving detection, tracking, and classification of UAV swarms [22]. Moreover, there are contributions made by RL-learned behaviors in UAV swarm methodologies toward enhancing search and rescue activities, intelligence acquisition, and disaster relief exercises within smart city settings [55].

To solve the growing safety issues in urban environments, the introduction of reinforcement learning (RL) to unmanned aerial vehicles (UAVs) has proven to be essential. RL enables drones to change behavior in real time based on dynamic data, increasing their ability to deal with new and unpredictable threats. In addition, in densely populated cities, it is necessary to prevent accidents in order to protect the operation of UAVs [56]. The RL algorithm can effectively train an unmanned aerial vehicle to navigate in a dark environment, reducing the risk of collisions with buildings and other aerial objects, thereby increasing safety generally in densely populated cities [57, 58]. Using RL capabilities, drones can coordinate their flight path strategies, contributing to efficient and safe operations in urban environments.

Deep Learning for Enhanced Capabilities

Deep learning (DL) is an important subfield of machine learning (ML) that uses artificial neural networks (ANN) to learn complex patterns from data [59]. Artificial neural networks are computational models inspired by the human brain model, consisting of networks that operate and transmit information. Deep neural network is a multilayer neural network that is good at extracting advanced features from raw data to solve optimization problems [59, 60]. By adjusting the weight and bias, these networks can improve time and learn from big data; so they can excel at tasks such as image recognition, natural language processing, and language skills. The depth of deep neural networks allows them to model the relationship between input data and predictive output, demonstrating their ability to solve complex problems across multiple locations [61].

Convolutional neural networks (CNNs) are very important in image recognition, especially in the detection and classification of unmanned aerial vehicles (UAVs). Several studies have demonstrated the use of CNNs to analyze aerial imagery, demonstrating their effectiveness in detecting features of interest in complex situations such as obstruction and protection [9, 62]. In addition, advances such as the integration of DCNN models with UAV devices and the development of new modules such as TA-GRU have significantly improved object detection in UAV-captured images and overcome issues such as image degradation and background interference [10, 63]. In addition, the development of new models such as YOLOv5 shows improvements in training speed and accuracy, demonstrating the continuous development and optimization of CNN architectures for spatial applications [64].

Regular neural networks (RNNs) show great potential in tasks involving structured data, especially in abstract situations such as monitoring high-frequency music readings over time. RNN architectures such as normalized short-term memory (Bi-LSTM) and cascaded short-term memory (C-LSTM) have been shown to successfully capture spatial correlations and detect short-term errors [65–67]. In addition, the combination of RNN with other deep learning methods such as convolutional neural networks (CNN) can improve the recognition of spatial patterns in high-resolution datasets and provide a powerful framework for learning different pattern recognition in drone music [68]. Leveraging the ability of RNNs to model data and related problems, these networks can play an important role in improving the accuracy and efficiency of fault detection in dynamic environments such as drones.

Applications of Machine Learning in Smart City Protection

Machine learning (ML) is playing a key role in turning UAV swarms into effective tools for smart city defense. Using different ML metrics, these multiples gain the ability to understand, analyze, and respond to their environment, improving safety and improving the overall performance of the city.

Surveillance and Monitoring

Special surveillance models, real-time threat detection, and monitoring capabilities of surveillance and monitoring systems have been greatly enhanced by the integration of advanced technologies such as artificial intelligence, drones, and sensors. Train-guided mobile robots with intelligent functions for autonomous surveillance [69], new models of human perception for conducting social journeys [70], and the use of UAVs

CHAPTER 4   MACHINE LEARNING APPLICATIONS IN UAV SWARMS

for surveillance and tracking of intruders, goods, smoke, and fire [71, 72] describe the use of various technologies to ensure safety and efficiency. In addition, the development of new space sensors, such as the CHEIA SST radar, will contribute to the understanding of the spatial and temporal characteristics of space objects orbiting the Earth and will increase more than the importance of surveillance systems in different areas [72].

Search and Rescue

To improve search and rescue (SAR) operations using UAVs, it is important to focus on rapid response times, optimal victim location, and resource allocation. The use of UAVs and object detection systems such as YOLO can improve victim detection recovery rates [73]. The implementation of resource allocation and stochastic optimization methods can improve mission performance and real-time resource allocation in SAR scenarios and ensure computational efficiency and accuracy [73]. In addition, by using multiple UAVs in a multitasking system, considering vehicle safety as a priority and using reinforcement learning and attention methods, resources can be allocated to traffic requirements and ensure safety and meet delay requirements [74]. By sending drones into inaccessible areas and providing real-time information through images and videos, SAR missions can be conducted quickly, increasing the chance of victim survival and reducing the risk to search parties [75].

Traffic Management

Route optimization, problem prediction, and traffic management problem solving can be greatly improved by the introduction of unmanned aerial vehicles (UAVs). By using unmanned aerial vehicles, road systems can empower road operations based on future predictions [76], improve safety and integrate UAV systems into the unsafe conditions [77, 78], and disseminate information to road users after the incident to reduce severity and increase road capacity [79]. The use of unmanned aerial vehicles not only helps in incident response but also helps in predicting disruptions and regulating traffic flow, ultimately leading to a more efficient and stable management system.

Challenges and Future Directions

Machine learning offers important opportunities to improve the capabilities of UAV fleets in smart city defense, enabling tasks such as real-time surveillance and emergency response [80, 81]. However, issues such as data privacy concerns, communication costs, energy limitations, and the need for continuous connectivity pose barriers to successful and responsible implementation [80, 82]. To overcome this problem, federated learning (FL) has emerged as a promising solution, which allows UAVs to train local models without sharing raw data, thereby reducing privacy risk and communication overload [83]. In addition, new technologies such as hierarchical personalized federated learning (HN-PFL) and UAV resource sharing also improve energy consumption, model performance, and network utilization of UAV groups, making ML more efficient [84]. Using this, integrating machine learning into drone systems can effectively protect smart cities.

Access to sensitive information collected by drones, including video and audio recordings, is important due to significant privacy and security concerns [3, 85]. To achieve this, strong security protocols (such as simple real-time algorithms such as AES) are required to prevent unauthorized access and ensure efficient use of data during delivery [81]. Furthermore, the development of security architectures, such as centralized systems and various attacks and methods, is necessary to improve the data security of UAS [86]. In addition, clear data management laws and regulations can enhance public trust in drone technology by ensuring compliance with privacy policies and promoting data management practices [87]. By using these methods, the protection and privacy of data collected by drones will be improved, encouraging efficient and reliable use of the technology.

On-board drone operators face computational limitations and must balance algorithm optimization for efficient processing and uploading data to the cloud for complex machine learning, which can quickly drain an app's battery life. Recent studies have explored solutions to these challenges, such as using lightweight 1D LiDAR sensors with adaptive motion to improve performance while reducing power consumption and weight [88, 89]. Furthermore, the use of front-end computers and small machine learning models has shown promise in improving energy efficiency and latency compared to cloud-based approaches focusing on meta-inferences to adapt algorithms to environmental conditions [90]. In addition, advances in nanodrone planning are made possible by new sensor technologies and coordination, which enable effective coordination and reduced on-board downtime, thereby reducing the need for greater computing power [91].

Explanatory artificial intelligence (XAI) plays an important role in solving the lack of understanding and explanation of complex artificial intelligence models and promoting trust and accountability [92]. Although most machine learning models are black boxes, efforts have been made to develop interpretive models or model black box models with logical models to enable users to deeply understand the concepts behind complex models [93]. However, defining a black box model by assigning statistics to input parameters leads to conflicting goals and makes it difficult to achieve high fidelity, scalability, completeness, and initial consistency [94, 95]. To improve reliability and clarity, future research should focus on the development of explanatory models that provide the information needed to make predictions and encourage human oversight and accountability in AI systems [96].

Combining machine learning (ML) for drone operations with traditional and secure infrastructure is a significant challenge due to the inability to use real drone data [81]. To avoid this, research highlights the need for communication methods and data standards to facilitate the integration of UAVs in existing systems and improve the operational effectiveness of UAVs [97]. By using blockchain-based mechanisms, such as the Drone Trusted Self-Organizing Network (BC-UTSON), the internal security of the drone fleet can be improved, ensuring reliability and preventing data transmission through malicious websites [82]. In addition, the reactive handover coordination system (RHCRB) with regenerative blockchain principles provides a highly secure system designed for large numbers of drones, providing smart and secure procedures to defend against threats and enhance system security [98]. Based on these advances, the researchers aim to improve communication, data exchange, and integration strategies to improve the performance of machine learning-controlled drones in urban areas.

Federated learning is an attractive solution to privacy concerns and scalability challenges by enabling the decentralized training of machine learning models on surface devices such as drones, thereby reducing the transmission of sensitive data, to the main exchange server of the device [99–101]. This approach does not improve data analysis, but reduces communication and improves system efficiency [102, 103]. However, even if the group is well educated, it faces obstacles such as the risk of data loss due to the lack of privacy measures, thus showing the need for security measures in storage modeling, transmission, and sharing [99]. Future research opportunities should focus on developing innovative methods to ensure data privacy, scalability, and effectiveness of blended learning systems, to address the challenges associated with justice, role modeling, and client involvement [101, 102].

Artificial intelligence (XAI) plays a key role in increasing transparency and trust in machine learning processes that help investors understand the decision-making process of these systems and facilitate human control, as seen in various discussions [104, 105]. XAI techniques such as Shapley's incremental interpretation help to brighten the black light algorithm, ultimately promoting the use of artificial intelligence

(AI) by providing fast interpretation, which includes significant impact on trust and usefulness [106]. The growth of XAI research in information systems (IS) and electronic markets highlights the importance of designing transparent and interpretable AI systems to meet the needs of stakeholders and ensure the deployment of responsible AI [107]. Furthermore, the discussion of master language understanding (LLM) demonstrates the need to understand human ways in order to have a chance of understanding humans and solving problems posed by the development of artificial intelligence technology.

Discussion

This chapter discusses the exciting world of machine learning (ML) and its potential to revolutionize the drone industry working in modern cities. We saw how supervised learning can be used to equip drones with powerful capabilities such as object detection, image processing, and artificial intelligence. These capabilities are important for tasks such as suspicious applications a detection, precautionary monitoring of pedestrians and traffic, and identification of potentially dangerous illegal drones. Vehicle data analysts can use algorithms to analyze large datasets without known labels, revealing hidden patterns and relationships. This enables security measures such as crime predictions based on historical data and better drone traffic. Additionally, unsupervised learning can improve resource allocation by examining security measures and areas where access time or response may be slow. Although supervised and unsupervised learning provide powerful tools, activity-based learning (RL) takes a different approach. Imagine flying a drone in a dangerous environment – a complex and dangerous environment with constantly changing obstacles. Traditional methods can struggle to change. Academic promotion is the best way in this regard. Drones equipped with AR algorithms can adjust their behavior through trial and error, achieving desired results like real-time learning and adjustment to find perpetrators of violence in dangerous places. This technology is particularly useful in autonomous vehicles (UAVs) and machine learning applications.

Nevertheless, the training of a drone delivery system is not a headache-free process. And data security and privacy are definitely among the key safety rules. Some security protocols and data encryption systems should be included to make sure that the drone, which has collected video and audio recordings, will be saved from unauthorized access and also used as it was initially designed. Another problem is that the drone cannot even contact the central server. The use of these devices will further deplete their battery life when feeding them with complex machine learning algorithms. In the case of intricate algorithms, it is necessary to take into account possible online processing power and information requirements. Furthermore, trust building in technology requires us to overcome some of the "black boxes" of machine learning models. Artificial intelligence (XAI) is a branch that focuses on manufacturing trustworthy and realistic models for human users. When we implement XAI technology, we can peek into the decision-making process of ML models, which will be of use for the managers, and the trust technology will be gained.

On the other hand, the process of the drone to get integrated into the city infrastructure as well as with the existing security systems are the basis for ultimate drone performance. It has been proposed that the findings of the research shall be made for the establishment of communication and data protocols which can enable the integration with the existing manned/unmanned systems, thus enhancing the intellect of people "motonomy." The prospect of public education can be the best way to solve particular problems and encourage the growth of cities. AI supports machines to obtain data by drones directly, which can cancel the necessity for transportation to a main server. This will not only bring out improvements in data privacy but will also shrink communication and thus amplify the efficiency of the whole system. Artificial intelligence is a powerful tool for making drones able to make a decision and also for the management of smart cities. As we address the challenges, seek solutions through public education and research like AAI, and improve public relations, we can use the power of machine learning to create healthier societies, better and safer for everyone.

Challenges and Future Directions

It is worth noting that challenges remain when applying ML for smart city protection with the help of UAV swarms. Security of the data collected is critical, and this has been endorsed by different authors [108, 109]. To tackle these problems, strict measures of security implementation should be put in place to guarantee that unauthorized individuals do not intrude into the sensitive information collected by the drones and to guarantee that the information collected by the drones will be used ethically. Furthermore, restricted computational resources on drone, for entailing computational demanding algorithm and communication compromise, has been presented in the study by Pan et al. [80]. Safety and security measures should be put in place while efficient computational power should be put into practice to enhance the use of ML in operation of UAV swarm in smart cities.

With the concepts of smart city protection based on UAV swarms, issues of data protection and privacy take a central position as it is mandatory to develop strict measures adhering to the noninterference by the unauthorized parties and the proper use of the data obtained [108]. Also, computation in drones is also a challenge due to the limited computing capability which has the challenge of processing power and communication for the execution of algorithms [110]. To overcome these problems, federated learning (FL) which is proposed as a solution allows drones to train models without transfer of raw data which minimizes the data privacy concern and communication overheads [80]. Through the use of FL and effective aggregation approaches like commutative FL and alternate FL, the smart city environment can further improve operation reliability, minimize power use, and lower communication expenses, which, in turn, improves the advanced protective UAV swarm solutions.

Efficient data processing to be done on-board and battery management are key challenge areas, along with the consideration of cloud-based processing strategies which involve low communication overhead. Solving these issues is crucial to gaining the populace's trust and adoption of ML-driven UAV formations in smart cities. Future contributions are directed to the design of secure data processing mechanisms, lightweight and efficient ML algorithms, and efficient utilization of the computing resources. This will help to enhance the likely hood of attaining successful and responsible incorporation of this technology.

Concerning the field of ML for UAV swarms, they can use the approach described by Dong et al. [111], which is federated learning to conduct training at the edge devices such as drones, thereby solving the problem of privacy and scalability. Also, the incorporation of explainable AI (XAI) which Zeng et al. [6] talk about is significant in increasing the level of transparence and trustworthiness in ML models making it easier for human beings to supervise as well as have confidence in complex models. Together with the federated learning applied to the distributed training, which is combined with the XAI in the given algorithms, researchers are able to advance the field of ML application in the UAV swarms while providing privacy, scalability, transparency, and trustworthiness for the continuously developing field of AI technologies.

Conclusion

Machine learning (ML) has proven to be a good candidate for smart city drones. This chapter explores how supervised learning enables object recognition, image segmentation, and artificial intelligence capabilities in drones – key tools for applications ranging from vehicle control to safety settings. We also discuss the role of unsupervised learning in discovering hidden patterns in data, supporting preventive measures, and improving resource allocation. Improved learning capabilities enhance a drone's adaptability and effectiveness in search, rescue, threat response, and collision prevention in a changing environment. However, effective management approaches present many challenges. Data security and privacy require strong protocols and clear governance rules. It is important to compare planning and communication requirements and build trust with models that can be interpreted through XAI. Insufficient integration with existing infrastructure requires further research into communications and data standards. Cooperative learning is a viable solution to privacy and security concerns, allowing decentralized machine learning models to be trained on the drone itself. By understanding these challenges and finding solutions through research and open communication, we can pave the way for machine learning-controlled drones. Ultimately, this will lead to a safe, clean, and secure city for all.

References

[1] K. AL-Dosari and N. Fetais, "A New Shift in Implementing Unmanned Aerial Vehicles (UAVs) in the Safety and Security of Smart Cities: A Systematic Literature Review," *Saf. 2023, Vol. 9, Page 64*, vol. 9, no. 3, p. 64, Sep. 2023, doi: https://doi.org/10.3390/SAFETY9030064.

[2] N. Mohamed, J. Al-Jaroodi, I. Jawhar, A. Idries, and F. Mohammed, "Unmanned aerial vehicles applications in future smart cities," *Technol. Forecast. Soc. Change*, vol. 153, p. 119293, Apr. 2020, doi: https://doi.org/10.1016/J.TECHFORE.2018.05.004.

[3] Y. Mekdad *et al.*, "A survey on security and privacy issues of UAVs," *Comput. networks*, vol. 224, pp. 109626–109626, Apr. 2023, doi: https://doi.org/10.1016/J.COMNET.2023.109626.

[4] "A Comprehensive Review of Classification and Application of Machine Learning in Drone Technology," Jun. 2023, doi: 10.20944/PREPRINTS202306.1901.V1.

[5] A. Javaid, M. A. Syed, and U. Baroudi, "A Machine Learning Based Method for Object Detection and Localization Using a Monocular RGB Camera Equipped Drone," *2023 Int. Wirel. Commun. Mob. Comput. IWCMC 2023*, pp. 1-6, Jun. 2023, doi: https://doi.org/10.1109/IWCMC58020.2023.10182369.

[6] L. Zeng, W. Wang, and W. Zuo, "A Federated Learning Latency Minimization Method for UAV Swarms Aided by Communication Compression and Energy Allocation," *Sensors*, vol. 23, no. 13, pp. 5787-5787, Jun. 2023, doi: https://doi.org/10.3390/S23135787.

[7] W. He, H. Yao, F. Wang, Z. Wang, and Z. Xiong, "Enhancing the Efficiency of UAV Swarms Communication in 5G Networks through a Hybrid Split and Federated Learning Approach," *2023 Int. Wirel. Commun. Mob. Comput. IWCMC 2023*, pp. 1371–1376, Jun. 2023, doi: https://doi.org/10.1109/IWCMC58020.2023.10183145.

[8] S. Wang, S. Hosseinalipour, M. Gorlatova, C. G. Brinton, and M. Chiang, "UAV-assisted Online Machine Learning Over Multi-Tiered Networks: A Hierarchical Nested Personalized Federated Learning Approach," *IEEE Trans. Netw. Serv. Manag.*, vol. 20, no. 2, pp. 1847–1865, Jun. 2023, doi: https://doi.org/10.1109/TNSM.2022.3216326.

[9] R. M. Bernardo, L. Claudio Batista Da Silva, and P. F. Ferreira Rosa, "UAV Embedded Real-Time Object Detection by a DCNN Model Trained on Synthetic Dataset," *2023 Int. Conf. Unmanned Aircr. Syst. ICUAS 2023*, pp. 580–585, Jun. 2023, doi: 10.1109/ICUAS57906.2023.10156134.

[10] R. Dey, B. K. Pandit, A. Ganguly, A. Chakraborty, and A. Banerjee, "Deep Neural Network Based Multi-Object Detection for Real-time Aerial Surveillance," *2023 11th Int. Symp. Electron. Syst. Devices Comput. ESDC 2023*, vol. 1, pp. 1–6, May 2023, doi: https://doi.org/10.1109/ESDC56251.2023.10149866.

[11] M. Kanevski, "On Unsupervised Learning from Environmental Data," *EGU23*, Feb. 2023, doi: https://doi.org/10.5194/EGUSPHERE-EGU23-9437.

[12] J. Talukdar, T. P. Singh, and B. Barman, "Unsupervised Learning," pp. 87–107, 2023, doi: 10.1007/978-981-99-3157-6_5.

[13] K. Jose Triny, J. Gowri, and S. Padmaja, "A Survey on Prediction of Risk Related to Theft Activities in Municipal Areas using Deep Learning," *Proc. 2023 2nd Int. Conf. Electron. Renew. Syst. ICEARS 2023*, pp. 1321–1326, 2023, doi: 10.1109/ICEARS56392.2023.10085123.

[14] N. Ashush, S. Greenberg, E. Manor, and Y. Ben-Shimol, "Unsupervised Drones Swarm Characterization Using RF Signals Analysis and Machine Learning Methods," *Sensors 2023, Vol. 23, Page 1589*, vol. 23, no. 3, p. 1589, Feb. 2023, doi: https://doi.org/10.3390/S23031589.

[15] Y. Wang, Y. He, F. R. Yu, Q. Lin, and V. C. M. Leung, "Efficient Resource Allocation in Multi-UAV Assisted Vehicular Networks with Security Constraint and Attention Mechanism," *IEEE Trans. Wirel. Commun.*, vol. 22, no. 7, pp. 4802–4813, Jul. 2023, doi: https://doi.org/10.1109/TWC.2022.3229013.

[16] H. Xiang, Y. Han, N. Pan, M. Zhang, and Z. Wang, "Study on Multi-UAV Cooperative Path Planning for Complex Patrol Tasks in Large Cities," *Drones 2023, Vol. 7, Page 367*, vol. 7, no. 6, p. 367, Jun. 2023, doi: https://doi.org/10.3390/DRONES7060367.

[17] G. Duflo, G. Danoy, E. G. Talbi, and P. Bouvry, "Learning to Optimise a Swarm of UAVs," *Appl. Sci. 2022, Vol. 12, Page 9587*, vol. 12, no. 19, p. 9587, Sep. 2022, doi: https://doi.org/10.3390/APP12199587.

[18] Y. Shen, Y. Qu, C. Dong, F. Zhou, and Q. Wu, "Joint Training and Resource Allocation Optimization for Federated Learning in UAV Swarm," *IEEE Internet Things J.*, vol. 10, no. 3, pp. 2272–2284, Feb. 2023, doi: https://doi.org/10.1109/JIOT.2022.3152829.

[19] S. Feng, X. Li, L. Ren, and S. Xu, "Reinforcement learning with parameterized action space and sparse reward for UAV navigation," *Intell. Robot.*, vol. 3, no. 2, pp. 161–75, Jun. 2023, doi: https://doi.org/10.20517/IR.2023.10.

[20] N. Zhang, C. Liu, and J. Ba, "Decomposing FANET to Counter Massive UAV Swarm Based on Reinforcement Learning," *IEEE Commun. Lett.*, vol. 27, no. 7, pp. 1784–1788, Jul. 2023, doi: https://doi.org/10.1109/LCOMM.2023.3269221.

[21] S. S. Carley, S. R. Price, X. M. D. Hadia, S. R. Price, and S. J. Butler, "Initial investigation of UAV swarm behaviors in a search-and-rescue scenario using reinforcement learning," p. 15, Jun. 2023, doi: 10.1117/12.2663629.

[22] W. Khawaja, Q. Yaqoob, and I. Guvenc, "RL-Based Detection, Tracking, and Classification of Malicious UAV Swarms through Airborne Cognitive Multibeam Multifunction Phased Array Radar," *Drones*, vol. 7, no. 7, Jul. 2023, doi: https://doi.org/10.3390/DRONES7070470.

[23] L. M. Da Silva, I. G. Ferrao, C. Dezan, D. Espes, and K. R. L. J. C. Branco, "Anomaly-Based Intrusion Detection System for In-Flight and Network Security in UAV Swarm," *2023 Int. Conf. Unmanned Aircr. Syst. ICUAS 2023*, pp. 812–819, Jun. 2023, doi: https://doi.org/10.1109/ICUAS57906.2023.10155873.

[24] "Comparative Study on Generating and Predicting Swarm Satellite Data by Deep Neural Networks," May 2023, doi: 10.5194/EGUSPHERE-EGU23-17550.

[25] D. N. M. Hoang, V. T. Truong, H. D. Le, and L. B. Le, "Clustered and Scalable Federated Learning Framework for UAV Swarms," *Authorea Prepr.*, Oct. 2023, doi: https://doi.org/10.36227/TECHRXIV.22730012.V1.

[26] X. Xu, J. Sun, and H. Hu, "Simulator Based mission optimization for swarm UAVs with Minimum Safety Distance Between Neighbors," Jun. 2023, doi: https://doi.org/10.2514/6.2023-4453.

[27] J. Correia, A. Bernardino, and R. Ribeiro, "Learning Performance Models of Distributed Computer Vision Methods for Decision Making in Detection and Tracking Algorithms in UAVs," *IEEE Internet Things J.*, vol. 10, no. 14, pp. 12486–12495, Jul. 2023, doi: https://doi.org/10.1109/JIOT.2023.3247589.

[28] V. Chava, S. S. Nalluri, S. H. Vinay Kommuri, and A. Vishnubhatla, "Smart Traffic Management System using YOLOv4 and MobileNetV2 Convolutional Neural Network Architecture," *Proc. 2nd Int. Conf. Appl. Artif. Intell. Comput. ICAAIC 2023*, pp. 41–47, May 2023, doi: 10.1109/ICAAIC56838.2023.10141268.

[29] P. Devadhas Sujakumari and P. Dassan, "Generative Adversarial Networks (GAN) and HDFS-Based Realtime Traffic Forecasting System Using CCTV Surveillance," *Symmetry (Basel).*, vol. 15, no. 4, pp. 779–779, Mar. 2023, doi: https://doi.org/10.3390/SYM15040779.

[30] B. Yang, H. Shi, and X. Xia, "Federated Imitation Learning for UAV Swarm Coordination in Urban Traffic Monitoring," *IEEE Trans. Ind. Informatics*, vol. 19, no. 4, pp. 6037–6046, Apr. 2023, doi: https://doi.org/10.1109/TII.2022.3192675.

[31] Y. Zheng, J. Luo, Y. Qiao, and H. Gao, "UAV-Assisted Traffic Speed Prediction via Gray Relational Analysis and Deep Learning," *Drones*, vol. 7, no. 6, pp. 372–372, Jun. 2023, doi: https://doi.org/10.3390/DRONES7060372.

[32] R. Hamadi, H. Ghazzai, and Y. Massoud, "Image-based Automated Framework for Detecting and Classifying Unmanned Aerial Vehicles," *2023 IEEE Int. Conf. Smart Mobility, SM 2023*, pp. 149–153, Mar. 2023, doi: 10.1109/SM57895.2023.10112531.

[33] M. Y. Zhu, X. Y. Hong, Z. Chen, J. X. Zhou, and N. Lv, "Transfer learning-based Traffic Identification for UAV-Assisted IoT," *Proc. - 2022 4th Int. Conf. Intell. Inf. Process. IIP 2022*, pp. 28–36, Oct. 2022, doi: 10.1109/IIP57348.2022.00013.

[34] I. Bisio, H. Haleem, C. Garibotto, F. Lavagetto, and A. Sciarrone, "Performance Evaluation and Analysis of Drone-Based Vehicle Detection Techniques From Deep Learning Perspective," *IEEE Internet Things J.*, vol. 9, no. 13, pp. 10920–10935, Jul. 2022, doi: https://doi.org/10.1109/JIOT.2021.3128065.

[35] N. K. Valappil and Q. A. Memon, "Vehicle Detection in UAV Videos Using CNN-SVM," *Adv. Intell. Syst. Comput.*, vol. 1383 AISC, pp. 221–232, Dec. 2020, doi: 10.1007/978-3-030-73689-7_22.

[36] I. Bisio, C. Garibotto, H. Haleem, F. Lavagetto, and A. Sciarrone, "Traffic Analysis Through Deep-Learning-Based Image Segmentation From UAV Streaming," *IEEE Internet Things J.*, vol. 10, no. 7, pp. 6059–6073, Apr. 2023, doi: https://doi.org/10.1109/JIOT.2022.3223283.

[37] J. Hong, Z. Gao, T. Mei, Y. Li, and C. Zhao, "UAV-based Traffic Flow Estimation and Analysis," *IEEE Int. Conf. Control Autom. ICCA*, vol. 2019-July, pp. 812–817, Jul. 2019, doi: https://doi.org/10.1109/ICCA.2019.8899679.

[38] X. Le, J. Jo, S. Youngbo, and D. Stantic, "Detection and Classification of Vehicle Types from Moving Backgrounds," *Rev. Informática Teórica E Apl.*, vol. 751, pp. 491–502, Dec. 2017, doi: https://doi.org/10.1007/978-3-319-78452-6_39.

[39] O. Kainz, M. Dopiriak, M. Michalko, F. Jakab, and I. Nováková, "Traffic Monitoring from the Perspective of an Unmanned Aerial Vehicle," *Appl. Sci.*, vol. 12, no. 16, pp. 7966–7966, Aug. 2022, doi: https://doi.org/10.3390/APP12167966.

[40] M. Islam, A. S. Dukyil, S. Alyahya, and S. Habib, "An IoT Enable Anomaly Detection System for Smart City Surveillance," *Sensors*, vol. 23, no. 4, pp. 2358–2358, Feb. 2023, doi: https://doi.org/10.3390/S23042358.

[41] Z. K. Abbas and A. A. Al-Ani, "DETECTION OF ANOMALOUS EVENTS BASED ON DEEP LEARNING-BILSTM," *Iraqi J. Inf. Commun. Technol.*, vol. 5, no. 3, pp. 34–42, Dec. 2022, doi: https://doi.org/10.31987/IJICT.5.3.207.

[42] R. Salama, F. Al-Turjman, and R. Culmone, "AI-Powered Drone to Address Smart City Security Issues," *Lect. Notes Networks Syst.*, vol. 655 LNNS, pp. 292–300, Jan. 2023, doi: https://doi.org/10.1007/978-3-031-28694-0_27.

[43] D. Protic, L. Gaur, M. Stankovic, and M. A. Rahman, "Cybersecurity in Smart Cities: Detection of Opposing Decisions on Anomalies in the Computer Network Behavior," *Electronics*, vol. 11, no. 22, pp. 3718–3718, Nov. 2022, doi: https://doi.org/10.3390/ELECTRONICS11223718.

[44] G. Liu, L. Shu, Y. Yang, and C. Jin, "Unsupervised video anomaly detection in UAVs: a new approach based on learning and inference," *Front. Sustain. cities*, vol. 5, Jun. 2023, doi: https://doi.org/10.3389/FRSC.2023.1197434.

[45] M. Altaweel, A. Khelifi, and M. M. Shana'ah, "Monitoring Looting at Cultural Heritage Sites: Applying Deep Learning on Optical Unmanned Aerial Vehicles Data as a Solution," *Soc. Sci. Comput. Rev.*, vol. 42, no. 2, pp. 480–495, Jul. 2023, doi: https://doi.org/10.1177/08944393231188471.

[46] H. Kocabas, C. Allred, and M. Harper, "Divide and Survey: Observability Through Multi-Drone City Roadway Coverage," *ISC2 2022 - 8th IEEE Int. Smart Cities Conf.*, Sep. 2022, doi: https://doi.org/10.1109/ISC255366.2022.9922207.

[47] A. A. Saadi, A. Soukne, Y. Meraihi, A. B. Gabis, and A. Ramdane-Cherif, "A Hybrid Improved Manta Ray Foraging Optimization With Tabu Search Algorithm for Solving the UAV Placement Problem in Smart Cities," *IEEE Access*, vol. 11, pp. 24315–24342, Jan. 2023, doi: https://doi.org/10.1109/ACCESS.2023.3255793.

[48] A. Roy and R. Naskar, "Improving Smart Cities Safety using Sanity-Check Deep Neural Network Algorithm," *Proc. 2022 9th Int. Conf. Comput. Sustain. Glob. Dev. INDIACom 2022*, pp. 108–114, Mar. 2022, doi: 10.23919/INDIACOM54597.2022.9763146.

[49] T. T. Khoei, A. Gasimova, M. A. Ahajjam, K. Al Shamaileh, V. Devabhaktuni, and N. Kaabouch, "A Comparative Analysis of Supervised and Unsupervised Models for Detecting GPS Spoofing Attack on UAVs," *IEEE Int. Conf. Electro Inf. Technol.*, vol. 2022-May, pp. 279–284, May 2022, doi: https://doi.org/10.1109/EIT53891.2022.9813826.

[50] N. Syed, M. A. Khan, N. Mohammad, G. Ben Brahim, and Z. Baig, "Unsupervised Machine Learning for Drone Forensics through Flight Path Analysis," *10th Int. Symp. Digit. Forensics Secur. ISDFS 2022*, Jun. 2022, doi: https://doi.org/10.1109/ISDFS55398.2022.9800808.

[51] M. Shurrab, R. Mizouni, S. Singh, and H. Otrok, "Reinforcement learning framework for UAV-based target localization applications," *Internet of things*, vol. 23, pp. 100867–100867, Oct. 2023, doi: https://doi.org/10.1016/J.IOT.2023.100867.

[52] M. U. and S. H. S. Schindler, "A Mini Review on the utilization of Reinforcement Learning with OPC UA," in *IEEE 21st International Conference on Industrial Informatics (INDIN), Lemgo, Germany*, May 2023, pp. 1–6. doi: https://doi.org/10.48550/ARXIV.2305.15113.

[53] Y. Zhu and S. Wang, "Learning-Based Cooperative Aerial and Ground Vehicle Routing for Emergency Communications," *Proc. - IEEE Glob. Commun. Conf. GLOBECOM*, pp. 4953–4958, Dec. 2022, doi: https://doi.org/10.1109/GLOBECOM48099.2022.10001484.

[54] J. Kang and J. H. Kim, "Optimal Movements of UAV using Reinforcement Learning in Emergency Environment," *Int. Conf. ICT Converg.*, vol. 2022-October, pp. 1248–1250, Oct. 2022, doi: https://doi.org/10.1109/ICTC55196.2022.9952915.

[55] H. Bayerlein, M. Theile, M. Caccamo, and D. Gesbert, "UAV Path Planning for Wireless Data Harvesting: A Deep Reinforcement Learning Approach," *Proc. - IEEE Glob. Commun. Conf. GLOBECOM*, 2020, doi: https://doi.org/10.1109/GLOBECOM42002.2020.9322234.

[56] V. Ř. omáš Brázdil, David Klaška, Antonín Kučera, Vít Musil, Petr Novotný, "On-the-fly adaptation of patrolling strategies in changing environments," in *Thirty-Eighth Conference on Uncertainty in Artificial Intelligence*, Jun. 2022, pp. 244–254. doi: 10.48550/ARXIV.2206.08096.

[57] R. Caballero, P. Jimenez, F. J. Perez-Grau, A. Viguria, and A. Ollero, "Towards Safe Operations in Urban Environments with UAVs," *2023 Int. Conf. Unmanned Aircr. Syst. ICUAS 2023*, pp. 336–342, Jun. 2023, doi: https://doi.org/10.1109/ICUAS57906.2023.10156360.

[58] M. Hamadouche, C. Dezan, D. Espes, and K. Branco, "Online reward adaptation for MDP-based distributed missions," *2023 Int. Conf. Unmanned Aircr. Syst. ICUAS 2023*, pp. 1059-1066, Jun. 2023, doi: https://doi.org/10.1109/ICUAS57906.2023.10156131.

[59] M. Almasi, "Deep Learning and Neural Networks: Methods and Applications," *Cutting-Edge Technol. Innov. Comput. Sci. Eng.*, Jun. 2023, doi: https://doi.org/10.59646/CSEBOOKC8/004.

[60] A. Hashemi and M. B. Dowlatshahi, "Neural Networks and Deep Learning," *Handb. Form. Optim.*, pp. 1-20, Jan. 2023, doi: https://doi.org/10.1007/978-981-19-8851-6_13-1.

[61] S. E, J. S, S. C. Mana, N. V. R., and A. Singh K, "An approach to deep learning," *AN APPROACH TO Deep Learn.*, Nov. 2022, doi: https://doi.org/10.47716/MTS.B.978-93-92090-12-7.

[62] L. Dabbiru, C. T. Goodin, D. W. Carruth, and J. Boone, "Object detection in synthetic aerial imagery using deep learning," in *Object detection in synthetic aerial imagery using deep learning*, Jun. 2023, vol. 12540, pp. 1254002-1254002. doi: https://doi.org/10.1117/12.2662426.

[63] Z. Zhou, X. Yu, and X. Chen, "Object Detection in Drone Video with Temporal Attention Gated Recurrent Unit Based on Transformer," *Drones*, vol. 7, no. 7, Jul. 2023, doi: https://doi.org/10.3390/DRONES7070466.

[64] N. V Gawali, K. Pagare, S. Daundkar, and D. Wagh, "Image Classification Using Convolutional Neural Networks," *Int. J. Sci. Technol. Eng.*, vol. 10, no. 12, pp. 586–591, Dec. 2022, doi: https://doi.org/10.22214/IJRASET.2022.47085.

[65] T. Xie, Q. Xu, and C. Jiang, "Anomaly detection for multivariate times series through the multi-scale convolutional recurrent variational autoencoder," *Expert Syst. Appl.*, vol. 231, pp. 120725–120725, Nov. 2023, doi: https://doi.org/10.1016/J.ESWA.2023.120725.

[66] M. Ducoffe, I. Haloui, and J. Sen Gupta, "ANOMALY DETECTION ON TIME SERIES WITH WASSERSTEIN GAN APPLIED TO PHM," *Int. J. Progn. Heal. Manag.*, vol. 10, no. 4, Jun. 2023, doi: https://doi.org/10.36001/IJPHM.2019.V10I4.2610.

[67] X. Zhang, J. Zhao, and Y. Lecun, "Character-level Convolutional Networks for Text Classification," *Adv. Neural Inf. Process. Syst.*, vol. 2015-January, pp. 649–657, Sep. 2015, doi: https://doi.org/10.48550/arxiv.1509.01626.

[68] A. Alos and Z. Dahrouj, "Using MLSTM and Multioutput Convolutional LSTM Algorithms for Detecting Anomalous Patterns in Streamed Data of Unmanned Aerial Vehicles," *IEEE Aerosp. Electron. Syst. Mag.*, vol. 37, no. 6, pp. 6–15, Jun. 2022, doi: https://doi.org/10.1109/MAES.2021.3053108.

[69] T. Botsis and K. Kreimeyer, "Improving drug safety with adverse event detection using natural language processing.," *Expert Opin. Drug Saf.*, vol. 22, no. 8, pp. 659–668, Jun. 2023, doi: https://doi.org/10.1080/14740338.2023.2228197.

[70] "Monitoring Social Distancing With Real-Time Detection and Tracking," in *Applied Big Data Analytics and Its Role in COVID-19 Research*, IGI Global eBooks, 2022, pp. 113–141. doi: https://doi.org/10.4018/978-1-7998-8793-5.ch005.

[71] A. Mariam, M. Mushtaq, and M. M. Iqbal, "Real-Time Detection, Recognition, and Surveillance using Drones," *2022 Int. Conf. Emerg. Trends Electr. Control. Telecommun. Eng. ETECTE 2022 - Proc.*, Dec. 2022, doi: 10.1109/ETECTE55893.2022.10007285.

[72] I. Tramandan and A. Rusu-Casandra, "Real-Time Monitoring and Command System for the Cheia Space Surveillance and Tracking Radar," *2022 15th Int. Symp. Electron. Telecommun. ISETC 2022 - Conf. Proc.*, Nov. 2022, doi: https://doi.org/10.1109/ISETC56213.2022.10010167.

[73] "Simulation Analysis of Exploration Strategies and UAV Planning for Search and Rescue," *arXiv. org*, vol. abs/2304.05107, Apr. 2023, doi: 10.48550/ARXIV.2304.05107.

[74] S. Razzaq, C. Xydeas, A. Mahmood, S. Ahmed, N. I. Ratyal, and J. Iqbal, "Efficient optimization techniques for resource allocation in UAVs mission framework," *PLoS One*, vol. 18, no. 4, pp. e0283923–e0283923, Apr. 2023, doi: https://doi.org/10.1371/JOURNAL.PONE.0283923.

[75]   F. Steinhausler and H. V. Georgiou, "Detection of victims with UAVs during wide area Search and Rescue operations," *SSRR 2022 - IEEE Int. Symp. Safety, Secur. Rescue Robot.*, pp. 14–19, Nov. 2022, doi: 10.1109/SSRR56537.2022.10018756.

[76]   J. Darko and H. Park, "A Proactive Dynamic-Distributed Constraint Optimization Framework for Unmanned Aerial and Ground Vehicles in Traffic Incident Management," *Lect. Notes Electr. Eng.*, vol. 901 LNEE, pp. 716–729, Jan. 2022, doi: https://doi.org/10.1007/978-981-19-2259-6_64.

[77]   J. Zhou, L. Jin, X. Wang, and D. Sun, "Resilient UAV Traffic Congestion Control Using Fluid Queuing Models," *IEEE Trans. Intell. Transp. Syst.*, vol. 22, no. 12, pp. 7561–7572, Dec. 2021, doi: https://doi.org/10.1109/TITS.2020.3004406.

[78]   K. S. Rekha, S. M. Sivapriyaa, M. D. Tanmayee, and B. L. Shree Raghav, "Dynamic Traffic Management System," *Proc. 2023 3rd Int. Conf. Innov. Pract. Technol. Manag. ICIPTM 2023*, Feb. 2023, doi: https://doi.org/10.1109/ICIPTM57143.2023.10118070.

[79]   K. D. Bilimoria, D. Helton, and G. Price, "Air Traffic Management Technology Demonstration – 3 (ATD-3) Multi-Flight Common Route (MFCR) Concept of Operations," California, Jun. 2018. Accessed: Jun. 30, 2024. [Online]. Available: https://ntrs.nasa.gov/citations/20180003519

[80] T. Pan, D., Khoshkholghi, M.A., Mahmoodi, "Decentralized Federated Learning Methods for Reducing Communication Cost and Energy Consumption in UAV Networks," *Mob. Comput. Appl. Serv. MobiCASE 2022*, vol. 495, pp. 18–30, Apr. 2023, doi: 10.48550/ARXIV.2304.06551.

[81] S. Rajasoundaran, S. V. N. S. Kumar, M. Selvi, and A. Kannan, "Reactive handover coordination system with regenerative blockchain principles for swarm unmanned aerial vehicles," *Peer-to-Peer Netw. Appl.*, vol. 17, no. 1, pp. 89–114, Jul. 2024, doi: https://doi.org/10.1007/s12083-023-01572-z.

[82] Y. Ding, Z. Yang, Q. V. Pham, Y. Hu, Z. Zhang, and M. Shikh-Bahaei, "Distributed Machine Learning for UAV Swarms: Computing, Sensing, and Semantics," *IEEE Internet Things J.*, vol. 11, no. 5, pp. 7447–7473, Jan. 2023, doi: https://doi.org/10.1109/JIOT.2023.3341307.

[83] S. Rangaraju, "SECURE BY INTELLIGENCE: ENHANCING PRODUCTS WITH AI-DRIVEN SECURITY MEASURES," *EPH - Int. J. Sci. Eng.*, vol. 9, no. 3, pp. 36–41, Dec. 2023, doi: https://doi.org/10.53555/EPHIJSE.V9I3.212.

[84] A.-R. Ottun *et al.*, "TOWARD TRUSTWORTHY AND RESPONSIBLE AUTONOMOUS DRONES IN FUTURE SMART CITIES," *Authorea Prepr.*, Oct. 2023, doi: https://doi.org/10.36227/TECHRXIV.21444102.V1.

[85] N. Cecchinato, A. Toma, C. Drioli, G. Oliva, G. Sechi, and G. L. Foresti, "Secure Real-Time Multimedia Data Transmission from Low-Cost UAVs with A Lightweight AES Encryption," *IEEE Commun. Mag.*, vol. 61, no. 5, pp. 160–165, May 2023, doi: https://doi.org/10.1109/MCOM.001.2200611.

[86] E. Basan, A. Basan, A. Nekrasov, C. Fidge, E. Abramov, and A. Basyuk, "A Data Normalization Technique for Detecting Cyber Attacks on UAVs," *Drones*, vol. 6, no. 9, Sep. 2022, doi: https://doi.org/10.3390/DRONES6090245.

[87] H. J. Hadi and Y. Cao, "Cyber Attacks and Vulnerabilities Assessment for Unmanned Aerial Vehicles Communication Systems," *Proc. - 2022 Int. Conf. Front. Inf. Technol. FIT 2022*, pp. 213–218, Dec. 2022, doi: 10.1109/FIT57066.2022.00047.

[88] S. Nayhouse, S. Chadha, P. Hourican, C. Moore, and N. Bezzo, "A General Framework for Human-Drone Interaction under Limited On-board Sensing," *2023 Syst. Inf. Eng. Des. Symp. SIEDS 2023*, pp. 308–313, Apr. 2023, doi: https://doi.org/10.1109/SIEDS58326.2023.10137774.

[89] J. Yao and N. Ansari, "QoS-Aware Machine Learning Task Offloading and Power Control in Internet of Drones," *IEEE Internet Things J.*, vol. 10, no. 7, pp. 6100–6110, Apr. 2023, doi: https://doi.org/10.1109/JIOT.2022.3222968.

[90] Q. Zhang, F. MacHida, and E. Andrade, "Performance Bottleneck Analysis of Drone Computation Offloading to a Shared Fog Node," *Proc. - 2022 IEEE Int. Symp. Softw. Reliab. Eng. Work. ISSREW 2022*, pp. 216–221, Oct. 2022, doi: https://doi.org/10.1109/ISSREW55968.2022.00070.

[91] B. Wang, J. Xie, S. Li, Y. Wan, S. Fu, and K. Lu, "Enabling High-Performance Onboard Computing with Virtualization for Unmanned Aerial Systems," *2018 Int. Conf. Unmanned Aircr. Syst. ICUAS 2018*, pp. 202–211, Jun. 2018, doi: https://doi.org/10.1109/ICUAS.2018.8453368.

[92] N. Thalpage, "Unlocking the Black Box: Explainable Artificial Intelligence (XAI) for Trust and Transparency in AI Systems," *J. Digit. Art Humanit.*, vol. 4, no. 1, pp. 31–36, Jun. 2023, doi: https://doi.org/10.33847/2712-8148.4.1_4.

[93] L. Adilova, M. Kamp, G. Andrienko, and N. Andrienko, "Re-interpreting rules interpretability," *Int. J. Data Sci. Anal.*, pp. 1–21, Jul. 2023, doi: https://doi.org/10.1007/S41060-023-00398-5/FIGURES/19.

[94] K. Jia, P. Saowakon, L. Appelbaum, and M. Rinard, "Sound Explanation for Trustworthy Machine Learning," *ArXiv*, Jun. 2023, Accessed: Jun. 30, 2024. [Online]. Available: http://arxiv.org/abs/2306.06134

[95] D. Lyu, F. Yang, H. Kwon, W. Dong, L. Yilmaz, and B. Liu, "TDM: Trustworthy Decision-Making Via Interpretability Enhancement," *IEEE Trans. Emerg. Top. Comput. Intell.*, vol. 6, no. 3, pp. 450–461, Jun. 2022, doi: https://doi.org/10.1109/TETCI.2021.3084290.

[96] C. Meske and E. Bunde, "Transparency and Trust in Human-AI-Interaction: The Role of Model-Agnostic Explanations in Computer Vision-Based Decision Support," *Lect. Notes Comput. Sci. (including Subser. Lect. Notes Artif. Intell. Lect. Notes Bioinformatics)*, vol. 12217 LNCS, pp. 54–69, 2020, doi: https://doi.org/10.1007/978-3-030-50334-5_4.

[97] J. Yang, X. Liu, X. Jiang, Y. Zhang, S. Chen, and H. He, "Toward Trusted Unmanned Aerial Vehicle Swarm Networks: A Blockchain-Based Approach," *IEEE Veh. Technol. Mag.*, vol. 18, no. 2, pp. 98–108, Jun. 2023, doi: https://doi.org/10.1109/MVT.2023.3242834.

[98] Y. Han, X. Wang, Y. Zhang, G. Yang, and X. Tan, "A UAV swarm communication network architecture based on consortium blockchain," *J. Phys.*, vol. 2352, no. 1, pp. 012008–012008, Oct. 2022, doi: https://doi.org/10.1088/1742-6596/2352/1/012008.

[99] B. Chhetri, S. Gopali, R. Olapojoye, S. Dehbashi, and A. S. Namin, "A Survey on Blockchain-Based Federated Learning and Data Privacy," in *Proceedings - International Computer Software and Applications Conference*, Jun. 2023, vol. 2023-June, pp. 1311–1318. doi: 10.1109/COMPSAC57700.2023.00199.

[100] H. Jin, "Advances in federal learning technology research," in *Proceedings Volume 12636, Third International Conference on Machine Learning and Computer Application (ICMLCA 2022)*, May 2023, vol. 12636, p. 97. doi: https://doi.org/10.1117/12.2675227.

[101] A. Blanco-Justicia, J. Domingo-Ferrer, S. Martínez, D. Sánchez, A. Flanagan, and K. E. Tan, "Achieving security and privacy in federated learning systems: Survey, research challenges and future directions," *Eng. Appl. Artif. Intell. Elsevier*, Dec. 2020, Accessed: Jun. 30, 2024. [Online]. Available: https://arxiv.org/abs/2012.06810v1

[102] M. Śmietanka, H. Pithadia, and P. Treleaven, "Federated Learning for Privacy-Preserving Data Access," *Int. J. Data Sci. Big Data Anal.*, vol. 1, no. 2, p. 1, May 2021, doi: https://doi.org/10.51483/IJDSBDA.1.2.2021.1-13.

[103] Q. Xia, W. Ye, Z. Tao, J. Wu, and Q. Li, "A survey of federated learning for edge computing: Research problems and solutions," *High-Confidence Comput.*, vol. 1, no. 1, p. 100008, Jun. 2021, doi: https://doi.org/10.1016/j.hcc.2021.100008.

[104] P. Hamm, M. Klesel, P. Coberger, and H. F. Wittmann, "Explanation matters: An experimental study on explainable AI," *Electron. Mark.*, vol. 33, no. 1, pp. 1–21, May 2023, doi: https://doi.org/10.1007/S12525-023-00640-9.

[105] Q. Vera Liao and J. Wortman Vaughan, "AI Transparency in the Age of LLMs: A Human-Centered Research Roadmap," *arXiv Prepr. arXiv*, pp. 5368–5393, Jun. 2023, doi: 10.48550/ARXIV.2306.01941.

[106] S. Rozario and G. Čevora, "Explainable AI does not provide the explanations end-users are asking for," *ArXiv*, Jan. 2023, doi: https://doi.org/10.48550/ARXIV.2302.11577.

[107] M. Kumari, A. Chaudhary, and Y. Narayan, "Explainable AI (XAI): A Survey of Current and Future Opportunities," *Stud. Comput. Intell.*, vol. 1072, pp. 53–71, Nov. 2022, doi: https://doi.org/10.1007/978-3-031-18292-1_4.

[108] S. Evangeline, A. Lenin, and V. B. Kumaravelu, "Blockchain System for Secure and Efficient UAV-to-Vehicle Communication in Smart Cities," *Int. J. Electron. Telecommun.*, vol. 69, no. 1, pp. 133–138, Jul. 2023, doi: https://doi.org/10.24425/IJET.2023.144342.

[109] S. Nazia Ashraf, S. Manickam, S. Saood Zia, A. Ahad Abro, M. Abdelhaq, and R. Alsaqour, "IoT empowered smart cybersecurity framework for intrusion detection in internet of drones," Jun. 2023, doi: 10.21203/RS.3.RS-3047663/V1.

[110] Y. Miao, K. Hwang, D. Wu, Y. Hao, and M. Chen, "Drone Swarm Path Planning for Mobile Edge Computing in Industrial Internet of Things," *IEEE Trans. Ind. Informatics*, vol. 19, no. 5, pp. 6836–6848, May 2023, doi: https://doi.org/10.1109/TII.2022.3196392.

[111] C. Dong *et al.*, "BDFL: A Blockchain-Enabled FL Framework for Edge-based Smart UAV Delivery Systems," *ACM Int. Conf. Proceeding Ser.*, Jul. 2023, doi: https://doi.org/10.1145/3591365.3592948.

CHAPTER 5

5G Integration for Enhanced UAV Swarm Connectivity

Faisal Rehman[1,2] Muhammad Hammad Nawaz[1], Shanza Gul[1]
[1]Department of Statistics & Data Science, University of Mianwali, Mianwali, Pakistan
[2]Department of Robotics & Artificial Intelligence, National University of Sciences & Technology, NUST, Islamabad, Pakistan
Correspondence: Faisal Rehman[1,2] (faisalrehman0003@gmail.com)

Introduction

The advancement in wireless communication technology has created a platform to introduce the new-generation fifth generation or 5G to replace existing communication technologies. However, when used in multiple parallel with a formation called swarm, it has proven to be an ideal solution for various sectors to enhance performance [1]. Assimilation of 5G technological solution with UAV low altitude system contributes immensely toward enhancing the capability, coordination, and the intelligence of the swarming system in function. Despite the fact that it is still considered to be an emerging analyzation area, it is by no means

CHAPTER 5 5G INTEGRATION FOR ENHANCED UAV SWARM CONNECTIVITY

a small project, and this article explores the potential of 5G technology in enhancing the interaction between UAV swarms, highlighting both technological advancements and potential concerns [2].

The paramount distinctive feature of 5G is the higher network throughput and minimum synthesis time. All these are significant for UAV swarms since they enable real-time data transfer and tackling within the swarm [2]. A feature that has been enhanced by 5G is the sending/receiving of high volumes of data; in this case, the drones coordinate themselves on the information shared among themselves. The low latency further extended to a few milliseconds in 5G networks implies that UAVs are capable of responding to changes in their environment and produces prompt actions in terms of control [3].

AI and edge computing are enabled by 5G necessary for the control of the multiple UAVs, given the great amount of data they produce. The concept of edge computing allows data to be processed near the source to counter the excessive use of bandwidth and huge delays. This capability is beneficial for UAV operations that demand almost immediate processing of information and decision-making such as in cases of detection of obstacles on the field, tracking of targets, and with regard to operations of FISO or other UAV navigation systems. The high bandwidth of 5G enables the AI processor to process the data from numerous UAVs operating in real time, enhancing operation effectiveness during swarm behavior [1, 2]. Next-gen 5G networks are developed to be more reliable and have a wider coverage range as compared with earlier ones. This is especially relevant in cases of swarming of UAVs working in conditions that may be hazardous or in areas not easily accessible [3, 4]. Evaluation of 5G networks for UAVs is also desirable as it enables them to have stable connections even in highly interfered or weak signal zones [4]. This makes communication and control persistent and reliable – a feature that is important for life and death scenarios like search and rescue missions or disasters, or for crop and livestock surveillance.

The Evolution of 5G Technology

5G defines the sixth generation of wireless technology, which is expected to reshape the landscape of wireless connectivity with unprecedented data transfer rates, capabilities, and reliability. This paper aims at elaborating on the historical progression of 5G technology as seen in Figure 5-1, in addition to the intrinsic enhancements that have been made in the past and the expectations of future changes that 5G is poised to bring about [1, 3].

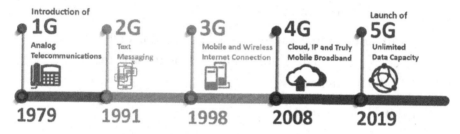

Figure 5-1. Evaluation of Technology over Time

Historical Context

1G to 4G: The Predecessors of 5G

1G (First Generation)

Introduction: 1G which was second generation of mobile networks was first developed in the 1980s.

Technology: Analog transmission.

Features: Simple voice translation without codes sending, low traffic channel, and low quality of signal.

2G (Second Generation)

Introduction: Early 1990s.

Technology: This signal can be transmitted by digital techniques such as the Global System for Mobile telecommunications (GSM) or the Code Division Multiple Access (CDMA).

Features: Improved voice, SMS (Short Message Service), and restricted data solutions are the key features of the 2G.

Advancements: These include the addition of the encryption security and general efficient use of the radio frequency.

3G (Third Generation)

Introduction: Early 2000s.

Technology: UMTS, CDMA2000.

Features: The background to the problem lies in the fact that people in the modern world use their mobile phones not only for calls but also for the Internet connection, video calls, and multimedia messaging.

Advancements: Scalability: With the IMLE data rates increased up to 2 Mbps, new applications that generate more data can now be implemented.

4G (Fourth Generation)

Introduction: Late 2000s.

Technology: LTE (long-term evolution).

CHAPTER 5   5G INTEGRATION FOR ENHANCED UAV SWARM CONNECTIVITY

Features: Internet services, data transfer rates, trough HDTV video, and voice over Internet protocol (VoIP) services.

Advancements: Higher data transmission rates up to one gigabit per second, lower end-to-end latency, and better overall network utilization.

Key Features of 5G

5G technology introduces several groundbreaking features that distinguish it from previous generations, including the following:

- **Ultralow Latency:** The latency has been very low, even to as low as 1 millisecond, to let entities communicate almost immediately.

- **Enhanced Mobile Broadband (eMBB):** Provides increased bandwidth for data transmission necessary to deliver high-quality video and audio and transport oversized files.

- **Massive Machine-Type Communication (mMTC):** The implementation of this technology can facilitate connectivity with a large number of devices which is crucial for IoT or UAV swarm networks.

- **Network Slicing:** Enables the instantiation of multiple logically isolated 5G networks, propositioned for certain needs and use cases.

CHAPTER 5   5G INTEGRATION FOR ENHANCED UAV SWARM CONNECTIVITY

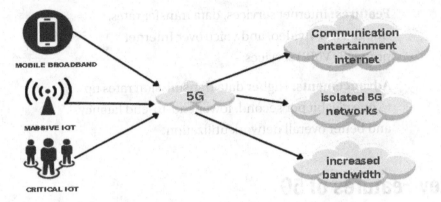

Figure 5-2. *5G Network Slicing*

5G Deployment and Infrastructure: Key Components and Their Roles in UAV Swarm Operations

The setting up of 5G technology is much more challenging and requires much more than previous generations of mobile networks [5]. It entails several key components that support the realization of high-speed, low-latency, and reliable connectivity which is necessary in the operations of systems such as the UAV. Here's an in-depth look at these components.

Small Cell Networks

Small cells, with limited capabilities, serve as low-capacity radio access points, aiding in traffic funneling from macrocellular networks, thus expanding network capacity. In general, small cells assist in the traffic funneling from the macrocellular network, hence expanding the general capacity of the network. It is imperative to this end in order to facilitate the compact as well as the heavy data traffic forecasted for 5G networks. 5G will incorporate higher frequency bands such as millimeter wave

CHAPTER 5 5G INTEGRATION FOR ENHANCED UAV SWARM CONNECTIVITY

to achieve the aforesaid but at the cost of achieving them over shorter distances [1, 4]. Small cells play the role of occupying these frequencies by offering coverage in limited geographical locations. Due to their closeness to the end users (or other devices such as UAVs), small cells prevent long-haul transmissions which contribute to high latency. This is important in real-time control and communication for managerial application in UAV swarm operations.

Distributed Antenna Systems (DAS)

Distributed antenna systems (DAS) are arrangements of connection points of at least two nodes employing multiple antennae to deliver wireless communication within specific geography. These nodes are linked together to a single point through a means of transportation medium, such as fibers in fiber-optic cables [5, 6]. DAS can supply equal coverage for big and cluttered areas such as cities, sports venues, or airports where it might be bulky to have cell tower installations. It can be quickly expanded in terms of user capacity and device handling, which is important based on an increasing number of connected devices (UAVs). DAS can also serve to rebroadcast the signal to ensure the unmanned aerial vehicles have a steady connection which is critical for multiple drone systems.

Cloud and Edge Computing

Cloud computing include data services, storage, hosting, servers, databases, networking, and software infrastructures offered over the Internet, which means that computation and data storage are moved to the location where it is needed to reduce response time demand and to save the network bandwidth. An analysis of the UAV swarms reveals that the coordination of multiple unmanned aerial vehicles requires large amounts of data processing in real time. Through using cloud computing, these data loads are dealt with efficiently due to the computational power that is required. Edge computing entails processing of data or data processing that happens at the endpoint or at the network's boundaries rather than at a centralized cloud. This is desirable in various areas such as in quadcopter navigation and collaboration. Cloud platforms can hence move resources up or down to meet requirements in terms of demand and usage. This is something that is particularly beneficial for UAV operations that have rather variable needs [4, 5].

Integration for UAV Swarm Operations

The integration of small cell networks, DAS, and cloud/edge computing creates a robust 5G [6] infrastructure capable of supporting UAV swarm operations in several ways:

> **Seamless Connectivity:** Small cells and DAS ensure end-to-end coverage, which will enable the UAVs to access connectivity wherever they are stationed. This serves a significant function of monitoring and leadership in a group of many individuals or robots.

> **Real-Time Data Processing:** The integration of edge computing allows processing of data to be done at the edge of the network through quick and real-time analysis of data. For this reason, the work is important given that UAV swarms are characterized by dynamic and unpredictable behavior.

> **High Bandwidth and Low Latency:** A system of small cells together with the edge computing is capable of meeting high bandwidth and low latency required for the transmission of high-resolution video, telemetry data, and control signals between UAVs and ground stations.

Enhanced Reliability: This distributed characteristic offers protection to the network, due to the fact that the technologies are spread out. If one node is down, others are able to take up operations to ensure that operation continues, which is important in critical UAV use.

UAV Swarms: Current State and Challenges

UAVs are referred to as swarms, which is a new and advanced tool that uses numerous unmanned aerial vehicles (UAVs) or drones to perform intricate operations. Ant kingdom, as the concept is called, is based on the similar principles to the principles of the swarms in the nature, to be precise, in the behavior of bird flocks, or fish schools [6]. Here's an in-depth look at the current state of UAV swarms and the challenges they face.

Coordination Algorithms

Consensus Algorithms: Originally, needed to coordinate the UAVs on specific parameters necessary for performances which require aerial formation such as flying in formation.

Flocking Algorithms: The fundamental principle: swarms should mimic natural swarming behaviors in order that they can continuously observe the original space cohesion, separation, and alignment among the swarm elements.

Leader–Follower Models: Label some UAVs as leaders for the swarm; this helps reduce the number of decisions that have to be made for the swarm to function optimally.

CHAPTER 5   5G INTEGRATION FOR ENHANCED UAV SWARM CONNECTIVITY

Communication Systems

Ad Hoc Networks: To guarantee that UAVs can freely talk to each other without having to check on the predetermined pedestal [1, 6].

Mesh Networks: Provide avenues of more indefatigable and elastic routes for business in the swarm which is important in maintaining series linkage.

Autonomous Capabilities

By incorporating sophisticated sensors in the UAVs and state-of-the-art embedded algorithms, the UAVs are now capable of safely navigating their surroundings to avoid any potential obstacles. Computing the routes, which a UAV should follow in a mission, the algorithms establish paths that do not lead to collisions among the UAVs. Strategies for efficiently allocating tasks to one or multiple UAVs based on their abilities and state on the fly.

Coordinated Search Patterns

The specific subsection of the area of interest can be given to each UAV in the swarm, and this way, the large area is covered with high speeds. This is much faster compared to the use of a single UAV, which solves the problem of time-consuming in searching for the target location [4, 5]. This is the case because multiple UAVs can work at the same time, allowing the swarm to perform multiple search missions in the same areas, which contributes to reducing the overall time needed to find missing individuals or assess the damages in disaster-stricken areas. Compared to fixed-wing UAVs, rotary-wing vehicles have the advantage of flexibility, especially searching for their target since they can adapt the mode of the search depending on the data received from the system and the conditions on the ground [7, 8]. For instance, if one UAV during surveillance identifies some object of interest, other UAVs can be guided to focus on the same area due to higher suspicion of enemy presence.

Real-Time Communication and Coordination

The UAVs are constantly in communication with one another, enabling seamless cooperation and the sharing of information as it becomes available which can include imagery, thermal scans, and any other data from the sensors. Organized models of swarm can have a centralized hub to control them, although other models might have UAVs that coordinate their decision-making process, making them diverse and robust.

Real-Time Data Capturing and Processing

Each UAV in the swarm can accommodate distinct sensors such as cameras, LiDAR, and multispectral sensors allowing the simultaneous collection of different types of data. Several UAVs can cover vast regions at high resolutions and furnish a ton of details of region's geographical characteristics such as ground topography, vegetation, water body, and more.

Applications in Ecological Studies

- **Wildlife Monitoring:** By flying such UAV swarms, animals and wildlife populations can be observed and followed, behavior can be monitored, and habitats evaluated without much interference with the creatures.

- **Pollution Detection:** Airborne platforms carrying air quality instrumentation can measure the presence of pollutants and environmental factors, to delineate the source of pollutive agents and their cumulative effects.

- **Forest Management:** Swarms can capture multimedia information on the health of forests and track violations of the directive, as well as foresee the effects of calamities like fires and storms.

Agriculture

Crop Monitoring: Multiple UAVs can fly over fields for crop and height monitoring, disease or infection check, and growth cycles check at frequent intervals. Standard satellite imagery in the high-resolution model and multispectral sensors offer resolutions that offer plant and soil heath information.

Variable Rate Application: From the real-time data, the UAV swarms can fertilize, spray pesticides, and water the crops in the exact areas where it is required, with maximum efficiency and least strain on the scarce resources and the environment. From surveillance activities to crop infections and sprays, the swarms of UAVs can save time and human resource compared to manual methods. Some of the advantages of precision farming practices include the following: they minimize the over application of resources such as water, fertilizers, and pesticides, hence reducing the cost while preventing adverse impacts on the environment.

CHAPTER 5   5G INTEGRATION FOR ENHANCED UAV SWARM CONNECTIVITY

Technical Challenges

The UAV swarms therefore have a great application potential that is however faced with a number of technical issues that require solution in order to optimize the operation of the swarms fully [4, 5].

Here's a detailed exploration of these challenges:

Connectivity and Communication

Signal interferences appear to be a common issue that UAVs may encounter, especially if they are flying in areas that have obstructions to the signal, harsh weather, or in close proximity to other electronic equipment. It is, therefore, difficult to foster very effective communication as we have seen, under these conditions [7, 8]. The information transferred between the UAVs and the ground control station, mainly through high-resolution pictures and sensors data, is abundant. To ensure this flow is managed without over burdening the communication channels will be key. For each UAV, it is likely to adopt ad hoc or mesh networks to make sure it can directly or indirectly communicate with other UAVs. It is challenging to design these networks so that they can be business continuity solutions that are also effective [7, 8].

Solutions and Approaches

- **Advanced Communication Protocols:** It is also important to devise processes and procedures which depend on the conditions that surround them and perform optimally if the transfer of information is to be considered smooth.
- **5G and Beyond:** Utilizing new advanced technology for achieving higher data rates increased reliability with low latency [7, 8] such as 5G.

- **Redundancy:** During its design, it is important to employ multiple connections, which will allow not being deprived of a connection to the swarm if one of the links is disrupted.

Coordination and Control

Coordinating UAV movement in a swarm ensures uniformity and harmony among them [6, 8], optimizing performance and ensuring mission objectives are achieved. This approach maintains suitable configurations for various activities like region monitoring or search and rescue.

Solutions and Approaches

Using algorithms that allow the UAVs to make decision at their level with information they share with other UAVs, this will help to reduce the control by a main controller. Consensus-based UAV methods, allow for simple local communication rules mimicking natural swarms, enabling the creation of complex global algorithms and behaviors.

Autonomy

UAVs have to receive a massive amount of data during the flight whereas they must respond quickly to it and make decision on navigational control and the avoidance of obstacles while performing the mission assigned to it [7, 8]. This takes a lot of computation time and hence needs a large number of iterations to be run. The UAVs have to be able to interact autonomously and perform properly especially when they encounter obstacles. The integration of artificial intelligence will enable UAVs to monitor behaviors and patterns, forming a model to predict the future environment.

Addressing the Challenges

To effectively tackle challenges, it is crucial to develop technical solutions that enhance technology, partner with stakeholders, and invest in research in technology-enhanced learning [8, 9]. Here's how these challenges can be tackled effectively:

> **Research and Development:** The sustained funding for more research and development and more and better communication protocols, better coordination in algorithms, and better system of decentralization and autonomy.
>
> **Collaboration:** Integration with stakeholders across various settings such as academic, industrial, and governmental environments with a mutual aim of developing standardization frameworks and providing information exchange points.
>
> **Field Testing:** The last one is elaborating the numerous problems encountered during comprehensive field testing to make improvements and adjust the systems.
>
> **Simulation and Modeling:** To train UAV swarm formations before a particular task, it is possible to use sophisticated software to simulate several types of situations.

CHAPTER 5   5G INTEGRATION FOR ENHANCED UAV SWARM CONNECTIVITY

5G Enhancements for UAV Swarm Connectivity

The connectivity given by 5G technology provides newer advancements for UAV swarm to relay data by promising higher speed, availability, and low latency [1, 2]. Here's an in-depth exploration.

Increased Data Transmission Speed

Features of 5G networks include enhanced data services with much higher data rate than the previous ones. This increased speed enables UAV swarms to relay a significant amount of data back to a central control point with ease. For instance, with higher data rates, high-resolution images, real-time video streams, and sensor data shared between UAVs and ground control stations are achievable with faster data rates in order to get increased detail in monitoring and analysis [3, 4].

Improved Reliability

Reliability is paramount especially for UAV swarm missions, because absence of control signals interrupting the communication is fatal for such operations. 5G networks can support high levels of reliability offered through features like parallel redundant paths, forward error corrections, and quality of service (QoS). These attributes guarantee the seamless reception of all data packets, preventing any errors due to interferences or signal loss.

Low Latency

The importance of latency minimization to fractional microseconds is significant in determining the success of a UAV swarm operation mission. The 5G technology is about to afford a much lower latency rate than the preceding technologies with the latency rate as low as a few milliseconds. This near real-time responsiveness can help the UAVs to respond quickly to the new scenarios and avoid obstacles that may arise in their flight paths, plan efficient flight paths, and coordinate their tasks effectively.

Massive Device Connectivity

5G networks have an ability to accommodate high number of connected devices, which makes 5G networks to be appropriate for UAV swarms where multiple UAVs have to transmit data between them as well as with GCS. This capability makes it possible for multiple UAVs to work in the required swarms, even within crowded towns or in cases where other wireless appliances are densely packed.

Network Slicing

Network slicing is a distinctive characteristic of 5G networks that permit operators to share a single physical network and split it virtually into numerous new ones corresponding to definite needs. In the context of UAV swarm connectivity, network slicing allows operators to provide guaranteed spectrum for mission-critical communications while the residual spectrum can be used for different kinds of other traffic, thus providing high reliability and low latency for operations.

Edge Computing Integration

The integration of 5G with edge computing can bring computational resources closer to the UAV swarm. This integration lowers the latency, and the reason lies in the fact that most computations are performed locally, at the edge of the network, not at cloud servers. The decision-making, control, and analysis in edge computing all occur in real time, thereby increasing the self-sufficiency and efficiency of UAV swarm operations.

CHAPTER 5   5G INTEGRATION FOR ENHANCED UAV SWARM CONNECTIVITY

Enhanced Data Transmission

5G global connectivity capabilities contribute significantly toward solely endowing UAV swarms with the monumental task of data handling, processing, and communication, such as HD video feeds and sensor data, thus enhancing situational awareness and decision-making systems. At the heart of this capability is 5G's capability to transmit data at unprecedented speed, something that is superior to that of the previous generations of mobile networks [7, 8]. This means that UAV swarms are capable interface and integrate at speed that was highly improbable in the past to enable a continuous flow of real-time data between individual UAVs and ground control stations. In the operational context, it means that UAVs in swarm should be assigned with complex sensors as well as cameras and other reconnaissance tools capable of recording high-definition video stream, images, and sensor data. These data streams are then sent over 5G networks back to ground control stations or to other UAVs in the swarm, for processing and decision-making [6, 7]. With the increased data transmission due to 5G, UAV swarms can deal with these data streams and maintain a comprehensive real-time understanding of the environment in real time.

This increased situational awareness is particularly important to support the decision-making process for UAV swarm formation and cooperation in various scenarios, such as surveillance, search and rescue operations, the environment, and precision agriculture. For example, in the search and rescue application, high-definition camera mounted on UAVs is capable of taking aerial images, which will be transmitted through 5G to the ground control center for further analysis. It enhances localization of possible danger-enhancing zones, search for lost individuals, and management of the rescue mission in real time. Furthermore, the capacity to process and manage complex sensor

CHAPTER 5 5G INTEGRATION FOR ENHANCED UAV SWARM CONNECTIVITY

data is crucial to ensuring UAV swarms can take intelligent decisions independently. Just like the typical UAVs, the predictive mode of operation can facilitate the identification of patterns from the streams of data received from different sensors, enabling the UAVs to do so without automation.

This level of delegation is very constructive in environments that are constantly in transition and where decisions concerning the high stakes are envisaged to occur. To sum up, through improving the data transmission rate of 5G, UAV swarms are capable of effectively collecting, transmitting, and analyzing a plethora of data for immediate use which greatly improve their awareness and decision-making. The rise of this capability creates new opportunities for UAV swarm activities as it now offers them an ability to both perform and coordinate as a single unit to better complete the tasks they are assigned with more effectiveness and accuracy and without much human interference or intervention [9, 8].

Massive IoT Integration for UAV Swarms

The opportunity to expand the concept of IoT communication capabilities using the capabilities of the fifth-generation network is critical to ensuring logical control and interaction between a huge number of UAVs and sensors that make up the network. Essentially, massive IoT can be defined as the use of 5G in Figure 5-3.

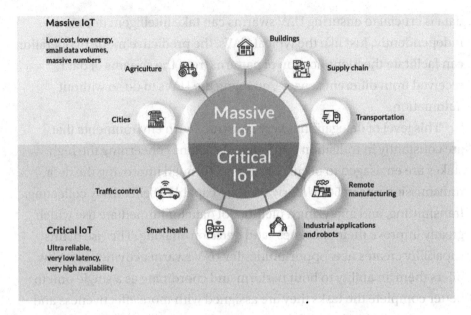

Figure 5-3. Massive IoT Integration

Networks are designed to efficiently connect multiple UAVs and sensors to other IoT devices, ensuring minimal packet loss [11, 12], low latency, and high efficiency.

When it comes to UAV swarm operation, the communication aspects where massive IoT facilities have to be incorporated are fundamental for extended UAV swarm coordination elaborate in their informational exchange. Here's how it works:

Connectivity for a Large Number of Devices:
MM IoT enables the connectivity of a massive number of UAVs and sensors in a single network solution that is represented by 5G. Hence, it becomes possible to guarantee connectivity between the many hundreds or thousands of UAVs that may be organized as a swarm [10, 11], but at the same time, the individual devices can always be connected and have clear communication and synchronization.

CHAPTER 5   5G INTEGRATION FOR ENHANCED UAV SWARM CONNECTIVITY

Efficient Data Exchange: Specifically, with the connections under the heading of massive IoT integration, the 5G network has the ability to support these interconnections of data between UAVs, sensors, or the related devices. This covers the live streaming of telemetry data, the sensors, high-definition videos, and other intelligence data that is vital for swarm action. Thus, the 5G guarantees reliable and fast data transfer rates with minimal end-to-end delay for data exchange empowering UAVs to sustain awareness of the surroundings and make accurate decisions in real time.

Scalability and Flexibility: The adversarial opponent in this regard is the massive IoT that is enabled by the 5G System, providing the necessary scalability and flexibility in order to accommodate the topical swarm operations in the sense that the system can range up from a few nodes to thousands as and when required. The use of the Internet allows it to achieve the connectivity required by the swarm whether it is a small swarm that has been dispatched for localized observation or a large number of swarms across an expansive region because 5G networks provide for the requisite connectivity for the swarm without a decline in its efficiency or effectiveness.

CHAPTER 5 5G INTEGRATION FOR ENHANCED UAV SWARM CONNECTIVITY

Centralized Management and Control:
A significant aspect of the 5G architecture is that it allows for the massive IoT network's integration to centralize and control the UAV swarm. Command centers as well as network administrators have the ability to manage the overall swarm, issuing instructions, receiving status information, and scheduling tasks in the real time. This effectively solves the issue of decentralized decision-making by one agent affecting the swarm, increasing safety and efficiency in operation.

Advanced Applications and Capabilities:
Appealing to this aspect of 5G technology, massive IoT integration in UAV swarms allows for approximately a rich array of enhanced opportunities and functions. These are cooperative perception and navigation, distributed simultaneous localization and mapping, decentralized sensing and detection, self-arming and path finding, and dynamic task planning. These capabilities make UAV swarms better at performing complex missions, operating in a broader spectrum of environments and missions, and increase their level of autonomous operations and amplified efficiency.

CHAPTER 5 5G INTEGRATION FOR ENHANCED UAV SWARM CONNECTIVITY

Applications of 5G-Enhanced UAV Swarms

Figure 5-4. *Applications of Enhanced UAV Swarms*

- **Search and Rescue Operations:** Dedicated to the 5G gorgeous, fast data communication, and quick response time, multiple UAV groups can quickly sweep the area, find victims, and provide supplies in the disaster area [12, 13].

- **Surveillance and Security:** With real-time video transmission and incorporating mandated sensors, 5G UAV swarms can offer huge surveillance services for security and surveillance at borders and event leakage of crucial structures.

- **Precision Agriculture:** Multimedia with high definition allows the use of precision agriculture with 5G enabling the utilization of UAV swarms that perform crop monitoring, irrigation, and yield control.

- **Environmental Monitoring:** With sensors and cameras mounted on them, 5G-enabled UAV swarms can collect data on temperature, wind speed in danger areas, censuses of wildlife, and levels of pollution.

- **Infrastructure Inspection:** Through enhancements achieved with 5G technology and modem control, numerous UAV swarms driven by imaging and control signals can provide precise and quick assessment of infrastructural assets such as bridges, power lines, and pipelines.

- **Emergency Response:** Wearing several infrared cameras and communication equipment, a number of 5G-connected UAV cooperatives can help lost people find their way, determine the extent of the destruction, and direct the actions of rescuers in cases of natural disasters or accidents.

- **Public Safety:** Serving the policemen and the emergency services control, 5G-supercharged UAV formations could include aerial observation, crowd observations, and traffic surveillance for the increased public safety and security.

- **Wildlife Conservation:** Advancements in UAV technology have enabled cities to effectively detect animal movements and perform regular monitoring of habitats, thereby assisting in wildlife conservation through the protection of endangered kinds and the maintenance of species diversity.

- **Urban Planning and Infrastructure Development:** Rapid and accurate mapping due to 5G enabled high mobility; 5G UAV swarms in urban planning and development: suitability evaluation of land use, infrastructure requirement, and impact of earth surface usage.

- **Delivery and Logistics:** Benefiting from optimum and timely navigation and tracking, a large number of UAVs combined with 5G networking can deliver items, such as medicine, parcels, and relief goods for rapid and reliable transport to the intended locations, particularly in mountainous or desolate regions.

- **Disaster Management:** The integration of 5G in UAV swarms brings into focus advanced multimodal disaster response by offering immediate viewing of the situation, search for victims, and supply of necessities to the affected areas.

- **Precision Agriculture:** In the field of agriculture, the 5G UAV swarms can also be effective to monitor the health of crops and efficiently apply water and fertilizers because they are highly accurate.

Technological Challenges and Solutions

Several technological issues that would need to be solved in order to facilitate the proper deployment of UAV swarms for various applications include the following: the first is the issue of connectivity and communication. The UAVs in a formation have to be seamlessly communicating with each other as well as with GCS and must be sharing and receiving real-time information. However, while operating in the dynamically changing and rather unpredictable settings, such as large city locales, or disaster areas, it is often difficult to ensure the stable signal management and optimal bandwidth usage as well as to avoid potential network overloads. To address this issue, current researchers and engineers in the field of UAV are working on more intricate communicational patterns like the mesh networking and ad hoc for UVAs to self-configure associations that can be updated depending on the prevailing conditions. Moreover, optimization of the available technologies such as 5G achieves higher data rate and throughput and commensurate low latency and reliability in the communication process which helps to strengthen the connectivity and command of UAV swarms [7, 8].

The other important difficulty is the disconnection and improvement of organizational consolidation. When several UAVs have to perform specific operations and movements in a coordinated manner, they also require complex control tactics and decentralized decision-making systems. Altogether, the task of both leading and following an agent and coordinating movements and actions of multiple agents in swarm fashion, for such purposes as formation flight or distributed task, ethically and safely and sufficiently quickly, is challenging. This challenge is fueling the development of swarm intelligence algorithms, rooted in observations of nature, proving those local interactions outsmart complex global ones.

These algorithms allow UAVs to "talk" with each other, work in tandem, and decide upon their actions on their own, with minimal interference, given that the overall leader and focus remains, thus making the UAV swarms more flexible and robust in their operations [19]. Another important issue for consideration in managing UAV swarms is the question of autonomy. Human operators will only give high level commands and interventions, and it is required for the UAVs to be semiautonomous and capable of learning about the surroundings and making necessary decisions within these environments. New technologies in sensing, perception, and decision-making enable unmanned aerial vehicles to embed multiple sensors like cameras, 5G integration, LiDAR, GPS, and others for environmental identification and decision-making [8, 9]. Another type of "intelligence" that can be employed in the development of smart UAVs is that of learning, which can be broadly interpreted as the ability of the UAVs to improve with experience that they gain over time, using learning algorithms such as deep learning and reinforcement learning. So, when integrated, these technologies can help increase the flexibility and future responsiveness of the UAV swarm to operate under tenuous and threatening circumstances.

Thus, it becomes easy to conclude that in order to manage UAV swarms, it will be important to try and solve the following technological problems: connectivity and communication issues, coordination and control issues, and autonomy problems. Through the promotion of coherent communication languages, cooperation strategies, and self-controlled mechanisms, the academics and technologists are free to utilize the benefits of collective UAV performing and apply them in different fields of tendentious civil use such as search and rescue operations, agricultural techniques, and so on.

Network Interference and Congestion

Due to higher connectivity of devices to the network, more interference and congestion can occur in 5G networks. The following is the summary of the findings: strengthening network topology techniques: applying intricate network management procedures and adaptive communication [13, 14] platforms can solve these problems.

Cybersecurity

It remains critical to secure the 5G-based UAV swarms from hacking and subsequent unauthorized accesses to it. This demands the use of secure encoding mechanisms for information exchange and early detection of weakness [15].

Energy Consumption

This is because managing the energy resource consumed by a UAV swarm is important in order to sustain the operations of the flying robots for long periods. Some ways are using energy-efficient communication protocols and path planning to reduce the amount of energy consumed frequently necessary for charging batteries.

Future Prospects

Implementing 5G networks in drones of UAV swarms is a major advancement in the field of unmanned aerial systems, and it is likely to bring new innovations across different sectors. When one looks toward future work in this area [16, 17], there are several promising pathways in technology and engineering research that will only continue to expand the applicability and ability of UAV swarms.

Advancements in AI Capabilities

Based on the success of UAV swarms for 5G, the improvement of AI capabilities is key to expanding its potential in the future. Distributed AI tools and techniques are necessary to assist UAVs to continually capture and process data regarding their environment and learn about the environment and optimally control them. Current research efforts have thus shifted toward breakthrough capabilities especially the deep learning and reinforcement learning techniques that will improve the autonomy and intelligence of UAV swarms. These enhancements will extend the capabilities of UAVs to its function in more challenging and specified operations such as UAVs operating in urban environments where they are to navigate on their own or work with other UAVs in responding to emerging situations.

Improvements in Swarm Intelligence

Another area of interest for future work pertains to improving the so-called swarm intelligence that occurs when a number of UAVs act in unison. Swarm intelligence is widely used in many fields, in part based on nature and its observation of the collective behaviors of animal herds or flocks to formulate computational models and control algorithms with the goal of creating UAV swarms capable of exhibiting emergent behavior and self-organizing in real-time adaptive responses to specific environments [19]. The improvement in swarm intelligence shall enhance the UAV swarms to perform the duties that are complicated more efficiently such as the distributed sensing, mapping, and other major decisions leading to new and efficient usages and opportunities.

Expansion of Applications

The combination of UAV swarms with the 5G network is perceived to open up numerous possibilities and opportunities in the different industry branches and economic spheres. From inspection of infrastructures, monitoring environments, responding to disasters, and so much more, the specific uses of 5G UAV swarms can go further than the imagination [17]. Based on such a roadmap, as the overall 5G infrastructure and others supporting UAV swarm activities become more developed and sophisticated, UAV swarm operations will also escalate to perform even more demanding, essential, and more massive missions. Furthermore, improvements in the miniaturization of UAVs along with the development of proper sensors and energy systems will extend the utility of UAV swarms as unique platforms suited for various tasks, crucial to accommodate to numerous opportunities and threats [18]. Furthermore, the advancement of AI has revolutionized various sectors, including IoT, the energy sector, quantum computing, and the fields of image and signal analysis [20–22].

CHAPTER 5 5G INTEGRATION FOR ENHANCED UAV SWARM CONNECTIVITY

On balance, it is possible to affirm that the integration of the 5G technology with UAV swarms has impressive perspectives for further development of autonomy, intelligence, and media scaling. Therefore, as further research and innovative advancements are made in the field of AI and swarm intelligence and its applications, higher roles of the UAV swarms could be projected throughout the various sectors ranging from disaster relief and environmental surveillance to agriculture and others in the future [18]. If we integrate 5G technology into UAV swarms and push forward the effectiveness of these systems, then it will be possible to improve the perspectives of creation of further innovations in this field as well as increase the contribution of the unmanned aerial systems into furthering the benefits for society and making it a better place in the years to come. The advancement of 5G technology with UAVs also signifies a great promise for a better future. The advancement of 5G technology with UAVs holds great promise for a better future, aiming to enhance artificial intelligence, improve swarm intelligence, and expand application fields. Due to the fact that 5G is being developed as a never-ending process, the potential of the respective UAV swarms will be enhanced for performing even more significant and significant tasks.

Conclusion

5G technology integrated into UAV swarms brings a new level of progress in terms of autonomous systems as well as the wireless sector. This integration relies on the features that distinguish the 5G network from the earlier technologies in terms of the data transfer rate, reliability, and response time required by applications or end users. These improvements are crucial for UAV swarms, where the communication between the devices and the utilization of certain channels represents the key factor. Regarding 5G technology, even if a large number of connecting UAVs

CHAPTER 5 5G INTEGRATION FOR ENHANCED UAV SWARM CONNECTIVITY

form a swarm, the technology guarantees that each individual UAV can perform the data transmission and reception quickly and in a stable manner. This continuous and integrated coordination is extremely important for the exchange of real-time information about members' positions, status, and environment they are working in. The enhanced connectivity to lettuce plant also provides constant information sharing which makes the swarm to be a single entity instead of bunch of different drones. Furthermore, the latency required in the 5G system is low, and this is very important when it comes to the coordination of many UAVs in a particular geographical area. High definition means that there is low delay that occurs within the drones and the control systems used in the process. This makes the swarm very quick in responding to any signals that are relayed by the operator or any changes in the surrounding environment. For instance, in case there is an unexpected event such as an obstacle on the way or change in the mission's requirements, then the entire swarm can respond correspondingly. This capability can be especially relevant in environments that rapidly change and is critical for prompt and synchronous response. The other is the efficiency which comes as a result of adopting a 5G technology. Each drone spends less time and energy on operational evaluation and decision-making due to well-coordinated signaling. This not only prolongs the swarm lifetime but also enables it to accomplish more elaborate tasks. The result is that the overall performance of the swarm is increased. The application of UAV swarms can be done in the field of disaster response as soon as they create maps of the affected areas, search for survivors, and provide necessary supplies to them. They can do the same tasks in parallel fashion such as crop checking, spraying, or even planting at very precise and efficient manner while little human interaction is required. In logistics, UAV swarms are useful in picking the right path and ensuring delivery of merchandise at the right time.

References

[1] Gupta, L., Jain, R., & Vaszkun, G. (2016). Survey of Important Issues in UAV Communication Networks. IEEE Communications Surveys & Tutorials, 18(2), 1123-1152.

[2] Zhang, S., Xu, H., & Gursoy, M. C. (2018). 5G Enabled UAV-to-UAV Communications. IEEE Transactions on Wireless Communications, 17(9), 5726-5739.

[3] Cao, Y., Yang, L., & Zhang, H. (2020). Integration of 5G and UAV Swarms for Enhanced Connectivity. IEEE Network, 34(5), 160-167.

[4] Fotouhi, A., Qiang, H., Ding, M., Hassan, M., Giordano, L. G., Garcia-Rodriguez, A., & Yuan, D. (2019). Survey on UAV Cellular Communications: Practical Aspects, Standardization Advancements, Regulation, and Security Challenges. IEEE Communications Surveys & Tutorials, 21(4), 3417-3442.

[5] Liu, C., Jiang, H., Han, Z., & Jiang, H. (2020). Low-Latency Communication for UAV Swarms with 5G NR and Integrated Access and Backhaul. IEEE Network, 34(6), 158-165.

[6] Choudhary, G., Sharma, S., & Zhang, Y. (2019). 5G Network Slicing for UAV Communication: A Survey. Sensors, 19(9), 2045.

[7] Xie, L., & Hu, R. Q. (2020). Energy-Efficient 5G UAV Swarm Networks. IEEE Transactions on Green Communications and Networking, 4(3), 702-715.

[8] Motlagh, N. H., Bagaa, M., & Taleb, T. (2016). UAV-Based IoT Platform: A Crowd Surveillance Use Case. IEEE Communications Magazine, 55(2), 128-134.

[9] Bithas, P. S., Kapsalis, A., Kanatas, A. G., & Samdanis, K. (2019). 5G UAV Networks: End-to-End Slicing and Service Optimization. IEEE Internet of Things Journal, 6(4), 6922-6935.

[10] Rimal, B. P., & Van, D. P. (2018). Impact of 5G on UAV Deployment: Challenges and Opportunities. IEEE Internet of Things Journal, 5(4), 1239-1246.

[11] Wang, X., & Zhang, Z. (2020). 5G and UAV Swarm Networks for Public Safety and Emergency Communications. IEEE Access, 8, 108032-108044.

[12] Choudhury, A., & Misra, S. (2017). Blockchain-Based Framework for 5G-Enabled UAV Communications. IEEE Communications Magazine, 55(9), 100-106.

[13] Galkin, B., Kovenko, A., & DaSilva, L. (2019). A Survey on Unmanned Aerial Vehicle Networks for Civil Applications: A Communications Viewpoint. IEEE Communications Surveys & Tutorials, 21(3), 2541-2563.

[14] Zhang, Y., & Wang, L. (2018). UAV Swarm Networking

[15] Riaz, N., Shah, S. I. A., Rehman, F., & Khan, M. J. (2021). An Intelligent Hybrid Scheme for Identification of Faults in Industrial Ball Screw Linear Motion Systems. *IEEE Access, 9*, 35136-35150.

[16] Humayoun, M., Sharif, H., Rehman, F., Shaukat, S., Ullah, M., Maqsood, H., ... & Chandio, A. H. (2023, March). From Cloud Down to Things: An Overview of Machine Learning in Internet of Things. In *2023 4th International Conference on Computing, Mathematics and Engineering Technologies (iCoMET)* (pp. 1-5). IEEE

[17] Riaz, N., Shah, S. I. A., Rehman, F., Gilani, S. O., & Udin, E. (2020). A Novel 2-D Current Signal-Based Residual Learning With Optimized Softmax to Identify Faults in Ball Screw Actuators. *IEEE Access, 8*, 115299-115313

[18] K. Wang, C. Pan, H. Ren, W. Xu, L. Zhang and A. Nallanathan, "Packet Error Probability and Effective Throughput for Ultra-Reliable and Low-Latency UAV Communications," in *IEEE Transactions on Communications*, vol. 69, no. 1, pp. 73-84, Jan. 2021, doi: https://doi.org/10.1109/TCOMM.2020.3025578

[19] Riaz, N., Shah, S. I. A., Rehman, F., & Gilani, S. O. (2020). An Intelligent Approach to Detect Actuator Signal Errors Based on Remnant Filter. In *Intelligent Technologies and Applications: Second International Conference, INTAP 2019, Bahawalpur, Pakistan, November 6–8, 2019, Revised Selected Papers 2* (pp. 675-683). Springer Singapore.

[20] Riaz, N., Shah, S. I. A., Rehman, F., & Gilani, S. O. (2020). An Approach to Measure Functional Parameters for Ball-Screw Drives. In *Intelligent Technologies and Applications: Second International Conference, INTAP 2019, Bahawalpur, Pakistan, November 6–8, 2019, Revised Selected Papers 2* (pp. 398-408). Springer Singapore.

[21] H. Sharif, F. Rehman and A. Rida, "Deep Learning: Convolutional Neural Networks for Medical Image Analysis - A Quick Review," 2022 2nd International Conference on Digital Futures and Transformative Technologies (ICoDT2), Rawalpindi, Pakistan, 2022, pp. 1-4, doi: https://doi.org/10.1109/ICoDT255437.2022.9787469.

[22] A. Ashfaq, M. Kamran, F. Rehman, N. Sarfaraz, H. U. Ilyas and H. H. Riaz, "Role of Artificial Intelligence in Renewable Energy and its Scope in Future," 2022 5th International Conference on Energy Conservation and Efficiency (ICECE), Lahore, Pakistan, 2022, pp. 1-6, doi: https://doi.org/10.1109/ICECE54634.2022.9758957.

[23] I. Manan, F. Rehman, H. Sharif, N. Riaz, M. Atif and M. Aqeel, "Quantum Computing and Machine Learning Algorithms - A Review," 2022 3rd International Conference on Innovations in Computer Science & Software Engineering (ICONICS), Karachi, Pakistan, 2022, pp. 1-6, doi: https://doi.org/10.1109/ICONICS56716.2022.10100452.

CHAPTER 6

Augmented Reality (AR) and Virtual Reality (VR) for UAV Swarm Visualization

Manish Thakral* (Senior Consultant, Ernst & Young (EY), Mumbai, India,
manishthakra@gmail.com)
Anita Devare, (Amity University, Mumbai, India,
devare.anita@gmail.com)
Manoj H Devare, (Professor & HOI AIIT-AUM, Mumbai, India,
mhdevare@mum.amity.edu)

Introduction

In recent years, unmanned aerial vehicle (UAV) swarms have emerged as a transformative technology with vast potential in various domains, including military surveillance, disaster response, environmental monitoring, and industrial inspections [1]. These swarms, consisting of multiple autonomous drones operating collaboratively, offer unparalleled capabilities for data collection, situational awareness, and task execution in dynamic and complex environments. However, effectively visualizing

CHAPTER 6 AUGMENTED REALITY (AR) AND VIRTUAL REALITY (VR) FOR UAV SWARM VISUALIZATION

and managing UAV swarm operations present significant challenges due to the sheer volume of data, the complexity of coordinating multiple drones, and the need for real-time decision-making.

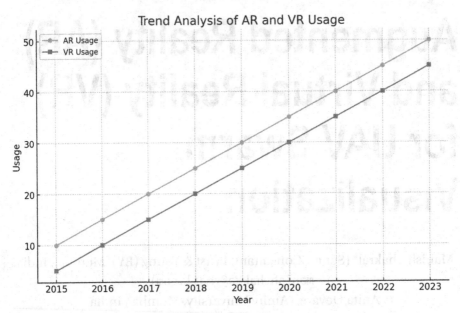

Figure 6-1. *Representation of trend analysis of AR and VR usage against the years 2015–2023*

Augmented reality (AR) and virtual reality (VR) technologies have garnered increasing attention as promising solutions to these challenges, offering immersive, interactive, and intuitive interfaces for UAV swarm visualization and control [2]. AR overlays digital information onto the real-world environment, providing operators with real-time visualizations of UAV positions, trajectories, sensor data, and environmental conditions. VR, on the other hand, creates immersive, simulated environments where operators can interact with virtual representations of UAV swarms, experiment with different control strategies, and simulate complex scenarios.

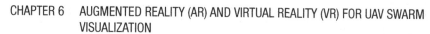
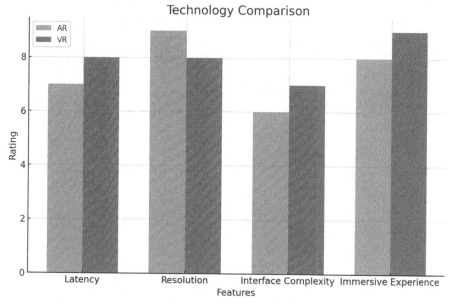

Figure 6-2. Representation of technology comparison with ratings

Challenges in UAV Swarm Visualization

Before delving into the applications of AR and VR in UAV swarm visualization, it's crucial to understand the challenges inherent in managing and visualizing UAV swarm operations:

Data Overload: UAV swarms generate vast amounts of data, including telemetry, sensor readings, video feeds, and environmental data. Effectively managing and processing this data in real time is essential for maintaining situational awareness and making informed decisions during swarm operations [3-5].

Coordination Complexity: Coordinating multiple drones in a swarm to perform collaborative tasks while avoiding collisions and maintaining formation poses significant coordination challenges. Operators need intuitive interfaces to monitor and control the swarm in real time and adjust mission parameters dynamically [6-7].

CHAPTER 6   AUGMENTED REALITY (AR) AND VIRTUAL REALITY (VR) FOR UAV SWARM VISUALIZATION

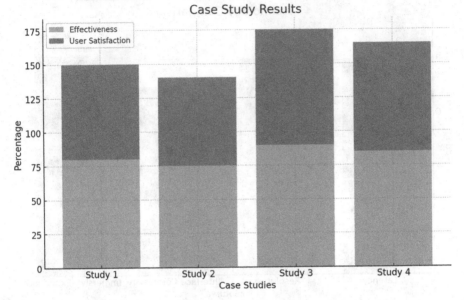

Figure 6-3. *Case study analysis in alignment with effectiveness and user satisfaction*

Human–Machine Interaction: Human operators play a crucial role in supervising and directing UAV swarm operations. Providing operators with intuitive, user-friendly interfaces that facilitate seamless interaction with the swarm is essential for optimizing human–machine collaboration and enhancing operational efficiency [8].

The Role of AR and VR in UAV Swarm Visualization

AR and VR technologies offer innovative solutions to address the challenges of UAV swarm visualization and control:

CHAPTER 6 AUGMENTED REALITY (AR) AND VIRTUAL REALITY (VR) FOR UAV SWARM VISUALIZATION

Enhanced Situational Awareness: AR overlays digital information onto the operator's field of view, providing real-time visualizations of UAV positions, trajectories, sensor data, and environmental conditions. This enhanced situational awareness enables operators to monitor the swarm's status, identify potential threats or obstacles, and make timely decisions to optimize mission performance [9-10].

Immersive Training and Simulation: VR creates immersive, simulated environments where operators can interact with virtual representations of UAV swarms, experiment with different control strategies, and simulate complex scenarios. VR-based training and simulation allow operators to develop and refine their skills in a safe and controlled environment, reducing the risk of accidents and improving operational readiness.

Source of Data

In the realm of augmented reality (AR) and virtual reality (VR) for UAV swarm visualization, the source of data serves as a cornerstone for creating immersive and informative visualizations. This chapter explores the diverse sources of data that contribute to AR and VR applications in UAV swarm visualization, ranging from telemetry and sensor data to environmental information and real-time video feeds [11-15].

CHAPTER 6 AUGMENTED REALITY (AR) AND VIRTUAL REALITY (VR) FOR UAV SWARM
VISUALIZATION

Telemetry and Sensor Data

Telemetry and sensor data collected from individual drones within the UAV swarm are fundamental sources of information for AR and VR visualization. Telemetry data includes real-time updates on drone position, altitude, speed, orientation, and battery status, providing operators with insights into the spatial dynamics and operational status of each drone. Sensor data, such as inertial measurement unit (IMU) readings, GPS coordinates, and environmental sensor measurements, offers additional context about the drone's surroundings, including weather conditions, air quality, and terrain topology. Integrating telemetry and sensor data into AR and VR interfaces enables operators to monitor the swarm's behavior, detect anomalies, and make informed decisions in real time [16].

Environmental Information

Environmental information, including topographical maps, satellite imagery, and weather forecasts, enriches AR and VR visualizations by providing contextual awareness of the operating environment. Topographical maps and satellite imagery offer detailed representations of terrain features, landmarks, and points of interest, enabling operators to plan and execute UAV swarm missions more effectively. Weather forecasts provide critical insights into atmospheric conditions, wind patterns, precipitation, and temperature variations, helping operators anticipate potential hazards and adjust flight parameters accordingly. By integrating environmental information into AR and VR interfaces, operators can enhance situational awareness, optimize mission planning, and mitigate risks during UAV swarm operations [17-20].

CHAPTER 6   AUGMENTED REALITY (AR) AND VIRTUAL REALITY (VR) FOR UAV SWARM VISUALIZATION

Real-Time Video Feeds

Real-time video feeds captured by onboard cameras mounted on UAVs offer valuable visual insights into the surrounding environment and mission-specific tasks. Live video feeds provide operators with a first-person perspective of the drone's field of view, enabling them to identify targets, navigate obstacles, and assess situational dynamics in real time. By streaming live video feeds to AR and VR headsets, operators can immerse themselves in the UAV's perspective, enhancing spatial awareness and decision-making capabilities during complex missions. Integrating real-time video feeds with AR overlays and VR simulations enables operators to augment the visual information with additional contextual data, such as mission objectives, waypoints, and threat alerts, enhancing the overall effectiveness of UAV swarm visualization [21-25].

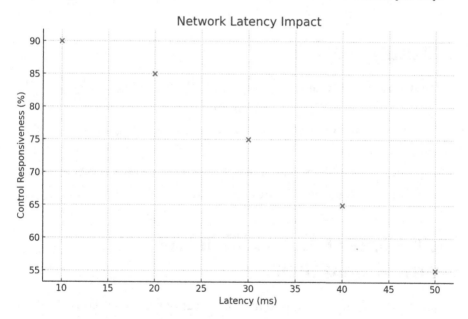

Figure 6-4. *Representation of control responsiveness against the latency (ms)*

CHAPTER 6 AUGMENTED REALITY (AR) AND VIRTUAL REALITY (VR) FOR UAV SWARM VISUALIZATION

Communication Networks

Communication networks serve as the backbone for transmitting data between UAVs, ground control stations, and remote operators in AR and VR applications for UAV swarm visualization. Communication protocols, such as Wi-Fi, cellular networks, and satellite links, enable real-time data exchange, command and control, and telemetry streaming between distributed nodes in the UAV swarm ecosystem. Robust and reliable communication networks are essential for maintaining connectivity, synchronizing data streams, and ensuring seamless interaction between AR and VR interfaces and the UAV swarm. By leveraging communication networks, operators can remotely monitor and control UAV swarm operations from anywhere in the world, facilitating collaborative decision-making and mission coordination.

Literature Review

Augmented reality (AR) and virtual reality (VR) technologies have gained significant attention in recent years for their potential to revolutionize the visualization and control of unmanned aerial vehicle (UAV) swarms. This literature review examines existing research and scholarship on the applications, challenges, and advancements of AR and VR in UAV swarm visualization, highlighting key findings, trends, and gaps in the literature.

Applications of AR and VR in UAV Swarm Visualization

Numerous studies have explored the diverse applications of AR and VR technologies in UAV swarm visualization across various domains:

CHAPTER 6 AUGMENTED REALITY (AR) AND VIRTUAL REALITY (VR) FOR UAV SWARM VISUALIZATION

Military and Defense: Research by demonstrates how AR and VR interfaces enhance situational awareness, mission planning, and target acquisition in military UAV swarm operations. VR simulations enable soldiers to train in realistic scenarios, practice tactical maneuvers, and coordinate UAV swarm missions in complex urban environments.

Civilian and Commercial: Studies by showcase the applications of AR and VR in civilian and commercial UAV swarm operations, including surveillance, monitoring, and infrastructure inspection. AR overlays provide utility workers with real-time visualizations of power lines, pipelines, and other critical infrastructure, enhancing safety and efficiency during inspection and maintenance tasks.

Disaster Response and Emergency Management: Research by investigates the use of AR and VR for disaster response and emergency management with UAV swarms. AR interfaces enable first responders to visualize disaster areas, assess damage, and identify survivors in real time, while VR simulations facilitate training and preparedness exercises for disaster scenarios.

CHAPTER 6   AUGMENTED REALITY (AR) AND VIRTUAL REALITY (VR) FOR UAV SWARM VISUALIZATION

Challenges and Limitations

Despite the promising applications of AR and VR in UAV swarm visualization, several challenges and limitations persist:

> Technological Constraints: The immersive and interactive nature of AR and VR interfaces requires robust hardware devices, high-resolution displays, and low-latency tracking systems. However, current technologies face limitations in terms of computational power, battery life, and sensory input, which can impact the user experience and effectiveness of UAV swarm visualization.

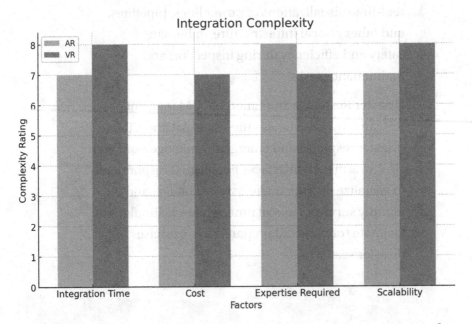

Figure 6-5. *Representation of factors such as integration time and cost in alignment with the complexity*

Human Factors: Human factors, such as motion sickness, fatigue, and cognitive overload, pose challenges for AR and VR users, particularly during prolonged use or intense visualization tasks. Designing user-friendly interfaces, optimizing display ergonomics, and incorporating adaptive feedback mechanisms are essential for mitigating human factors and enhancing user comfort and performance.

Integration and Interoperability: Integrating AR and VR systems with existing UAV platforms, communication networks, and data management systems requires interoperability standards, compatibility testing, and integration frameworks. Ensuring seamless integration and interoperability is essential for facilitating data exchange, command and control, and collaborative decision-making in UAV swarm operations.

Advancements and Future Directions

To address these challenges and unlock the full potential of AR and VR in UAV swarm visualization, researchers are exploring several advancements and future directions:

Hardware Innovations: Advances in display technologies, tracking sensors, and haptic feedback devices are enhancing the immersion, realism, and usability of AR and VR interfaces for UAV swarm visualization. Lightweight, portable devices with improved performance and battery life are enabling more accessible and practical applications in diverse environments.

Software Developments: Software developments in AR and VR platforms, simulation engines, and spatial computing frameworks are expanding the capabilities and flexibility of UAV swarm visualization applications. Open-source software libraries, development toolkits, and application programming interfaces (APIs) are empowering developers to create custom solutions tailored to specific use cases and requirements.

Interdisciplinary Collaboration: Interdisciplinary collaboration between researchers, practitioners, and stakeholders from fields such as computer science, human–computer interaction, aerospace engineering, and psychology is fostering innovation and cross-pollination of ideas in UAV swarm visualization. Collaborative research projects, industry partnerships, and academic–industry consortia are driving advancements and accelerating the adoption of AR and VR technologies in real-world applications.

Experimental Setup

The experimental setup for studying augmented reality (AR) and virtual reality (VR) in UAV swarm visualization plays a critical role in evaluating the effectiveness, usability, and performance of these technologies in real-world scenarios. This chapter explores the key components, considerations, and methodologies involved in designing an experimental setup for AR- and VR-based UAV swarm visualization studies.

CHAPTER 6 AUGMENTED REALITY (AR) AND VIRTUAL REALITY (VR) FOR UAV SWARM VISUALIZATION

Hardware Infrastructure

The hardware infrastructure forms the foundation of the experimental setup, encompassing the devices and equipment necessary to support AR and VR experiences. Key components of the hardware infrastructure include the following:

> Head-Mounted Displays (HMDs): HMDs serve as the primary interface for delivering AR and VR content to users. High-resolution displays, accurate tracking sensors, and comfortable ergonomic design are essential features of HMDs to ensure an immersive and comfortable user experience.
>
> Motion Tracking Systems: Motion tracking systems, such as optical, inertial, or magnetic trackers, capture the user's movements and gestures in real time, enabling natural interaction with virtual objects and environments. Precise tracking accuracy and low-latency performance are critical for maintaining immersion and presence in AR and VR simulations.
>
> Input Devices: Input devices, such as handheld controllers, gesture recognition sensors, and haptic feedback devices, allow users to interact with virtual objects and manipulate the virtual environment. Intuitive and responsive input devices enhance user engagement and facilitate task performance in AR and VR applications.

CHAPTER 6 AUGMENTED REALITY (AR) AND VIRTUAL REALITY (VR) FOR UAV SWARM VISUALIZATION

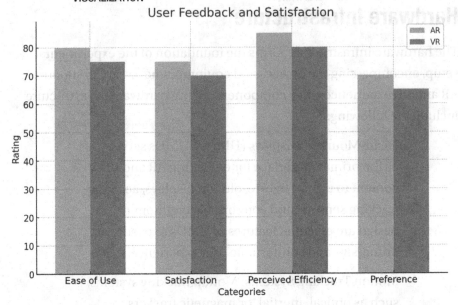

Figure 6-6. *Representation of rating against different categories starting from the use till the preferences*

Software Platforms

Software platforms play a crucial role in creating, rendering, and interacting with AR and VR content in the experimental setup. Key software components include the following:

>AR and VR Development Tools: AR and VR development tools, such as Unity3D, Unreal Engine, and Vuforia, provide a framework for creating immersive virtual environments, integrating 3D models and animations, and implementing interactive user interfaces. These development tools offer a wide range of features and functionalities for building custom AR and VR applications tailored to specific research objectives.

Simulation Engines: Simulation engines, such as Gazebo, AirSim, and Unity Simulation, enable realistic modeling and simulation of UAV swarm behavior, environmental dynamics, and mission scenarios. These simulation engines provide a platform for testing and validating AR- and VR-based UAV swarm visualization techniques in controlled virtual environments before deployment in real-world settings.

Data Acquisition and Analysis Tools: Data acquisition and analysis tools, such as MATLAB, Python, and ROS (Robot Operating System), facilitate the collection, processing, and visualization of telemetry, sensor data, and user interactions during AR and VR experiments. These tools enable researchers to analyze user behavior, evaluate system performance, and derive insights from experimental data to inform iterative design improvements.

Experimental Methodologies

Experimental methodologies define the procedures and protocols for conducting AR- and VR-based UAV swarm visualization studies. Key considerations include the following:

User Recruitment and Training: Recruiting participants with diverse backgrounds, expertise, and familiarity with AR and VR technology ensures a representative sample for the study. Providing training sessions or tutorials on using AR and

CHAPTER 6 AUGMENTED REALITY (AR) AND VIRTUAL REALITY (VR) FOR UAV SWARM VISUALIZATION

VR equipment and interfaces helps familiarize participants with the experimental setup and minimizes learning curves during the study.

Task Design and Scenarios: Designing realistic and engaging tasks or scenarios for participants to perform in the AR and VR environment ensures relevance and validity of the study. Task scenarios may include UAV swarm coordination, mission planning, target tracking, or environmental monitoring, depending on the research objectives and application domains [26-30].

Data Collection and Analysis: Collecting observational data, user feedback, and performance metrics during AR and VR experiments enables quantitative and qualitative analysis of user interactions, task performance, and system usability. Analyzing collected data using statistical methods, qualitative coding techniques, and user experience metrics provides insights into the effectiveness and usability of AR- and VR-based UAV swarm visualization techniques.

CHAPTER 6   AUGMENTED REALITY (AR) AND VIRTUAL REALITY (VR) FOR UAV SWARM VISUALIZATION

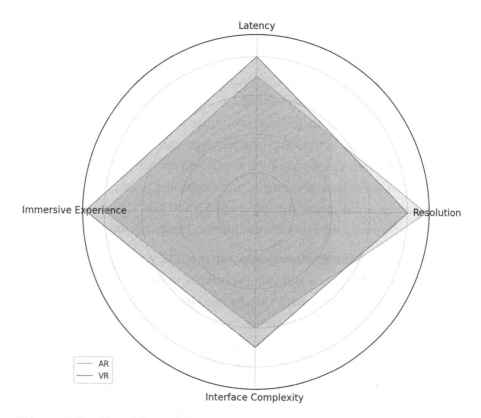

Figure 6-7. Cloud-based demonstration of different AR and VR

Validation and Evaluation

Validation and evaluation of the experimental setup are essential to ensure the reliability, validity, and generalizability of research findings. Key validation and evaluation methods include the following:

> Usability Testing: Usability testing involves observing participants' interactions with the AR and VR system, collecting feedback on user experience, and identifying usability issues or areas

for improvement. Usability metrics, such as task completion time, error rates, and subjective ratings, are used to evaluate the effectiveness and efficiency of the experimental setup.

Performance Evaluation: Performance evaluation assesses the effectiveness of AR- and VR-based UAV swarm visualization techniques in achieving research objectives and task requirements. Performance metrics, such as mission success rate, situational awareness, and decision-making accuracy, quantify the impact of AR and VR technology on user performance and task outcomes.

Comparative Studies: Comparative studies compare the performance and user experience of AR- and VR-based UAV swarm visualization techniques against alternative methods or control conditions. Comparing AR and VR interfaces with traditional 2D displays, physical models, or manual control interfaces provides insights into the relative advantages and limitations of each approach.

Legal Considerations

The integration of augmented reality (AR) and virtual reality (VR) technologies into UAV swarm visualization brings about various legal considerations that researchers, developers, and operators must navigate to ensure compliance, safety, and ethical use. This chapter examines the legal aspects surrounding the deployment and operation of AR and VR systems in UAV swarm visualization, including regulatory frameworks, privacy concerns, liability issues, and intellectual property rights.

Regulatory Frameworks

UAV swarm operations are subject to a complex regulatory landscape governed by national aviation authorities, such as the Federal Aviation Administration (FAA) in the United States, the European Union Aviation Safety Agency (EASA) in Europe, and the Civil Aviation Administration of China (CAAC) in China. Regulatory requirements for UAV swarm operations encompass aspects such as airworthiness certification, pilot licensing, flight planning, and airspace restrictions. AR and VR technologies used in UAV swarm visualization must comply with applicable regulations governing UAV operations, as well as emerging regulations specific to AR and VR applications in aviation.

Privacy Concerns

AR and VR systems collect and process vast amounts of data, including telemetry, sensor readings, video feeds, and user interactions, raising concerns about privacy, data protection, and surveillance. Privacy laws and regulations, such as the General Data Protection Regulation (GDPR) in Europe and the California Consumer Privacy Act (CCPA) in the United States, impose strict requirements on the collection, storage, and use of personal data. Operators of AR and VR systems must implement privacy-by-design principles, obtain informed consent from data subjects, and establish robust data protection measures to safeguard privacy rights and comply with regulatory requirements.

Liability Issues

The deployment of AR and VR systems in UAV swarm visualization introduces liability risks for operators, manufacturers, developers, and users. In the event of accidents, property damage, or personal injury resulting from AR- or VR-related malfunctions or errors, liability may arise due to negligence, product defects, or failure to adhere to safety standards. Establishing clear liability frameworks, insurance coverage, and risk mitigation strategies is essential for addressing liability concerns and ensuring accountability in UAV swarm operations involving AR and VR technologies.

Intellectual Property Rights

AR and VR technologies encompass a wide range of intellectual property rights, including patents, copyrights, trademarks, and trade secrets. Developers and manufacturers of AR and VR systems must navigate intellectual property laws and regulations to protect their innovations, inventions, and creative works from infringement, misappropriation, and unauthorized use. Licensing agreements, nondisclosure agreements, and intellectual property policies play a crucial role in safeguarding intellectual property rights and fostering innovation in the AR and VR ecosystem.

Ethical Considerations

In addition to legal requirements, AR and VR applications in UAV swarm visualization raise ethical considerations related to autonomy, transparency, accountability, and societal impact. Ethical guidelines, such as the IEEE Ethically Aligned Design framework and the ACM Code of Ethics and Professional Conduct, provide principles and guidelines for responsible and ethical use of AR and VR technologies. Addressing ethical considerations requires a holistic approach that balances technological innovation with ethical values, human rights, and societal welfare.

CHAPTER 6 AUGMENTED REALITY (AR) AND VIRTUAL REALITY (VR) FOR UAV SWARM VISUALIZATION

Limitation and Conclusion

Despite the transformative potential of augmented reality (AR) and virtual reality (VR) technologies in UAV swarm visualization, several limitations and challenges must be addressed to realize their full benefits. This chapter examines the key limitations of AR and VR in UAV swarm visualization and concludes by outlining future directions and opportunities for innovation in this field.

Limitations of AR and VR in UAV Swarm Visualization

Technological Constraints: AR and VR technologies face limitations in terms of display resolution, field of view, tracking accuracy, and latency, which can impact the effectiveness and usability of UAV swarm visualization. Addressing technological constraints requires advancements in hardware devices, software algorithms, and integration techniques to improve immersion, realism, and performance in AR and VR systems.

Human Factors: Human factors, such as motion sickness, fatigue, and cognitive overload, pose challenges for AR and VR users, particularly during prolonged use or intense visualization tasks. Designing user-friendly interfaces, optimizing display ergonomics, and incorporating adaptive feedback mechanisms are essential for mitigating human factors and enhancing user comfort and performance in UAV swarm visualization.

CHAPTER 6 AUGMENTED REALITY (AR) AND VIRTUAL REALITY (VR) FOR UAV SWARM VISUALIZATION

Regulatory Compliance: The integration of AR and VR technologies into UAV swarm operations must comply with regulatory requirements governing UAV operations, airspace management, data privacy, and safety standards. Navigating regulatory compliance challenges requires collaboration with aviation authorities, legal experts, and industry stakeholders to ensure adherence to applicable laws and regulations.

Cost and Accessibility: AR and VR systems can be costly to procure, deploy, and maintain, limiting their accessibility and affordability for research, development, and deployment in UAV swarm visualization. Addressing cost and accessibility barriers requires investment in affordable hardware solutions, open-source software platforms, and collaborative research initiatives to democratize access to AR and VR technology.

Conclusion

In conclusion, augmented reality (AR) and virtual reality (VR) technologies hold immense promise for enhancing situational awareness, decision-making, and coordination in UAV swarm visualization. Despite the limitations and challenges, AR and VR offer innovative solutions to complex problems, enabling operators to visualize and interact with UAV swarms in immersive and intuitive ways. By addressing technological constraints, human factors, regulatory compliance, and cost barriers, stakeholders can unlock the full potential of AR and VR in transforming UAV swarm operations across military, civilian, and emergency response domains.

CHAPTER 6   AUGMENTED REALITY (AR) AND VIRTUAL REALITY (VR) FOR UAV SWARM VISUALIZATION

Future Directions and Opportunities

Looking ahead, several future directions and opportunities for innovation in AR- and VR-based UAV swarm visualization emerge:

> Advancements in Hardware and Software: Continued advancements in hardware devices, software algorithms, and integration techniques will enhance the performance, realism, and usability of AR and VR systems for UAV swarm visualization. Innovations in display technology, tracking sensors, and immersive interfaces will enable more immersive and intuitive user experiences.
>
> Interdisciplinary Collaboration: Interdisciplinary collaboration between researchers, practitioners, and stakeholders from fields such as computer science, human–computer interaction, aerospace engineering, and psychology will foster innovation and cross-pollination of ideas in UAV swarm visualization. Collaborative research projects, industry partnerships, and academic-industry consortia will drive progress and accelerate the adoption of AR and VR technologies in real-world applications.
>
> Ethical and Societal Considerations: Addressing ethical and societal considerations related to autonomy, transparency, accountability, and societal impact is essential for responsible and ethical use of AR and VR technologies in UAV swarm operations. Ethical guidelines, regulatory frameworks, and public engagement initiatives will promote the responsible development and deployment of AR and VR systems that prioritize safety, privacy, and human welfare.

CHAPTER 6 AUGMENTED REALITY (AR) AND VIRTUAL REALITY (VR) FOR UAV SWARM VISUALIZATION

Future Scope

The future scope of augmented reality (AR) and virtual reality (VR) in UAV swarm visualization holds immense promise for transforming the way we perceive, interact with, and manage unmanned aerial systems. This chapter explores the evolving landscape of AR and VR technologies and envisions future opportunities for innovation, research, and application in UAV swarm visualization.

Advancements in Hardware and Software

Next-Generation HMDs: Continued advancements in head-mounted displays (HMDs) will lead to lighter, more ergonomic, and higher-resolution devices with improved field of view, tracking accuracy, and display quality. Innovations in display technology, optics, and sensor integration will enhance immersion, realism, and comfort for AR and VR users.

Sensor Fusion and AI Integration: Sensor fusion techniques, combined with artificial intelligence (AI) algorithms, will enable seamless integration of data from multiple sensors, cameras, and IoT devices for enhanced situational awareness and decision-making in UAV swarm visualization. AI-driven analytics and predictive modeling will enable autonomous control and adaptive behavior in AR- and VR-based UAV swarm operations.

CHAPTER 6   AUGMENTED REALITY (AR) AND VIRTUAL REALITY (VR) FOR UAV SWARM VISUALIZATION

Emerging Applications and Use Cases

Urban Air Mobility (UAM): AR and VR technologies will revolutionize urban air mobility (UAM) by enabling real-time visualization and management of aerial traffic, airspace congestion, and drone delivery services in urban environments. AR overlays will provide pilots and operators with dynamic airspace information, flight paths, and collision avoidance alerts, while VR simulations will enable stakeholders to plan, simulate, and optimize UAM operations.

Telepresence and Remote Collaboration: AR and VR systems will facilitate telepresence and remote collaboration in UAV swarm operations, enabling operators and stakeholders to remotely monitor, control, and coordinate drone missions from anywhere in the world. Immersive virtual environments will enable real-time communication, visualization, and decision-making, fostering collaboration and synergy among distributed teams.

Ethical and Societal Implications

Ethical Design and Responsible Innovation: Addressing ethical and societal implications of AR and VR technologies will become increasingly important as these technologies become more pervasive in UAV swarm visualization. Ethical design principles, human-centered approaches, and inclusive design practices will prioritize safety, privacy, transparency, and accountability in AR- and VR-based UAV swarm operations.

Community Engagement and Stakeholder Consultation: Engaging communities, stakeholders, and end users in the development and deployment of AR and VR systems will foster trust, acceptance, and adoption of these technologies in UAV swarm visualization. Public consultation, participatory design workshops, and user feedback mechanisms will ensure that AR and VR solutions address real-world needs and concerns.

Research and Innovation Ecosystem

Interdisciplinary Collaboration and Knowledge Exchange: Interdisciplinary collaboration between researchers, practitioners, and stakeholders from diverse fields will drive innovation and knowledge exchange in UAV swarm visualization. Collaborative research projects, industry-academic partnerships, and open innovation platforms will accelerate the development and adoption of AR and VR technologies in real-world applications.

Education and Workforce Development: Training the next generation of AR and VR, professionals, researchers, and practitioners will be essential for advancing the field of UAV swarm visualization. Educational programs, workshops, and training initiatives will equip students and professionals with the skills, knowledge, and expertise needed to design, develop, and deploy AR and VR solutions for UAV swarm operations.

References

[1] Kaushik, K. (2022). A Novel Approach to Secure Files Using Color Code Authentication. In: Sugumaran, V., Upadhyay, D., Sharma, S. (eds) Advancements in Interdisciplinary Research. AIR 2022. Communications in Computer and Information Science, vol 1738. Springer, Cham. https://doi.org/10.1007/978-3-031-23724-9_4.

[2] Kaushik, K., Singh, V., Manikandan, V.P. (2022). A Novel Approach for an Automated Advanced MITM Attack on IoT Networks. In: Sugumaran, V., Upadhyay, D., Sharma, S. (eds) Advancements in Interdisciplinary Research. AIR 2022. Communications in Computer and Information Science, vol 1738. Springer, Cham. https://doi.org/10.1007/978-3-031-23724-9_6.

[3] K. Kaushik and V. Naik, "Making Ductless-split Cooling Systems Energy Efficient using IoT," 2023 15th International Conference on Communication Systems & Networks (COMSNETS), Bangalore, India, 2023, pp. 471-473, doi: 10.1109/COMSNETS56262.2023.10041408.

[4] Porwal, P., & Devare, M. (2024, August 1). Scientific impact analysis: Unraveling the link between linguistic properties and citations. Journal of Informetrics. https://doi.org/10.1016/j.joi.2024.101526.

[5] Porwal, P., & Devare, M. (2023, September 20). Citation count prediction using weighted latent semantic analysis (wlsa) and three-layer-deep-learning paradigm: a meta-heuristic approach. Multimedia Tools and Applications. https://doi.org/10.1007/s11042-023-16957-8.

[6] P. Porwal and M. Devare, "Citation Classification Prediction Implying Text Features Using Natural Language Processing and Supervised Machine Learning Algorithms," Communications in computer and information science, Jan. 01, 2021. https://doi.org/10.1007/978-981-16-0507-9_46.

[7] K. A. Tatkare and M. Devare, "A Novel Region Duplication Detection Algorithm Based on Hybrid Approach," Social Science Research Network, Jan. 01, 2019. https://doi.org/10.2139/ssrn.3425340.

[8] Fuzzy Probability Model for Quantifying the Effectiveness of the MSW Compost," IEEE Conference Publication I IEEE Xplore, Jan. 01, 2019. https://ieeexplore.ieee.org/document/8945830.

A. Shitole and M. Devare, "Machine Learning Supported Statistical Analysis of IoT Enabled Physical Location Monitoring Data," Springer eBooks, Jan. 01, 2020. https://doi.org/10.1007/978-3-030-41862-5_13.

[9] M.. Devare and M.. Thakral, "Enhancing Automatic Speech Recognition System Performance for Punjabi Language through Feature Extraction and Model Optimization", Int J Intell Syst Appl Eng, vol. 12, no. 8s, pp. 307–313, Dec. 2023.

[10] "Designing an Automatic Speech Recognition System for the Minor Age Group," IEEE Conference Publication I IEEE Xplore, Dec. 15, 2023. https://ieeexplore.ieee.org/abstract/document/10431311.

[11] M. Thakral, R. K. Singh, and K. Kaushik, "Integration of Blockchain Technology and Intelligent System for Potential Technologies," Chapman and Hall/CRC eBooks, Aug. 23, 2022. https://www.taylorfrancis.com/chapters/edit/10.1201/9781003193425-6/integration-blockchain-technology-intelligent-system-potential-technologies-manish-thakral-rishi-raj-singh-keshav-kaushi.

[12] M. Thakral and S. Singh, "A Secure Bank Transaction Using Blockchain Computing and Forest Oddity," Blockchain technologies, Jan. 01, 2022. https://link.springer.com/chapter/10.1007/978-981-19-1960-2_6.

[13] Thakral, M., Singh, R.R., Singh, S.P. (2022). An Extensive Framework Focused on Smart Agriculture Based Out of IoT. In: Choudhury, A., Singh, T.P., Biswas, A., Anand, M. (eds) Evolution of Digitized Societies Through Advanced Technologies. Advanced Technologies and Societal Change. Springer, Singapore. https://doi.org/10.1007/978-981-19-2984-7_12.

[14] M. Thakral, R. K. Singh, and B. V. Kalghatgi, "Cybersecurity and Ethics for IoT System: A Massive Analysis," Transactions on computer systems and networks, Jan. 01, 2022. https://link.springer.com/chapter/10.1007/978-981-19-1585-7_10.

[15] Singh, R.R., Thakral, M., Kaushik, S., Jain, A., Chhabra, G. (2022). A Blockchain-Based Expectation Solution for the Internet of Bogus Media. In: Hemanth, D.J., Pelusi, D., Vuppalapati, C. (eds) Intelligent Data Communication Technologies and Internet of Things. Lecture Notes on Data Engineering and Communications Technologies, vol 101. Springer, Singapore. https://doi.org/10.1007/978-981-16-7610-9_28.

[16] M. Thakral, A. Jain, V. Kadyan and A. Jain, "An Innovative Intelligent Solution Incorporating Artificial Neural Networks for Medical Diagnostic Application," 2021 Sixth International Conference on Image Information Processing (ICIIP), Shimla, India, 2021, pp. 529-532, doi: 10.1109/ICIIP53038.2021.9702631.

[17] M. Thakral, R. R. Singh, A. Jain and G. Chhabra, "Rigid Wrap ATM Debit Card Fraud Detection Using Multistage Detection," 2021 6th International Conference on Signal Processing, Computing and Control (ISPCC), Solan, India, 2021, pp. 774-778, doi:10.1109/ISPCC53510.2021.9609521.

[18] I. Girish, A. Kumar, A. Kumar and A. M, "Driver Fatigue Detection," 2020 IEEE 17th India Council International Conference (INDICON), New Delhi, India, 2020, pp. 1-6, doi: 10.1109/INDICON49873.2020.9342456.

[19] Prajeesha and A. M, "EDGE Computing Application in SMART GRID-A Review," 2021 Second International Conference on Electronics and Sustainable Communication Systems (ICESC), 2021, pp. 1-6, doi: 10.1109/ICESC51422.2021.9532792.

CHAPTER 6   AUGMENTED REALITY (AR) AND VIRTUAL REALITY (VR) FOR UAV SWARM VISUALIZATION

[20] D. M. Udeshi, S. G. L. Divakarla, N. C. Rajdev and A. M, "Wind Speed Forecasting using Hybrid Model," 2022 IEEE 7th International conference for Convergence in Technology (I2CT), Mumbai, India, 2022, pp. 1-5, doi: 10.1109/I2CT54291.2022.9823995.

[21] Tammana, A., Amogh, M.P., Gagan, B., Anuradha, M., Vanamala, H.R. (2021). Thermal Image Processing and Analysis for Surveillance UAVs. In: Kaiser, M.S., Xie, J., Rathore, V.S. (eds) Information and Communication Technology for Competitive Strategies (ICTCS 2020). Lecture Notes in Networks and Systems, vol 190. Springer, Singapore. https://doi.org/10.1007/978-981-16-0882-7_50.

[22] S. Balasubramanian, R. Kashyap, S. T. CVN and M. Anuradha, "Hybrid Prediction Model For Type-2 Diabetes With Class Imbalance," 2020 IEEE International Conference on Machine Learning and Applied Network Technologies (ICMLANT), Hyderabad, India, 2020, pp. 1-6, doi: 10.1109/ICMLANT50963.2020.9355975.

[23] I. Girish, A. Kumar, A. Kumar and A. M, "Driver Fatigue Detection," 2020 IEEE 17th India Council International Conference (INDICON), New Delhi, India, 2020, pp. 1-6, doi: 10.1109/INDICON49873.2020.9342456.

[24] M. Yakasiri, A. M and K. B. K, "Comparative Analysis of Markov Chain and Polynomial Regression for the Prognostic Evaluation of Wind Power," 2020 IEEE International Conference for Innovation in Technology (INOCON), Bangluru, India, 2020, pp. 1-5, doi: 10.1109/INOCON50539.2020.9298374.

[25] P. J, V. M, V. G. Pai and A. M, "Comparative Analysis of Marker and Marker-less Augmented Reality in Education," 2020 IEEE International Conference for Innovation in Technology (INOCON), Bangluru, India, 2020, pp. 1-4, doi: 10.1109/INOCON50539.2020.9298303.

[26] M. Yakasiri, J. Avrel, S. Sharma, M. Anuradha and B. K. Keshavan, "A Stochastic Approach for the State-Wise Forecast of Wind Speed Using Discrete-Time Markov Chain," TENCON 2019 - 2019 IEEE Region 10 Conference (TENCON), Kochi, India, 2019, pp. 575-580, doi: 10.1109/TENCON.2019.8929529.

[27] Gunjan Chhabra, Ajay Prasad & Venkatadri Marrabenta (2022) Comparison and performance evaluation of human bio-field visualization algorithm, Archives of Physiology and Biochemistry, 128:2, 321-332, DOI: 10.1080/13813455.2019.1680699.

[28] D. M. Udeshi, S. G. L. Divakarla, N. C. Rajdev and A. M, "Wind Speed Forecasting using Hybrid Model," 2022 IEEE 7th International conference for Convergence in Technology (I2CT), Mumbai, India, 2022, pp. 1-5, doi: 10.1109/I2CT54291.2022.9823995.

[29] Tammana, A., Amogh, M.P., Gagan, B., Anuradha, M., Vanamala, H.R. (2021). Thermal Image Processing and Analysis for Surveillance UAVs. In: Kaiser, M.S., Xie, J., Rathore, V.S. (eds) Information and Communication Technology for Competitive Strategies (ICTCS 2020). Lecture Notes in Networks and Systems, vol 190. Springer, Singapore. https://doi.org/10.1007/978-981-16-0882-7_50.

[30] S. Balasubramanian, R. Kashyap, S. T. CVN and M. Anuradha, "Hybrid Prediction Model For Type-2 Diabetes With Class Imbalance," 2020 IEEE International Conference on Machine Learning and Applied Network Technologies (ICMLANT), Hyderabad, India, 2020, pp. 1-6, doi: https://doi.org/10.1109/ICMLANT50963.2020.9355975.

CHAPTER 7

Surveillance and Monitoring in Smart Cities with UAV and Machine Learning Integration

Zeeshan Ali Haider[1], Taj Rahman[1], Asim Zeb[2,*], Inayat Khan[3], Amin Sharafian[4], Inam Ullah[5]

[1]Department of Computer Science, Qurtuba University of Science & Information Technology, Peshawar, Pakistan
[2]Department of Computer Science, Abbottabad University of Science & Technology, Abbottabad, Pakistan
[3]Department of Computer Science, University of Engineering & Technology, Mardan, Pakistan
[4]Department of Mechatronics and Control Engineering, Shenzhen University, Guangdong Province, China
[5]Department of Computer Engineering, Gachon University, Seongnam 13120, Republic of Korea;
[1]zeeshan.ali9049@gmail.com; [1]tajuom@gmail.com; [2]asimzeb1@gmail.com;
[3]inayatkhan@uetmardan.edu.pk; [4]aminsharafian@szu.edu.cn;
[5]inam.fragrance@gmail.com
Corresponding Author: Asim Zeb (asimzeb1@gmail.com)

CHAPTER 7 SURVEILLANCE AND MONITORING IN SMART CITIES WITH UAV AND MACHINE LEARNING INTEGRATION

Introduction

Surveillance and monitoring are major aspects of smart city projects, which are implemented to boost public safety capabilities, streamline urban planning functions, and foster better living standards. From CCTV cameras to sensors embedded in public infrastructures or even drones and data analytics platforms that monitor the routine duties of municipalities, there is a plethora of technologies that cities can implement. These technologies manage transportation infrastructure, such as vehicle traffic, and monitor environmental conditions like temperature or light levels (again to detect disease outbreaks), all of which regulate the spread of pathogens. By using the collected data toward these technologies, city officials can use the information to inform intelligent decisions in real-time and improve responsiveness when it comes to emergencies or resource optimization.

A smart city is a new urban construction and planning idea for solving the resulting problems of rapid urbanization, such as too much energy consumed, traffic congestion on the roads, and the efficiency with which many urban mechanisms work. IoT and smart cities utilize advanced technology to enhance residents' lives, integrate city mechanisms, and provide integrated services. The IoT makes data collection almost real time in certain areas like medicine, transportation, energy management, and environmental data logging. The IoT makes information use via monitoring and usage-based analytics to create better data for resource use. It will have the potential to create economic viability, societal welfare, and planet sustainability via the integrated method of smart infrastructure, sustainable resource use, and efficient publishing services in integration [1].

Smart city technologies have a myriad of advantages once deployed. For example, in the realm of transportation, smart traffic management systems could alleviate congestion and increase public transit ridership. These systems can respond to actual on-road performance, adjust the timing of traffic signals, and supply drivers with real-time information about traffic-impacting situations. By using connected medical devices and telehealth services, smart cities in healthcare can provide better patient care more easily, especially in underserved areas. Smart cities could obtain the benefits of smart grids and RE sources to help their economy and clean the environment healthy, avoiding emissions of carbon by using efficient substitutes. It also means that IoT for environmental monitoring makes the city capable of regulating resources in a sustainable approach compared to the previous years in issues such as water conservation and air quality. Nonetheless, the use of surveillance and monitoring technologies in smart cities has the following problems despite the given benefits. The amount of data that is now possible to gather and the detailed levels of analysis to distill these live fire sensor outcomes make it a tragic threat to civil liberties. Thus, accumulating facts and stickers about the desirable sites or staking out and observing the activity of people can lead to oppressive surveillance [2]. Examining this privacy issue, authorities can exploit that data, and on the same note, there can be compromised access or data intrusion. Also, computation-based decision and analysis of data might employ an algorithm that results in the formation of bias in the results, thereby resulting in discrimination. Such biases are from flawed data, insensitive algorithms, or structural injustices that increase society's inequality instead of minimizing it.

Based on all foregoing, one of the main challenging areas is the protection of data, and significant measures must be established. The final crucial aspect is safeguarding the data from unauthorized access through proper encryption, appropriate access control mechanisms, and regular data vulnerability checks. One should also note how data are being obtained and retained, as well as their purpose for utilization. Local authorities have to be honest with the residents about what data they are gathering, for the purpose stated in this article (or any other noble cause), the kind of information being gathered, and what they will be used for. Also important is the involvement of the citizens in the decisions regarding the utilization of data and surveillance. Public consultations, feedback loops, and a move toward participatory governance models can help in community engagement, allowing the local citizens to trust their government that deploying smart city technologies is done to benefit them rather than subduing personal rights.

Smart cities should also be supported by a framework illustrating the ethical use of AI (data analytics) technologies. Such focuses feature preventing algorithmic biases, supporting fairness and accountability, and promoting inclusivity. However, the risks of biased decision-making can be mitigated by developing and deploying AI systems that take into account ethical considerations in smart city projects. Global standards and best practices for the ethical deployment of AI deserve to be spearheaded by the government, technology providers, and civil society in a truly collaborative approach around smart cities.

Smart cities represent a radical departure in the way we live, where state-of-the-art technology helps bring about more efficient and responsive urban solutions that support new forms of sustainability. IoT is being integrated with other smart technologies to collect data in real time, which can then be easily analyzed and used for system-wide improvements. Yet, the usefulness of such systems would depend on how effectively privacy and ethical issues are balanced against surveillance methods to safeguard justice. In sum, smart cities can live up to their potential in improving urban life only when they develop through transparency and citizen participation backed by strong data protection safeguards. To ensure that smart cities are not only innovative but inclusive and equitable as well, the potential of technological terms mustn't exceed societal ones.

Importance of Surveillance and Monitoring

The end-to-end security mechanisms result in an increase in safety provision and much more cost-effective wide-use cases possible with appropriate surveillance. Video surveillance systems and surveillance drones are used for urban and roadway surveillance, crime prevention, and emergency response. Advanced technologies like deep learning and blockchain analyze visual data for real-time anomaly detection, human action recognition, and data integrity, enhancing public safety and health. Visual data from fixed and moving monitoring devices (CCTV cameras, drones) is analyzed with advanced technologies like deep learning and blockchain for real-time anomaly detection, human action recognition, and integrity of the data. In addition to driving public safety and health with remote patient monitoring, emergency response is available 24/7, which aids in the good management of urban structures by supporting traffic patterns or crowd behavior. However, any implementation of these surveillance systems must overcome challenges to be effective as well as unobtrusive, especially in smart cities, such as privacy, data storage, and scalability [3].

The conventional top-down "god-executed" way to deal with brilliant urban areas is progressively advancing toward a more adjusted, iterative cycle that incorporates both the national and private parts. This, in turn, means the ability to better integrate public needs, especially transport-wise. However, accessing and using the data that a smart city collects about residents are problematic from a privacy perspective, among other perspectives. The use of data for commercial purposes by economic actors can outweigh public policy goals. This underscores the importance of an approach to privacy that is as principled, deontological, and law-and-society theoretic as it is grounded in its pragmatism. The findings signal a need for careful calibration of digital innovation and data protection to secure the future sustainability of smart urban communities [4].

Role of UAV Swarms in Smart Cities

An UAV swarm is a fundamental object in smart city management and evolution that encompasses various aspects and capabilities, as well as total utility to improve people's city-based existence. In Figure 7-1, these swarms encompass multiple drones for creating several use cases. Surveillance and security is one of the primary missions of UAV swarms and enables perpetual surveillance for police by simply "plugging in" on the integrated skyhook structures nestled on lampposts for recharging. These cameras also enable monitoring of a larger area than the fixed security cameras, and due to their ability to move, they can effectively address any ongoing events, thus increasing the security.

In traffic monitoring, UAV swarms can observe traffic patterns, find areas with heavy traffic, and input information into algorithms to be used to minimize traffic density. It also allows real-time processing, which reduces traffic jams, increases the efficiency of networks' work, etc. In terms of environmental surveillance, drones that are fitted with sensors can assess the quality of the air, the intensity of the noise, and other factors that are of the essence to the enforcement of environmental protection agencies and the maintenance of public health.

To be precise, UAV swarms become relevant and useful in natural disasters for searching and rescuing people as well as evaluating the extent of damage in the disaster-struck regions. In physical infrastructure inspection, UAV swarms are a more efficient and less risky solution than traditional ways to examine infrastructure; this can give a detailed report and conduct repairs on bridges, buildings, and power lines before catastrophes occur.

On the same note, the UAV swarms are used in delivery services where drones can carry loads and consignment and medical supplies among them, effectively negotiating around obstacles on the road. Last, in agriculture, the UAV swarms help in monitoring the crop health and water demand and pest attacks in urban agriculture for efficient and sustainable farming in cities.

These applications prove that UAV swarms can increase different aspects of the quality of life in urban environments and turn cities into smart, secure, and efficient environments. UAV technology integration into the concept of smart cities means a new level of advancement of city management and development to provide a favorable basis for modern urban issues' solutions.

CHAPTER 7 SURVEILLANCE AND MONITORING IN SMART CITIES WITH UAV AND MACHINE LEARNING INTEGRATION

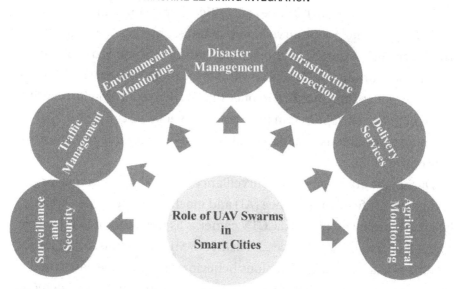

Figure 7-1. *Role of UAV Swarms in Smart Cities*

Current Technologies in Smart City Surveillance

Smart city surveillance is also using several high-tech technologies to keep its surroundings safe and secure. Figure 7-2 illustrates the current technologies used in smart city surveillance. The main surveillance infrastructure backbone systems are CCTV and sensor networks. These videos are further processed and analyzed to spot any security threat or incident that may have occurred in the city through CCTV cameras positioned at various strategic places throughout the different parts of a country. Typically, these cameras may be associated with multiple sensors, such as motion detectors and environmental sensors, to provide a complete monitoring solution [5].

CHAPTER 7 SURVEILLANCE AND MONITORING IN SMART CITIES WITH UAV AND MACHINE LEARNING INTEGRATION

Most importantly, IoT devices are critical in providing much more functional and integrated surveillance systems. Smart cameras, drones, and connected sensors: a city could have tens of thousands or millions of these devices producing data from various locations in the city. This data is then compiled and analyzed to offer the city live reports on its security status. IoT devices deployed in the area of surveillance can help a faster and more autonomous system take immediate reactions to any abnormal issue or threat [6].

During the era of smart city surveillance, it is an important player influencing artificial intelligence (AI) and machine learning (ML). The data produced by mobile and CCTV cameras as well as from IoT devices are big data that is being processed using AI/ML algorithms. It can locate patterns of usage, detect anomalous behaviors, and in some cases predict future security issues before they even happen. Automated fraud detection AI then can automatically highlight such behavior allowing for less human interaction to monitor continuously and quicker reprisal. AI and ML technology can improve the overall efficiency of resource utilization by making it possible to target surveillance efforts in areas or situations that are at higher risk [7].

CHAPTER 7   SURVEILLANCE AND MONITORING IN SMART CITIES WITH UAV AND MACHINE LEARNING INTEGRATION

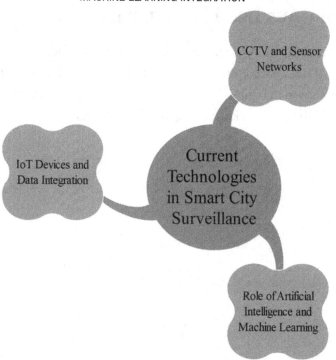

Figure 7-2. *Current Technologies in Smart City*

Unmanned Aerial Vehicles (UAVs) in Surveillance

Drones, also known as unmanned aerial vehicles (UAVs), have established themselves as a key part of modern surveillance systems that excel at tasks traditional methods cannot. The flexibility, long-range, and powerful technical features of UAVs also allow them to be a great ally in a wide range of applications, from urban areas with only confined scenarios up to large unoccupied terrains [8]. This chapter will look at the various types of UAVs used in surveillance, their areas of best utilization, and technological enhancements improving operational capabilities.

CHAPTER 7 SURVEILLANCE AND MONITORING IN SMART CITIES WITH UAV AND
            MACHINE LEARNING INTEGRATION

Types of UAVs Used in Surveillance

The three main drone types for surveillance are multirotor, fixed-wing, and hybrid VTOL, each with unique attributes for specific surveillance jobs. [9], as shown in Figure 7-3, Which type is the best for certain surveillance jobs, as each kind has specific attributes.

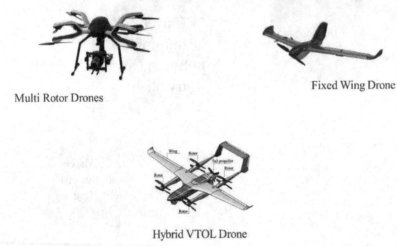

Figure 7-3. *Types of Drones*

Multirotor Drones

Quadcopters and hexacopters are examples of multirotor drones that function as UAVs for surveillance purposes; they have more than one rotor. Small and lightweight, these drones can hover practically on the spot, fly with a level of precision unmatched among their larger cousins, and get in or out where only birds could once go. Their vertical take-off and landing enable them to be conveniently deployed in more densely populated areas. They are ideal for urban surveillance, crowd monitoring, and infrastructure inspection. This includes high-resolution cameras, thermal imaging sensors, and more on-board payloads capable of snapping detailed photos and gathering data. The prime benefits of using multicopter drones are their mobility and stability, which will make them highly flexible in detail capture as well as focus surveillance. They can stay over an area for a very long time, allowing monitoring and transmission of data in real time to the control centers.

Fixed-Wing Drones

These are a lot like regular planes and can be designed to fly long distances for many hours or miles. Compare this to multirotor drones, which can hover in place, albeit not nearly as fast and efficient at covering large areas quickly due partially to a less aerodynamic design. The fixed-wing drones are best for border security, long-distance area scanning, and surveying farms. Similarly, they are employed in search and rescue operations as well as disaster response when wide-area coverage is required. One of the key differences between fixed-wing and multirotor drones is endurance. This allows it to fly up and patrol for hours, performing a long endurance mission over such an expanse of terrain. Flying is more efficient and can cover much greater distances with less energy expended compared to rechargeable multirotor drones.

Hybrid VTOL Drones

These types of VTOL drones are known as hybrids because they take the best from multirotor and fixed-wing. It can lift off and touch down vertically like a multirotor, but then move to more efficiently fly forward with the speed of a fixed-wing drone. Given these two functions, you can see why load cells are so flexible. The hybrid VTOL drones are used for surveillance with emphasis on both the urban and the rural regions. Due to their maneuverability, especially regarding take-off and landing, as well as the coverage area for which they are employed during surveillance missions, they are ideal for infrastructural, environmental, and tactical surveillance. The characteristic that may be derived from the hybrid VTOL drone's design is flexibility. They incorporate both features of multicopter drones as well as the ones of fixed-wing planes as they can fly. This makes it possible to fit the different operation needs, hence making it suitable for a wide range of applications in surveillance.

Technological Advancements in UAV Surveillance

The formation of body-built UAVs and their miniaturization have been quite well supported by the new technologies; wager, UAV-based surveillance has enhanced because of new sensors, data handling, and communication apparatus [10]. These innovations help the UAVs to do the surveillance more expansively. Figure 7-4 shows the technological advancement in UAV surveillance. Modern UAVs have photographic equipment, zoom lenses, and thermal imaging in such situations as in darkness or any unfavorable weather. Also, LiDAR sensors can be mounted on the UAVs for instance in categorizing terrain and environment in three dimensions. AI and machine learning (ML) must be integrated to analyze the data that drones gather since they can capture enormous amounts of data. The core of these technologies helps in semiautomated analytics, anomaly detection, as well as deductive modeling that enables UAVs to identify and adapt to new threats swiftly.

CHAPTER 7   SURVEILLANCE AND MONITORING IN SMART CITIES WITH UAV AND MACHINE LEARNING INTEGRATION

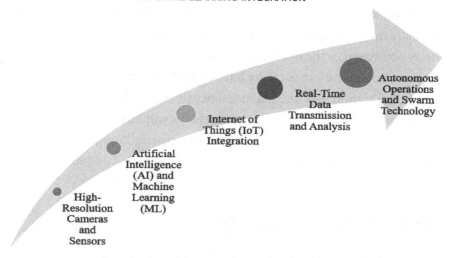

Figure 7-4. Technological Advancements in UAV Surveillance

Furthermore, the UAVs to be used in the place need to be integrated with the IoT devices making it possible for the surveillance systems to communicate with each other by sharing information. It delivers a clear view of the area being monitored or surveilled to enhance the decision-making view and the coverage area. The implementation of the 5G network also supports real-time channels between UAVs and control centers that are crucial in decision-making and receptive action. It facilitates immediate recommendation on dubious things; surveillance activities are as much preventive as they are responsive.

Two core concepts, autonomy and swarm technology, are constituents of the future from a surveillance UAV's standpoint. Autonomous UAVs have the capabilities of self-navigation, that is, they can be flown without remote control by any human operator or following a predesignated flight route, or the UAV can set its flight path at any given time. In swarm technology, many UAVs work together to share information and synchronize their movements to provide better coverage. It enhances radiation coverage, decreases operational expenses, and fosters surveillance networks, which are more tenacious [11]. UAV's use specifically in surveillance operations has greatly revolutionized how even the simplest surveillance of either the densely populated or the open country is conducted. Trends in AI, IoT, and machine-to-machine connectivity have made UAVs more operational in safeguarding smart cities and extended ocean areas. Thus, it may be stated that with the development of technology, UAVs will be playing a significant role in the sphere of surveillance offering effective and progressive solutions for the management of the urban environment and its security.

UAV Swarm Technology

With UAV swarm technology, several UAVs collaborate and form an intelligent entity toward achieving collective objectives.

Definition and Concept

UAV swarm technology refers to a system wherein multiple drones or UAVs work in tandem with each other toward the fulfillment of an end goal. As opposed to single UAV operations, multiple UAV swarms perform tasks that require collective behavior to improve overall performance and efficiency [12]. The idea is inspired by interactions in nature, from swarms of birds to schools of fish and colonies of insects, which exhibit complex group behaviors through simple individual rules.

There are several key advantages to UAV swarms:

- **Redundancy and Reliability:** If the failure of one UAV does not affect the mission as there are multiple UAVs.

- **Scalability:** Swarm systems can scale up by simply adding more UAVs to the group.

- **Flexibility and Adaptability:** Swarms should be able to change reconfiguration on the fly, based on changing environments and mission requirements.

- **Increased Coverage:** Several UAVs can cover a wider area in less time and more efficiently than one single UAV.

Communication and Coordination Among UAVs

For a UAV swarm to operate as desired, effective communication and coordination are essential. These systems normally use one of each or all of the following:

- **Direct Communication:** The UAVs may communicate directly with one another using wireless communication protocols in the swarm. This approach is designed for low-latency and point-to-point information exchange, but in practice, it will be severely limited in range and performance.

- **Indirect Communication:** This method has the UAVs communicating through their environment rather than directly to one another. One instance could be a UAV leaving some marker or signal at a certain location that other UAVs can follow and then act upon.

- **Centralized Coordination:** The swarm is managed centrally by a central controller or leader UAV. This can simplify decision-making and reduce disputes; however, there is a natural single failure point in this configuration.

- **Decentralized Coordination:** All UAVs make decisions only considering local information and interactions with neighbors. This increases resilience and adaptability at the expense of being much harder to code for when keeping groups together.

Algorithms and Protocols for Swarm Intelligence

In the case of UAV swarms, there must be self-organization as well as coordination among the multiple UAVs the priority of which is on the use of swarm intelligence algorithms and protocols. Algorithms are mainly based on the modeling of specific behavior of self-organizing swarms that perform tasks like foraging, hunting, or even simple flying in formation. Some of these are the Boids Algorithm which is a form of population biology situated at the level of simple rules like alignment, cohesion, and separation and the Ant Colony Optimization or ACO which is based on foraging of ants to find the probable paths in the environment. Another optimization approach, similar to the category of the genetic algorithms, is particle swarm optimization (PSO) which simulates birds' flocking or fish schooling; every UAV is regarded as a particle with the position determined by the particle's own and other particles' knowledge.

Consensus algorithms are vital to coordinate all the values of a swarm of UAVs like direction, speed, or even certain tasks. For instance, there exists the leader–follower consensus in which a specific UAV behaves as the leader and other UAVs replicate their behaviors from the leader; the average consensus involves the update of the UAVs' states in cycles with the average of the states of other UAVs in the swarm. While the task assignment method establishes the generic form of the work that is to be done by each UAV in the swarm, the task allocation algorithms define the UAVs and the specific jobs or roles that these UAVs have to undertake based on their capability and state. Market-based algorithms allow UAVs to tender for tasks where the task spots the UAV based on task ability bids. A good example is the threshold-based algorithm that assigns one or several tasks when a specific threshold is crossed increasing the efficiency of CPU resources.

There is a need to put in place correct communication strategies that would enable the swarm to be in harmony and passing of information. TDMA stands for time division multiple access. It breaks the time into slots for individual UAVs so that collisions and interferences are common. Carrier sense multiple access (CSMA) enables UAVs to check the communication channel before sending or transmitting an information/RF signal to avoid interferences. Such basic algorithms and protocols allow multiple UAVS to manage and perform complex coordinated operations in several applications, hence improving the multiple UAVS operation and effectiveness in smart city applications.

Applications of UAV Swarms in Smart Cities

In these areas, numerous commercial opportunities have opened up recently due to advances in drone technology. Typically, drones vary depending on the task or issue that needs to be resolved. Because of the development of scientific technology, public and marketable drones have also manifested as superior. UAVs are expected to provide great value in many areas, such as public safety, rescue, and product delivery [13]. The emerging technology will potentially use drones to usher medical supplies into the arms of rare, regrown, and perhaps retreated places. Sensors, cameras, and other payloads offer a variety of tasks that UAVs can perform [14]. Figure 7-5 presents examples of UAV applications in smart cities. UAVs, acting as an infrastructure layer, can provide relief in situations where other communication methods have failed by acting as a communication network capacity. This section presents the related work

CHAPTER 7   SURVEILLANCE AND MONITORING IN SMART CITIES WITH UAV AND
            MACHINE LEARNING INTEGRATION

on deploying UAV applications for smart cities. Drones offer efficient data management solutions due to reduced response times and increased infrastructure. [15].

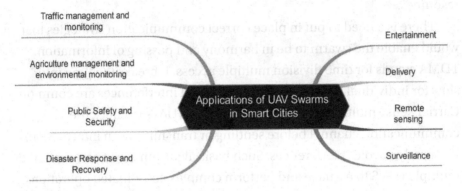

Figure 7-5. Applications of UAV Swarms in Smart Cities

Entertainment

High-resolution aerial photos and videography are widely used by the film and entertainment industry with UAVs. The entertainment category encompasses the average user interested in using some of today's newest technology to photograph and video aerial stunts, map landmarks, or simply fly a small drone around for personal enjoyment. As prices on drones escalate, the need for drone tourism and, hence, new tourists and customers grows with an identical increase in aerial photography film products. Agricultural, environmental monitoring and leisure use also demand smaller drones. Drones have had another purpose in entertainment and advertising, where they carry banners to perform light displays. Drone-based image and video capture tend to be more complex than regular video and photography approaches [16]. They often fly to far-flung or difficult-to-reach sites. Aerial footage and photography for advertising commercials, product launches, or events are captured using

UAVs. They can directly post promotional material (such as flyers, banners, and other kinds of promotional materials). Drones can access hard-to-reach areas [17].

Delivery

Drone delivery systems are specifically suited for smart cities due to their sophisticated technology and infrastructure. Relatively, drones have not been used for delivery in the city, but today, with the development of smart cities, using a drone to deliver pizza is becoming common and considerably faster as they can fly above traffic. A drone can be used to deliver supplies faster and with less effort. Drones can reach areas like high-rise buildings and zones limited to foot traffic only that are difficult for ground-based delivery vehicles. In [18], it was reported that smart cities leverage their advanced technology and infrastructure to facilitate the operation of drones, which then enables efficient delivery networks on a large scale, as depicted. Drones are used for delivery demand forecasting in last-mile logistics, but safety and privacy concerns necessitate compliance with regulations to ensure public well-being. At the same time, safety and privacy concerns make it necessary for city drone use to comply with a certain set of regulations and procedures, so their operation is responsible while ensuring public well-being. Rescue missions have been using drones as well to provide rapid and reliable relief activities in disaster areas, contributing greatly to the welfare of people.

Remote Sensing

Drones might transport items rapidly and safely, resulting in faster shipping situations and more satisfied prospects. This enables UAVs to deliver high-resolution data collected through remote sensing at this scale and timeframe [19]. The unique ability of UAVs to provide data on hazards and features from any perspective in almost any topography cannot be duplicated by virtually every other single technology [20]. Various firms are working on developing and testing UAV taxi services in the region. Vole Copter, a company that has overcome flight over urban canopies, appears to have created an air taxi for travel of up to 60 km with two passengers and offers quick-battery recharge and speedy transport service for shorter-flight occasions. UAVs are essential for surveying the disaster-affected zone and estimating the damage caused by disasters. UAVs are essential for emergency supplies delivery, power line monitoring in smart cities, and inspections without human staff. They can detect cracks and corrosion before they cause accidents, ensuring safety and efficiency [21]. We need UAVs to monitor the power lines in smart cities. The IoT devices can include UAVs that, for instance, could be used to detect cracks in power lines or corrosion on bridges before they fail and lead to potentially deadly accidents. Inspections can be conducted without the need for human staff.

Surveillance

Recent technological developments suggest that UAVs may offer a substitute for surveillance, at least in some scenarios typically met by ground robots. UAVs can fly faster than ground vehicles or robots and reduce the turnaround time between target points [22]. The traffic monitoring system is one of the key applications for UAVs. Drones give a dynamic overview of traffic monitoring, providing necessary data to help optimize road traffic systems due to their mobility and ability to cover various points. This is possible by capturing ordinary methods of surveillance and infiltration [23]. They also inspect industrial premises and power transmission infrastructure. Localizing UAVs to inspect pipelines, the inspection of power lines from a UAA perspective has long been based on the same principle. Based on the above information, UAV-based crowd-monitoring systems are feasible and cost-effective solutions, which will be a part of technology for crowd monitoring. UAVs can perform surveillance on wide portions of land, for example, perimeter areas (borders or coastal zones) and industrial sites during security checks. They can be over an incident scene quickly, collect information, and conduct situation assessments, leading to a faster response time. The design of infrastructure inspection by UAVs is expensive [24]. The spatial tool utilizes high-resolution imaging and sensor payloads on UAVs to accurately inspect and analyze infrastructure health, providing clear images and live data from remote locations. Every city must focus on the availability of basic life-saving and emergency healthcare services. The services people expect from cities need to be there when they are needed, irrespective of where and at what time. Healthcare emergency service delivery or support using UAVs is a real solution [25]. UAVs can act fast by reliably carrying critical medical supplies and services to patients, making basic life support systems instantly available in the new realm of healthcare privacy. For example, a drone can deliver a defibrillator at great speed and other prehospital equipment to the cardiovascular patient who needs it most by facilitating immediate cardiopulmonary resuscitation.

Traffic Management and Monitoring

UAV swarms are used for traffic monitoring and management in smart cities. With high-resolution cameras and sensors carried on small drones, traffic situations can be monitored in real time by local police. UAV swarms could rapidly identify traffic congestion, accidents, and other incidents for immediate response and efficient traffic diversion. The collected data and analysis can be performed to optimize traffic flow by adjusting traffic signals and providing real-time information for drivers [26], thereby alleviating congestion and further enhancing overall transportation efficiency. In addition, UAV swarms are more capable of surveying larger regions while using fewer resources than a traditional ground-based surveillance system, so they can provide the comprehensive understanding we need to restructure our urban environment.

Agriculture Management and Environmental Monitoring

Everything from fertilizers to pesticides, seeds, and water can be drone-deposited onto fields. Regular inspection of crops helps in the estimation of crop damage. UAVs may also be used to identify the maturity state of any crop for choosing a suitable time for harvesting or applying preventive remedies. For example, sudden weather changes can destroy certain crops. UAVs are used to watch and measure the environment, like, previously said, CO_2 and air pollution reaching high limits. This capability of early detection provides for a prompt response to the information and an improved ability to assess risks [27].

Public Safety and Security

The swarming of UAVs results in increased safety and security overall, leading to swifter surveillance and response. These drones can be used in routine patrolling, crowd monitoring during events, and emergency-response scenarios like accident investigations. UAV swarms equipped with facial recognition technologies can identify suspicious activity, track individuals of interest, and provide real-time video feeds for law enforcement purposes. This enhances the degree of situational awareness and reduces reaction time to critical situations. Furthermore, UAV swarms can be deployed in search and rescue missions to support aerial imaging for locating a missing person, even in challenging terrain [28]. The drone can autonomously detect and fly over low- and high-altitude areas to expedite response times to evolving security threats.

CHAPTER 7   SURVEILLANCE AND MONITORING IN SMART CITIES WITH UAV AND MACHINE LEARNING INTEGRATION

Disaster Response and Recovery

The range, speed, and approach of UAV swarms can bring a wealth of advantages to disaster response and recovery. UAV swarms can be used to assess disaster damage and search for survivors in inaccessible areas such as earthquakes, floods, and wildfires. Well if UAVs are used, they can fly over disaster areas and provide real-time aerial imagery on the scene that will help emergency responders to see what type of disaster they have in their hands so that appropriate response measures can be implemented effectively. UAV swarms can be used for mapping the disaster area, monitoring ongoing threats, and aiding in civil response coordination to gather much-needed situational awareness.

In the reconstruction phase, UAVs can help with infrastructure inspection as they can constantly monitor how fast the construction work is going on and ensure that recovery is going according to plan. Unmanned drones are of utmost necessity during disasters as they can function in perilous circumstances and supply critical information, which is vital for disaster management and recovery.

Technical Challenges and Solutions

As shown in Table 7-1, UAVs have the potential to be highly beneficial for smart city services, but significant challenges need attention before their effective use. Technical issues are related to safety, security, and reliability concerns, as well as communication problems, while nontechnical matters focus on licensing and legislation aspects of driving autonomous vehicles; the cost-benefit structure in comparison with conventional transport means ethical questions regarding automation features taking (life or death) decisions for humans, privacy-related considerations, etc. [29]. There are multiple advantages to UAVs, most of which have become more pronounced over time as technology matures. Although some are low on running resources, others represent security challenges. The absence of standards and regulations for UAVs subjects them to security and privacy issues. Because of the design restrictions and hostile operation areas, security experts face difficulties in building an all-in-UAV-security module [30]. Even though UAVs are capable of being a part of smart city services successfully, using them also brings some challenges that should be solved. This chapter explains the challenges faced by UAVs used in smart cities.

Speed and Security

UAV drones may be able to change the way deliveries are carried out and allow for accelerated and more efficient delivery times. Yet, speed with drones can also be dangerous when flying fast in areas where many people are present, as it threatens public safety [31]. The discussion can extend to security concerns, specifically surveillance and privacy concerns, which could potentially extend to UAV drone discussions. Data collection drones with cameras and other sensors on them can gather all kinds of information; some of them sensitive; a hacked drone could be pointed at an enemy target above or below the ground to gather GPS data. The adoption of appropriate regulations and best practices is paramount to addressing speed and security-related issues; this will significantly help in their smooth urban deployment.

Protection of Communication Channels

A reliable communication channel in intelligent cities allows UAVs to be protected against possible security threats and provides secure, safe (the core purpose of the current research study) operation among drones. This data must be protected from unauthorized access and transmitted securely, for example, through a secure communication channel [32]. UAS also must be operated safely to prevent mishaps that could endanger public safety. Ensuring UAVs are appropriately managed, controlled, and safe through secure communication that makes real-time continuous two-way data transfer between the drone and its operator impossible to intercept or manipulate by a third party.

Inventory Resource Management System

One of the most severe utilization constraints in commercial operations is related to drones. Some drones have limited flight time, which makes it tough for extensive inventory management jobs. Not ideal for all large items, as we mentioned earlier, drones have limited space and do not work well with larger things. Using drones for inventory management is a more complex and time-consuming process, possibly preventing many organizations from using the technology due to a lack of technical knowledge [33].

Global Mediums for Connecting and Communicating

Smart cities include the use of UAVs to ensure that everything works well in air traffic management; they are also essential and need global communication. Real-time communication and information exchange between various drone systems is vital to overall operations such as GCS (ground control station) and ATC (air traffic control) since drones are used in different services [34]. This guarantees no blind zones, compliance with regulations, and optimized flight paths. Additionally, these communication channels are used worldwide, providing open lines of communication between the truck and other air-based vehicles as well as road-based vehicles, while also delivering data such as weather conditions or traffic information.

Interservice Operational Capabilities

Drone interservice operation capabilities relate to the ability of drones in smart cities to interface and collaborate with other technologies, systems, or services. Interoperability in UAVs can be challenging due to differing communication logics among services, despite the vast amount of data they provide. [35]. While UAVs provide a wealth of data, sharing information between the services streaming this data can be complex. Services may have different procedures in place, complicating matters during UAV operations. These headwinds can slow down a UAV operation and make it less efficient, making cooperation among the services key to getting past those problems.

Limited Storage and Processing Power

The storage and processing power of UAVs is limited. These restrictions limit the potential of a drone, making you unable to get exactly what you are trying. Additionally, drones lack strong processing prowess; hence, they cannot process data in real time (especially for functions like object recognition, avoiding obstacles, and real-time decision-making). The limitation of storage and processing resources on drones may affect the flight time due to this system of battery-powered consumption, where a drone will still be in the air for a limited period [36].

Self-Controller Security Governance System

Deploying an autonomous network security system for UAVs in smart cities presents challenges due to potential interference issues, potentially disrupting normal operations. While UAV deployment increases, the communication system should be scalable enough to support a large number of devices and have enough coverage to adapt to traffic patterns. Part of the challenge UAVs pose is requiring a lot of bandwidth, which can be difficult to allocate and manage effectively.

Securely Sending Data to End Users

Unauthorized parties can intercept data transmitted between UAVs and end users, thus exposing the security vulnerability. Of course, UAVs create a lot of data, which can overload the network, primarily in heavily populated urban areas [37]. Since UAV communication gets disturbed by other wireless systems in an urban environment, it causes errors and loss of data. However, there are new doubts about the security and privacy of end user data: UAV communication systems have to ensure that any critical financial statement is sent securely in an encrypted way, so unauthorized people cannot access it.

Line of Sight Blockade

In smart cities, line of sight (LOS) is commonly the main concern because buildings and bridges impede communication from one entity to another in a UAV link. This might lead to the UAV losing contact with the ground station or end user, whose consequences will impair real-time data transmission [38]. LOS blocks communication in the UAV system, increasing latency and decreasing reliability. In those regions where LoS is blocked, the coverage provided by UAV communication systems may be significantly degraded and create dead communication zones.

UAV Surveillance Network

UAVs need to cover a relatively large and timely portion of the area, which implies agility toward an essentially wide-coverage, low-latency communication network. Energy resources on the UAV are limited, requiring a focus on low power consumption in communication and data processing prediction systems. UAV surveillance platforms need to adhere to privacy rules and protect individual information. This proposed surveillance system should be interlinked with the existing system of police and security programs to establish robust operations [39].

UAV Networks for Edge Computing

Edge computing UAV networks can be applied in surveillance, environmental monitoring, agriculture, and emergency response. The UAV networks for edge computing require the data to be collected and processed in situ, which means that cannot rely on centralized (cloud) processing. It allows for real-time decision-making and less data to be sent over long distances, resulting in a reduction in latency, therefore shortening the time it takes to process data. The increase in the number of UAVs within a network introduces challenges for the management and coordination of communication among devices [40].

Limited Battery Issue

A UAV flight time is from 20 minutes to several hours, depending on the size of the drone itself. If it carries cargo or uses batteries; these obligations require a lot of energy. Batteries also limit the length of time drones can serve. One of the great challenges of the great challenges of UAVs in smart cities is their short battery life [41]. The increased prevalence of UAVs in urban areas for tasks such as surveillance, traffic monitoring, environmental assessments, and emergency response further compounds the issue of limited battery life.

Table 7-1. *Technical Challenges and Solution*

Sr. no.	Challenges	Problem description	Solution
1	Speed and security	If flying in a populated area, UAVs can travel at speeds that are too high for public safety and be viewed as a security risk	Establish appropriate regulations and guidelines for drone safety protocols
2	Protection of communication channel	Security threats in the data and the need for secure communication to ensure safe operation	Secure communication protocols and encryption
3	Inventory resource management system	Less flight time and storage, need for high skills in UAV operations	Specialized drones developed for inventory tasks supported by comprehensive training programs
4	Global channel for connect and communicate	Common communication for efficient UAV coordination and safety in air traffic management	Adoption of international standards for communication and full integration into the ATC network
5	Interservice operation capabilities	Interfacing and data sharing across diverse UAV services	Standardizing communication protocols and collaborative environments among services as common platforms

(*continued*)

CHAPTER 7 SURVEILLANCE AND MONITORING IN SMART CITIES WITH UAV AND
MACHINE LEARNING INTEGRATION

Table 7-1. (*continued*)

Sr. no.	Challenges	Problem description	Solution
6	Limited storage and processing power	This has a dramatic impact on the performance and flight time of all kinds of UAVs suffering from severe constraints in terms of storage and processing capabilities	Improving the hardware of UAVs and streamlining data processing techniques
7	Self-controller security governance system	Secure and efficient autonomous network security system implementation for UAVs	Advanced security algorithms and infrastructure that is equipped to handle failures
8	Securely sending data to end users	To transmit information securely despite network congestion and interference	Use of encryption and secure communication protocols
9	Line of sight blockade	The reliability and latency of UAV communication are notably influenced by physical obstructions	Relay station deployment, alternative communication methods
10	UAV surveillance network	A strong and reliable communication network to receive on-time UAV surveillance operated by the Military as well as To ensure the protection of that information	Compatible with current surveillance systems and privacy law

(*continued*)

Table 7-1. (*continued*)

Sr. no.	Challenges	Problem description	Solution
11	UAV networks for edge computing	Problems of managing and coordinating wireless communication, and data processing among many UAVs	Efficient network management systems and edge computing frameworks – RPMA facilitates efficient implementation of such medium
12	Limited battery issue	Even though the battery life of some UAS is limited, it also affects UAVs' utilization time and capabilities	Advances in the battery technologies and energy management systems

Regulatory and Ethical Considerations

The legal frameworks that surround the use of UAVs differ with country and region of the world, the main aspects being regulation and compliance, privacy, and safety. The UAV activities are under the control of governments with civil aviation departments' agencies and other organizations that lay down rules on licensing of UAV pilots, geographical areas that they are allowed to fly, and the need for UAVs to be registered. Concerning the protection of the people's resultant observation from the data accrued through aerial surveillance, legal standards are in place for the accumulation, storage, and usage of the gathered data. Such laws make sure that the practices of surveillance, mostly by the police, are as clear as possible. Safety concerns include UAV construction, pilot qualifications, operations, collisions, and ensuring the safety of surrounding areas and nearby objects from the UAV.

CHAPTER 7 SURVEILLANCE AND MONITORING IN SMART CITIES WITH UAV AND MACHINE LEARNING INTEGRATION

Surveillance using UAVs brings to the fore several ethical issues the major of them being privacy, data protection, and liability. UAVs with cameras are a menace to the right to privacy because they can capture people going about their private business, thus the need for UAV operators to adhere to the ethical standards regarding intrusion into people's privacy. When such a situation is inevitable, the transparency of the process has to be observed. Another important factor is data security; any information collected by UAVs must be protected against leakage and particular attention to the exclusion of personal data as well as limitations concerning data access. Also, ethical issues uphold that UAV surveillance should be accountable and responsible to the public whereby the public has the right to know about UAV surveillance and where to report misuse of the UAVs.

Acceptance and adoption of UAV surveillance in smart cities mainly depend on the perception of the public. Trust is essential to attain and it must be established through the communication of organs and agencies conducting surveillance activities through enlightening the stakeholders, the Wayans, of the usefulness and indispensability of surveillance programs and also to educate citizens on the measures taken to uphold their rights and freedoms. The benefits which tend to be considered by members of the public include improved security and the capability of responding to an incident as soon as possible against the likelihood of infringement of an individual's privacy and misuse of data. It is noteworthy that risk communication and public consultation can change such perceptions for the betterment. Integration of the community in the development and formulation of UAV surveillance program plans will help to increase the uptake and reduce concerns once and for all. This engagement makes it possible for the stakeholders to give input on the changes that should be made to the proposed plans, thus enhancing public support and confidence in the UAV surveillance programs.

This chapter reveals that ensuring the compliance of UAVs with the current legislation, escalating ethical questions, and taking into consideration public opinion are decisive in popularizing the use of smart city surveillance through UAVs. Thus, the challenge is to find the right level of innovative UAV surveillance that would still be acceptable to the public, with their privacy violated to the extent necessary to make surveillance a more effective activity.

Future Trends in UAV Swarm Surveillance

Several future trends are likely to influence UAV swarm surveillance as technology is likely to grow further, as demonstrated in Figure 7-6.

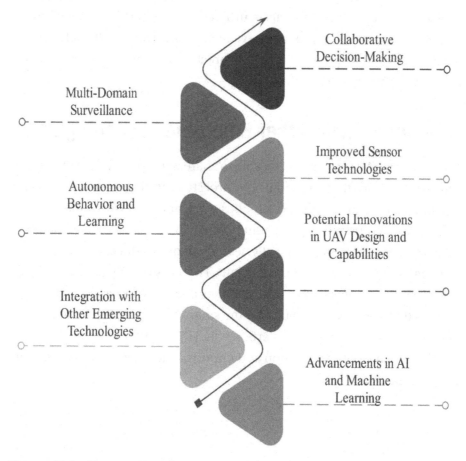

Figure 7-6. Future Trends in UAV Swarm Surveillance

CHAPTER 7 SURVEILLANCE AND MONITORING IN SMART CITIES WITH UAV AND MACHINE LEARNING INTEGRATION

Advancements in AI and Machine Learning

A UAV swarm operation will be made possible by AI and machine learning to reduce the level of interference by human beings. This comprises functionalities such as real-time decision-making capacity, route planning, selection capacity, and the capacity to modify its behavior based on changes in the environment. The better algorithms will also help in the identification of objects of interest as well as tracking them more efficiently through UAVs. Some examples of video data surveillance (VDS) applications are the recognition of abnormal situations and crowds, as well as the recognition of threats in urban spaces. Smart UAVs, together with AI technology, can process surveillance data, which will enable the government and responding teams to forecast future mishaps and devise subsequent moves effectively.

Integration with Other Emerging Technologies

With 5G, the UAV swarms will be able to share high-quality data in real time with one another and ground control stations. It also allows quicker reactions and improves the possibilities of expanding observation activities. Operationally, edge computation can be used to enable UAVs to analyze and process data locally, which not only reduces latency but also makes the procedure more efficient. This is especially the case when the immediate processing of large volumes of data is necessary. IoT devices that are present throughout the smart city structures, including sensors and smart grids, can interface with UAVs. This integration helps in achieving big data acquisition and improves the view of observation for city surveillance.

Potential Innovations in UAV Design and Capabilities

It will be possible to produce mini-advanced UAVs built of composite materials that will be lighter and able to accommodate a variety of high-tech loads, such as enhanced endowment cameras, complicated sensors, and sophisticated communication equipment. Advancements in batteries and energy control systems will increase the time UAVs remain airborne and the distances that they can cover. This is very important in cases of surveillance operations and assignments undertaken in harsh conditions. Next-generation UAV swarms will have better algorithms used for the formation of UAVs, data link integrity, and distributing the workload among different members of the swarm.

These UAV designs may contain elements for functionality in adverse weather and complex city environments, accompanied by efficient GPS and other extremely helpful avoidance systems.

Autonomous Behavior and Learning

Swarms of UAVs will be smarter, capable of making their own decisions, and capable of responding to new conditions. The use of artificial intelligence will help the UAVs adapt to the environment by integrating new knowledge of their surroundings into their practices; the effectiveness of surveillance will rise along with the capacity of the UAVs to recognize signs of odd behavior.

Improved Sensor Technologies

The discovery of better sensors will help UAV swarms have better surveillance abilities. Thus, it can be expected that better cameras, thermal sensors, and LiDAR will more efficiently collect and analyze data for better results. This will also create the ability for UAVs to identify targets and follow them, especially under worse environmental conditions.

CHAPTER 7 SURVEILLANCE AND MONITORING IN SMART CITIES WITH UAV AND
          MACHINE LEARNING INTEGRATION

Multidomain Surveillance

Future UAV swarm surveillance systems will move to different domains of surveillance other than aerial surveillance. For submersible surveillance, there are undercover UAVs, and for overland surveillance, there are ground-based UAVs. Thus, integrating these various domains will result in improving overall surveillance strategies.

Collaborative Decision-Making

UAV swarm surveillance systems will employ artificial intelligence and a decision-making approach that allows multiple UAVs to make collective decisions. This will permit the swarm to make massive analyses and real-time interpretations of the gathered data to make the right choice. These future trends of UAV swarm surveillance will build upon the strengths and effectiveness of the surveillance activities and make these activities more precise, safe, and dynamic for various applications like security and surveillance, natural calamity control, ecological monitoring, and building structural assessment. Changemakers in UAV swarm surveillance are defined by the incorporation of AI/ML, adaptability with new technologies such as 5G and edge computing, and improvement in UAV construction. All these advancements seem to boost the efficiency, usability, and security of UAV surveillance in smart cities, hence the management of the urban setting and security.

Case Studies and Real-World Examples

With smart city technologies, there have been recent and remarkable changes in surveillance and monitoring in urban areas. This section focuses on the use of UAVs, IoT devices, artificial intelligence, and other related technologies in real-life scenarios and the successful implementation of some of the strategies for the safety of city operations and support for the people. Singapore has been among the leading countries considered for the integration of UAVs into the traffic system. These UAVs acquire an awareness of the stream, density, and frequency of the traffic as well as the occurrence of accident information that enables the authorities to act as well in the regulation of traffic during such an occasion. The integration of UAVs has been observed to reduce traffic density and enhance overall transportation capabilities. Air quality and noise level monitoring are major areas where Amsterdam makes use of UAVs. These are equipped with several sensors for the collection of chemicals and other parameters of pollutants. The data collected makes it easier to make decisions over environmental policies and programs to be undertaken in the city. Such an approach has resulted in improved aeration and a better environment within the urban setting for the occupants.

In Tokyo, the UAVs are useful, mainly when it comes to the management and occurrence of disasters. In natural disasters like earthquakes or tsunamis, drones are used to evaluate areas, search for victims, and transport necessities to hard-to-reach places. UAVs have provided an efficient, fast, and safe way of responding to unfortunate incidents, which include the loss of lives and, hence, assisting in quick recovery. New York City is still adding to the large smart surveillance system, which includes the larger surveillance CCTV, IoT, and AI. Such concepts can include virtually any aspect of individuals' lives in cities, including safety, construction, and maintenance of buildings and facilities. Other patterns within this domain are, for instance, using machine learning algorithms to identify such activity that may be considered malicious or likely to cause some risks that would necessitate police action. It is clear that there has been an improvement in this integration, and at the same time, acts of criminal activity have been reduced in society, thereby boosting overall security.

Therefore, various carnivals, like the Rio Carnival, for example, use UAVs for the surveillance of the crowd in terms of security. Each one of these swarms also performs the duty of tracking the movement of mobs and, in matters of vice, isolating undesirable behavior and pointing it out to the security departments. The collected data is processed concurrently to enable the right decisions for the enforcement authorities as well as the participants. By observing the effectiveness of UAV swarms in the accomplishment of crowd management, the same has been observed in other immense gatherings in other parts of the world.

Conclusion

One of the directions for the development of smart cities is the use of UAVs and other modern technologies to manage urban processes. Thus, these technologies foster the improvement of safety measures and city functioning while also making cities better places to live. With the nature of cities changing and developing with time, the further development of these technologies is extremely vital in the fight against modern-day urban issues and in raising the standard of living.

Further research should be carried out to work on the issues arising from the implementation of these technologies, such as privacy, security of the information, and the current laws governing the issue. Descriptive improvements in AI, machine learning, and self-managed operations will refine the competencies of UAVs and other intelligent surveillance systems. Furthermore, there is the viability of integrating UAVs with other developing technologies, such as 5G and blockchain, for smart city security purposes.

References

[1] M. Zaman, N. Puryear, S. Abdelwahed, and N. Zohrabi, "A Review of IoT-Based Smart City Development and Management," *Smart Cities,* vol. 7, no. 3, pp. 1462-1500, 2024.

[2] A. Gohari, A. B. Ahmad, R. B. A. Rahim, A. S. M. Supa'at, S. Abd Razak, and M. S. M. Gismalla, "Involvement of Surveillance Drones in Smart Cities: A Systematic Review," *IEEE Access,* vol. 10, pp. 56611-56628, 2022.

[3] B. F. Fabrègue and A. Bogoni, "Privacy and Security Concerns in the Smart City," *Smart Cities,* vol. 6, no. 1, pp. 586-613, 2023.

[4] Y. Myagmar-Ochir and W. Kim, "A Survey of Video Surveillance Systems in Smart City," *Electronics,* vol. 12, no. 17, p. 3567, 2023.

[5] H. Kim, L. Mokdad, and J. Ben-Othman, "Designing UAV Surveillance Frameworks for Smart City and Extensive Ocean with Differential Perspectives," *IEEE Communications Magazine,* vol. 56, no. 4, pp. 98-104, 2018.

[6] N. A. Ghoniem, S. Hesham, S. Fares, M. Hesham, L. Shaheen, and I. T. A. Halim, "Intelligent Surveillance Systems for Smart Cities: A Systematic Literature Review," *Smart Systems: Innovations in Computing: Proceedings of SSIC 2021,* pp. 135-147, 2022.

[7] K. Al-Dosari and N. Fetais, "A New Shift in Implementing Unmanned Aerial Vehicles (UAVs) in the Safety and Security of Smart Cities: A Systematic Literature Review," *Safety*, vol. 9, no. 3, p. 64, 2023.

[8] N. Mohamed, J. Al-Jaroodi, I. Jawhar, A. Idries, and F. Mohammed, "Unmanned aerial vehicles applications in future smart cities," *Technological forecasting and social change*, vol. 153, p. 119293, 2020.

[9] A. A. Mugheri, M. A. Siddiqui, and M. Khoso, "Analysis on Security Methods of Wireless Sensor Network (WSN)," *Sukkur IBA Journal of Computing and Mathematical Sciences*, vol. 2, no. 1, pp. 52-60, 2018.

[10] A. Mosaif and S. Rakrak, "A New System for Real-time Video Surveillance in Smart Cities Based on Wireless Visual Sensor Networks and Fog Computing," *J. Commun.*, vol. 16, no. 5, pp. 175-184, 2021.

[11] N. Wahab, T. Seow, I. Radzuan, and S. Mohamed, "A Systematic Literature Review on the Dimensions of Smart Cities," in *IOP Conference Series: Earth and Environmental Science*, 2020, vol. 498, no. 1: IOP Publishing, p. 012087.

[12] T. Mazhar *et al.*, "Analysis of IoT Security Challenges and Its Solutions Using Artificial Intelligence," *Brain Sciences*, vol. 13, no. 4, p. 683, 2023.

[13] R. Chatterjee, A. Chatterjee, M. R. Pradhan, B. Acharya, and T. Choudhury, "A Deep Learning-Based Efficient Firearms Monitoring Technique for Building Secure Smart Cities," *IEEE Access*, vol. 11, pp. 37515-37524, 2023.

[14] Y. Said and A. A. Alsuwaylimi, "AI-based outdoor moving object detection for smart city surveillance," *AIMS Mathematics,* vol. 9, no. 6, pp. 16015-16030, 2024.

[15] T. Mazhar *et al.*, "Analysis of Challenges and Solutions of IoT in Smart Grids Using AI and Machine Learning Techniques: A Review," *Electronics,* vol. 12, no. 1, p. 242, 2023.

[16] M. Islam, A. S. Dukyil, S. Alyahya, and S. Habib, "An IoT Enable Anomaly Detection System for Smart City Surveillance," *Sensors,* vol. 23, no. 4, p. 2358, 2023.

[17] A. Zahra, M. Ghafoor, K. Munir, A. Ullah, and Z. Ul Abideen, "Application of region-based video surveillance in smart cities using deep learning," *Multimedia Tools and Applications,* vol. 83, no. 5, pp. 15313-15338, 2024.

[18] I. H. Sarker, "Machine Learning: Algorithms, Real-World Applications and Research Directions," *SN computer science,* vol. 2, no. 3, p. 160, 2021.

[19] H. U. Khan, M. Sohail, F. Ali, S. Nazir, Y. Y. Ghadi, and I. Ullah, "Prioritizing the multi-criterial features based on comparative approaches for enhancing security of IoT devices," *Physical Communication,* vol. 59, p. 102084, 2023.

[20] S. Bhoyar, P. Bhoyar, A. Kumar, and P. Kiran, "Enhancing applications of surveillance through multimedia data mining."

[21] D. C. Nguyen, M. Ding, P. N. Pathirana, A. Seneviratne, J. Li, and H. V. Poor, "Federated Learning for Internet of Things: A Comprehensive Survey," *IEEE Communications Surveys & Tutorials*, vol. 23, no. 3, pp. 1622-1658, 2021.

[22] T. Mazhar, H. M. Irfan, S. Khan, I. Haq, I. Ullah, M. Iqbal, and H. Hamam, "Analysis of Cyber Security Attacks and Its Solutions for the Smart Grid Using Machine Learning and Blockchain Methods," *Future Internet*, vol. 15, no. 2, p. 83, 2023.

[23] I. Rafiq, A. Mahmood, S. Razzaq, S. H. M. Jafri, and I. Aziz, "IoT applications and challenges in smart cities and services," *The Journal of Engineering*, vol. 2023, no. 4, p. e12262, 2023.

[24] M. Qaraqe *et al.*, "PublicVision: A Secure Smart Surveillance System for Crowd Behavior Recognition," *IEEE Access*, vol. 12, pp. 26474-26491, 2024.

[25] P. Y. Ingle and Y.-G. Kim, "Real-Time Abnormal Object Detection for Video Surveillance in Smart Cities," *Sensors*, vol. 22, no. 10, p. 3862, 2022.

[26] R. Miranda *et al.*, "Revolutionising the Quality of Life: The Role of Real-Time Sensing in Smart Cities," *Electronics*, vol. 13, no. 3, p. 550, 2024.

[27] G. Mohi ud din dar *et al.*, "A Novel Framework for Classification of Different Alzheimer's Disease Stages Using CNN Model," *Electronics*, vol. 12, no. 2, p. 469, 2023.

[28] D. Alahakoon, R. Nawaratne, Y. Xu, D. De Silva, U. Sivarajah, and B. Gupta, "Self-Building Artificial Intelligence and Machine Learning to Empower Big Data Analytics in Smart Cities," *Information Systems Frontiers,* pp. 1-20, 2023.

[29] A. Medjdoubi, M. Meddeber, and K. Yahyaoui, "Smart City Surveillance: Edge Technology Face Recognition Robot Deep Learning Based," *International Journal of Engineering,* vol. 37, no. 1, pp. 25-36, 2024.

[30] N. K. Dewi, "Review of Vehicle Surveillance Using IoT in the Smart Transportation Concept," *International Journal of Engineering and Manufacturing,* vol. 11, no. 1, p. 29, 2021.

[31] H. Khalil *et al.*, "A UAV-Swarm-Communication Model Using a Machine-Learning Approach for Search-and-Rescue Applications," *Drones,* vol. 6, no. 12, p. 372, 2022.

[32] I. Alrashdi and A. Alqazzaz, "Synergizing AI, IoT, and Blockchain for Diagnosing Pandemic Diseases in Smart Cities: Challenges and Opportunities," *Sustainable Machine Intelligence Journal,* vol. 7, pp. (6): 1-28, 2024.

[33] M. Shorfuzzaman, M. S. Hossain, and M. F. Alhamid, "Towards the sustainable development of smart cities through mass video surveillance: A response to the COVID-19 pandemic," *Sustainable cities and society,* vol. 64, p. 102582, 2021.

[34] M. Ponnusamy and A. Alagarsamy, "Traffic monitoring in smart cities using internet of things assisted robotics," *Materials Today: Proceedings,* vol. 81, pp. 290-294, 2023.

[35] I. Ullah, A. Noor, S. Nazir, F. Ali, Y. Y. Ghadi, and N. Aslam, "Protecting IoT devices from security attacks using effective decision-making strategy of appropriate features," *The Journal of Supercomputing*, vol. 80, no. 5, pp. 5870-5899, 2024.

[36] M. Dai, Z. Su, Q. Xu, and N. Zhang, "Vehicle Assisted Computing Offloading for Unmanned Aerial Vehicles in Smart City," *IEEE Transactions on Intelligent Transportation Systems*, vol. 22, no. 3, pp. 1932-1944, 2021.

[37] N. Abbas, Z. Abbas, X. Liu, S. S. Khan, E. D. Foster, and S. Larkin, "A Survey: Future Smart Cities Based on Advance Control of Unmanned Aerial Vehicles (UAVs)," *Applied Sciences*, vol. 13, no. 17, p. 9881, 2023.

[38] H. Sharma and N. Kanwal, "Video surveillance in smart cities: current status, challenges & future directions," *Multimedia Tools and Applications*, pp. 1-46, 2024.

[39] A. Kumar, "Secure Surveillance in Smart Cities: A Comprehensive Framework," in *Green Computing for Sustainable Smart Cities*: CRC Press, pp. 100-111.

[40] H. Verma, "IoT Empowerment in Urban Development: Using Surveillance," *Available at SSRN 4822527*, 2024.

[41] A. Haque, K. Bharath, M. Amir, and Z. Khan, "Role and applications of power electronics, renewable energy and IoT in smart cities," in *Smart Cities: Power Electronics, Renewable Energy, and Internet of Things*: CRC Press, 2024, pp. 66-95.

CHAPTER 8

Artificial Intelligence-Enhanced IoT and UAV Integration in Ad Hoc Networks for Smart Cities

Abdullah Akbar[1], Waqar Hussain[2]

[1]Department of Electrical Engineering, National University of Computer and Emerging Sciences, Peshawar, Pakistan,
Email: abdullahakbar0209@gmail.com

[2]Department of Computer Science Shaheed Zulfikar Ali Bhutto Institute of Science and Technology Islamabad, Pakistan,
Email: waqarabaaj@gmail.com

Introduction

Today's the era of technological innovation to facilitate humans. Monitoring data traffic in internet must be practiced properly to ensure security using artificial intelligence [1]. Ad hoc networks are formed by the combination of sink node, workstation, and satellite. Ad hoc networks come in a variety of configurations which include wireless ad hoc networks

CHAPTER 8 ARTIFICIAL INTELLIGENCE-ENHANCED IOT AND UAV INTEGRATION IN AD HOC NETWORKS FOR SMART CITIES

(WANETs), mobile ad hoc networks (MANETs), ship ad hoc networks (SANETs), flying ad hoc networks (FANETs), vehicular ad hoc networks (VANETs), robot ad hoc networks (RANETs), and aeronautical ad hoc network (AANET) [2, 3]. Wireless communication systems, for example, IEEE 802.11, IEEE 802.12, and Zigbee, can be utilized as backbone for ad hoc networks [4]. The mentioned subject is having applications both in civil and military domains. AI-enabled wireless ad hoc networks are having vast applications in transportation, smart cities, smart grid, smart homes, drone-taxi, smart phones, etc. Also, AI is having many other applications in healthcare industry, fake news detection, digital marketing, agriculture, smart government, and education as well [5, 6]. Wireless control ad hoc networks are having many problems which include energy, communication, management, routing, localization, decentralization, network operations, dynamic topology, and line of sight difficulties that must also be addressed. Locating node positioning in ad hoc networks is considered critical problem. Signal strength from base station to wireless enabled ad hoc nodes is also a problem which needs to be given solution. However, in smart cities, many ad hoc networks like MANET, VANET, FANET, and IoT connected nodes will improve communication standard. Therefore, due to low signal strength, the intruder will attack easily on the network. Attacks such as Denial of Service (DoS), Distributed Denial of Service (DDoS), and Ping of Death (PoD) can have a significant impact on the overall network [7–11]. AI/ML techniques use to provide facility directly from environment. Machine learning techniques can be applied on huge amount of data. Therefore, due to AI/ML techniques, decision-making process will be easily made possible. The majority of the time, MANETs lack the infrastructure necessary to address security issues. AI/ML methods can secure communication links, while metaheuristic search algorithms like AntHocNet can optimize communication in wireless connected flying ad hoc networks [12, 13]. Ad hoc networks have the potential to revolutionize smart cities by combining AI-enhanced IoT with unmanned aerial vehicles (UAVs). Unmanned aerial vehicles (UAVs) function as mobile nodes in ad hoc networks, offering flexible and dynamic communication capabilities that are crucial for efficient administration of urban environments. Unmanned aerial vehicles (UAVs) powered by artificial intelligence (AI) can be used in smart cities for a variety of purposes, such as real-time traffic monitoring, emergency response, surveillance, and environmental monitoring. Because AI algorithms are being used, these UAVs are able to evaluate massive amounts of data instantaneously, make intelligent decisions, and optimize their courses for maximum efficiency and coverage. Additionally, the collaboration of AI, IoT, and UAVs enhances the adaptability and resilience of ad hoc networks, ensuring reliable communication even in challenging urban settings. This relationship makes it possible to design infrastructure and resources more wisely, which improves city peoples' quality of life and encourages sustainable urban expansion. Using AI and UAVs can help smart cities become more automated and networked, which can lead to innovative solutions for urban issues. In addition, other ad hoc networks like MANET, SANET, RANET, VANET, FANET, and WSN can utilize artificial intelligence techniques in near future to improve quality of experience metrics within networks. Major contribution points of this book chapter are as under:

- Detailed limitations of ad hoc networks are discussed.

- Comprehensive study regarding artificial intelligence for ad hoc networks like MANET, VANET, WPIoT, FANET, UWSN, RANET, and SANET.

- AI and machine learning applications for ad hoc networks are discussed properly.

Figure 8-1 shows of variations in ad hoc networks and demonstrates their connection mechanism. It demonstrates the many ways that ad hoc networks establish connections between devices as well as the kinds of devices that are a part of each form of ad hoc network.

CHAPTER 8 ARTIFICIAL INTELLIGENCE-ENHANCED IOT AND UAV INTEGRATION IN AD HOC NETWORKS FOR SMART CITIES

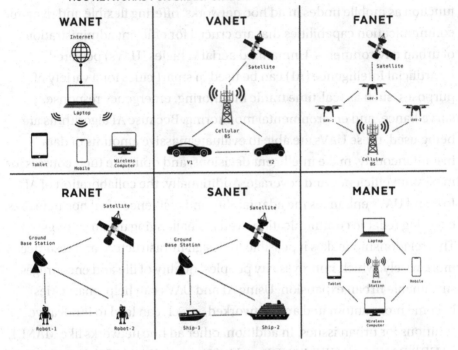

Figure 8-1. Different Types of Ad Hoc Networks

Literature Study

Wireless ad hoc networks have offered promising wireless communication between nodes without involving a centralized infrastructure. This section discusses the literature review that was done to identify the following limitations in ad hoc networks using various techniques.

Ad hoc networks are dependent on the nodes and ground base station but lack a stable infrastructure. These nodes can be stable or moveable. In VANET, the nodes are either roadside units (RSUs) or vehicles that are wirelessly connected. These nodes (vehicles) are moving at a speed range from zero to 200 kilometers per hour. These high-speed cars do not move only in one direction, but they move on both sides of the road. VANET's 300m wireless range and 90km/h car speed limit communication for 12 seconds, highlighting the drawbacks of ad hoc networks in VANET. These limitations present the drawbacks of ad hoc networks in VANET [14].

Information-centric networking (ICN) is among one of the finest communication structures for VANET. Research on named data networking (NDN) driven VANET is in its initial stages and has limited scalability. However, to put connected automobile technologies into practice, NDN can be quite important. A brief survey where they described the possibility of implementing the CCN model for communication between vehicles, known as vehicular CCN (VCCN). The survey described challenges related to caching, naming, routing, and security in VCCN, but their analysis doesn't cover all the recent solutions available in NDN-based VANETs [15]. Mobile ad hoc networks are formed by wirelessly connecting different nodes like mobile, tablets, or laptops. In MANET, every node is honest and cooperative which makes the design to access any node in a particular path. This design makes it easier to make connections between nodes and use routing protocols, but it also makes it open to attack. One of these attacks is called packet dropping attack where a corrupt node gets rid of some or all the data passing through it. These types of attacks are a serious risk to MANET because MANET doesn't have physical protection where it can control the flow of data between nodes or a reliable way to control access to data. In packet-dropping attacks, a node can be quickly taken over by an attacker who joins the network and corrupts it. It gets rid of the data that is supposed to pass the node to make a routing connection between the source and the receiver. Many surveys have been conducted to deal with security threats in MANETs. Most of them have provided useful overviews of various risks and protective measures. However, neither of these studies focuses on a specific assault and outlines all its features across various routing approaches [16].

CHAPTER 8 ARTIFICIAL INTELLIGENCE-ENHANCED IOT AND UAV INTEGRATION IN AD HOC NETWORKS FOR SMART CITIES

Unmanned aerial vehicles (UAVs) are used for many purposes in agriculture, healthcare, traffic monitoring, search and rescue, and other applications. These single UAVs are combined to make a network known as FANET. These UAVs are connected where they exchange data with each other and a nearby base station. The design infrastructure and working mechanism of multiple UAVs is vulnerable to attacks which is harmful to the network security and the data transmission between these nodes. An attacker can easily connect and compromise a node. It can listen to control data and manipulate information. Wi-Fi-connected UAVs are more open to attacks than cellular networks because of weak security and unreliable connections. Moreover, Wi-Fi-connected UAVs when flying over a restricted area are more attractive to attackers because they can access all their base data through ports like USB. These UAVs can cause serious damage to nature and humans if it crashes in a remote area due to extreme weather or any technological issue [17]. Transportations for passengers and transport of goods by airplanes are expected to grow at a larger scale in the coming decades. According to studies in 2035, there will be 14.4 million flights which will result in a 1.8% average annual growth in revenue compared to that of 2012. These airplanes are one of few places where internet data cannot be available at low cost and high speed. Many researchers have worked to overcome this problem by using the AANETs concept. AANET provides an exchange of communication between nodes using multi-hop A2A radio communication. It is difficult to secure AANET from every threat and attack. These attacks can lead an airplane to any incident which can endanger the lives of many passengers. The AANET can be hacked both internally and externally, causing service impairments and flight control issues, and hacking a node can lead to serious security issues [18]. Table 8-1 presents different types of ad hoc networks with drawbacks and limitations.

Table 8-1 *Different types of ad hoc networks with limitations*

References	Types of Networks	Description	Limitations
[19]	Vehicular Ad Hoc Networks	VANETs are now considered a part of advanced automobiles that help passengers, drivers and road systems communicate and exchange information.	As VANETs advanced it faced numerous challenges like Sybil Assault, Node Impersonation Assault, Black Hole Assault, Gray Hole Assault, and many more. These solutions are not discussed to overcome these challenges.
[20]	Vehicular Ad Hoc Networks with Cloud Computing Concept	A new concept of VANETs with cloud computing concepts is presented. The new design inherits the characteristics, applications, and challenges of both the parent clouds.	New challenges arise due to the emergence of two different concepts which lead to data loss, data breaching, Account or Service Hijacking, Denial of Service, and Location Sharing and Privacy Issues.

(*continued*)

Table 8-1. (*continued*)

References	Types of Networks	Description	Limitations
[21]	Mobile Ad Hoc Networks	MANETs deploy radio technology to transfer data between devices such as Mobile, Laptops, and printers without any fixed infrastructure. Two different routing techniques in MANETs "Proactive and Reactive" are proposed.	Proactive methods are good for quality but not suitable for larger networks as they use a lot of energy. Reactive methods save energy but may cause delays in finding the best route.
[22]	Mobile Ad Hoc Networks with Multipath Routing	New routing method called Mobility, Contention window, and LInk quality sensitive multipath Routing (MCLMR) has introduced which finds the multiple route for data to prevent congestion.	The MCLMR method can also be extended for large- scale networks and 5G IoT applications.
[23]	Flying Ad Hoc Network	A survey is proposed which classify the description and comparing important number	Technical issues like frequent disconnections, higher packet latency,

CHAPTER 8   ARTIFICIAL INTELLIGENCE-ENHANCED IOT AND UAV INTEGRATION IN AD HOC NETWORKS FOR SMART CITIES

Findings

An analysis of the literature shows that in ad hoc network enabled IoT and UAV applications for smart cities, artificial intelligence (AI) strategies outperform traditional approaches significantly. Artificial intelligence (AI) enhances urban environment management and anomaly detection speed and accuracy. Examples of research requirements include the need for lifelong learning methodologies and cross-domain knowledge transfer in dynamic IoT and UAV scenarios. Applying machine learning algorithms to smart city use cases improves situational awareness, predictive maintenance, and resource allocation.

Limitations

Despite AI's great promise, there are a number of disadvantages, including interpretability problems with important vital to safety algorithms like DL, ML, and RL. The large-scale implementation of federated learning and blockchain is complicated by the constantly evolving structure of UAV and IoT networks. Managing personal IoT and UAV data raises security and privacy issues. Unreliable connections make VANET operations difficult, and poor Wi-Fi connectivity leaves UAVs vulnerable to assaults. Maintaining ALANTE security is difficult because of both external and internal threats. The study's overall findings emphasize the necessity of strong privacy, security, and interpretability protections for AI-enhanced IoT and UAV systems in smart cities [19–32].

AI for Ad Hoc IoT Networks

Artificial intelligence (AI) and the Internet of Things (IoT) can collaborate closely to evaluate data from linked devices and derive meaningful knowledge that can enhance people's lives. A wealth of information about how physical objects are used and the environment around them may be regularly gathered and shared by devices that are outfitted with sensors and connected to the internet through the Internet of Things.

CHAPTER 8 ARTIFICIAL INTELLIGENCE-ENHANCED IOT AND UAV INTEGRATION IN AD HOC NETWORKS FOR SMART CITIES

Artificial intelligence (AI) systems can then examine the large and diverse data streams from IoT devices to identify patterns and forecast outcomes. This enables IoT and AI to work together, with IoT providing the data and AI powering the applications to offer services ranging from basic automation of household appliances to more advanced features like automated driving, personalized health monitoring, and enhanced medical diagnosis. Ultimately, by combining their routine and innovative capabilities, these two technologies enhance people's quality of life through innovative solutions. Ad hoc networks enable Internet of Things devices to communicate autonomously and gather data, enabling flexible communication and analysis. This, combined with AI's ability to draw conclusions from large datasets, opens up new possibilities for innovative applications in IoT.

Deep Learning for IoT in Ad Hoc Networks

Promising outcomes have been observed in deep learning for a range of Internet of Things applications, such as anomaly detection, activity recognition, and predictive maintenance. Convolutional neural networks (CNNs), recurrent neural networks (RNNs), and long short-term memory (LSTMs) are examples of deep learning approaches that have shown useful for processing picture data and IoT time series gathered on ad hoc networks. Deep learning models are particularly ideal for processing Internet of Things data from ad hoc networks because they are capable of complete training straight from raw sensor inputs. This function allows the models to automatically identify relevant patterns in the data without requiring human-defined features. In time-series data, temporal relationships between subsequent readings can be captured by RNNs and LSTMs. CNNs are effective at analyzing picture and video streams because they can identify spatial trends in pixel values. When applied to data dispersed over ad hoc IoT networks, deep learning techniques can attain high accuracy, allowing for real-time on-device processing for anomaly detection, crowd behavior analysis, and predictive maintenance. This distributed processing method is effective for analyzing large amounts of unstructured data and prevents bottlenecks [33–34].

Federated Learning for Distributed IoT Devices in Ad Hoc Topology

Federated learning allows machine learning models to be trained on scattered, IoT devices without requiring data transfer by keeping training localized. While preserving privacy, this technique is effective for distributed learning from data created over ad hoc IoT networks that are evolving dynamically [35]. By leveraging locally collected data from ad hoc-connected IoT devices, [36] offered a federated learning framework for collaborative predictions. Due to their dispersed and dynamic character, ad hoc Internet of Things network applications substantially benefit from federated learning. Because they would require a constant transfer of all raw data to a central server, which would be exceedingly costly in terms of latency and bandwidth, standard centralized training methods are not feasible in these settings. Federated learning gets around this by enabling IoT devices to cooperate to train a shared machine-learning model while preserving the localization of each device's unique private data. Federated learning makes it possible for IoT devices connected via dynamic ad hoc networks to collaborate with machine intelligence. Without disclosing personal information, it gains knowledge from collective experiences, and updates are federated to continuously enhance the common model.

CHAPTER 8   ARTIFICIAL INTELLIGENCE-ENHANCED IOT AND UAV INTEGRATION IN AD HOC NETWORKS FOR SMART CITIES

Reinforcement Learning for Dynamic Resource Management in Ad Hoc IoT Networks

In dynamic wireless sensor networks, reinforcement learning is being used to optimize routing, scheduling tasks, and allocation of resources. Because reinforcement learning can learn efficient actions through trial-and-error interactions, it is the perfect tool for optimizing behaviors in scattered ad hoc IoT networks. Because it doesn't need labeled samples, dynamic networks can use it. Agents search action spaces using reward and punishment signals to find rules that maximize long-term benefits. This allows IoT devices to work together to optimize operations in real time [37]. Raghupathi et al. [38] established the theoretical framework for the use of reinforcement learning models to acquire the best possible behaviors for the coordination and distributed decision-making of Internet of Things devices connected via ad hoc networks.

Blockchain Technologies for Secure Data Sharing in Ad Hoc IoT Systems

Blockchain technology improves privacy, safety, stability, and trusted data sharing in unmanaged ad hoc Internet of Things applications with its distributed ledger architecture. Peer-to-peer exchanges of information are enabled, maintaining openness and responsibility. Smart contracts enable fine-grained access control, and a distributed structure lessens the chance of having one point of failure. This enhances supply chain traceability by recording the origin when goods and data are exchanged between network participants [39–40].

CHAPTER 8 ARTIFICIAL INTELLIGENCE-ENHANCED IOT AND UAV INTEGRATION IN AD HOC NETWORKS FOR SMART CITIES

AI-Enabled Ad Hoc Networks

With the progress in ad hoc networks, artificial intelligence (AI) is crucial for enhancing the effectiveness of the network and improving networking decisions in different types of ad hoc networks. AI is applied in VANETs, MANETs, FANETs, and RANETs to enable self-control and intelligent networking choices, leading to improved system performance through learning-based networking techniques [41]. Like other ad hoc networks, the use of AI has significantly improved the network outcome of VANETs. Vehicular networks aim to enhance the safety and efficiency of transport systems by sharing information among cars and roadside base stations [42]. Next, we discuss the influence of artificial intelligence (AI) in various types of ad hoc networks:

Artificial Intelligence-Enabled Mobile Ad Hoc Networks

AI has a significant impact in advancing network security, routing protocols, Quality of Service (QoS), and self-organization of ad hoc networks. Due to rapid development in the area of MANET, there are some significant hazards which threaten the network security. One of the most impactful threats is the Black Hole Attack (BHA) in which the attacker uses an attack node to stop the movement of data between the destination and source node which weakens the network performance. This issue caused by the attacking node within the discovered route is prevented by using ad hoc on-demand distance vector (AODV) along with artificial bee colony (ABC), artificial neural network (ANN), and support vector machine (SVM) methods. The combination of artificial intelligence with these techniques in the AODV mechanism discovers and identifies the attacker node within the path. This approach of applying AI with SVM and AODV helps in supplying a better route for data transmission between the destination and

CHAPTER 8   ARTIFICIAL INTELLIGENCE-ENHANCED IOT AND UAV INTEGRATION IN AD HOC NETWORKS FOR SMART CITIES

source node [43]. Figure 8-2 demonstrates how a Black Hole Attack on a node in a MANET works. In a Black Hole Attack scenario, a rogue mobile device can appear to be the best path, causing packets to be dropped into a "black hole" instead of being transported to their intended destination after passing through the attacker's device. Consequently, the corrupted intermediary mobile node may consume packets, disrupting the link between legal nodes.

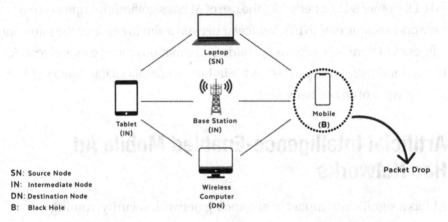

Figure 8-2. Black Hole Attack in Mobile Ad Hoc Networks

As we have already indicated, MANETs are formed by individual nodes which are connected to each other without any fixed infrastructure to form a temporary connection between nodes. These nodes in MANETs are not fixed and sometimes move continuously during data transmission. MANETs use many different ways for the transfer of data between the destination and source node, but it makes communication difficult when the nodes are constantly moving. It is because the devices don't have much power and the network structure is changing continuously. This problem of moving nodes is overcome by using artificial intelligence with smart computers and those that learn from data. These methods can make the network work better by using less power, reducing extra stuff, and more. This helps in creating a better way for nodes to communicate with each other [44].

CHAPTER 8   ARTIFICIAL INTELLIGENCE-ENHANCED IOT AND UAV INTEGRATION IN AD HOC NETWORKS FOR SMART CITIES

To enable communication between nodes in MANET from the origin to the destination, we need to establish and maintain connections between these nodes. To improve its performance, MANETs are required to select the best possible route to make the communication possible. As we know, AI has influenced all fields of technologies around the world to improve the system's performance. An AI-based Ant-OBS technique is used here to improve the system performance to work faster and save energy, like when you find the quickest way to get somewhere. Ants are clever insects when they reach a crossroad; they decide which path to follow to reach their destination. The same concept is used here for mobile nodes to select the best possible path using the Ant-OBS mechanism. This mechanism makes sure that your messages don't go back and forth and makes the connection faster because it reduces the extra effort that slows down the performance of the system. Ant-OBS uses ant's behavior and smart technology which enhances its operation and shows that its work is better than the previous methods like ADV and LAR1 [45].

Artificial intelligence (AI) is used in more complicated and dynamic situations in mobile ad hoc networks than in standard setups. AI approaches make wireless communications far more reliable and efficient in contexts with high mobility and rapid changes, such as UAV networks and urban infrastructure. Such networks can swiftly detect and eliminate risks by using techniques like AODV with ABC, ANN, and SVM, guaranteeing safe and trustworthy data transmission. Furthermore, artificial intelligence (AI)-driven algorithms like Ant-OBS, which improve resource distribution and route choice and enable dependable and energy-efficient communication in the face of constantly changing network topologies. Ultimately, this makes it easier to implement the complex functions and quick responses needed in such dynamic environments.

CHAPTER 8 ARTIFICIAL INTELLIGENCE-ENHANCED IOT AND UAV INTEGRATION IN AD HOC NETWORKS FOR SMART CITIES

Vehicular Ad Hoc Networks Using Artificial Intelligence

Vehicular ad hoc network (VANET) is a sub-class of Mobile ad hoc network (MANET). In VANETs, vehicles communicate and share information with each other and roadside unit (RSU) to improve the transportation system. As the concept of smart and intelligent cities continues to gain momentum, intelligent transportation systems (ITS) need to be improved to increase comfort, safe driving, and driver pleasure. Considering that wireless communication is exposed in VANETs, it is susceptible to several kinds of attacks. The attacks slow down the performance of VANETs and cause serious problems like when attackers may spread false messages to cause road accidents. To prevent VANETs from such dangers, artificial intelligence is used in the form of machine learning algorithms to optimize support vector machine (SVM) so that the system possesses the capability to defend itself against such cybersecurity threats. Three different machine learning algorithms such as genetic algorithm (GA), particle swarm optimization (PSO), and ant colony optimization (ACO) are used. These techniques are compared with each other and found that GA worked best as compared to the other two for making SVM capable of detecting these attacks [46].

VANET plays an important part in improving driving enjoyment and road safety. In VANETs, the nodes are in constant motion, sharing information and roadside units. However, the population has grown, and there are more vehicles as well leading to various challenges such as traffic congestion and increased road accidents. To overcome these issues effectively, a large amount of data must be exchanged within the network. The current version of VANET is not capable of managing such massive data sets. For this purpose, machine learning (ML) methods are used to overcome this problem. ML algorithms are smart and make quick decisions to improve systems' functionality and performance. Machine learning methods with cognitive radio (CR) are used to manage large amounts of data in VANET and make CR more intelligent. Cognitive radio (CR) provides a promising way to address this challenge. CR-based VANETs with ML are used to tackle this problem. This new method of ML is still in its early days but has already shown great potential as a solution for handling the vast amount of data exchanged within VANET [47].

AI integration into vehicle ad hoc networks (VANETs) improves communication efficiency and robustness in the context of building smart cities. Through the use of machine learning techniques such as GA, PSO, and ACO, VANETs may enhance traffic flow and protect data transfers. This meets infrastructure criteria for smart cities and enhances traffic safety.

Artificial Intelligence for Flying Ad Hoc Networks

Flying ad hoc network (FANET) is made of a network of unmanned aerial vehicles (UAVs) which are connected to each other and ground base stations wirelessly to communicate without any fixed infrastructure. FANETs are used in different range of domain which includes civil and military areas. In FANETs, the connection between the nodes is frequently disconnecting due to constant motion of UAVs at high and low altitude. UAVs can travel between 30 and 460 km/h which make it difficult for FANET to keep a constant connection between the nodes. These UAVs are equipped with batteries which have limited power. It is quite challenging in FANET to make sure that the route between nodes has fast communication and quick route setup which ensure a reliable, long-term, and fixed connection between the nodes. To address this issue, we use an artificial intelligence-based approach called reinforcement learning which helps the connection last longer. This technique makes UAVs to learn from its environment and make intelligent decision based on previous data. They try different ways to determine the most direct path between the nodes. This route uses less power and includes stable drones which keep the network stable and last longer [48].

CHAPTER 8 ARTIFICIAL INTELLIGENCE-ENHANCED IOT AND UAV INTEGRATION IN AD HOC NETWORKS FOR SMART CITIES

5G technology has laid the foundation for FANET in ad hoc networks. As we have discussed above, in FANET, the networks are formed by flying nodes which are communicating and exchanging information. These flying nodes are used for search and rescue, border monitoring, agriculture inspection, etc. Networks formed by UAVs find it difficult to communicate due to the continuous mobility of the network. To solve this problem, we take notes from nature like how insects behave in a network can help overcome this issue. We use an artificial-based intelligent cluster routing scheme (CRSF) for FANET. Artificial intelligence allows a system to demonstrate intelligence comparable to human performance. CRSF chooses a head within the network called cluster head (CH) which is selected based on how fit they are, considering their position and how much power they have left. We also find a way to manage a large group of drones by getting the idea from how moths behave in a network. To make sure that these groups are stable and communicate without any difficulty. The CRSF changes the cluster head when needed and makes a plan for the nodes to choose the ideal communication route. The effectiveness of the CRSF shows a significant improvement in the efficiency of the drones. The findings demonstrate that the suggested CRSF performs better than previous nature-based technologies while having less complexity [49].

Artificial intelligence significantly enhances the efficacy and reliability of flying ad hoc networks (FANETs), which are networks of interconnected unmanned aerial vehicles (UAVs) used in both military and commercial purposes. Maintaining stable connectivity when UAVs move quickly and have limited battery life is an issue. In particular, reinforcement learning facilitates AI's optimization of UAV communication and route planning, leading to dependable and long-lasting network connectivity. This technology plays a major role in smart city applications, which simplify and improve the dependability of jobs like emergency response, infrastructure inspection, and urban monitoring.

CHAPTER 8   ARTIFICIAL INTELLIGENCE-ENHANCED IOT AND UAV INTEGRATION IN AD HOC NETWORKS FOR SMART CITIES

Figure 8-3 shows how UAVs in a cluster are connected to share information with a distant control center in times of emergency. A peer-to-peer wireless mesh network for emergency response is depicted in the picture, linking a number of unmanned aerial vehicles (UAVs). The drones communicate location information and video to a remote-control center. In order to ensure prompt action, responders can monitor the situation and give commands to coordinate the UAV cluster.

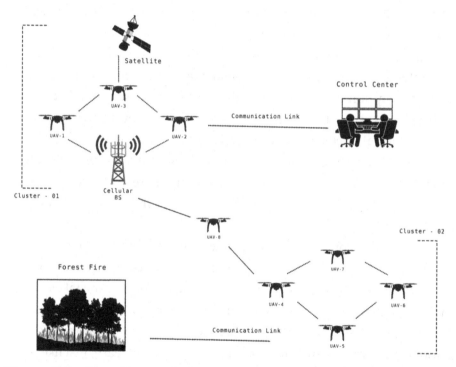

Figure 8-3. *Data Transmission to a Distant Control Center in Case of Any Disaster*

Artificial Intelligence for Robot Ad Hoc Networks

Robot ad hoc networks are formed by network of robots which are connected to each other without any fixed infrastructure. These robots perform different task quickly and with more accuracy. They are like buddies who help each other out like a group of robots used at a warehouse to move items from one place to another. These robots are smart, make their own decisions, and learn from their previous experience. These robots can also communicate with each other wirelessly and share information. Imagine if a group of robots need to find a treasure in a maze. Initially, they don't have any map, so they just start moving randomly on the board. With each movement, they use a mix of learning and planning. They learn from their mistakes and successes as they explore the maze. To make this setup smoother and faster, we introduce a more advanced way of communication between the robots. These robots communicate with each other in an order like each robot takes a turn to communicate, so in this way, they don't interrupt each other and share information like maps and tips quickly which helps them work more smoothly. For this purpose, we use artificial intelligence (AI)-based reinforcement learning (RL) which shows how communication helps robots to perform tasks quickly. It helps them to know where they are better and share secret information. These technologies are used in vehicles and smart industries for better performance [50].

To make this communication advanced, we use another AI technique to enhance robotic interaction. AI is used to teach a computer how to think like humans. It makes the computer to learn from its mistakes and improve its performance. It is being utilized in different fields like engineering, science, medicine, and finance. AI techniques are employed in robotics communication to teach them to think and learn, just like people. They can get better by learning from the things they do wrong. This helps them do tasks even better over time. Also, they can share helpful information to work together. Artificial neural network (ANN), fuzzy logic, neuro-fuzzy interference system (ANFIS), genetic algorithms, pattern recognition, clustering, machine learning (ML), deep learning (DL), etc. help robots to better learn about each other and their environment that keep changing; the robots can work together. They plan tasks for each robot, avoid crashing into each other and other obstacles, make their team performance better, and do their tasks faster and more easily [51].

Artificial intelligence has changed robot ad hoc networks by enabling unmanned aerial vehicles (UAVs) to perform coordinated tasks and effortlessly interact with each other. Such networks of autonomous devices can successfully collaborate to enhance urban logistics, emergency response, and infrastructure maintenance in smart city environments. Thanks to the use of AI methods like reinforcement learning and complex robotic interaction methods, robots can adjust to changing urban environments, increasing their efficiency and ability to do challenging tasks independently.

Under Water Sensor Networks Using Artificial Intelligence

Under Water Wireless sensor network (UWSNs) are used to monitor under water conditions such as pressure of water and temperature. It is also used for search and rescue, exploration, and security like monitoring under water pipelines and cables and detecting if any unexpected situation occurs. The underwater world is quite different and unique from our world which makes it difficult for the devices like sensors to work there. This made it hard to create networks between the sensors. Artificial intelligence (AI)-based technologies are used to build strong networks between these sensors, so they can share information and send data to each other without any interruption. They build a system called marine integrated communication network system (MICN) which works with another system called maritime wireless mesh network (MWMN) which makes it possible for the networks to work effectively underwater. A special computer program distributed hybrid fish swarm optimization algorithm (FSOA) is formed. It is like a large group of fish that works together to find their way. This program guides underwater devices on how to work which enables them to share information and explore underwater. Furthermore, they create intelligent ships that predict the upcoming situation and make useful decisions to make sure that the ship runs well. The AI-based systems help to send data underwater which also worked well. These AI technologies enhance the underwater sensor networks and improve their performance [52]. An underwater wireless sensor network (UWSN) system for environmental monitoring is depicted in Figure 8-4. It is made up of three mesh network topologically connected clusters of underwater sensors. There is a local sink for gathering and combining data in each cluster. For redundancy transmission, the data is routed to an onshore ground station and a central sink buoy.

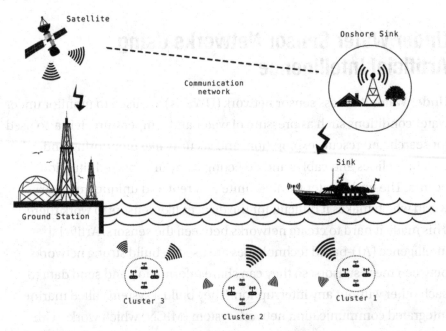

Figure 8-4. *Mesh network topologically connected clusters for underwater sensors*

Artificial intelligence (AI) has enabled computers and wireless networks to think and learn like humans. AI has played a great role to improve the Quality of Service of these networks by teaching computers to understand, reason, and solve problems. It is transforming wireless communications by helpings them to deal with large amount of data learn from it and make proper decisions. AI finds application across different domains like medicine, science, engineering, robotics, and many more. AI-enabled underwater wireless communication systems have attracted a lot of attention because they are used in different security and military areas. These systems can be attacked by intruders who they might change the original information, disappear it, or mess up the whole data. They can also waste the power the sensor needs to run the network. One common attack they used is Denial of Service where the intruder messes the whole network and tries to stop the system from working. A network security system is crucial for detecting and preventing unauthorized intrusions, which can be categorized into four types of attacks. There are four types of attacks: black holes, gray holes, flooding, and scheduling. For this purpose, an AI-based artificial neutral is using a network to educate the system to understand and find these attacks. The conclusion demonstrates that the system recognizes and comprehends the upcoming attacks and helps in improving the system security of the underwater sensor [53].

Artificial Intelligence Assisted Wireless Powered Internet of Things (IoT)

Internet of Things (IoT) was initially introduced by "Ashton" where he expands the network connectivity to physical objects around us. IoT enables these objects to interact with each other and share real-time information. These devices communicate without any fixed infrastructure and can be controlled and monitored wirelessly. IoT has influenced many sectors like healthcare, smart transportation, smart cities, agriculture, energy management, and many more [54]. IoT collects huge amounts of data from the physical objects connected and sharing real-time information. This large amount of data can be used to enable our system to learn. Artificial intelligence and algorithms use this huge amount of data to learn and make accurate and quick decisions which improve the outcome of IoT devices. The growing use of AI is improving the utilization of IoT by using advanced data analytics. With AI, we can enable home monitoring and control system where the devices learn from the previous data and respond according to your behavior. The thermostat can sense the temperature and take steps according to it. Lights can turn ON and OFF according to your schedule. It controls and automates most of the home appliances like TVs, fans, lights, AC, etc., which makes the home appliances work more efficiently. AI is used to make the network or system intelligent. It's like giving machines a brain, so they can think and make their own choices. AI-enabled IoT devices are used in smart hotels where customers can use smart booking systems, suitable temperature control, and helpful information based on previous data. AI has influenced the oil and gas industry where the sensors are used to measure the oil temperature, pressure, and other factors. AI is also used to observe patterns that show when machinery might fail. These early warnings can prevent the industry from costly breakdowns of machines [55].

IoT has emerged as a ground-breaking application for many different problems. But because of the wireless communication between the objects and openness of network connection, the system is open and suffers from various types of attacks. These attacks include Denial of Service (DoS), Distributed Denial of Service (DDoS), and device authentication. The system security is enhanced using AI methods and algorithms to detect any security breach and prevent it from causing damage to the system. AI uses machine learning methods which are further divided into reinforcement learning (RL), deep learning (DL), ensemble learning, unsupervised learning, and supervised learning. The normal behavior of IOT devices is predictable and fixed. It is like a group of students who are well-behaved and mannered. This group of well-behaved students is acting as the predictable behavior of the system. To recognize abnormal behavior of the system, machine learning techniques including supervised learning and unsupervised learning are applied. The system uses supervised learning to detect abnormal activity in a group of well-behaved students, separating them from normal data and identifying unauthorized activities using two different colored points [56].

Artificial Intelligence-Enabled Ship Ad Hoc Networks

Ship ad hoc networks (SANET) is another type of ad hoc network where the nodes are ships, submarines, base station, and satellite connected wirelessly with each other without any fixed infrastructure. These nodes are used to enhance sea connectivity. The node density in SANETs is medium as compared to MANET which is low, and the topology changes in SANET are slow as compared to VANET and FANET where the topology changes are very high [57].

In SANETs, the nodes use the available frequencies around them for navigation, calls, search and rescue communication, etc. These frequencies which are selected by SANETs are sometimes in use of another important system which causes signals interference and disruptions in the network. To avoid this problem, we use an artificial-based cognitive radio (CR) which uses AI algorithms to analyze the available radio spectrum and select the best possible frequency band for communication and other factors. It learns from the data and makes its own decisions about which band of frequencies to use based on the current conditions. CR helps to sense the frequencies and make sure if it is not in use by another important system [58].

SANET send and receive signals to share information with other nodes. To make SANET able to share data quickly, we introduced a new method called ITU-R M.184-1. This technology makes sure that the communication is well understood by the receiver and the overall information is shared quickly. It helps engineers to understand the nature of signals and how they should be shared between the nodes. It is like making sure that all the nodes are speaking the same language (signal) for better communication. They used the technology and found out it worked well near the shore where the ships were sailing. To avoid signal interference, we use ITU-R M.184-1 to figure out what signals to use and how to understand them [59].

In smart city contexts, artificial intelligence is revolutionizing ship ad hoc networks (SANET) and improving UAV navigation and communication. AI-driven cognitive radio (CR) is used by SANETs, which are composed of satellites, ships, and submarines, to efficiently manage radio spectrum utilization, minimizing interference and maximizing communication performance. With the use of this technology, UAVs may choose and adjust frequency bands on their own depending on the circumstances, improving communication dependability in busy metropolitan settings. Furthermore, developments such as ITU-R M.184-1 enable quick and efficient data transfer between SANET nodes, supporting a range of marine operations and infrastructure management duties in smart cities.

AI and Machine Learning Applications for Ad Hoc Networks

The development of wireless communication was significantly aided by ad hoc networks. They are used in disasters, battlefields, search and rescue, wildlife tracking and different industries, etc. The amazing feature is that the ad hoc networks formed by different nodes can communicate and share data without any fixed infrastructure. Due to their wireless communication nature, these networks are open for attackers to change the information, slow down the system, or send malicious data to the system. Several different protocols and techniques are used for secured data transfer, but the security concerns are still increasing day by day. AI and ML techniques played a vital role to enhance the network infrastructure and improve system security. Vehicular ad hoc networks (VANETs) are facing the same security concerns due to the openness of network communication. The goal is to create a secured communication strategy where the system needs authentication and authorization of the node before it is involved in network communication. These security models ask several questions from the node to prove itself. They gather and save data about every node and then check if the data is correct and if the nodes want to join the network. AI-enabled logistic regression enables these techniques to overcome the problem's complexity. Logistic regression analysis is used for predicting two choices and provides a result based on the given data [60–62].

As we have discussed, the VANETs are formed by several vehicles connected and roadside units. The data transmission in VANET is difficult as compared to that of MANET; it is because of the frequent mobility of nodes which creates continuous changes within the network structure and topology of the network. The constant changes in the network increased the transmission delays and reduced the stability of the system. Several techniques were used where the researchers' observed roads with high traffic and used them to detect traffic flow and road selection. However, it reduces the transmission delays between the nodes, but real-time traffic statistics cannot be gathered using these approaches. It is because the traffic flow or traffic density changes over time which can change the data transmission rate between the nodes due to which the system might not work very well all the time. To find a solution, we need to develop a model that can keep track of traffic density and improve the system's accuracy for road selection. For this purpose, we use DRL-based machine learning architecture model which will provide essential data for VANET. Roadside units monitor the traffic density using DRL. This model predicts the movement of the vehicles and proposes the best possible route for communication between nodes. This model determines the data required to send information from roadside units. It divides the area into multiple parts and then chooses the best possible route based on how many cars, how close the roadside units, and how fast the message should go to avoid any delay in transmission [63]. Figure 8-5 shows the system model of VANET and how nodes are connected. Reinforcement learning is used by the vehicular ad hoc network (VANET) system model to manage vehicle-to-vehicle communications. Two vehicles are linked to Base Station 1 through a VANET, while two other vehicles are linked to Base Station 2 through a different VANET. A satellite link is used to establish additional communication between the two base stations. The network is capable of dynamically optimizing routes for information sharing between vehicles by employing reinforcement learning agents in both base stations and each vehicle.

Figure 8-6 illustrates the deep reinforcement learning (DRL) working mechanism. The operation of a deep reinforcement learning system is shown in Figure 8-6. The environment state, which depicts the system or issue under consideration, is on the right side. The agent will monitor the current state of the environment from its location in the feature state on the left. The agent will choose what to do based on this observed state. The environment is then used to carry out this action. After that, the environment will update its condition and give back an instant reward. The agent notices this new state, and the cycle continues. Reinforcement learning involves an agent's interaction and feedback loop, allowing them to improve their actions through trial and error over numerous learning iterations.

CHAPTER 8 ARTIFICIAL INTELLIGENCE-ENHANCED IOT AND UAV INTEGRATION IN AD HOC NETWORKS FOR SMART CITIES

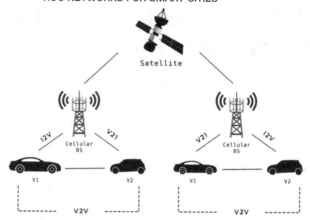

Figure 8-5. *Underwater Sensor Networks Under Dynamic Monitoring*

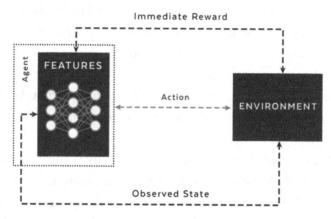

Figure 8-6. *Deep Reinforcement Learning (DRL)*

CHAPTER 8   ARTIFICIAL INTELLIGENCE-ENHANCED IOT AND UAV INTEGRATION IN AD HOC NETWORKS FOR SMART CITIES

AI-Enabled Cyberattack Detection System for Ad Hoc Networks and IoT

Ad hoc networks exist where devices communicate without a fixed infrastructure. AI is crucial in detecting network issues before they cause damage, addressing security vulnerabilities posed by potential attackers entering the network. We utilize SA-IDPS, an intelligent intrusion detection and prevention system, to enhance MANET security, optimize energy efficiency, and minimize false alarms. Initially, we make sure that all the nodes available in the network are registered in network authority, so no new node can join the network without registration. This thing can be done using a method called One Way Hash Chain Function. When any node sends data, the packet analyzer reads the data and matches it with any known attack patterns. Furthermore, the processing unit and extraction unit select only important data and keep it in the standard form. The classification unit employs a number of techniques, including bootstrapped optimistic algorithms and artificial neural networks, to decide if the data is normal or malicious. These smart methods enable us to identify and stop any upcoming threat that could damage our network architecture [64–65].

As we have already discussed, network security is always a problem in ad hoc networks. Many technologies and algorithms are used to minimize the attacks and secure the network. Detection systems use different learning methods called supervised and deep learning because deep learning is good at finding patterns in data. This method of learning is very great, but it always needs a lot of data to learn, and this type of system is not applicable for new attacks. For this reason, we need to create such system which can detect and predict upcoming threats. We use unsupervised learning with these aspects in order to fix this issue.

CHAPTER 8   ARTIFICIAL INTELLIGENCE-ENHANCED IOT AND UAV INTEGRATION IN AD HOC NETWORKS FOR SMART CITIES

To do this, we provide a novel technique called deep embedded median clustering (DEMC). It consists of two key steps; first one is organizing the area where features are stored and grouping those features using a technique called K-median clustering. This aids in the discovery of abnormalities or strange occurrences inside the network [66].

Results from Literature Review on AI Applications in Ad Hoc Networks

Table 8-2 shows how well various ad hoc network types can be served by implementing artificial intelligence algorithms. The use of AI techniques, such as reinforcement learning, deep learning, federated learning, and convolutional neural networks, enhanced significant metrics and efficiency across the networks, including mobile ad hoc, vehicular ad hoc, flying ad hoc, and underwater wireless sensor networks, in comparison to traditional non-AI approaches.

CHAPTER 8   ARTIFICIAL INTELLIGENCE-ENHANCED IOT AND UAV INTEGRATION IN AD HOC NETWORKS FOR SMART CITIES

Table 8-2. *Results of applying AI in ad hoc IoT networks*

Reference	Ad hoc network	AI technique used	Key results
[67]	Mobile ad hoc networks (MANETs)	Decision trees, SVM	Implemented machine learning methods to increase routing efficiency, such as SVM and decision trees, increased the packet delivery ratio by up to 30%
[68]	Vehicular ad hoc networks (VANETs)	Deep reinforcement learning	Using deep reinforcement learning, the best routes were chosen by current traffic circumstances, decreased transmission delays, and improved efficiency by 20–30%
[69]	Flying ad hoc networks (FANETs)	Cognitive radio	Developed an AI-powered cognitive radio technology that dynamically chooses the optimal frequency ranges for communication. A 15–20% increase in the lifetime of networks
[69]	Underwater wireless sensor networks	Distributed hybrid fish swarm optimization algorithm	Produced a resilient underwater sensor network by utilizing AI to build a combined network of communication, increased connection, and a 25–30% decrease in packet loss

CHAPTER 8   ARTIFICIAL INTELLIGENCE-ENHANCED IOT AND UAV INTEGRATION IN AD HOC NETWORKS FOR SMART CITIES

Future Directions

Ad hoc networks have experienced several problems recently. One issue is that communication between devices has a short range which means these devices need to be close to each other for connections [70]. Lack of centralized control, which makes it challenging to organize the network, is another problem [71]. Ad hoc networks may also experience signal loss and interference, especially in busy areas [72]. Security is another issue because it might be difficult to keep the network safe from unauthorized or intruder access [73]. Finding answers to these issues can be addressed in terms of AI, if we want to increase the efficiency and dependability of ad hoc networks [74]. Deep reinforcement learning techniques can optimize routing protocols for mobile ad hoc networks (MANETs), improving throughput and routing efficiency by considering factors like overloading levels and bandwidth availability [75]. Federated techniques for learning provide training for models among ad hoc network nodes in a secured private manner. Federated learning allows nodes to collaboratively create a common prediction model while preserving the decentralized character of learning information from local devices. In addition to helping with the identification of errors and network intrusions, this could protect user privacy [76].

Artificial intelligence (AI)-powered wireless ad hoc networks might significantly improve UAV operations in smart cities. Two critical areas of resource management for sustainable urban development are data and energy efficiency management. Improving AI algorithms to better adapt and arrange UAV networks can greatly improve these areas. It may be possible to create smarter UAV networks with increased operating efficiency and connection by fusing AI with IoT and 5G technologies. Enhancing security in urban airspace with AI-powered threat detection guarantees a strong defense against changing cyber threats. By taking ethical issues into account, AI decision-making processes are made fair and transparent, which promotes acceptance and confidence in smart city efforts. Increasing the security, efficacy, and efficiency of UAV networks in urban settings is the aim of these ensuing pathways.

Conclusion

The applications of machine learning and artificial intelligence approaches to various problems in different kinds of wireless ad hoc networks were covered in this research. An extensive review of the literature revealed that issues with routing, intrusion detection, spectrum sensing, and network optimization for MANETs, VANETs, FANETs, RANETs, SANETs, AANETs, UWSNs, and IoT networks can be successfully resolved with methods like deep learning, reinforcement learning, and other machine learning and AI algorithms. A summary of these network types and the main AI applications that were looked into for each are shown in Table 8-2. High accuracy was attained by the suggested methods in tasks such as device coordination, path selection, and anomaly detection. Adopting AI, however, also brings up new concerns about energy efficiency, security flaws, and data privacy that need to be further researched, particularly in contexts with limited resources. The integration of artificial intelligence and machine learning algorithms in wireless ad hoc networks enables improved UAV operations in smart cities. Research has indicated that these technologies are effective in tackling network optimization, spectrum sensing, intrusion detection, and routing problems for a range of ad hoc network topologies. Techniques like deep learning and reinforcement learning have shown to be particularly accurate in tasks like path selection and anomaly recognition, which might help UAVs operate better in urban environments. However, adding AI to these networks also raises concerns about data security, privacy, and energy efficiency—especially in resource-constrained situations. Further research is needed to fully realize AI's potential benefits and solve the challenges presented by dynamic, decentralized network operations. Innovative real-time applications in smart city domains such as transport and IoT might result from this research. Even though the study showed how AI allows for independent communication and the analysis of large datasets across dispersed ad hoc networks, it still needs more research to fully realize AI's potential benefits and address issues arising from dynamic, decentralized network operations. These challenges could lead to creative real-time applications in domains like transportation, smart cities, and the Internet of Things.

Major Contribution Points

1) A thorough analysis of machine learning and artificial intelligence applications for several kinds of wireless ad hoc networks, such as MANETs, FANETs, VANETs, UWSNs, and IoT networks.

2) An explanation of how problems with network optimization, spectrum sensing, intrusion detection, routing, and various other machine learning approaches, such as deep learning and reinforcement learning, can be successfully solved.

3) An analysis of the possible advantages of AI-enhanced UAVs for applications related to smart cities, specifically with regard to reliable communication networks, effective resource management, and real-time data processing.

4) A review of the main uses of AI/ML that have been studied for every kind of network and the high levels of accuracy attained in tasks such as device coordination, path selection, and anomaly detection.

5) The recognition of fresh problems with privacy of data, security flaws, and efficiency of energy that need more investigation and are brought about by the use of AI/ML techniques.

6) An example of how AI allows for independent dataset processing and interaction via dispersed ad hoc networks.

7) A suggestion of chances for creative real-time applications in areas like smart cities, transportation, and the Internet of Things by overcoming difficulties with dynamic, decentralized network operations.

References

[1] Khan, InamUllah, AsrinAbdollahi, Abdul Jamil, BismaBaig, Muhammad Adnan Aziz, and FazalSubhan. "A Novel Design of FANET Routing Protocol Aided 5G Communication Using IoT." Journal of Mobile Multimedia (2022): 1333-1354.

[2] Khan, InamUllah, IjazMansoor Qureshi, Muhammad Adnan Aziz, Tanweer Ahmad Cheema, and Syed Bilal Hussain Shah. "Smart IoT Control-Based Nature Inspired Energy Efficient Routing Protocol for Flying Ad Hoc Network (FANET)." IEEE Access 8 (2020): 56371-56378.

[3] Bilen, Tuğçe, BerkCanberk, Vishal Sharma, Muhammad Fahim, and Trung Q. Duong. "AI-Driven Aeronautical Ad Hoc Networks for 6G Wireless:

Challenges, Opportunities, and the Road Ahead." Sensors 22, no. 10 (2022): 3731.

[4] Khan, InamUllah, Ryan Alturki, Hasan J. Alyamani, Mohammed AbdulazizIkram, Muhammad Adnan Aziz, Vinh Truong Hoang, and Tanweer Ahmad Cheema. "RSSI-Controlled Long-Range Communication in Secured IoT-Enabled Unmanned Aerial Vehicles." Mobile information systems 2021 (2021): 1-11.

[5] Kreutzer, Ralf T., Marie Sirrenberg, Ralf T. Kreutzer, and Marie Sirrenberg. "Fields of Application of Artificial Intelligence—Energy Sector, Smart Home, Mobility and Transport." Understanding Artificial Intelligence: Fundamentals, Use Cases and Methods for a Corporate AI Journey (2020): 195-210.

[6] Ferrão, Isadora Garcia, David Espes, Catherine Dezan, and Kalinka Regina Lucas Jaquie Castelo Branco. "Security and Safety Concerns in Air Taxis: A Systematic Literature Review." Sensors 22, no. 18 (2022): 6875.

[7] Younis, Mohamed, and SebnemZorlu Ozer. "Wireless ad hoc networks: technologies and challenges." Wireless Communications and Mobile Computing 6, no. 7 (2006): 889-892.

[8] Khan, InamUllah, Syed Bilal Hussain Shah, Lei Wang, Muhammad Adnan Aziz, Thompson Stephan, and Neeraj Kumar. "Routing protocols & unmanned aerial vehicles autonomous localization in flying networks." International Journal of Communication Systems (2021): e4885.

[9] Shah, Syed Bilal, Chen Zhe, Fuliang Yin, InamUllah Khan, Seema Begum, Muhammad Faheem, and FakhriAlam Khan. "3D weighted centroid algorithm & RSSI ranging model strategy for node localization in WSN based on smart devices." Sustainable cities and society 39 (2018): 298-308.

[10] Khan, InamUllah, AsrinAbdollahi, Ryan Alturki, Mohammad DahmanAlshehri, Mohammed AbdulazizIkram, Hasan J. Alyamani, and Shahzad Khan. "Intelligent detection system enabled attack probability using Markov chain in aerial networks." Wireless Communications and Mobile Computing 2021 (2021): 1-9.

[11] Abdollahi, Asrin, and Mohammad Fathi. "An Intrusion Detection System on Ping of Death Attacks in IoT Networks." Wireless Personal Communications 112 (2020): 2057-2070.

[12] Popli, Renu, Monika Sethi, IshaKansal, Atul Garg, and Nitin Goyal. "Machine Learning Based Security Solutions in MANETs: State of the art Approaches." In Journal of Physics: Conference Series, vol. 1950, no. 1, p. 012070. IOP Publishing, 2021.

[13] Khan, InamUllah, Muhammad Abul Hassan, Muhammad Fayaz, JeonghwanGwak, and Muhammad Adnan Aziz. "Improved Sequencing Heuristic DSDV Protocol Using Nomadic Mobility Model for FANETS." Comput. Mater. Contin 70, no. 2 (2022): 3653-3666.

[14] Hamdi, M.M., Audah, L., Rashid, S.A., Mohammed, A.H., Alani, S. and Mustafa, A.S., 2020, June. A Review of Applications, Characteristics and Challenges

in Vehicular Ad Hoc Networks (VANETs). In 2020 international congress on human-computer interaction, optimization and robotic applications (HORA) (pp. 1-7). IEEE.

[15] Khelifi, H., Luo, S., Nour, B., Moungla, H., Faheem, Y., Hussain, R. and Ksentini, A., 2019. Named Data Networking in Vehicular Ad Hoc Networks: State-of-the-Art and Challenges. IEEE Communications Surveys & Tutorials, 22(1), pp. 320-351.

[16] Djahel, S., Nait-Abdesselam, F. and Zhang, Z., Mitigating Packet Dropping Problem in Mobile Ad Hoc Networks: Proposals and Challenges. IEEE communications surveys & tutorials, 13(4), pp. 658-672.

[17] Noor, F., Khan, M.A., Al-Zahrani, A., Ullah, I. and Al-Dhlan, K.A., 2020. A Review on Communications Perspective of Flying Ad-Hoc Networks: Key Enabling Wireless Technologies, Applications, Challenges and Open Research Topics. Drones, 4(4), p. 65.

[18] Zhang, J., Chen, T., Zhong, S., Wang, J., Zhang, W., Zuo, X., Maunder, R.G. and Hanzo, L., 2019. Aeronautical $ Ad~ Hoc $ Networking for the Internet-Above-the-Clouds. Proceedings of the IEEE, 107(5), pp. 868-911.s

[19] Mahmood, J., Duan, Z., Yang, Y., Wang, Q., Nebhen, J. and Bhutta, M.N.M., 2021. Security in Vehicular Ad Hoc Networks: Challenges and Countermeasures. Security and Communication Networks, 2021, pp. 1-20.

[20] Sharma, S. and Kaul, A., 2021. VANETs Cloud: Architecture, Applications, Challenges, and Issues. Archives of Computational Methods in Engineering, 28, pp. 2081-2102.

[21] Korir, F. and Cheruiyot, W., 2022. A survey on security challenges in the current MANET routing protocols. Global Journal of Engineering and Technology Advances, 12(01), pp. 078-091.

[22] Tilwari, V., Maheswar, R., Jayarajan, P., Sundararajan, T.V.P., Hindia, M.N., Dimyati, K., Ojukwu, H. and Amiri, I.S., 2020. MCLMR: A Multicriteria Based Multipath Routing in the Mobile Ad Hoc Networks. Wireless Personal Communications, 112, pp. 2461-2483.

[23] Oubbati, O.S., Atiquzzaman, M., Lorenz, P., Tareque, M.H. and Hossain, M.S., 2019. Routing in Flying Ad Hoc Networks: Survey, Constraints, and Future Challenge Perspectives. IEEE Access, 7, pp. 81057-81105.

[24] Tareque, M.H., Hossain, M.S. and Atiquzzaman, M., 2015, September. On the routing in Flying Ad Hoc Networks. In 2015 federated conference on computer science and information systems (FedCSIS) (pp. 1-9). IEEE.

[25] Raja, L. and Baboo, S.S., 2014. An Overview of MANET: Applications, Attacks and Challenges. International journal of computer science and mobile computing, 3(1), pp. 408-417.

[26] Hui, B., Jeon, K., Chang, K., Kim, S., Park, J. and Lim, Y., 2011, September. Design of radio transmission technologies for VHF band ship ad-hoc network. In ICTC 2011 (pp. 626-629). IEEE.

[27] Su, X., Hui, B. and Chang, K., 2016. Multi-hop clock synchronization based on robust reference node selection for ship ad-hoc network. Journal of Communications and Networks, 18(1), pp. 65-74.

[28] Bilen, T. and Canberk, B., 2022. Three-phased clustered topology formation for Aeronautical Ad-Hoc Networks. Pervasive and Mobile Computing, 79, p. 101513.

[29] Sheikh, M.S., Liang, J. and Wang, W., 2020. Security and Privacy in Vehicular Ad Hoc Network and Vehicle Cloud Computing: A Survey. Wireless Communications and Mobile Computing, 2020, pp. 1-25.

[30] Khan, I.U., Qureshi, I.M., Aziz, M.A., Cheema, T.A. and Shah, S.B.H., 2020. Smart IoT control-based nature inspired energy efficient routing protocol for flying ad hoc network (FANET). IEEE Access, 8, pp. 56371-56378.

[31] Tahboush, M. and Agoyi, M., 2021. A Hybrid Wormhole Attack Detection in Mobile Ad-Hoc Network (MANET). IEEE Access, 9, pp. 11872-11883.

[32] Mukhtaruzzaman, M. and Atiquzzaman, M., 2020. Clustering in vehicular ad hoc network: Algorithms and challenges. Computers & Electrical Engineering, 88, p. 106851.

[33] Chen, M., Hao, Y., Hwang, K., Wang, L. and Wang, L., 2017. Disease Prediction by Machine Learning Over Big Data from Healthcare Communities. Ieee Access, 5, pp. 8869-8879.

[34] Wang, S. and Cao, J., 2021. AI and Deep Learning for Urban Computing. Urban Informatics, pp. 815-844.

[35] Konečný, J., McMahan, H.B., Ramage, D. and Richtárik, P., 2016. Federated Optimization: Distributed Machine Learning for On-Device Intelligence. arXiv preprint arXiv:1610.02527.

[36] Yang, Q., Liu, Y., Chen, T. and Tong, Y., 2019. Federated Machine Learning: Concept and Applications. ACM Transactions on Intelligent Systems and Technology (TIST), 10(2), pp. 1-19.

[37] Sutton, R.S. and Barto, A.G., 1998. Reinforcement learning: an introduction MIT Press. Cambridge, MA, 22447.

[38] Raghupathi, W. and Raghupathi, V., 2014. Big data analytics in healthcare: promise and potential. Health information science and systems, 2, pp. 1-10.

[39] Dorri, A., Kanhere, S.S., Jurdak, R. and Gauravaram, P., 2017, March. Blockchain for IoT security and privacy: The case study of a smart home. In 2017 IEEE international conference on pervasive computing and communications workshops (PerCom workshops) (pp. 618-623). IEEE.

[40] Viriyasitavat, W., Da Xu, L., Bi, Z. and Hoonsopon, D., 2019. Blockchain Technology for Applications in Internet of Things—Mapping from System Design Perspective. IEEE Internet of Things Journal, 6(5), pp. 8155-8168.

[41] Rovira-Sugranes, A., Razi, A., Afghah, F. and Chakareski, J., 2022. A review of AI-enabled routing protocols for UAV networks: Trends, challenges, and future outlook. Ad Hoc Networks, 130, p. 102790.

[42] Mchergui, A., Moulahi, T. and Zeadally, S., 2022. Survey on Artificial Intelligence (AI) Techniques for Vehicular Ad-hoc Networks (VANETs). Vehicular Communications, 34, p. 100403.

[43] Rani, P., Kavita, Verma, S., Kaur, N., Wozniak, M., Shafi, J. and Ijaz, M.F., 2021. Robust and Secure Data Transmission Using Artificial Intelligence Techniques in Ad-Hoc Networks. Sensors, 22(1), p. 251.

[44] Safari, F., Savić, I., Kunze, H., Ernst, J. and Gillis, D., 2023, June. A Review of AI-based MANET Routing Protocols. In 2023 19th International Conference on Wireless and Mobile Computing, Networking and Communications (WiMob) (pp. 43-50). IEEE.

[45] Jazyah, Y.H., 2021. Enhancing the Performance of Wireless Routing Protocols of MANET using AI. Journal of Computer Science, 17(10), pp. 953-959.

[46] Alsarhan, A., Alauthman, M., Alshdaifat, E.A., Al-Ghuwairi, A.R. and Al-Dubai, A., 2021. Machine Learning-driven optimization for SVM-based intrusion detection system in vehicular ad hoc networks. Journal of Ambient Intelligence and Humanized Computing, pp. 1-10.

[47] Hossain, M.A., Noor, R.M., Yau, K.L.A., Azzuhri, S.R., Z'aba, M.R. and Ahmedy, I., 2020. Comprehensive Survey of Machine Learning Approaches in Cognitive Radio-Based Vehicular Ad Hoc Networks. IEEE Access, 8, pp. 78054-78108.

[48] Khan, M.F. and Yau, K.L.A., 2020, August. Route Selection in 5G-Based Flying Ad-Hoc Networks using Reinforcement Learning. In 2020 10th IEEE international conference on control system, computing and engineering (ICCSCE) (pp. 23-28). IEEE.

[49] Khan, A., Khan, S., Fazal, A.S., Zhang, Z. and Abuassba, A.O., 2021. Intelligent cluster routing scheme for flying ad hoc networks. Science China Information Sciences, 64(8), p. 182305.

[50] Chen, K.C. and Hung, H.M., 2019, May. Wireless Robotic Communication for Collaborative Multi-Agent Systems. In ICC 2019-2019 IEEE International Conference on Communications (ICC) (pp. 1-7). IEEE.

[51] Alsamhi, S.H., Ma, O. and Ansari, M.S., 2019. Survey on artificial intelligence based techniques for emerging robotic communication. Telecommunication Systems, 72, pp. 483-503.

[52] Lv, Z., Chen, D., Feng, H., Wei, W. and Lv, H., 2022. Artificial Intelligence in Underwater Digital Twins Sensor Networks. ACM Transactions on Sensor Networks (TOSN), 18(3), pp. 1-27.

[53] Ahmad, B., Jian, W., Enam, R.N. and Abbas, A., 2021. Classification of DoS Attacks in Smart Underwater Wireless Sensor Network. Wireless Personal Communications, 116, pp. 1055-1069.

[54] Ahmad, I., Shahabuddin, S., Sauter, T., Harjula, E., Kumar, T., Meisel, M., Juntti, M. and Ylianttila, M., 2020. The Challenges of Artificial Intelligence in Wireless Networks for the Internet of Things: Exploring Opportunities for Growth. IEEE Industrial Electronics Magazine, 15(1), pp. 16-29.

[55] Mohamed, E., 2020. The Relation of Artificial Intelligence with Internet of Things: A Survey. Journal of Cybersecurity and Information Management, 1(1), pp. 30-24.

[56] Wu, H., Han, H., Wang, X. and Sun, S., 2020. Research on Artificial Intelligence Enhancing Internet of Things Security: A Survey. Ieee Access, 8, pp. 153826-153848.

[57] Yogarayan, S., 2021. Wireless Ad Hoc Network of MANET, VANET, FANET and SANET: A eRview. Journal of Telecommunication, Electronic and Computer Engineering (JTEC), 13(4), pp. 13-18.

[58] Huang, Y., Su, X. and Chang, K., 2013, October. Notification-based cooperative spectrum sensing for ship ad-hoc network. In 2013 International Conference on ICT Convergence (ICTC) (pp. 199-200). IEEE.

[59] Su, X., Yu, H., Chang, K., Kim, S.G. and Lim, Y.K., 2015. Case Study for Ship Ad-hoc Networks under a Maritime Channel Model in Coastline Areas. KSII Transactions on Internet & Information Systems, 9(10).

[60] Wu, J. and Stojmenovic, I., 2004. Ad Hoc Networks. Computer, 37(2), pp. 29-31.

[61] Sekhar, B.V.D.S., Udayaraju, P., Kumar, N.U., Sinduri, K.B., Ramakrishna, B., Babu, B.R. and Srinivas, M.S.S.S., 2023. Artificial neural network-based secured communication strategy for vehicular ad hoc network. Soft Computing, 27(1), pp. 297-309.

[62] Zou, X., Hu, Y., Tian, Z. and Shen, K., 2019, October. Logistic Regression Model Optimization and Case Analysis. In 2019 IEEE 7th international conference on computer science and network technology (ICCSNT) (pp. 135-139). IEEE.

[63] Saravanan, M. and Ganeshkumar, P., 2020. Routing using reinforcement learning in vehicular ad hoc networks. Computational Intelligence, 36(2), pp. 682-697.

[64] Duraipandian, M., 2019. PERFORMANCE EVALUATION OF ROUTING ALGORITHM FOR MANET BASED ON THE MACHINE LEARNING TECHNIQUES. Journal of trends in Computer Science and Smart technology (TCSST), 1(01), pp. 25-38.

[65] Islabudeen, M. and Kavitha Devi, M.K., 2020. A Smart Approach for Intrusion Detection and Prevention System in Mobile Ad Hoc Networks Against Security Attacks. Wireless Personal Communications, 112, pp. 193-224.

[66] Rajendran, A., Balakrishnan, N. and Ajay, P., 2022. Deep embedded median clustering for routing misbehaviour and attacks detection in ad-hoc networks. Ad Hoc Networks, 126, p. 102757.

[67] Duraipandian, M., 2019. Performance evaluation of routing algorithm for Manet based on the machine learning techniques. Journal of trends in Computer Science and Smart technology (TCSST), 1(01), pp. 25-38.

[68] Saravanan, M. and Ganeshkumar, P., 2020. Routing using reinforcement learning in vehicular ad hoc networks. Computational Intelligence, 36(2), pp. 682-697.

[69] Lv, Z., Chen, D., Feng, H., Wei, W. and Lv, H., 2022. Artificial intelligence in underwater digital twins sensor networks. ACM Transactions on Sensor Networks (TOSN), 18(3), pp. 1-27.

[70] Obaidat, M., Khodjaeva, M., Holst, J. and Ben Zid, M., 2020. Security and Privacy Challenges in Vehicular Ad Hoc Networks. Connected Vehicles in the Internet of Things: Concepts, Technologies and Frameworks for the IoV, pp. 223-251.

[71] Abdel-Fattah, F., Farhan, K.A., Al-Tarawneh, F.H. and AlTamimi, F., 2019, April. Security Challenges and Attacks in Dynamic Mobile Ad Hoc Networks MANETs. In 2019 IEEE jordan international joint conference on electrical engineering and information technology (JEEIT) (pp. 28-33). IEEE.

[72] RahnamaeiYahiabadi, S., Barekatain, B. and Raahemifar, K., 2019. TIHOO: An Enhanced Hybrid Routing Protocol in Vehicular Ad-hoc Networks. EURASIP Journal on Wireless Communications and Networking, 2019(1), pp. 1-19.

[73] Manivannan, D., Moni, S.S. and Zeadally, S., 2020. Secure authentication and privacy-preserving techniques in Vehicular Ad-hoc NETworks (VANETs). Vehicular Communications, 25, p. 100247.

[74] Behbahani, H., MohammadianAmiri, A., Nadimi, N. and Ragland, D.R., 2020. Increasing the efficiency of vehicle ad-hoc network to enhance the safety status of highways by artificial neural network and fuzzy inference system. Journal of Transportation Safety & Security, 12(4), pp. 501

CHAPTER 8 ARTIFICIAL INTELLIGENCE-ENHANCED IOT AND UAV INTEGRATION IN AD HOC NETWORKS FOR SMART CITIES

[75] Jazyah, Y.H., 2021. Enhancing the Performance of Wireless Routing Protocols of MANET using AI. Journal of Computer Science, 17(10), pp. 953-959.

[76] Lv, Z., Chen, D., Feng, H., Wei, W. and Lv, H., 2022. Artificial intelligence in underwater digital twins sensor networks. ACM Transactions on Sensor Networks (TOSN), 18(3), pp. 1-27.

CHAPTER 9

Blockchain Application in UAV Swarm Security

Faisal Rehman[1,2], Junaid Akbar[1], Muhammad Abid Mehmood[3]
[1]Department of Statistics & Data Science, University of Mianwali, Mianwali, Pakistan
[2]Department of Robotics & Artificial Intelligence, National University of Sciences & Technology, NUST, Islamabad, Pakistan
[3]Department of Computer Science, Lahore Leads University, Lahore, Pakistan
Correspondence: Faisal Rehman[1,2] (faisalrehman0003@gmail.com)

Introduction

Within the context of the UAVs, the phenomenon of swarming has appeared as a more novel approach that reassures the overall effectiveness and effectiveness of a given number of tasks that can be applied in shelves such as surveillance, reconnaissance, disaster relief, and agriculture [29]. Envisioned uses of UAV swarms signify an innovative horizon of collective aerial routines, where several UAVs work collectively with others to perform various tasks in unison. However, amidst the excitement surrounding the potential of UAV swarms, there lies a critical challenge that cannot be overlooked: maintaining the security of such ever-evolving as well as integrated systems.

CHAPTER 9   BLOCKCHAIN APPLICATION IN UAV SWARM SECURITY

Unveiling the Significance of Security in UAV Swarms

This is especially so in the case of multiple UAV swarms that are becoming more common in various operational theaters the world over; the matter of security becomes rather pertinent. Swarm operation is generally defined as multiple robots operating successively and consequently while sharing information in a coordinated way, and just as can be anticipated, multiple coordinated networks imply numerous risks [1]. UAV swarm communication systems are highly vulnerable to network security threats such as cyberattacks, unauthorized access, and data breaches, compromising operational integrity and mission success. As a result, it is very important and indeed the next big challenge that cannot be overlooked to ensure that protection is offered to the UAV swarms and that they are secure and safe.

Harnessing Blockchain: A Revolutionary Approach to Security

Enter blockchain technology – a distributed and highly secure, digital ledger technology widely promoted for its versatility to solve a wide range of industries' problems of which the safety of UAV swarm systems is one. With greater consideration, I can state that it can be suggested that blockchain is a rather original and novel attempt to create a basis of trust, openness, and responsibility in terms of the cryptographic apparatus and decentralized consensus. These features make blockchain a potential solution for UAV systems so they could enhance system security by reducing dependence on a single control center, prohibiting the control by one party, and excluding data tampering [2].

CHAPTER 9 BLOCKCHAIN APPLICATION IN UAV SWARM SECURITY

Navigating the Objectives and Scope of Exploration

Therefore, this chapter aims to provide further in-depth analysis of how best the concepts of blockchain and UAV swarm security books can be intertwined [3]. The primary objectives include the following:

- Defining the key ideas behind UAV swarms and blockchain, as well as the possibilities of their combination and collaboration

- Explaining the specific risks and security concerns associated with swarms of UAVs and pointing to possible consequences of the use of the blockchain system

- Discussing the potential usages of the blockchain approach in promoting efficient interaction, credible data exchange, and proper identification of UAV units

- Exploring the potential and challenges and exploring the opportunities for application of blockchain-based security in the context of UAV swarms

In this chapter, we present the main idea of the book, which is to integrate a UAV swarm with the blockchain technology with the focus on the concepts related to the UAV swarms and the blockchain technology as well as their benefits in terms of security. We consider certain risks and security threats related to UAV swarms; we identify possible implications of applying blockchain systems. Furthermore, we discuss how blockchain can help in proper communication among the units with proper identification of UAV and correct data exchange. Finally, we outline the discussion that focuses on the prospects and issues related to the implementation of blockchain security to UAV swarms before proceeding to the forthcoming chapters filled with detailed investigations and analyses that will help guide researchers, engineers, policymakers, and managers in improving UAV swarm security and performance.

Background
Key Concepts
UAV Swarms

Swarms, relative to UAVs, show the UAVs that are synchronized to accomplish a task and are often referred to as unmanned aerial vehicle swarms. These swarms operate multiple UAVs together to work in a coordinated manner, which increases their efficiency, flexibility, and effectiveness across applications [4]. Compared to single UAVs that can be constrained by factors such as distance, coverage, or computation ability, swarms are capable of allowing various tasks to be divided among different vehicles, meaning that a swarm can perform a larger area search or display formations, complex aerobatics, or coordinated operations. Figure 9-1 represents UAV swarms' architecture by presenting how numerous UAVs can be controlled to perform various tasks efficiently. It amplifies the coupling of nodes in UAVs; this stresses the fact that to achieve highly coordinated movement and implementation, swarm functionality must open lines of communication to effectively multiply or divide functionality to the strain needed.

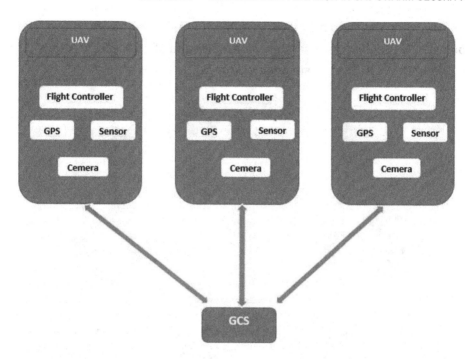

Figure 9-1. UAV swarms architecture

Blockchain

Blockchain is a decentralized distributed ledger technology that allows records of digital assets exchange to be processed and documented without the use of central authorities. In the blockchain, units are arranged in a hierarchical manner where related transactions are strung blocks; each block has a timestamp and block reference, constituting an unalterable ledger of all the transactions. Participants in the network and the transactional flow operate through consensus approaches such as proof of work or proof of stake, which enables them to validate and approve the transactions and create the distributed ledger without using a central controller. Figure 9-2 represents conceptual model for secure communication in UAV swarms using a blockchain layer. It shows the movement of data and how through blockchain the safe swarm transaction between UAVs occurs so that the communication of the swarm is not vulnerable to cyberterrorism and hacking [5].

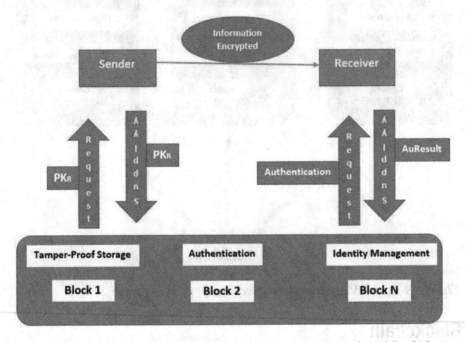

Figure 9-2. *Secure communication architecture based on blockchain*

Security in UAV Swarms

Security within UAV swarms is the protection of the ecosystem of the swarms from attack or influence by unauthorized entities and also protection of the packaged data and information from theft or invasion [6]. These include but are not limited to the following: ensuring the integrity of the interactions between distinct UAVs, guaranteeing the exchanged or processed data by the swarm integrity, validating and enforcing swarm member's access rights, and ensuring swarm solutions resilience to malicious actions, hence making security similar to UAV swarms imperative for enhancing the swarm dependability, dependability, and operational efficiency in various applications and environments [31].

Need for Secure Communication
The Collaborative Nature of UAV Swarms

UAVs act in groups and need to have effective message exchange mechanisms for the overall management of the UAV system, or for relaying information between the UAVs [7]. They differ from lone UAVs that can act independently or near others, as swarms need uninterrupted communication to align their movements, exchange data gathered by their sensors, and perform tasks in unison. It is imperative to establish secure communication channels through which to compose, transmit, and synchronize instructions, information updates, and data exchange that should circulate within the swarm and be supported by all the engaged UAVs.

Vulnerabilities in Communication

In the absence of the proper measures, current UAV swarm communication has some drawbacks and prohibitions. It derives that these vulnerabilities can be exploited by terrorists to reduce the integrity and confidentiality or by terrorists to affect the availability of swarm operations [8]. For example:

- Hijacking of UAVs: Such compromise of communication means may enable the attacker to intercept or falsify the control signals, that is, to seize control or manipulate control individual UAVs in the swarm maliciously [9].

- Data Tampering: Given insecure communication channels may allow the adversary to inject dangling/offending/unwanted messages into the communication flow or alter the message content altogether, which may cause a wrong decision to be made or the swarm control signal to be altered [10].

- Loss of Control: Inability to maintain control over the specific UAV or all of the UAVs involved in a mission or even the entire swarm can lead to misfortune and mishaps and pose risks to safety [11].

That is why solving these vulnerabilities calls for the addition of strong encryption, authentication, and integrity checks where all the flows that happen within the UAV swarms will have to be secured. Secure communication protocols in achieving the confidentiality, integrity, and availability of data exchanges contribute to the robustness, reliability, and functionality of UAV swarm engagements in various mission settings and cluttered geo space milieu.

Challenges in Ensuring Security

Some of the issues that need to be solved when protecting the security of UAVs include securing links in the multiple UAV formation, safeguarding information, and recognizing the authorized identity. He found that in an attempt to improve the reliability and performance of these systems, he proposed the development of an intelligent hybrid scheme that effectively addresses the identification of faults in industrial ball screw linear motion systems. This integration mandates that UAV swarms will be able to perform better against threats and malicious actions that may exist [23].

Scalability

Some of the problems arising while controlling a swarm of UAVs are how multiple UAVs can be managed at once and how secure communication can be established between them. The data show that as more UAVs are deployed, the challenge to achieve secure and reliable communication between the individual participants of the swarm also grows. Some limitations of the proposed system include scalability, with a large amount of information being processed, timely update of information, and the privacy and security of the communicated information [12]. The management of conventional security procedures to support swarm operational environments is a vital part of UAV swarm management; in special, the scalability of the security measures should not impair the optimal performance of swarm operation.

Resource Constraints

UAVs inherently have low computation capabilities and power supply which creates a challenge on which security functionalities to include. Cryptographic mechanisms aimed at providing confidentiality, integrity, nonrepudiation, and other necessary features for communications consume significant computational power and energy. Having modest battery capacities and processing power, the issue of having satisfactory security while maintaining an optimal load is one of the biggest concerns in developing UAVs. Due to the dysfunctions that we have already described, lightweight cryptosystems and efficient energy securing become effective in preventing the decrease of opportunities for UAV swarm by the security systems [13].

Dynamic Network Topology

Frequently, a UAV swarm modifies the network topology because of its movement and variability in its operations. This constant reconfiguring of the networks means that the security can be challenging to manage. In such a complex, dynamic network, it is possible that preexisting security solutions that are developed for a static type or slowly evolving network environment may not suffice [12]. Widespread and high mobility also means that the dynamic topology of the system demands fast and agile security solutions to address, for example, dynamic membership of the swarm, joining or leaving UAV, or changes in the connectivity of the links due to motion or other factors. On the drone network environment fluctuation, guaranteeing smooth and safe communication must be achieved to provide the proper functionality of the UAV swarm.

Security of UAV Swarms Using Blockchain System

Overview of Blockchain Technology

Chain of Blocks

It is implemented in the form of a blockchain, blocks which in turn contain records of all the operations made. These are any communication or exchange of information or other value flowing through the network. Each block is tied to another block with a complicated mathematical algorithm that creates a chain of blocks. By formatting the structure in this way, it becomes impressive to adjust a block to all the other subsequent blocks, thereby making it almost impossible for manipulation to be done [4].

Immutable Ledger

An important focus of blockchain is maintaining the data integrity of the stored information, that is, its immutability [15, 16]. Once the block is added to the chain, it becomes a permanent record that cannot be altered or removed because of the use of cryptography. This makes the record immutable and unchangeable, thus affording a reliable source of record of the undertaking of transactions, hence acting as a check and ensuring no alterations are made to any information that is recorded.

Decentralized Network of Nodes

Blockchain is a distributed open network where each node stores the entire data of the blockchain and no central hub operates it. These nodes operate to authenticate and make consensus on the transactions done through consensus like proof of work or proof of stake [17]. Decentralization removes the dependency on a central control point, which thereby decreases the probability of the occurrence of a downfall point leading to a more secure system. Before transactions are added to the blockchain, the nodes themselves check the transactions entered, including checking if it is a valid transactions or not.

Enhancing Security with Blockchain

Decentralization

It is still questionable whether or not there are completely shielded blockchains, but the decentralized architecture reduces central points of vulnerability by default. When data is distributed across different centers, there is no particular authority managing the whole network and this exposes the decentralized network to fewer attacks [18]. In UAV swarms, decentralization involves spreading the control and decision of the formation among individual subsystems to improve the general security and performances of the swarms.

Immutability

This is because, once data is inputted into a block at a given time, the data cannot be altered or deleted in the future. This has the effect of providing data authenticity since no one can alter its content after it has been created: it cannot be changed [19]. For UAV swarms in particular, this can be interpreted as indicating that all the information, interactions, and transactions of actions and communications are time-stamped and encrypted, which precludes their alteration once they have been desynchronized.

Transparency

Blockchain offers the attribute of transparency since the information recorded in the block affects all the participants who are allowed to scrutinize them. This makes accountability much easier to achieve because each transaction is easily seen and tracked because of the transparency created by electronic money. In the context of UAV swarms, transparency always allows all communications and the exchange of data to be visible to those who are entitled to inspect the records, thus making it easier to establish the necessary levels of trust between swarm members [20].

Benefits of Blockchain in UAV Swarms

Improved Security

To the degree that these swarms engage in data sharing and coordination, blockchain provides a better safeguard against fraud and unauthorized modification [4]. The techniques of cryptography that have been adopted in blockchain guarantee that all data that is input is well encrypted and that a system is not subject to fraudulent interfaces by wrong individuals. This protection is required for the purposes of preserving and safeguarding the propriety and secrecy of the messages and information exchanged within the swarm formation. For instance, innovative methods such as machine learning have been incorporated to improve both the security and functionality of the Internet of Things in detecting unwanted entrances and potential risks [30].

Transparency

It further ensures that every transaction or communication made within the UAV swarm is clearly accounted for and observable by only the parties that have been predesignated in the blockchain system. This traceability adds to the accountability and reliable nature of the swarm because every action done can be proven and tracked within the swarm [20]. Transparency again enables the provision of quick identification of any aberrations or infringements as and when they occur.

Scalability

Using the insights mentioned above, we can conclude that implementing a decentralized blockchain-based control platform is a feasible solution to manage the dynamic and scalable environment of UAV swarms. The decentralized and distributed character of blockchain is effective for the massive control of numerous UAVs with the same robust safety and performance level. This scalability is crucial for the increasing scale and the increasing number of UAVs that will be included in a swarming operation, to be able to securely and efficiently manage all the UAVs that are included in the operation [12].

Applications of Blockchain in UAV Swarm Security

Secure Communication Protocols

Enhancing Security with Blockchain

Security measures that may be employed within UAV swarms include provision of security to data transmitted and communication channels and determining access rights. To increase the dependability and effectiveness of these systems, a conventional I scheme known as an intelligent hybrid

CHAPTER 9   BLOCKCHAIN APPLICATION IN UAV SWARM SECURITY

system has been developed that increases the chances of diagnosing faults in industrial ball screw linear motion systems tenfold. This integration makes it possible for the UAV swarms to function with further escalation against threats or inadequate actions. Some new techniques are also used to quantify functional parameter quantities in ball screw drives, helping the industrial application to be far more accurate and efficient [29, 31].

Encrypted Data and Identity Verification

With regard to safety and data protection, blockchain can help safeguard communication links between UAV swarms by enhancing data security in the messages that are sent and received by the UAVs and authenticating the sending and receiving UAVs [12]. Each UAV may include the ability to employ private blockchain-based cryptographic keys to ensure that the messages transmitted are encrypted from end to end, thus preventing the messages from being intercepted by any other unauthorized party. The use of blockchain-based identity checks assures that only UAVs that have valid access keys obtained from the blockchain network can engage in the communication, hence minimizing the incidences of spoofing and impersonation.

Smart Contracts for Secure Message Exchanges

Using smart contracts, self-contained contracts where the conditions of the contract whether in the form of coded instructions are carried out by the machine automatically, secure message exchanges can be performed within the swarm [14]. These contracts can also mandate that specific security guidelines are followed, for instance, encryption protocols and message authentication, and this would ensure that all messages that flow through the system are compliant with standard security requirements. Through such means, it is possible to eliminate the presence of human factors that can negatively affect swarm communications by increasing the dependability and security provided by smart contracts.

Data Storage and Access Control
Decentralized Storage

Blockchain as explained thus offers a system of decentralized storage and hence would help in the storage of data collected by UAVs across a large number of nodes. This decentralization means that the data is not anchored on one server that can be easily messed up and secured against any form of manipulation or intrusion [21]. The ledger in the blockchain is permanent and can therefore not be changed once created; this assures users of the reliability of the data stored.

Smart Contracts for Access Control

The access to the data within a UAV swarm is best managed by deploying smart contracts to only allow access to data in accordance with roles and privileges assigned to the users. These contracts can specify who or what can get into an individual set of data or control a particular operation of the UAVs, which would only be accessible or executable by those who are allowed. Because smart contracts can be operational, they can be configured to change the access permission severally and then offer a flexible means of access control [22].

Identity Management and Authentication
Unique Blockchain Identity for UAVs

For instance, in the case of numerous UAVs forming a swarm, a distinct blockchain identity can be assigned to each UAV and the identity registration and management is ensured and done within the blockchain environment. This specific identity aids in reducing the vulnerability of the drone to be impersonated and ensures that every UAV that is in the vicinity can be recognized and not be confused with other UAVs. The addition of a blockchain-based identity management system to the IdP bolsters the IdP's capacity to present authenticated credentials with a record of each UAV [7].

Blockchain-Based Authentication

By using techniques of blockchain-based authentication, it may be possible to allow only UAVs which have access rights, granted by the blockchain, to be included in the swarm and perform an activity. Thus, in this authentication, the cryptographic credentials effectively validate the legitimacy of every UAV in the blockchain register and permit only those possessing the specific credentials to share their data [24]. This assists in enhancing the general security of the network since the first element of facetious UAV that would attempt to gain access to the network is refused entry.

Inconsistency: Consensus Mechanisms and Smart Contracts

Consensus Mechanisms for Trust

In blockchain, consensus mechanisms ensure the validity of a given transaction through the coordination of all nodes belonging to a given network to authenticate the transaction before being written into a database. More related to consensus mechanisms, these can confirm the command issued for their operations, data to each other UAVs, and others, enhancing the degree of trust and reliability in the UAV swarms. Thus, it is agreed that all UAVs are running in a certain state that is capable of avoiding conflicts or mischief from the network [6].

Smart Contracts for Operational Protocols

Smart contracts are also capable of performing various number of logical operations that are associated with the UAV swarm's working mechanisms as well as its protection measures. Such contracts can contain instructions for the behavior, interaction, or security which can ensure that all unmanned aerial vehicles used will perform in the same manner [25]. For example, for a certain task, a smart contract can ensure that only a UAV with corresponding credentials can execute the said task. This kind of automation helps to secure and make it beyond reasonable doubt accurate as it is not subjected to human intervention in as much as human errors or intention to open the doors to criminals are likely to cause havoc.

Challenges and Future Directions
Scalability and Performance

- A UAV swarm entails the use of several drones operating concurrently with the capability of executing tasks such as surveillance, delivery, or producing comprehensive maps. The advantages include decentralized control of the database, preserving the authenticity of records and transactions, and secure transactions. However, there would still be limitations on the real-time drone operation by UAV swarms due to the computational and communication overhead of blockchain operations above a certain threshold of operations within a specific time frame to enable quick decisions and coordinated actions [26].

- To meet this challenge, knowledge and analysis are necessary to create a separate lightweight blockchain environment that utilizes UAV constraints. Such protocols should reduce computational and communication complexity to the required levels, but at the same time include the features that ensure the system's secure and dependable operation [26]. Possible interventions include enhancing consensus algorithms, minimizing the needed data storage, and implementing proper communication strategies peculiar to UAV networks.

Energy Efficiency

- Delving into consensus mechanisms, such as PoW or PoS, it is prevalent for blockchain operations to be associated with significant energy consumption. Although UAV swarms remain relatively efficient due to their structure, for swarms that fly in far-off areas or areas that have no access to power sources, power management is significant [27]. The energy used in conducting the blockchain operations may have some buck on the flight time and functioning capacity of the UAVs.
- Possible optimizations include energy efficiency: consensus solutions include energy-efficient consensus algorithms like proof of authority and practical Byzantine fault tolerance. These algorithms have less computational demands as compared to others like

PoW or PoS and therefore utilize less power. Moreover, the physical hardware layer contains optimizations, like the low energy processors or energy harvesting systems, which might help minimize the energy consumption of blockchain-based UAVs [32].

Interoperability with Existing Systems

- Combining the elements of blockchain technology with conventional UAV platforms and control networks poses certain difficulties resulting from the differences in the systems' architectures, modes of data exchange, and data representation modes [28]. Data from unmanned aerial vehicles has usually prebuilt kernel software and impermeable standard interfaces which do not make it easy to integrate with blockchain systems.

- Thus, it is crucial to establish standards and frameworks that will integrate blockchain-UAV systems with the infrastructure of current systems and facilities. These standards should cover areas such as common data formats for storage and sharing, communication protocols between UAVs and the blockchain infrastructure, and the interfaces through which the UAVs can communicate with the blockchain [28]. Moreover, there are opportunities to create middleware for UAVs that would allow integrating them into existing blockchains and creating an efficient system for sharing and exchanging information between platforms.

CHAPTER 9   BLOCKCHAIN APPLICATION IN UAV SWARM SECURITY

Future Research Directions

As the field of blockchain-enabled UAV systems continues to evolve, several promising research directions can be explored:

- From a technical perspective, I would like to design new, efficient, and performant blockchain platforms, as well as to consider applying them to the operations of drone swarms because of their extraordinary differences [27].

- Assessing the feasibility of using other highly efficient consensus mechanisms adaptive to situations where targets are constantly shifting and resources are scarce, including lighter versions of existing algorithms or completely different consensus protocols designed particularly with UAV networks in mind [26].

- Exploring the subject of the interaction of blockchain with AI and ML, as applied to the advanced decision-making, and optimizing autonomy of UAV swarms. Blockchain may be effectively used to ensure secure data storage and effective sharing of UAV data, while AI/ML may analyze the information to make effective decisions in real time and thus control swarm behavior [30]. Furthermore, the advancement of AI has revolutionized various sectors, including IoT, the energy sector, quantum computing, and the fields of image and signal analysis [33–35].

Conclusion

Enabling the functionality of blockchain and UAV swarms paves the way for enhanced security, reliability, and functionality. Although these are promises portended in data analytics, there are several defining factors of major hurdles that must be crossed in order to unlock this potential. Challenges such as scalability and performance are related to the amount of time and effort required in computing and communicating the blockchain functionality, which makes it important to establish low-complexity blockchain protocols specifically designed for UAVs with limited resources. There is also the matter of efficiency or lack thereof as most blockchains consume large amounts of energy due to the different consensus algorithms involved. To address this issue, more advanced thinking is needed including the use of efficient consensus algorithms and hardware. Furthermore, integration with other existing UAVs and the protocols that are already set also pose many challenges in that standards and frameworks can work as a solution. Larger studies should endeavor to relate to the generation of energy-effective blockchain solutions, examine the high-level consensus mechanisms relevant to volatile environments, and incorporate the blockchain with artificial intelligence and machine learning to improve decision-making and self-control in UAV formations. Thus, further successful management of these challenges will create the foundation for the application of blockchain technology in UAV swarms and enhance their functionality, thus expanding their applicability to different fields of industries, logistics, energy, environmental monitoring, and more.

References

[1] Hassan Jalil Hadi, Yue Cao, Khaleeq Un Nisa, Abdul Majid Jamil, Qiang Ni, A comprehensive survey on security, privacy issues and emerging defence technologies for UAVs, Journal of Network and ComputerApplications, Volume213, 2023, 103607, ISSN1084-45.

[2] F. Zidan, D. Nugroho, and B. A. Putra, "Securing Enterprises: Harnessing Blockchain Technology Against Cybercrime Threats", *IJCITSM*, vol. 3, no. 2, pp. 167–172, Oct. 2023.

[3] Manikandan, K.; Sriramulu, R. Optimized Path Planning Strategy to Enhance Security under Swarm of Unmanned Aerial Vehicles. *Drones* 2022, 6, 336. https://doi.org/10.3390/drones6110336.

[4] J. Jensen, D. F. Selvaraj and P. Ranganathan, "Blockchain Technology for Networked Swarms of Unmanned Aerial Vehicles (UAVs)," *2019 IEEE 20th International Symposium on "A World of Wireless, Mobile and Multimedia Networks" (WoWMoM)*, Washington, DC, USA, 2019, pp. 1-7, doi: 10.1109/WoWMoM.2019.8793027.

[5] W. Xiao, M. Li, B. Alzahrani, R. Alotaibi, A. Barnawi and Q. Ai, "A Blockchain-Based Secure Crowd Monitoring System Using UAV Swarm," in *IEEE Network*, vol. 35, no. 1, pp. 108-115, January/February 2021, doi: 10.1109/MNET.011.2000210.

[6] S. H. Alsamhi *et al.*, "Blockchain-Empowered Security and Energy Efficiency of Drone Swarm Consensus for Environment Exploration," in *IEEE Transactions on Green Communications and Networking*, vol. 7, no. 1, pp. 328-338, March 2023, doi: 10.1109/TGCN.2022.3195479.

[7] Xueping Zhu, Zhengchun Liu, Jun Yang, Model of Collaborative UAV Swarm Toward Coordination and Control Mechanisms Study, Procedia Computer Science,Volume 51, 2015, Pages 493-502, ISSN 1877-0509.

[8] L. CHAARI, S. CHAHBANI and J. REZGUI, "Vulnerabilities Assessment for Unmanned Aerial Vehicles Communication Systems," *2020 International Symposium on Networks, Computers and Communications (ISNCC)*, Montreal, QC, Canada, 2020, pp. 1-6, doi: 10.1109/ISNCC49221.2020.9297293.

[9] Subbarayalu, V.; Vensuslaus, M.A. An Intrusion Detection System for Drone Swarming Utilizing Timed Probabilistic Automata. *Drones* 2023, 7, 248. https://doi.org/10.3390/drones7040248.

[10] Islam and S. Y. Shin, "BUS: A Blockchain-Enabled Data Acquisition Scheme With the Assistance of UAV Swarm in Internet of Things," in *IEEE Access*, vol. 7, pp. 103231-103249, 2019, doi: 10.1109/ACCESS.2019.2930774.

[11] D. M. Hart and P. A. Craig-Hart, "Reducing swarming theory to practice for UAV control," *2004 IEEE Aerospace Conference Proceedings (IEEE Cat. No.04TH8720)*, Big Sky, MT, USA, 2004, pp. 3050-3063 Vol.5, doi: 10.1109/AERO.2004.1368111.

[12] G. Bansal and B. Sikdar, "S-MAPS: Scalable Mutual Authentication Protocol for Dynamic UAV Swarms," in *IEEE Transactions on Vehicular Technology*, vol. 70, no. 11, pp. 12088-12100, Nov. 2021, doi: 10.1109/TVT.2021.3116163.

[13] R. Dong, B. Wang and K. Cao, "Security enhancement of UAV swarm enabled relaying systems with joint beamforming and resource allocation," in *China Communications*, vol. 18, no. 9, pp. 71-87, Sept. 2021, doi: 10.23919/JCC.2021.09.007.

[14] Manikandan K, Sriramulu R. ASMTP: Anonymous secure messaging token-based protocol assisted data security in swarm of unmanned aerial vehicles. *Int J Network Mgmt*. 2024; e2271. doi:10.1002/nem.2271.

[15] Riaz, N., Shah, S. I. A., Rehman, F., Gilani, S. O., & Udin, E. (2020). A Novel 2-D Current Signal-Based Residual Learning with Optimized Softmax to Identify Faults in Ball Screw Actuators. *IEEE Access*, 8, 115299-115313.S. Hafeez et al., "Blockchain-Assisted UAV Communication Systems: A Comprehensive Survey," in *IEEE Open Journal of Vehicular Technology*, vol. 4, pp. 558-580, 2023, doi: 10.1109/OJVT.2023.3295208.

[16] G. Raja, S. Anbalagan, A. Ganapathisubramaniyan, M. S. Selvakumar, A. K. Bashir and S. Mumtaz, "Efficient and Secured Swarm Pattern Multi-UAV Communication," in *IEEE Transactions on Vehicular Technology*, vol. 70, no. 7, pp. 7050-7058, July 2021, doi: https://doi.org/10.1109/TVT.2021.3082308.

[17] Han, P.; Sui, A.; Wu, J. Identity Management and Authentication of a UAV Swarm Based on a Blockchain. *Appl. Sci.* 2022, 12, 10524. https://doi.org/10.3390/app122010524.

[18] Jian Wang, Yongxin Liu, Shuteng Niu, Houbing Song, Lightweight blockchain assisted secure routing of swarm UAS networking, Computer Communications,Volume 165, 2021, Pages 131-140, ISSN 0140-3664.

[19] Tejasvi Alladi, Vinay Chamola, Nishad Sahu, Mohsen Guizani, Applications of blockchain in unmanned aerial vehicles: A review, Vehicular Communications,Volume 23, 2020, 100249, ISSN 2214-2096.

[20] Wang, Z. Jiao, J. Chen, X. Hou, T. Yang, and D. Lan, "Blockchain-Aided Secure Access Control for UAV Computing Networks," in *IEEE Transactions on Network Science and Engineering*, doi: 10.1109/TNSE.2023.3324639.

[21] Y. Liu et al., "Lightweight Blockchain-Enabled Secure Data Sharing in Dynamic and Resource-Limited UAV Networks," in *IEEE Network*, doi: 10.1109/MNET.2024.3383237.

[22] Y. Tan, J. Wang, J. Liu and N. Kato, "Blockchain-Assisted Distributed and Lightweight Authentication Service for Industrial Unmanned Aerial Vehicles," in *IEEE Internet of Things Journal*, vol. 9, no. 18, pp. 16928-16940, 15 Sept.15, 2022, doi: 10.1109/JIOT.2022.3142251.

[23] Riaz, N., Shah, S. I. A., Rehman, F., & Khan, M. J. (2021). An Intelligent Hybrid Scheme for Identification of Faults in Industrial Ball Screw Linear Motion Systems. *IEEE Access*, 9, 35136-35150.

[24] R. Karmakar, G. Kaddoum and O. Akhrif, "A Blockchain-Based Distributed and Intelligent Clustering-Enabled Authentication Protocol for UAV Swarms," in *IEEE Transactions on Mobile Computing*, vol. 23, no. 5, pp. 6178-6195, May 2024, doi: 10.1109/TMC.2023.3319544.

[25] Y. Zhou, B. Rao and W. Wang, "UAV Swarm Intelligence: Recent Advances and Future Trends," in *IEEE Access*, vol. 8, pp. 183856-183878, 2020, doi: 10.1109/ACCESS.2020.3028865.

[26] S. Javed *et al.*, "State-of-the-Art and Future Research Challenges in UAV Swarms," in *IEEE Internet of Things Journal*, vol. 11, no. 11, pp. 19023-19045, 1 June1, 2024, doi: 10.1109/JIOT.2024.3364230.

[27] W. Chen, J. Liu, H. Guo and N. Kato, "Toward Robust and Intelligent Drone Swarm: Challenges and Future Directions," in *IEEE Network*, vol. 34, no. 4, pp. 278-283, July/August 2020, doi: 10.1109/MNET.001.1900521.

[28] M. Adil, M. A. Jan, Y. Liu, H. Abulkasim, A. Farouk and H. Song, "A Systematic Survey: Security Threats to UAV-Aided IoT Applications, Taxonomy, Current Challenges and Requirements With Future Research Directions," in *IEEE Transactions on Intelligent Transportation Systems*, vol. 24, no. 2, pp. 1437-1455, Feb. 2023, doi: https://doi.org/10.1109/TITS.2022.3220043.

[29] Riaz, N., Shah, S. I. A., Rehman, F., Gilani, S. O., & Udin, E. (2020). A Novel 2-D Current Signal-Based Residual Learning with Optimized Softmax to Identify Faults in Ball Screw Actuators. *IEEE Access, 8*, 115299-115313.

[30] Humayoun, M., Sharif, H., Rehman, F., Shaukat, S., Ullah, M., Maqsood, H., ... & Chandio, A. H. (2023, March). From Cloud Down to Things: An Overview of Machine Learning in Internet of Things. In *2023 4th International Conference on Computing, Mathematics and Engineering Technologies (iCoMET)* (pp. 1-5). IEEE.

[31] Riaz, N., Shah, S. I. A., Rehman, F., & Gilani, S. O. (2020). An Intelligent Approach to Detect Actuator Signal Errors Based on Remnant Filter. In *Intelligent Technologies and Applications: Second International Conference, INTAP 2019, Bahawalpur, Pakistan, November 6–8, 2019, Revised Selected Papers 2* (pp. 675-683). Springer Singapore.

[32] Riaz, N., Shah, S. I. A., Rehman, F., & Gilani, S. O. (2020). An Approach to Measure Functional Parameters for Ball-Screw Drives. In Intelligent Technologies and Applications: Second International Conference, INTAP 2019, Bahawalpur, Pakistan, November 6–8, 2019, Revised Selected Papers 2 (pp. 398-408). Springer Singapore.

[33] H. Sharif, F. Rehman and A. Rida, "Deep Learning: Convolutional Neural Networks for Medical Image Analysis - A Quick Review," 2022 2nd International Conference on Digital Futures and Transformative Technologies (ICoDT2), Rawalpindi, Pakistan, 2022, pp. 1-4, doi: https://doi.org/10.1109/ICoDT255437.2022.9787469.

[34] A. Ashfaq, M. Kamran, F. Rehman, N. Sarfaraz, H. U. Ilyas and H. H. Riaz, "Role of Artificial Intelligence in Renewable Energy and its Scope in Future," 2022 5th International Conference on Energy Conservation and Efficiency (ICECE), Lahore, Pakistan, 2022, pp. 1-6, doi: 10.1109/ICECE54634.2022.9758957.

[35] I. Manan, F. Rehman, H. Sharif, N. Riaz, M. Atif and M. Aqeel, "Quantum Computing and Machine Learning Algorithms - A Review," 2022 3rd International Conference on Innovations in Computer Science & Software Engineering (ICONICS), Karachi, Pakistan, 2022, pp. 1-6, doi: 10.1109/ICONICS56716.2022.10100452.

CHAPTER 10

Security and Privacy in AG-IoT-Enabled Unmanned Aerial Vehicles for Smart Cities

Yasir khan[1], Sadaf Manzoor[2]

[1]Department of Science and Technology & Information Technology (ST&IT), Peshawar, Pakistan Email: imyasir.308@gmail.com

[2]NED University of Engineering and Technology, Karachi, Pakistan Email: sadafmanzoor089@gmail.com

Introduction

Unmanned aerial vehicles (UAVs) are automatic vehicles which can perform their tasks without any assistance of human beings. For instance, agricultural UAVs perform operations for crop checking and applying fertilizer and chemicals, as well as for checking large pieces of land for precision farming. Likewise, the use of UAVs for disaster response entails using the drones to determine the extent of destruction in the disaster affected regions, as well as to offer assistance, and for the purpose of searching and rescuing stranded individuals. UAVs can be operated through several electronics advancement like microprocessors and sensors [1]. Currently, UAVs have slowly been expanded and have many qualities that can be used in various fields. For instance, the impact and adaptation of self-driving cars and drones [2] are drastically reshaping the transportation sector. Besides this, cooperative drones and robots [3] can perform tasks like assembling, joining, holding, and inspecting and can also work in quality assurance. This helps in making work easier, enables people to achieve their goals, and protects people by reducing accidents.

AG, or artificial general intelligence, is an AI technology that aims to make machines behave in a human-like manner, enabling them to reason, innovate, and learn like humans. Narrow AI on the other hand is designed to function on a specific task or need, but AG has the human-like broader intelligence that provides it with the ability of self-learning and solving new problems encountered. For example, in AG, UAVs are able to comprehend natural language unto itself, and production of natural language thereto, for interaction with humans and other intelligent systems. This capability makes UAVs capable of understanding instruction and even ask questions making the drones to be more useful and appropriate in many applications. Interestingly, the AG-based systems of UAVs has been more efficient and useful in carrying out the operations, as suggested and proved by Smart Tissue Autonomous Robot (STAR) [4]. The STAR supervisor-controlled interface offers two modes: a manual mode where a surgeon can define the position of each stitch and an automatic mode where an appropriate amount of stitches are placed evenly in reference to an indentation of an incision.

The Internet of Things (IoT) entails the interconnected environment created by devices dedicated to three key functions: this device performs the function of sending data, receiving data, and analyzing received data [5]. As stated by authors, IoT infrastructure is aimed "to connect

CHAPTER 10 SECURITY AND PRIVACY IN AG-IOT-ENABLED UNMANNED AERIAL VEHICLES FOR SMART CITIES

everything and everyone everywhere with everything and everyone else" [6, 7]. The area has grown progressively over the years due to the increasing numbers of connected IoT devices, which passed the 12 billion mark and marked an 8% growth compared to the previous year, according to a report done in May 2022. This is likely to grow to more than double to 27 billion by 2025 due to this trend [8]. As the great impact that IoT technologies could bring, its role could be realized in the vast majority of spheres of human activity. Systems known as "smart homes" became possible due to the growth in popularity of IoT technology and changed the way houses are managed through improved safety and energy use. Consumer IoT market has nine segments; smart home technology is the second largest segment having the total value of $108 billion USD, slightly lower than the value of smart manufacturing that leads this segment with $119 billion [9]. In the same way, human wearables such as fitness tracker, health monitoring devices, and networking healthcare devices are transforming the way healthcare is being delivered. Consequently, it has led to the emergence of an entirely new horizon recognized as "smart health." Continuing from that, Internet of Things includes networked automobile, smart traffic patterns, and the sensors integrated in roads and bridges. They lead toward the idea of smart cities which to avoid jams and wastage of energy.

Following are some of the major contributions of this book chapter:

- A taxonomy framework is introduced based on three key aspects: threat vectors and defense strategies, security challenges, and privacy considerations regarding AG-IoT-based UAVs. This taxonomy aims to highlight vulnerabilities in AG-powered and IoT-driven drones, offering a comprehensive understanding of these aspects.

- The security issues present in AG algorithms and IoT infrastructure, which form the fundamental building blocks of UAVs, are discussed. Moreover, the security measures employed to prevent these issues are underscored.

- Different privacy concerns identified over time in AG-IoT-based UAVs are also explored. These concerns are emphasized to promote responsible usage of UAVs for smart cities.

- Lastly, future research directions are suggested in these areas to highlight what needs further study by researchers and academicians.

The remaining book chapter is structured as follows: the "Literature Review" section provides a holistic review of the existing literature on AG-IoT-driven UAVs for smart cities. The "Security Issues in AG-IoT-Enabled UAVs for Smart Cities" section discusses the security challenges inherent in AG, IoT, and their integration into drones. In the "Privacy Concerns in AG-IoT-Enabled UAVs" section, critical privacy considerations associated with the use of AG-IoT-enabled UAVs are highlighted. The "Future Research Directions for Secure AG-IoT-Based UAVs" section identifies future research areas necessary to be explored for building robust and secure AG-IoT-enabled UAVs for smart cities. Finally, the "Conclusion" section concludes the chapter.

Literature Review

This section reviews the existing literature on privacy concerns and security issues in AG-IoT-enabled drones, highlighting their vulnerabilities and associated privacy issues.

de las Morenas et al. [10] discuss two case studies highlighting potential security risks to flying robots. In both cases, the identified threats were related to accessing information exchanged, whether it involved transmission of data for Monitoring systems remotely or communication between involved parties. The recommended course of action to handle these security threats included certificate for MQTT broker and client, the encryption of the TLS/SSL, firewalls to eliminate the DoS, and DDoS vulnerability.

Khan et al. [11] have developed a new vision for the Internet of flying cars incorporating the fastest routing protocol known as DSDV. On this basis, the concept of energy conservation of flying IoT can be resolved. As explained earlier, ISH-DSDV is an extension of the Bellman–Ford algorithm and uses routing update, information broadcasting, and subsequent stale data handling. Where DSDV stands compared to other modern routing protocols? Here, it coordinates as many as 32 trials and proves to demonstrate the optimal outcome. In any fly area, there is encompassing implementation of flying network, and therefore, the nomadic mobility model is applied in tests or assessments of these routing protocols.

Khan et al. [12] have provided routing protocol for FANET with the help of modified AntHocNet. In the present area of communication, one of the major issues which is gaining the considerable attention is the issue related to routing in ad hoc networks. One of the new developing fields is flying ad hoc network that is a new addition to mobile ad hoc networks. One of the significant issues in routing that is difficult to address when choosing the best optimal path in any network is the global optimum to be used by the routing protocols because the performance of the network depends on several performance parameters like throughput, quality of service (QoS), user experience, and response time of the system.

Arshi et al. [13] have tried to explain the classification of social analysis of IoT in the context of energy sector including characteristics, components, and protocols of IoT. It also covers the architecture and current advancement in the IoT and also the issues facing the field as well. Many topics were covered that are a part of IoT such as the IoT communications model, sensor boards, challenges faced by industries, the threats of IoT, and a few solutions to IoT threats.

CHAPTER 10 SECURITY AND PRIVACY IN AG-IOT-ENABLED UNMANNED AERIAL VEHICLES FOR SMART CITIES

Table 10-1 presents a summary of the related work that addresses privacy concerns and security challenges in smart cities that employed UAVs. It includes description of the techniques with an account of each technique briefly and the menace of each technique.

Table 10-1. *Tackling Privacy and Security Issues in AG-IoT-Enabled UAVs*

Methodology	Contribution	Limitation	Reference
Survey work	This work discusses faster delivery services by drones and their risk to public safety if operated at higher speed	Complicated UAV behavior and decision-making due to integration of AI techniques	[14]
Simulation and comparative analysis	This paper evaluated blockchain consensus methods for use in UAV networks	Scalability challenges with huge UAV deployments	[15]
Comparative analysis and cost-benefit study	Researchers examined the cost efficiency of UAVs used in smart cities using IoT and AG technology	Elevated transaction charges and network traffic congestion	[16]
Computer vision	This work empowers machines to interpret and understand visual data and involves analyzing data collected from IoT devices	Variations in lighting, angles, and image quality cause inaccuracies in object recognition and scene understanding	[17]

(continued)

CHAPTER 10   SECURITY AND PRIVACY IN AG-IOT-ENABLED UNMANNED AERIAL VEHICLES FOR SMART CITIES

Table 10-1. (*continued*)

Methodology	Contribution	Limitation	Reference
Field testing and performance evaluation	This paper includes experiments that evaluate the performance of UAVs with IoT and blockchain in smart cities	Limited flexibility owing to blockchain complexity	[18]
Knowledge representation	This work describes practical applications and scenarios for AGI-IoT-enabled UAVs for smart cities	Representing complex knowledge in a machine-readable format is difficult, limiting depth of understanding	[19]
Prevent-Model	This paper presents the Prevent-Model, which includes AG-IoT-based UAVs designed to ameliorate smart cities	The model might be empirically validated in future research	[20]

Security Issues in AG-IoT-Enabled UAVs for Smart Cities

Security issues in AG-IoT-based UAVs are multifaceted, involving vulnerabilities at both the hardware and software levels. These drones remain vulnerable to various cyberattacks, including data violations, unauthorized access, and malware threats, due to their interconnected nature. The integration of AI and IoT in drones increases these risks by creating more entry points for potential attackers. Figure 10-1 provides an overview of different categories of cyberattacks that often target various components of UAV systems. Each category comprises of a number of cyber threats that pose risk to these systems.

CHAPTER 10 SECURITY AND PRIVACY IN AG-IOT-ENABLED UNMANNED AERIAL VEHICLES FOR SMART CITIES

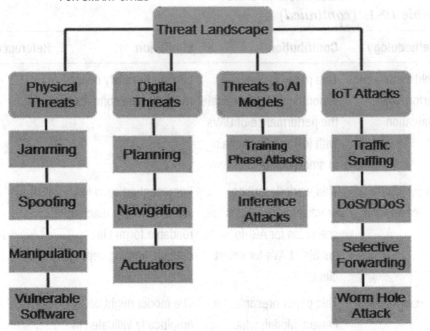

Figure 10-1. An overview of attacks on AG-IoT-driven UAVs

Threat landscape, also known as attack surface, refers to a combination of potentially vulnerable domains of a system, a part within a system, or its surrounding environment. These are points where adversaries may try to compromise or manipulate the drone's functions [21]. In the subsequent sections, each category of threats is explained in detail. First of all, physical threats, consisting of jamming, spoofing, manipulation, and vulnerabilities within software applications, are discussed. Then, detailed insights into potential digital threats, covering planning and navigation attacks, coupled with threats to actuators are explained. Furthermore, attacks targeting AI models, posing security risks to AG-IoT-based UAVs, which include training phase attacks and inference attacks, are also discussed. Lastly, this study examines other important contemporary threats, such as ransomware, backdoors, malware, and attacks on communication channels.

Physical Threats

Physical attacks consist of threats that directly manipulate the physical hardware or different parts of a system in a bid to disrupt their usual operation or gaining unauthorized access. These attacks comprise of various actions, such as tampering, destruction, removal, theft, or manipulation of integrated sensors. By targeting these components, threat actors want to disrupt or compromise the sensing capabilities of IoT components, potentially undermining their main functionality or security measures. Oruma et al. [22] have discussed how social and humanoid UAVs, which engage in interactions with humans, remain vulnerable to such attacks. On the contrary, the hardware-level Trojans, as outlined in the [23], possess the capacity to secretly manipulate integrated circuits within UAVs. This interference can lead to unexpected malfunctions, causing disruptions and potentially compromising the reliability and functionality of the attacked systems.

Jamming Attacks for UAVs

Jamming attacks are kind of attacks that involve the deliberate interference with wireless communication technologies, like Wi-Fi, cellular networks, and Bluetooth, using malicious ways. The fundamental aim of a jamming attack is to disrupt a network by interfering with communication, thereby causing damage to the device connectivity, disrupting crucial services, or potentially causing a network outage in a denial of service (DOS) scenario. Jamming attacks usually employ physical devices to overwhelm a network by sending powerful signals, thereby disturbing its normal functionalities. Figure 10-2 aptly demonstrates the way threat actor targets communication channels between source and destination, leading to jamming attack.

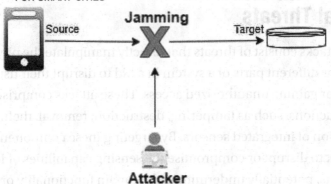

Figure 10-2. Explaining the concept of jamming attack

Yan et al. [24] investigated using lights whose intensity was too high, to disrupt camera sensors, which makes the images they capture much worse. This makes it hard for the camera to recognize objects in the pictures, showing how easily visual perception systems can be disrupted. Similarly, Shin et al. [25] talked about attacks that blind LiDAR sensors. These attacks involve powerful light matching LiDAR's wavelength, which makes the sensing detector unable to detect items facing the light. Additionally, Kar et al. [26] looked into how GPS signals can be interrupted by jamming devices. This can disrupt communication, making navigation and positioning inaccurate, and cause financial problems. To defend against jamming attacks, authors in [27] suggested using special filters and lenses to block out certain types of light.

Spoofing Threats for UAVs

Spoofing basically refers to the practice of tricking people by pretending to be someone or something they trust, like a friend or a device. The goal is to make them do actions that help the hacker but hurt them. Spoofing can happen through different ways of communicating and can range from simple to complicated methods. These attacks often involve tricking people through social engineering, which simply means that scam artists

CHAPTER 10 SECURITY AND PRIVACY IN AG-IOT-ENABLED UNMANNED AERIAL VEHICLES FOR SMART CITIES

exploit human vices such as greed, fear, or ignorance in cyberspace. Global positioning system (GPS) spoofing occurs when a GPS receiver is tricked into sending out fake signals that look like the real ones. This allows criminals pretend to be in one place when they're actually somewhere else. By doing this, they can make a car's GPS guide you to the wrong place, or even disrupt GPS signals for ships or airplanes. Numerous mobile applications depend on smartphone location data, making them susceptible to such spoofing attacks. In Figure 10-3, the scenario shows a GPS spoofing attack aimed at a UAV. In this scenario, the attacker sends a signal interference and/or jamming signal to mislead UAV about its position making it to follow a different GPS reading than the actual one. The second type of assault is shown in Figure 10-3, which describes spoofing, in which an attacker imitates another entity.

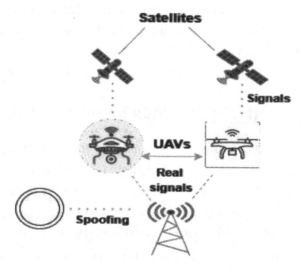

Figure 10-3. *Demonstrating spoofing attack on UAVs*

Zeng et al. [28] have studied a carefully planned GPS attempt aimed at controlling systems for navigating. Furthermore, recent research, such as [29], [30], and [31], suggests that attackers have learned to spoof LiDAR sensors using different methods, including advanced laser projection techniques. This technique involves directing misleading laser signals toward LiDAR sensors, tricking the sensors into detecting false environmental information. This could undermine the accuracy and dependability of LiDAR-based systems, particularly in crucial applications like autonomous vehicles and surveillance systems. Furthermore, spoofing LiDAR sensors through object manipulation methods involves taking advantage of weaknesses in the sensor's perception mechanisms [32]. One approach involves placing physical objects or reflective surfaces strategically within the LiDAR's range to create misleading readings. Attackers can manipulate the arrangement and characteristics of these objects, such as their size, shape, and reflectivity, to trick the LiDAR into detecting obstacles or features that either don't exist or don't accurately represent the surroundings. Such manipulation can mislead autonomous vehicles or surveillance systems that depend on LiDAR data, resulting in erroneous decision-making or compromised safety. To guard against spoofing attacks, several countermeasures have been proposed in various studies, including filtering [33], fusion [34], and randomization [35].

Manipulation Attacks for UAVs in Smart Cities

Manipulation attacks can target various aspects of UAV's operations. These attacks can include GPS spoofing, data tempering, and firmware attacks. Resultantly, safety, security, and reliability of UAV operations in smart cities are compromised. Several research studies have highlighted the susceptibility of inertial measurement units (IMUs) utilized in drone machines, to deliberate threats. These targeted threats are designed to control the accelerometer, magnetometer, and gyroscope sensors, potentially compromising the accuracy and reliability of the sensor data crucial for navigation, motion control, and orientation estimation in UAVs. Such vulnerabilities underscore the significance of putting string security measures in place to safeguard IMUs against malicious tampering and ensure the integrity of drone operations [36]. Further to that, these attacks are intended to interfere with the precise measurement of both linear and angular velocity, leading to the loss of control in UAV systems and, in certain instances, misleading object detection tasks [37].

Software Vulnerabilities in UAVs for Smart Cities

UAVs are vulnerable to software flaws existing in the applications in charge of interpreting and handling data collected by sensors. These weaknesses pose a risk to the overall functionality and security of drone operations, potentially leading to errors in data analysis, decision-making, and control processes. For instance, Bradski [38] introduces OpenCV, a comprehensive library for computer vision and image processing. Bradski's discussion emphasizes OpenCV's potential to facilitate the creation of applications for everything from basic image processing jobs to complex computer vision models. Importantly, Bradski points out specific vulnerabilities within the OpenCV framework that could be subject to exploitation. Most specifically, a buffer overrun based on heap, vulnerability (CVE-2018-5268) has been identified in OpenCV, which may be activated through interpreting a corrupted picture file while it is loaded [39]. Likewise, weaknesses may emerge due to inadequate or incomplete input parameter verification (CVE-2021-37650) within the TensorFlow. This vulnerability occurs when assuming that the analyzed dataset's records are all strings [40].

Digital Threats in UAVs

A digital threat refers to malicious activities targeting the software aspects of firmware, operating system, communication channels, and other software elements of a UAV, coupled with AI-driven entities. These threats aim to exploit vulnerabilities within these software elements to interfere with or manipulate various layers of UAV, particularly those responsible for navigation, planning, and control. Such attacks can compromise the integrity, reliability, and security of drone operations, potentially leading to unauthorized access, data manipulation, or disruption of critical functions.

CHAPTER 10   SECURITY AND PRIVACY IN AG-IOT-ENABLED UNMANNED AERIAL VEHICLES
             FOR SMART CITIES

In the following paragraphs, we present a detailed analysis regarding possible attacks on planning layer, navigation layer, and control layers such as actuators of an AG-IoT-driven UAV. Also, we discuss the security strategies adopted to cope up with the attacks.

Threat Planning in Smart Cities

Traffic signals and IoT devices play a crucial role in smart city's monitoring system. These components of smart city help in collecting data regarding traffic, waste, water supply, drainage, roads and bridges, energy, security breaches, and environmental factors. Further to that, the data is analyzed with the help of big data analytics, machine learning, and deep learning techniques, which gives out the patterns that are not seen and could be used for prediction. The predictions made using the data can prove useful to the administrator or city planner to deal with and repair the social ills infecting the residents. Similarly, with the increasing utilization of independent UAVs in domains such as delivery services, military tasks, supply chain, and industrial automation, it has become pivotal to concentrate on secure and private planning and navigation systems. It is due to the fact that systems are important for ensuring that drones function safely and efficiently in various environments. Any violation of security or privacy in navigation and planning systems may likely result in serious repercussions like safety risks or disrupted essential services. Consequently, it's important to safeguard these systems from digital attacks and vulnerabilities to maintain trust and confidence of people in UAVs. Hence, safeguarding these systems against potential digital risks and weaknesses is crucial for upholding public trust and confidence in the widespread deployment of autonomous drones.

Perception-targeted attacks concentrate on the sensors utilized by UAVs for perceiving and comprehending their surroundings in smart cities. These attacks seek to disrupt or manipulate these sensors, critical for providing the drone with environmental information. Manipulating these sensors enables attackers to feed the drone false or misleading data about its environment, which could cause errors in its decision-making process. This might hinder the drone's ability to navigate safely and interact efficiently with its surroundings. Conversely, navigation and planning attacks involve tampering with the navigation systems of AI-driven UAV systems by utilizing corrupted data from compromised sensors.

Attacks on Navigation Layer in Smart Cities

Possible interferences targeting navigation processes can be of a rather simple nature or contain combined and complex manipulations. One of the methods used in this case is when the attacker mitigates the UAV's sensory gadgets such as GPS, cameras, or LiDAR to provide incorrect information. This may help the drone to create a wrong perception about the environment, hence having issues such as collision or wrong judgment [41, 42]. Furthermore, it can also be noted that navigational systems could be highly exposed to data controlling threats [43]. These kinds of attacks involve changing of mappings or location details which the machines depend on, and this confuses the machines about the environment. For example, this might lead to a drone in a city dropping parcels at the wrong location or through poorly lit streets which are risky to people and cars. For the defensive purpose, Bianchin et al. [44] proposed a model which is designed to detect the fake sensor data and manipulated regulate inputs which are toward to design the path for the UAV. They mainly defined unknown threats targeting GNSS and RSSI sensors that are not easily identified by the conventional methods. Their detection method is based on control inputs that are standard. In the same way, Yang et al. [45] in their study aimed at exposed exclusive adversarial attacks on a navigation system concerning strategies of timing to impair it based on deep reinforcement learning (DRL). They jammed physical noise patterns into specific time frames to create disturbances. The attack involved three main scenarios: a pulsed zero-out attack using off-the-shelf hardware, adding noise to sensor fusion through a noisy system developed by a Gaussian filter, and generating adversarial patterns against trained deep Q-network (DQN) using the fast gradient sign method.

As far as attacks on planning layer are concerned, attacks during the planning phase can also be very disruptive. They focus on the algorithms that decide what actions UAVs should take based on the information it senses. Threat actors might make small alterations to the problem-solving of drone, making it ineffective for selection or even dangerous deeds [46, 47]. For instance, a UAV could be fooled into entering a prohibited zone or conducting unapproved monitoring. These efforts could occur through targeted signal interference, sensor distortions, or manipulation of data. Bhandarkar et al. [48] suggested using deep reinforcement learning (DRL) methods to execute Sybil attacks, where fake beacon signals are transmitted to disturb the path planning logic. In order to defend against planning attacks, Wang et al. [49] presented an attack utilizing infrared light to manipulate environmental perception outcomes and introduce errors in simultaneous localization and mapping (SLAM) for autonomous vehicles. To counter such attacks, the researchers devised a software-driven detection module. Likewise, Li et al. [50] showcased a threat leveraging time-sensitive multimodal information from various sensors to destabilize the SLAM.

Attacks on Control System of UAVs for Smart Cities

In the architecture of UAV, drones engage with the environment by employing their joint manipulators, such as arms or limbs for humanoid drones, wheels for ground-based mobile drones, or propellers for aerial and marine UAVs. These components allow drones to navigate, manipulate objects, and perform a variety of tasks within their environments. Actuators play a crucial role in these systems, functioning as mechanical devices that convert energy from sources such as hydraulics, pneumatics, electricity, or external signals into motion. This motion can then be precisely controlled based on specific parameters [51]. Commonly used actuators in UAVs include electric motors, which provide precise and adjustable control; hydraulic actuators, which offer high force and power; and pneumatic actuators, known for their simplicity and rapid response. Each type of actuator serves distinct purposes, depending on the requirements of the UAVs. Possible points of attack in relation to drone sensors encompass the hardware and firmware of the actuators themselves, communication protocols, and network components utilized by tele-operated drones. Attackers may block switches or hardware-level backdoors to get around control measures and obtain entry to the drone's control mechanisms [52]. Actuator firmware is another vulnerable area, primarily because of insecure communication methods, such as updating the firmware of a drone via open internet.

AI-Enabled Cyber Threats for UAVs

Training Phase Attacks

In an AI-based system, datasets are used to train the model so it can classify or recognize new data based on the training data. The model's precision and effectiveness heavily depend upon the training phase of the model development. In this context, training attacks involve intentionally altering, tainting, or modifying the training dataset. These attacks aim to manipulate the training procedure, compromising AI model's functionality and conduct. AI-based UAV systems are susceptible to various types of security risks targeting their training process or training data. For instance, model poisoning is one such attack, where the training data for AI models is compromised through label corruption. The VirusTotal poisoning incident [53] is a real-world example where altered samples uploaded to the platform tricked an ML model into making incorrect classifications. VirusTotal, a well-known tool for judging doubtful malware, shares results with the security societies. But the attackers uploaded fake samples, causing the platform to wrongly classify even harmless files as malicious.

There are several strategies to defend against model poisoning attacks. For example, if a model is poisoned, it might be feasible to use generative hostile networks to identify the inserted artifacts [54] and fix them through trimming and fine-tuning [55]. Similarly, another strategy is preventing model poisoning threats which includes fixing the compromised model or blocking the model from being poisoned [56].

Inference Attacks

An inference attack is a privacy violation in which an attacker tries to reveal private information regarding specific data points or the training set by questioning the AI model and examining its responses [57]. Inference attacks, such as model inversion, involve a threat actor retrieving sensitive information from training data and meticulously querying the inference API [58]. As a defensive strategy, a method called differential privacy (DP) determines a theoretical upper bound on how much an individual input can affect an output. This helps to resolve privacy concerns. In certain circumstances, this tactic can provide assurances that information won't be disclosed [59]. Learning with selective forgetting is another way to prevent data leaks. This involves using a synthetic signal specific to each class to neglect unwanted categories without relying on the actual information [60].

IoT Attacks for Smart Cities
Traffic Sniffing Attack

Traffic sniffing involves capturing and monitoring a network to observe and analyze data packets as they move between devices. In sniffing, unencrypted passwords, content of emails, voice and video calls, configuration information of networks, session tokens and login sessions, traffic patterns, browsing-related information are some of the activities that network administrators can perform. On the opposite hand, threat actors also take advantage of traffic sniffing and use it for malicious purposes. For instance, Address Resolution Protocol (ARP) spoofing is one of the sniffing techniques used to intercept data packets on a network, alter content, or stop the flow entirely. This assault is often used as a pretext for more sophisticated attacks such as session hijacking, denial of service, or man in the middle (MITM). Other examples of sniffing attack include IP spoofing and MAC spoofing [61]. Packet or traffic sniffing attack operates through four main stages: collection, conversion, analysis, and data theft [62]. In order to protect against sniffing attacks, several techniques, such as encryption, secure protocols, network monitoring, firewalls, access controls, and regular updates, are used.

DoS/DDoS Attacks in Smart Cities

The two most frequent cyber threats to IoT networks are thought to be denial of service (DoS) and distributed denial of service (DDoS) attacks. The network of physical items, or "things," that are outfitted with sensors, software, and other technological components is what makes up the Internet of Things (IoT). These objects are designed to connect and share data over the Internet with other systems and devices. Surprisingly, global Internet of Things (IoT) device count is expected to than double by 2020, from 15.1 billion in 2020 to over 29 billion by 2030 [63]. However, with such a rapid proliferation of IoT devices, there has been a notable surge in cyber assaults, with DoS and DDoS attacks emerging as prominent ones. Khader and Eleyan [64] suggest that in order to safeguard IoT networks in UAVs against DoS/DDoS, security solutions must be employed at all layers of IoT network which include software layer, middleware layer, and network layer. Similarly, Khan et al. [65] have proposed an intrusion detection system (IDS) based on attack probability using the Markov chain to detect DoS and DDoS attacks. This IDS employs an optimal threshold to balance false-positive and false-negative rates, utilizing Markov binomial and Markov chain distribution stochastic models. Figure 10-4 illustrates the underlying framework of DoS attack, making the server unavailable to the legitimate users.

CHAPTER 10 SECURITY AND PRIVACY IN AG-IOT-ENABLED UNMANNED AERIAL VEHICLES
FOR SMART CITIES

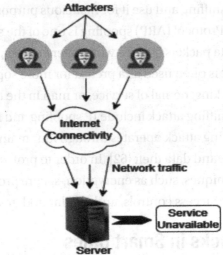

Figure 10-4. Demonstrating DoS attack on a network server

Selective Forwarding Attack in Smart Cities

A selective forwarding attack in smart cities refers to a type of cyberattack where an attacker manipulates the data transmitted between devices or systems in a smart city infrastructure. Consequently, it forwards only a selected portions of the data while discarding or modifying others. This threat can have significant consequences in smart cities, where data is relied upon to make critical decisions such as energy management, public safety, and waste management. Undermining the network's availability and integrity is the goal. The selective forwarding attack basically consists of three subclasses: black hole attack, gray hole attack, and on/off attack. One of the ways that could be used to combat the selective forwarding attacks is by using reliable communication channel through the use of authentication and encryption to prevent all forms of access and alteration of the forwarded messages. Another approach concerns the use of IDS/IPS to monitor the functioning of a network and prevent various attacks.

CHAPTER 10 SECURITY AND PRIVACY IN AG-IOT-ENABLED UNMANNED AERIAL VEHICLES FOR SMART CITIES

Wormhole Attack in Smart Cities

This kind of attack takes place at the network layer and targets a number of evil nodes. These nodes are greater in capability than normal nodes and help in establishing better communication channels over larger distances. The attack involves passing data from one compromised node to the other malicious node at the furthest end of the network through a tunnel. Therefore, other nodes in the WSN may be misled to believe that they are closer to some nodes than reality, which is disruptive in the routing algorithm. This attack can occur when the host is not compromised even when all the corresponding communication assurances are authentic and confidential [66]. Javed et al. [67] have implemented the PCC-RPL (Parental Change Control RPL) trust concept across communicating nodes in IoT networks to prevent unsolicited parent changes. This study aims to enhance the security of the RPL protocol by using a trust model based on the subjective logic framework to detect and mitigate wormhole attacks [67].

Privacy Concerns in AG-IoT-Enabled UAVs

In the current world which is teaming with technology, UAVs have tremendously transformed various sectors throughout the world. With a combination of AGI and IoT technologies integrated into it, drone systems have become the standard tools in various automation and service industries. The improvement of performance and invention, however, has created a problem of privacy leak. As these UAVs expand in their usage across the different sectors, it bristles with questions of privacy regarding individuals based on the information collection, processing, and data transfer capabilities of the UAVs. This section seeks to address some of the privacy considerations in UAVs by offering a discussion of the diverse nature of privacy threats according to different sectors. Thus, this section is

CHAPTER 10   SECURITY AND PRIVACY IN AG-IOT-ENABLED UNMANNED AERIAL VEHICLES
              FOR SMART CITIES

dedicated to explain how technological advancement in specific aspects [68] influences the evolution of legal, ethical, and regulating perspectives related to privacy protection [69].

In this section, moral concerns in the context of AG-IoT-enabled UAVs are explored in great detail.

Ethical Concerns Using UAVs for Smart Cities

The responsible use of data gathered by AG-IoT-enabled drone systems, maintaining accountability and transparency in their operations, and addressing potential social repercussions are ethical considerations in these systems [70]. It aims to understand and monitor the ethical ramifications and effects of UAV technology in our society, particularly with regard to talented and self-propelled drones. This covers a broad range of use cases where ethical issues play a critical role in directing the creation, implementation, and application of UAVs.

Healthcare UAVs for Smart Cities

Unmanned aerial vehicles (UAVs) have transformed the healthcare industry. AG-IoT-enabled UAVs could communicate with Internet of vehicles (IoV) to monitor health status such as epileptic seizure, heart failure, and panic attacks, of the drivers to prevent life-critical accidents [71, 72]. This can sometimes pose ethical issues concerning patient information, cost, and consent for drone procedures. For instance, there is an ethical question as to whether it is right to bring financial costs to patients or the healthcare system for the slim benefit that can be derived from the use of drones bearing in mind the high cost. Other ethical issues include patients' consent about the robot surgery and giving patients an option to choose whether they want the drone to be of assistance during the surgery. Another concern is privacy and data security. Examples include protection from hacking and safeguarding patient data gathered by medical UAVs.

CHAPTER 10   SECURITY AND PRIVACY IN AG-IOT-ENABLED UNMANNED AERIAL VEHICLES FOR SMART CITIES

Healthcare ethics deals with the development of AG-IoT-enabled UAVs that respects patient safety, privacy, and equal opportunities among others. Healthcare drones also have to adhere to certain privacy standards like the GDPR of Europe [73] and the HIPAA of the United States [74]. These rules help engineers in protecting the information of the users. These guidelines can help reduce the risks of data breaches, unauthorized access, and misuse of private health information for developers.

Autonomous UAVs for Smart Cities

A UAV can execute a designated task or mission autonomously, making all decisions onboard based on sensor observations and adapting to operational and environmental changes without human intervention. Autonomous UAVs are attracting growing attention because of their wide range of uses. They can deliver packages to customers, provide medical assistance at traffic accident scenes, track military targets, and support search and rescue missions, among many other applications. Self-propelled vehicles are another prime area of application of UAV systems [75]. The autonomy of cars has given rise to two major ethical questions: the moral dilemma and the question of who is safe from being killed by an autonomic car. An ethical decision-making scenario arises when an autonomous car is programmed to either deviate, hence running over one person or stuck to the current course leading to multiple deaths of pedestrians. This brings a number of ethical questions that involves how AGI should be designed to make such decisions and how the societal norms and values should have an impact on any autonomous car.

The controversy regarding on demand autonomous vehicle choice and priority between passengers and pedestrian safety is in question in the passenger safety argument. In an inevitable collision situation, the vehicle must perform trade-off such that it will minimize damage on the passengers at the expensive of maximum damages on the pedestrians, or maximize passengers' safety at the expensive of the pedestrians. What is the best course of action that raises this conflict as well as the way in which AI should be designed to handle situations such as this and how visible these decisions will need to be [76]?

These moral issues indicate why the development of AGI systems must occur within the context of the society's norms, rules, and ethos. This entails AVs ensuring that privacy regulations such as GDPR in the EU and the HIPAA Act in the United States are observed, especially concerning AG-IoT incorporative drone structures. These laws reduce the risk of someone "hacking" into UAVs and gaining unauthorized access to the users' sensitive information as well as create confidence in these technologies because data privacy and ethical use are incorporated into the design of autonomous vehicles [77].

Assistive UAVs in Smart Cities

Assistive unmanned aerial vehicles (UAVs) are designed to support and aid various tasks, often providing assistance in areas where human presence is limited, hazardous, or less efficient. For this reason, these UAVs can improve independence and quality of life, but their use raises a number of ethical issues such as dependency, risks associated with machines, costs, and opportunities. One of those concern is loneliness [78]. Issues arise, for instance, regarding how much authority people should exert over the choices and maneuvers of assistive drones. Another concern is determining who is responsible for accidents and blunders committed by these drones.

Compliance with such things as GDPR and HIPAA is crucial in addressing these issues, especially where UAVs empowered by AG-IoT are concerned. These laws ensure that ethical aspects are considered across the life cycle of assistive drones and provide additional safeguards of user information. While using these rules, developers may ensure greater levels of transparency and trust in these technologies and reducing the likelihood of data leakage and unauthorized access. Moreover, various moralities, laws, and culture may also be integrated into the development of assistive drones, which will go a long way in reducing the ills or ethical dilemmas and enhance the lives of the users.

UAVs for Construction and Infrastructure Inspection in Smart Cities

Small and remotely operated UAVs have served as an effective solution for as-built mapping, construction surveillance, and site inspections. They have been able to provide clear and high-quality images and real-time data that have, in turn, enhanced efficiencies and accuracy and has saved on time and manpower. Supervising construction right from the planning stage up to the stage the sites are actually built guarantees quality site work. This type of monitoring provides data and images in text, photo, video, and 3D mapping formats, making them easily sharable. This improves the level of transparency as well as timely corrective action that would address specific issues affecting the overall project calendar and cost projections. Drones are actually safer and more efficient than conventional methods for UAV inspections as they allow for the capture of high-quality images and instant data feedback. This technology subverts the need for manual inspection in restricted areas to perform inspection and reduces inspection time and increases the monitoring efficiency of the activities and maintenance. Thus, they have become integral part of various industries that seek improvement in operational performance and safety [79]. Figure 10-5 shows an example of how a UAV that is performing an inspection on a construction site will send data to the control station. This concerns the goals of the control station where end users evaluate the data and then come up with necessary recommendations.

CHAPTER 10 SECURITY AND PRIVACY IN AG-IOT-ENABLED UNMANNED AERIAL VEHICLES
FOR SMART CITIES

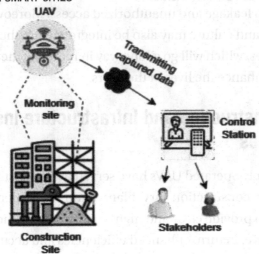

Figure 10-5. Depiction of UAV monitoring construction site

Manufacturing UAVs for Smart Cities

The following points form the important aspects that are under consideration in the ethics and that include issues related to replacement of workforce, human integration with UAVs and issues to dialogue with automatic machines in the manufacturing sector, and issues of responsible automation. This is even more essential now that there are great data privacy benchmarks such as GDPR and HIPAA that must be adhered to in the collection of the network's data. Combining privacy and security measures into the process of constructing and deploying UAV systems benefits these polices on two fronts: regarding the protection of its users' information and dealing with ethics issues [80]. These are important risk management checkpoints that will ensure people are not fired, encourage good practices with automation, and assist the automation developers to gain peoples' confidence and be more transparent. It should however be indicated that these systems are ethical when applied and helpful in a way that they enhance the workforce and the worth of the human capital consequently. Moreover, they need to be directed toward some definite social values, strict norms, and cultural norms.

Legal Concerns

Legal issues with AG-IoT-equipped UAVs include protection against data violation, illegitimate data access, and unauthorized usage of personal data. Due to these revelations, developers must implement strict privacy and security measures to mitigate risks and ensure that legal specifications are fulfilled in the designing of these systems. As effectively demonstrated on AG-IoT, the introduction of such innovative drone technologies means that the accountability for actions and the manner in which data and information are processed and manage call for specific set protocols with regard to laws that may dictate the course of legal action [81].

Thus, when considering the use of these surgical UAVs especially in the war-torn regions for treatments such as autonomous drone surgical operation or tele-surgery using drone assistants, the role of principles of accountability and responsibility comes into question. Accidents are likely to happen, more so in times of surgery, and if a drone actuator or the drone-based surgical system attacks a patient or causes harm to him/her due to a slight mishap, then legal consequences could follow. This gives rise to important issues in terms of liabilities, for instance, is it the turn of the creator of the system, the programmer, the autonomous machines, or the worker supervising the process ought to bear some or all the blames [82].

Secondly, the privacy and the subsequent processing of personal data are crucial legal issues of AI and UAVs. As a result, several legislations like the General Data Protection Regulation in Europe since 2016 and the CCPA in California, United States, in 2020 have been postulated to safeguard individual data. Offering a nod to embracing transparency and public trust in these forward-thinking UAV systems, these regulations aim to ensure robust privacy protections and data security standards, thereby imposing compliance by the makers in a bid to prevent data tampering and illicit access.

CHAPTER 10   SECURITY AND PRIVACY IN AG-IOT-ENABLED UNMANNED AERIAL VEHICLES
FOR SMART CITIES

Future Research Directions for Secure AG-IoT-Based UAVs

Since there has been a consistent increase in UAVs being employed in various industries, researchers need to focus on developing trustworthy AG-IoT-driven UAVs that can confront cyber threats, adhere to the ethical principles, and ensure privacy. First of all, spreading awareness and education regarding secure IoT-enabled drones is the most important area which can be explored further. Some researchers [83] suggest that there should be UAV-related curricula integrated in educational institutions so that students become capable of developing secure drone applications and systems. Moreover, regularly conducting practical hacking workshops or webinars can help spread awareness about emerging threats targeting UAV systems. In this regard, the authors [84] have reviewed existing hacking drills, evaluating their effectiveness, and problems based on a collection of teaching rules.

In the "Privacy Concerns in AG-IoT-Enabled UAVs" section, we discussed different types of cyberattacks that target AG-IoT-enabled UAVs. Adversaries may target embedded sensors that perceive the environment, either by physically damaging them or by jamming and spoofing signals. Future technologies for drones ought to be aimed at countering these threats. For example, strong access controls and secure perimeters can protect against physical attacks, while lightweight detection algorithms can address jamming and spoofing [85]. To protect attack-related actuators, study is focusing on developing systems for finding anomalies. These systems can monitor and identify anomalies during drone manipulation tasks [86].

CHAPTER 10 SECURITY AND PRIVACY IN AG-IOT-ENABLED UNMANNED AERIAL VEHICLES FOR SMART CITIES

Conclusion

In today's world, UAV systems have become increasingly integrated into our lives due to their usability and efficiency. On the other hand, these drones have integrated a number of components such as IoT devices, communication channels, data, algorithms, physical systems, and AG. As a result, each component is vulnerable to different types of security risks, leading to significant financial, social, and legal consequences. In this piece of work, we highlight in great detail the security risks consisting of four domains: physical threats, digital threats, threats to AI models, and IoT attacks. We specifically focused on the techniques used by threat actors to target and influence the performance of AI models. Furthermore, we discuss the privacy concerns that emerge when IoT and AG technologies are integrated into UAV systems. This integration poses significant risks to user data and privacy, necessitating robust legal frameworks to ensure protection. In this context, we highlight essential legal frameworks such as the GDPR and CCPA, which establish rigorous guidelines for protecting user data and ensuring privacy. We analyze how these regulations specifically apply to UAV systems, assessing their effectiveness in tackling privacy concerns and identifying potential areas for enhancement.

References

[1] Nourmohammadi, Amin, Mohammad Jafari, and Thorsten O. Zander. "A Survey on Unmanned Aerial Vehicle Remote Control Using Brain–Computer Interface." *IEEE Transactions on Human-Machine Systems* 48, no. 4 (2018): 337-348.

[2] Waymo, "The World's Most Experienced Driver". Available at: https://waymo.com/.

[3] F. Sherwani, M. M. Asad, and B. S. K. K. Ibrahim, "Collaborative Robots and Industrial Revolution 4.0 (IR 4.0)," in *Proc. International Conference on Emerging Trends Smart Technology (ICETST)*, Mar. 2020, pp. 1–5.

[4] S. Leonard, K. L. Wu, Y. Kim, A. Krieger, and P. C. W. Kim, "Smart Tissue Anastomosis Robot (STAR): A Vision-Guided Robotics System for Laparoscopic Suturing," *IEEE Transactions on Biomedical Engineering*, vol. 61, no. 4, pp. 1305–1317, Apr. 2014.

[5] Ashton K, "That Internet of Things Thing. RFID J (2009) 22:97–11

[6] L. Atzori, A. Iera, G. Morabito, "The Internet of Things: A survey", Computer. Networks 54 (15) (2010) 2787–2805.

[7] D. Miorandi, S. Sicari, F. De Pellegrini, I. Chlamtac, "Survey internet of things: vision, applications and research challenges", Ad Hoc Netw. 10 (7) (2012) 1497–1516.

[8] Hasan, M.: "State of IoT 2022: number of connected IoT devices growing 18% to 14.4 billion globally" (2022). Available: https://iot-analytics.com/number-connected-iot-devices/ Accessed: 15 April 2024.

[9] IDC: "Top 6 IoT use cases worldwide by share of spending" (2019). Available: https://www.idc.com/promo/customerinsights?modal=tile-IoT&modal-ytb=dzR6h_U4TrU&modal-ytb-api=1. Accessed: 15 April 2024.

[10] de las Morenas, Javier, et al. "Security Experiences in IoT based applications for Building and Factory Automation." *2020 IEEE International Conference on Industrial Technology (ICIT)*. IEEE, 2020.

[11] Khan, Inam Ullah, Asrin Abdollahi, Abdul Jamil, Bisma Baig, Muhammad Adnan Aziz, and Fazal Subhan. "A Novel Design of FANET Routing Protocol Aided 5G Communication Using IoT." *J. Mobile Multimedia* 18, no. 5 (2022): 1333-1354.

[12] Khan, Inam Ullah, Ijaz Mansoor Qureshi, Muhammad Adnan Aziz, Tanweer Ahmad Cheema, and Syed Bilal Hussain Shah. "Smart IoT Control-Based Nature Inspired Energy Efficient Routing Protocol for Flying Ad Hoc Network (FANET)." *IEEE Access* 8 (2020): 56371-56378.

[13] Arshi, Oroos, Akanksha Rai, Gauri Gupta, Jitendra Kumar Pandey, and Surajit Mondal. "IoT in energy: a comprehensive review of technologies, applications, and future directions." *Peer-to-Peer Networking and Applications* (2024): 1-40.

[14] N. Cherif, W. Jaafar, H. Yanikomeroglu and A. Yongacoglu, "3D Aerial Highway: The Key Enabler of the Retail Industry Transformation," in *IEEE Communications Magazine*, vol. 59, no. 9, pp. 65-71, September 2021

[15] Euchi, Jalel. "Do drones have a realistic place in a pandemic fight for delivering medical supplies in healthcare systems problems?." *Chinese Journal of Aeronautics* 34, no. 2 (2021): 182-190.

[16] Alam, Mehtab, Akshay Chamoli, and Nabeela Hasan. "Smart Cities and the Internet of Drones." In *The Internet of Drones*, pp. 295-322. Apple Academic Press, 2022.

[17] Singh, Prabhdeep & Singh, Kiran Deep, "Security and Privacy in Fog/Cloud-based IoT Systems for AI and Robotics. EAI Endorsed Transactions on AI and Robotics", 2023, 2. 10.4108/airo.3616.

[18] Pamnani, A., and V. Parvathi. "Innovations in contemporary marketing through artificial intelligence and robotic drones." *Innovations* 1, no. 1 (2021): 34-43.

[19] Gyrard, A., Tabeau, K., Fiorini, L. et al. "Knowledge Engineering Framework for IoT Robotics Applied to Smart Healthcare and Emotional Well-Being", International Journal of Social Robotics 15, 445–472 (2023), Available: https://doi.org/10.1007/s12369-021-00821-6.

[20] Christina Glasauer, "The PREVENT-Model: Human and Organizational Factors Fostering Engineering of Safe and Secure Robotic Systems", Journal of Systems and Software, Volume 195,2023, Available: https://doi.org/10.1016/j.jss.2022.111548.

[21] R. Ross, V. Pillitteri, K. Dempsey, M. Riddle, and G. Guissanie, *Protecting Controlled Unclassified Information in Nonfederal Systems and Organizations*. Gaithersburg, MD, USA: Nat. Inst. Standards Technol., 2019.

[22] S. O. Oruma, M. Sánchez-Gordón, R. Colomo-Palacios, V. Gkioulos, and J. K. Hansen, "A Systematic Review on Social Robots in Public Spaces: Threat Landscape and Attack Surface," *Computers*, vol. 11, no. 12, p. 181, Dec. 2022.

[23] S. Bhunia, M. S. Hsiao, M. Banga, and S. Narasimhan, "Hardware Trojan Attacks: Threat Analysis and Countermeasures," *Proc. IEEE*, vol. 102, no. 8, pp. 1229–1247, Aug. 2014.

[24] C. Yan, W. Xu, and J. Liu, "Can You Trust Autonomous Vehicles: Contactless Attacks against Sensors of Self-driving Vehicle," *Defcon*, vol. 24, no. 8, p. 109, 2016.

[25] H. Shin, D. Kim, Y. Kwon, and Y. Kim, "Illusion and Dazzle: Adversarial Optical Channel Exploits Against Lidars for Automotive Applications," in Proc. 19th Int. Conf. Cryptograph. Hardw. Embedded Syst., Taipei, Taiwan. Cham, Switzerland: Springer, 2017, pp. 445–467.

[26] G. Kar, H. Mustafa, Y. Wang, Y. Chen, W. Xu, M. Gruteser, and T. Vu, "Detection of on-road vehicles emanating GPS interference," in *Proc. ACM SIGSAC Conf. Comput. Commun. Secur.*, 2014, pp. 621–632.

[27] G. Kar, H. Mustafa, Y. Wang, Y. Chen, W. Xu, M. Gruteser, and T. Vu, "Detection of On-Road Vehicles Emanating GPS Interference," in *Proc. ACM SIGSAC Conf. Comput. Commun. Secur.*, 2014, pp. 621–632.

[28] K. C. Zeng, S. Liu, Y. Shu, D. Wang, H. Li, Y. Dou, G. Wang, and Y. Yang, "All Your GPS Are Belong to Us: Towards Stealthy Manipulation of Road Navigation Systems," in *Proc. 27th USENIX Secur. Symp.*, 2018, pp. 1527–1544.

[29] H. Shin, D. Kim, Y. Kwon, and Y. Kim, "Illusion and Dazzle: Adversarial Optical Channel Exploits Against Lidars for Automotive Applications," in *Proc. 19th Int. Conf. Cryptograph. Hardw. Embedded Syst.*, Taipei, Taiwan. Cham, Switzerland: Springer, 2017, pp. 445–467.

[30] Y. Cao, C. Xiao, B. Cyr, Y. Zhou, W. Park, S. Rampazzi, Q. A. Chen, K. Fu, and Z. M. Mao, "Adversarial Sensor Attack on LiDAR-based Perception in Autonomous Driving," in *Proc. ACM SIGSAC Conf. Comput. Commun. Secur.*, Nov. 2019, pp. 2267–2281.

[31] Z. Jin, X. Ji, Y. Cheng, B. Yang, C. Yan, and W. Xu, "PLA-LiDAR: Physical Laser Attacks against LiDAR-based 3D Object Detection in Autonomous Vehicle," in *Proc. IEEE Symp. Secur. Privacy (SP)*, May 2023, pp. 710–727.

[32] Y. Zhu, C. Miao, T. Zheng, F. Hajiaghajani, L. Su, and C. Qiao, "Can We Use Arbitrary Objects to Attack LiDAR Perception in Autonomous Driving?" in *Proc. ACM SIGSAC Conf. Comput. Commun. Secur.*, Nov. 2021, pp. 1945–1960.

[33] D. F. Kune, J. Backes, S. S. Clark, D. Kramer, M. Reynolds, K. Fu, Y. Kim, and W. Xu, "Ghost Talk: Mitigating EMI Signal Injection Attacks against Analog Sensors," in *Proc. IEEE Symp. Secur. Privacy*, May 2013, pp. 145–159.

[34] W. Xu, C. Yan, W. Jia, X. Ji, and J. Liu, "Analyzing and Enhancing the Security of Ultrasonic Sensors for Autonomous Vehicles," *IEEE Internet Things J.*, vol. 5, no. 6, pp. 5015–5029, Dec. 2018

[35] H. Shin, D. Kim, Y. Kwon, and Y. Kim, "Illusion and Dazzle: Adversarial Optical Channel Exploits Against Lidars for Automotive Applications," in *Proc. 19th Int. Conf. Cryptograph. Hardw. Embedded Syst.*, Taipei, Taiwan. Cham, Switzerland: Springer, 2017, pp. 445–467

[36] S. Nashimoto, D. Suzuki, T. Sugawara, and K. Sakiyama, "Sensor CON-Fusion: Defeating Kalman Filter in Signal Injection Attack," in *Proc. Asia Conf. Comput. Commun. Secur.*, May 2018, pp. 511–524.

[37] X. Ji, Y. Cheng, Y. Zhang, K. Wang, C. Yan, W. Xu, and K. Fu, "Poltergeist: Acoustic Adversarial Machine Learning against Cameras and Computer Vision," in *Proc. IEEE Symp. Secur. Privacy (SP)*, May 2021, pp. 160–175.

[38] G. Bradski, "The OpenCV library," *Dr. Dobb's J., Softw. Tools Prof. Programmer*, vol. 25, no. 11, pp. 120–123, Nov. 2000.

[39] CVE-2018-5268 Detail. [Online]. Available: https://nvd.nist.gov/vuln/detail/CVE-2018-5268

[40] *CVE-2021-37650 Detail.* [Online]. Available: https://nvd.nist.gov/vuln/detail/CVE-2021-37650

[41] Y. Xu, X. Han, G. Deng, J. Li, Y. Liu, and T. Zhang, "SoK: Rethinking Sensor Spoofing Attacks against Robotic Vehicles from a Systematic View," 2022, *arXiv:2205.04662.*

[42] J. H. Ryu, M. Irfan, and A. Reyaz, "A Review on Sensor Network Issues and Robotics," *J. Sensors*, vol. 2015, pp. 1–14, Aug. 2015.

[43] A. Bhardwaj, V. Avasthi, and S. Goundar, "Cyber security attacks on robotic platforms," *Netw. Secur.*, vol. 2019, no. 10, pp. 13–19, Oct. 2019.

[44] G. Bianchin, Y.-C. Liu, and F. Pasqualetti, "Secure Navigation of Robots in Adversarial Environments," *IEEE Control Syst. Lett.*, vol. 4, no. 1, pp. 1–6, Jan. 2020.

[45] C. H. Yang, J. Qi, P.-Y. Chen, Y. Ouyang, I. D. Hung, C.-H. Lee, and X. Ma, "Enhanced Adversarial Strategically-Timed Attacks against Deep Reinforcement Learning," in *Proc. IEEE Int. Conf. Acoust., Speech Signal Process. (ICASSP)*, May 2020, pp. 3407–3411.

[46] X. Peng and D. Xu, "Intelligent Online Path Planning for UAVs in Adversarial Environments," *Int. J. Adv. Robotic Syst.*, vol. 9, no. 1, p. 3, Mar. 2012.

[47] X. Wang and M. C. Gursoy, "Resilient UAV Path Planning for Data Collection Under Adversarial Attacks," in *Proc. IEEE Int. Conf. Commun.*, May 2022, pp. 625–630.

[48] A. B. Bhandarkar, S. K. Jayaweera, and S. A. Lane, "Adversarial Sybil attacks against Deep RL based drone trajectory planning," in *Proc. IEEE Mil. Commun. Conf. (MILCOM)*, Nov. 2022, pp. 1–6.

[49] W. Wang, Y. Yao, X. Liu, X. Li, P. Hao, and T. Zhu, "I Can See the Light: Attacks on Autonomous Vehicles Using Invisible Lights," in *Proc. ACM SIGSAC Conf. Comput. Commun. Secur.*, Nov. 2021, pp. 1930–1944.

[50] A. Li, J. Wang, and N. Zhang, "Chronos: Timing interference as a new attack vector on autonomous cyber-physical systems," in *Proc. ACM SIGSAC Conf. Comput. Commun. Secur.*, Nov. 2021, pp. 2426–2428.

[51] D. Herath and D. St-Onge, *Foundations of Robotics: A Multidisciplinary Approach With Python and ROS*. Berlin, Germany: Springer, 2022.

[52] X. Wang, T. Mal-Sarkar, A. Krishna, S. Narasimhan, and S. Bhunia, "Software exploitable hardware Trojans in embedded processor," in *Proc. IEEE Int. Symp. Defect Fault Tolerance VLSI Nanotechnol. Syst. (DFT)*, Oct. 2012, pp. 55–58.

[53] *How Data Poisoning Attacks Can Corrupt Machine Learning Models*. Accessed: May. 20, 2024. [Online]. Available: https://www.linkedin.com/pulse/how-data-poisoning-attacks-can-corrupt-machinelearning-misra

[54] L. Zhu, R. Ning, C. Wang, C. Xin, and H. Wu, "GangSweep: Sweep Out Neural Backdoors by GAN," in *Proc. 28th ACM Int. Conf. Multimedia*, Oct. 2020, pp. 3173–3181.

[55] K. Liu, B. Dolan-Gavitt, and S. Garg, "Fine-Pruning: Defending Against Backdooring Attacks on Deep Neural Networks," in *Proc. RAID*, 2018, pp. 273–294.

[56] M. Goldblum, D. Tsipras, C. Xie, X. Chen, A. Schwarzschild, D. Song, A. Madry, B. Li, and T. Goldstein, "Dataset Security for Machine Learning: Data Poisoning, Backdoor Attacks, and Defenses," 2020, *arXiv:2012.10544*.

[57] R. Shokri, M. Stronati, C. Song, and V. Shmatikov, "Membership Inference Attacks Against Machine Learning Models," in *Proc. IEEE Symp. Secur. Privacy (SP)*, May 2017, pp. 3–18.

[58] Y. Kaya and T. Dumitras, "When Does Data Augmentation Help With Membership Inference Attacks?" in *Proc. ICML*, 2021, pp. 5345–5355.

[59] S. Zanella-Béguelin, L. Wutschitz, S. Tople, V. Rühle, A. Paverd, O. Ohrimenko, B. Köpf, and M. Brockschmidt, "Analyzing Information Leakage of Updates to Natural Language Models," in *Proc. ACM SIGSAC Conf. Comput. Commun. Secur.*, Oct. 2020, pp. 1–13.

[60] T. Shibata, G. Irie, D. Ikami, and Y. Mitsuzumi, "Learning with Selective Forgetting," in *Proc. 30th Int. Joint Conf. Artif. Intell.*, Aug. 2021, pp. 989–996

[61] Tuli, Ruchi. (2020). Packet Sniffing and Sniffing Detection. *International Journal of Innovations in Engineering and Technology, 16*, 22. DOI:10.21172/ijiet.161.04.

[62] Siswanto, Apri & Syukur, Abdul & Abdul Kadir, Evizal & Suratin,. (2019). *Network Traffic Monitoring and Analysis Using Packet Sniffer.*10.1109/COMMNET.2019.8742369.

[63] Statista, "IoT Connected Devices Worldwide" Available: https://www.statista.com/statistics/1183457/iot-connected-devices-worldwide/, Accessed: May 2024

[64] Khader, R. and Eleyan, D., 2021. Survey of DoS/DDoS attacks in IoT. *Sustainable Engineering and Innovation, 3*(1), pp. 23–28.

[65] Khan, Inam Ullah, Asrin Abdollahi, Ryan Alturki, Mohammad Dahman Alshehri, Mohammed Abdulaziz Ikram, Hasan J. Alyamani, and Shahzad Khan. "Intelligent Detection System Enabled Attack Probability Using Markov Chain in Aerial Networks." *Wireless Communications and Mobile Computing* 2021 (2021): 1-9.

[66] Preeti, Yogesh Chaba, Yudhvir Singh, —Review of Detection and Prevention of DDOS attack in MANET‖, Proc. National Conference on Challenges &Opportunities in Information Technology (COIT – 2008), India, pp. 56-59

[67] Javed, Sarmad, Ahthasham Sajid, Tayybah Kiren, Inam Ullah Khan, Christine Dewi, Francesco Cauteruccio, and Henoch Juli Christanto. "A Subjective Logical Framework- Based Trust Model for Wormhole Attack Detection and Mitigation in Low-Power and Lossy (RPL) IoT-Networks." *Information* 14, no. 9 (2023): 478.

[68] Pinto, G.P.; Donta, P.K.; Dustdar, S.; Prazeres, C. A Systematic Review on Privacy- Aware IoT Personal Data Stores. Sensors 2024, 24, 2197. https://doi.org/10.3390/s24072197

[69] Eleshin, F., Iradukunda, P., Ruberamitwe, D. and Ishimwe, E., *Your Robot Might Be Inadvertently or Deliberately Spying on You: A Critical Analysis of Privacy Practices in the Robotics Industry*. DOI: 10.5220/0012422200003648

[70] Spyros G Tzafestas. Roboethics.A navigating overview. Heilberg: Springer, 2016.

[71] Min Chen, Yuanwen Tian, Giancarlo Fortino, Jing Zhang, Iztok Humar, Cognitive Internet of Vehicles, Comput. Commun. 120 (2018) 58–70

[72] Y. Zhang, M. Chen, N. Guizani, D. Wu, V.C.M. Leung, SOVCAN: Safety-Oriented Vehicular Controller Area Network, IEEE Commun. Mag. 55 (8) (2017) 94–99

[73] General Data Protection Regulation (GDPR)- Legal Text. Available at: https://gdpr-info.eu/

[74] ACCOUNTABILITY ACT HEALTH INSURANCE PORTABILITY AND ACCOUNTABILITY ACT OF 1996. Public law, 104:191, 1996.

[75] Hong Wang, Amir Khajepour, Dongpu Cao, and Teng Liu. "Ethical Decision Making in Autonomous Vehicles: Challenges and Research Progress." IEEE Intelligent Transportation Systems Magazine, 14(1):6–17, 2020.

[76] Patrick Lin. The ethics of saving lives with autonomous cars are far murkier than you think. Wired Opinion, 30, 2013.

[77] Erin A McDaniel. "Robot Wars: Legal and Ethical Dilemmas of Using Unmanned Robotic Systems in 21st Century Warfare and Beyond". PhD thesis, Fort Leavenworth, KS: US Army Command and General Staff College, 2008.

[78] Laurel Riek and Don Howard. A Code of Ethics for the Human-Robot Interaction Profession. Proceedings of we robot, 2014.

[79] Shakhatreh, H.; Sawalmeh, A.H.; Al-Fuqaha, A.; Dou, Z.; Almaita, E.; Khalil, I.; Othman, N.S.; Khreishah, A.; Guizani, M. Unmanned Aerial Vehicles (UAVs): A Survey on Civil Applications and Key Research Challenges. *IEEE Access* **2019**, *7*, 48572–48634.

[80] Bartneck, Christoph & Lütge, Christoph & Wagner, Alan & Welsh, Sean. (2021). Privacy Issues of AI. 10.1007/978-3-030-51110-4_8.

[81] D. Papp, Z. Ma, and L. Buttyan, "Embedded systems security: Threats, vulnerabilities, and attack taxonomy," in Privacy, Security and Trust (PST), 2015 13th Annual Conference on. IEEE, 2015, pp. 145–152.

[82] Shane O'Sullivan, Nathalie Nevejans, Colin Allen, Andrew Blyth, Simon Leonard, Ugo Pagallo, Katharina Holzinger, Andreas Holzinger, Mohammed Imran Sajid, and Hutan Ashrafian.Legal, regulatory, and ethical frameworks for development of standards in artificial intelligence (AI) and autonomous robotic surgery.The international journal of medical robotics and computer assisted surgery, 15(1):e1968, 2019.

[83] S. Mittal and J. Chen, "AI Security Threats Against Pervasive Robotic Systems: A Course for Next Generation Cybersecurity Workforce," 2023, *arXiv:2302.07953*.

[84] D. Votipka, E. Zhang, and M. L. Mazurek, "HackEd: A Pedagogical Analysis of Online Vulnerability Discovery Exercises," in *Proc. IEEE Symp. Secur. Privacy (SP)*, May 2021, pp. 1268–1285.

[85] S. Rivera, S. Lagraa, A. K. Iannillo, and R. State, "Auto-Encoding Robot State Against Sensor Spoofing Attacks," in *Proc. IEEE Int. Symp. Softw. Rel. Eng. Workshops (ISSREW)*, Oct. 2019, pp. 252–257.

[86] D. Park, Z. Erickson, T. Bhattacharjee, and C. C. Kemp, "Multimodal execution monitoring for anomaly detection during robot manipulation," in *Proc. IEEE Int. Conf. Robot. Autom. (ICRA)*, May 2016, pp. 407–414.

CHAPTER 11

The Essential Role of Cybersecurity in UAV Swarm Operations

Fida Muhammad Khan[1], Taj Rahman[1], Asim Zeb[2], Inam Ullah[3], Inayat Khan[4], Muhammad Shoaib Akhter[5]

[1]Department of Computer Science, Qurtuba University of Science & Information Technology, Peshawar, Pakistan.
[2]Department of Computer Science, Abbottabad University of Science and Technology, Abbottabad 22010, Pakistan
[3]Department of Computer Engineering, Gachon University, Seongnam 13120, Republic of Korea
[4]Assistant Professor of Computer Science, Department of Computer Science, UET Mardan, Pakistan
[5]Department of Electrical Engineering, University of Science & Technology, Bannu, Pakistan
[1]fida5073@gmail.com, [1]tajuom@gmail.com, [2]asimzeb1@gmail.com, [3]inam.fragrance@gmail.com, [4]inayatkhan@uetmardan.edu.pk, [5]engrshoaibkhanmrt@gmail.com

Corresponding Author: Asim Zeb (asimzeb1@gmail.com)

CHAPTER 11   THE ESSENTIAL ROLE OF CYBERSECURITY IN UAV SWARM OPERATIONS

Overview

UAV swarm operations and cybersecurity measures are crucial to protect them from cyber threats due to their high reliance on wireless communication and advanced software systems. Ensuring security in the realm of cybersecurity is crucial for maintaining the rapid, secure, and effective operation of UAV swarms that encompass a wide range of mission capabilities, such as military operations, short-term surveillance, and search and rescue environmental monitoring [1].

Cybersecurity in the context of UAV swarm operations is made to secure unmanned aerial vehicles (UAVs) from falling victim to or suffering as a result of cyber threats. These kinds of operations are necessary because UAV swarms may be used for some of the toughest and, in space or with people around them at least, safety-critical scenarios from military engagements to disaster response and monitoring natural resources. It is therefore essential to keep these systems away from cyberattacks. The utilities of UAV applications have drastically transformed, starting from individual to swarming systems. Swarming systems naturally have synergistic benefits compared to single UAVs, exploiting their biocomplexity for increased adaptability as well as self-governing and control behaviors. Besides that, they are prepared for shortcoming resilience, scalability, and difficult circumstances which makes them stronger in unanticipated situations. It also signifies the movement toward developing advanced and robust UAV operations to cater to intricate and transitioning environments [2, 3].

UAVs are useful in multiple variety of scenarios that could potentially pave the way for more widespread use cases and multiple robots. An example might be a multi-UAV swarm fulfilling tasks like mapping inaccessible sites and visual inspection of industrial plants (e.g., area surveillance), etc. Employing a mix of UAVs in coordination can make these operations more efficient and effective, especially in military, disaster response, and environmental monitoring missions that require

CHAPTER 11 THE ESSENTIAL ROLE OF CYBERSECURITY IN UAV SWARM OPERATIONS

the highest levels of precision, accuracy, as well as security. In military applications, UAV swarms can be used to carry out missions such as reconnaissance and surveillance or tactical interventions. These swarms need to operate in a distributed manner so that the data can be shared instantaneously with real speed and actions can be coordinated for better mission successes [4]. But at the same time, these interconnections also increase their exposure to cyber threats; as a result, robust cybersecurity measures are necessary to protect for instance interconnected UAVs from possible attacks. UAV swarms are also highly advantageous for disaster response operations. These systems are capable of making fast damage assessments, identifying trapped people, and delivering supplies to regions that have become so seriously challenging space-wise because of the disaster. One of its key features is that it can scale as the fault-tolerant design allows a failed UAV simply to continue from where another left off, enabling missions in such adverse environments (actuator failure or single-point drone loss, etc.). In this respect, cybersecurity measures are required to ensure the operational integrity of the UAVs performing their mission in the way they were designed.

UAV swarms are nothing less when it comes to environmental monitoring. Such systems are capable of large-scale coverage and continuous data acquisition across diverse parameters as well as offering real-time insights that are essential for conservation, pollution monitoring, resource management, etc. This increased operational capacity of the UAV swarms, backed by their agile and controlled interoperability, enhances performances over a singular UAS in tasks that would be difficult or impossible with one sole UAV. The last few years have seen a significant maturity of UAV swarm technology which offers promise for building fast, controllable, and interoperable systems designed to carry out complex tasks. Today, these systems are a part of many operational scenarios and are thus both efficient as well as effective. Nevertheless, being able to coordinate

CHAPTER 11 THE ESSENTIAL ROLE OF CYBERSECURITY IN UAV SWARM OPERATIONS

large groups of UAVs also makes them prime runners for advanced cyber threats. This, therefore, requires continual cyber advancements to prevent these kinds of emerging threats from attacking your networks.

A key issue in the security of UAV swarms is that both individual drones and cooperative ones must be protected as well. The crowd UAVs/UAS manifestation is a potential surface that can be used by malicious cyber activities, and it just takes one particular to make your entire group of drones vulnerable. As such, cybersecurity measures should be holistic including protection for individual UAVs and their communication links as well as the overall swarm network. This means using high-quality encryption technology, secure channels for communication, and a durable network architecture that can resist cyberattacks while maintaining its effectiveness. Cyber threats against UAV swarms are of different and ever-evolving types. Some of these threats include hacking, which refers to unauthorized infiltration leading the UAVs being controlled by nonlaw-abiding citizens; denial-of-service (DoS) attacks on communication systems used for operating these drones that are designed to overload their data link preventing them from working properly; and finally GPS spoofing where the ability deliver legit signals pretending it's an authorized source signal successfully misleading a drone into either landing in a predefined place or course. Any one of these threats can result in considerable loss, including the loss of UAVs themselves, mission failure, or even physical injury to personnel and resources. Thus, protection from such threats needs to be designed and implemented to reach the maximum level of effectiveness, including methods for identifying these threats in the process [5].

To advance the protection of swarms, scientists are beginning to focus on artificial intelligence and machine learning. Also, AI-based anomalous detection systems can monitor the UAV's actions and interactions which will aid in the identification of changes that may signify cyberattacks. To this end, many algorithms of machine learning can be utilized since they can be trained to process big amounts of information and make the types of predictions regarding what type of risks are to be expected which makes swarms less vulnerable to cyber threats. AI also makes it possible

CHAPTER 11 THE ESSENTIAL ROLE OF CYBERSECURITY IN UAV SWARM OPERATIONS

for decision-making in the swarm to be made independently because it enables a given UAV to recognize and deal with cases of cyber threats independently of help from humans. One can also identify opportunities for blockchain technology that will, in turn, enhance the security levels of UAV swarming systems. As far as other than tamper-proof flaws, it is possible to state that UAV communication and data exchange can be potentially secured by decentralization of blockchain. This implies that the UAV swarms can guarantee that commands and data are not tampered with and that any modification done to them by other people apart from the authorized personnel has been done also; it guarantees that all the UAVs in the swarm are operating from verified and correct information [6].

To enhance basic cybersecurity, it is crucial to incorporate edge computing and cyber-physical systems (CPS) in UAV swarm management. Edge computing therefore provides a means by which cloud services are extended into the network perimeter or different geographical areas from where they are hosted. As a result of this, local fast processing is vital to distinguishing and respond to the threat in an instant, which allows the drone to have capacities to counter cyber threats as soon as possible. As a result, the number of additional rules correlated with security standards associated with air space control as well as data security becomes proportional to the emergence of the new use cases based on UAVs. According to these frameworks, the procedural control of UAVs is provided with formal regulations, as well as operation laws to maintain public and ethical values throughout the time and, thus, prevent cyberattacks. Threats to individuals are just as poorly addressed during the UAV swarm operations as the impacts on human life are desperately needed during these operations. Of these, methods of minimizing human mistakes and insider threats are by educating UAV operators and personnel with the best cybersecurity practices [4, 5].

The cybersecurity in UAV swarm operations spans practices, technologies, and strategic measures to prevent cyber threats that target the UAV swarms. These systems are interconnected, often have sophisticated software, and communicate wirelessly with all parts of each system; therefore, they can be hacked. In conclusion, the key technical challenges such as cybersecurity using AI and blockchain technologies, performance improvements via edge computing integrations with CPS, and compliance with regulations including privacy laws will address how to correctly deploy UAV swarms in critical real-time applications [7, 8].

Importance of Cybersecurity in UAV Swarm Operations

As our world becomes more digitally connected, a growing number of industries and applications are utilizing drones in their services. We see many different ways to apply a drone with the cost being relatively low, very fast, dexterous, and flexible in how they can be used all of this means [9]. Today, drones are enabling e-commerce and various critical service deliveries often used for things like aerial video capture, geographically spanning surveillance, and observing natural environmental events. The importance of drone functionality increases dramatically as now continuous service deliveries depend on drones with this explosion of possibilities and potential for new applications with medical, military, and social services [10]. Drone (UAV) integration among defense-based incorporating drones into defense-specific use cases further promotes the urgency for cybersecurity. Military and defense departments use drones for electromagnetic battle management (EMBM) which includes communication systems, jammer snares, signal collectors, etc. [11]. Drones occupy a congested airspace, normal as they are being used globally in both governmental and commercial sectors. Therefore, the security and stability of these networks must be maintained to protect them from cyberattacks [12].

Remotely controlled drones and swarms pose cybersecurity threats, including malicious hacking, DoS attacks, and performance modification, due to their independent entities and software-defined nature. These

CHAPTER 11 THE ESSENTIAL ROLE OF CYBERSECURITY IN UAV SWARM OPERATIONS

attacks have the potential to endanger drone operations and result in a great threat to public safety [13]. GPS spoofing attacks enable bad actors to hijack drone navigation, manipulate flight routes, or even downgrade performance when critical maneuvers are conducted, even though most drones include an in-built return-to-home (RTH) program. This can be disastrous for drone swarms and pilots' control at the time of mission-critical exercises, by disrupting flight reliability [14]. Previously, drone operation was based on manned UAV navigation; however, modern operations involve AI-supported flight management for autonomous or semiautonomous and remotely piloted high-altitude, contra-rotating propellers, wing/kite drone. GPS spoofing tricks drones into thinking they're in one place, using software defined radios to send malicious signals. Most drones accept unencrypted GPS signals, leading to drone loss and misdirected payloads due to discrepancies between hijack and detection [15].

Although behavioral models, inertial monitoring, and real-time flight path analysis are used to reveal possible misuse of the system, an impotent or no effect at all occurs as with its current countermeasure capability, it is left powerless to prevent a control loss. Civilian and drone operators are vulnerable to malicious attacks, spoofing, or hijacking [16], which may have severe consequences. As an example, a drone under DoS attack can be forced to activate its RTH function or land immediately, potentially resulting in device loss and/or mission failure. Once drones have been hijacked, they can be used as weapons in their own right effectively. While operators may think they are sending the drone to a safe spot, location spoofing can send incorrect GPS data and cause the trigger for fly-off behavior which leads to vehicle loss or other system disruptions [17]. Drones face constant threats and stress onboard security features, requiring system protection and stability for flight performance. Experimentation and trial and error can lead to at-risk profiles requiring security hardening and operational protection. Drones are rife with

CHAPTER 11 THE ESSENTIAL ROLE OF CYBERSECURITY IN UAV SWARM OPERATIONS

exposure to any number of risks; most learned the hard way through trial and error which could result in a potentially at-risk profile requiring security hardening as well as operational protection. The drone industry is working to ensure drone users are responsible for their actions and air traffic, but loss of control remains a significant threat [18].

However, while work continues to support enhanced risk identification and mitigation modalities, MET research efforts will be required for fingerprinting hijacked drones and the data/informational needs that should support future recovery initiatives. Providing additional research with new proof on the latest trends in advanced machine learning may widen its application field and help adopt a risk-averse, control-driven approach toward drone operations.

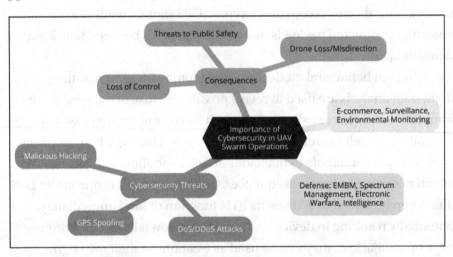

Figure 11-1. *Importance of Cybersecurity in UAV Swarm Operations*

The importance of cybersecurity in UAV swarm operations is illustrated in Figure 11-1. This figure discussed the importance of cybersecurity in securing UAV swarms against different cyberattacks. Such interconnected systems are ubiquitous in UAV swarms extensively used for military, surveillance, and disaster response operations with little

CHAPTER 11   THE ESSENTIAL ROLE OF CYBERSECURITY IN UAV SWARM OPERATIONS

consideration of their security features against GPS spoofing or denial-of-service (DoS) attacks. It is indeed crucial that strict cybersecurity protocols are in place to safeguard the dependability, integrity, and efficacy of swarm missions for a wide range of mission sets and modular environments. The use of AI/machine learning in flight management introduces new attack surfaces, hence the need to continuously research and develop advanced ways for mitigating these threats.

Evolution of UAV Technology

Unmanned aerial vehicles (UAVs) have come a very long way since they first appeared, with substantial technological advancements having occurred over the years and adopted to include practically every sector. Originally military gathering and surveillance weapons, drones have developed from little more than cameras in the sky into advanced self-sufficient analytics frameworks equipped for the execution of complex tasks. Crucial to this evolution has been innovation in areas such as battery technology, lightweight materials, miniaturized sensors, and advanced computing. UAVs have been developed over the years from simple single manually piloted drones to large teamwork coordinated swarms which may be controlled together either autonomously or by a human operator/virtual pilot.

Early Developments

Unmanned aircraft system (UAS) technology was first begun in the early 20th century; however, the development of unarmed aerial vehicles (UAV) has been largely driven by military purposes and applications. UAVs were originally developed for military target practice and reconnaissance and later became used in the 1980s as attack aircraft. Such early UAVs did provide the technology but at a very basic level compared to machines

CHAPTER 11   THE ESSENTIAL ROLE OF CYBERSECURITY IN UAV SWARM OPERATIONS

that today fly faster and higher and are capable of nearly autonomous operations. A key early moment was World War I, when the United States came up with the Kettering Bug, a cruise missile, like this prototype for an early military application. The Kettering Bug took this to a whole new level, being developed from the start as an autonomous "remote-controlled" bomb. The Kettering Bug began the development of UAVs for combat which would later lead to further developments on both military and civilian ends. These other technological advancements in the field better engine designs, simple autopilot systems, and even some early aerodynamic design principles of this time arguably could be said to have contributed toward advances that would come with UAVs throughout the 20th century [19].

Cold War Era

The Cold War period was perceived as a period of UAV technology with demand urged on by increased military conflict and aligned to national strategies. It was seen, captured, and exploited by both the United States and the Soviet Union as a proven concept of UAV, which influenced heavy manpower, and dollars on development for intelligence, surveillance, and reconnaissance (ISR). The Ryan Firebee was a major advancement in the history of UAV technology during this era, which was developed and deployed by the United States. The Ryan Firebee started work as a reconnaissance drone, but its service was also extended to the Vietnam War. The newer versions offered increased range and improved reconnaissance capabilities over the original, improving what had up to then been a lemon by most military standards.

The Soviet Union developed the Tupolev Tu-141, named "Strizh" (Swift), for ISR purposes. The Tu-141 was intended to fly at high altitudes and carry advanced cameras and sensors for aerial reconnaissance. It was instrumental in Soviet military operations during the Cold War, offering valuable wide-area imaging of large parts. The Cold War promoted the

CHAPTER 11 THE ESSENTIAL ROLE OF CYBERSECURITY IN UAV SWARM OPERATIONS

development of aerodynamics, propulsion systems, and sensors that spilled over to later applications for UAV technology. These changes in technology enabled UAVs to execute broader ranges of missions autonomously which began taking some risks away from human pilots and expanding the military reconnaissance capabilities [20].

Technological Advancements

The 1980s and early to mid-1990s brought on a lot of technological advancements. Several key innovations in this period revolutionized UAV capabilities and performance across military, civilian, and economic spheres. The miniaturizing of electronics was a major leap in technology. The invention allowed smaller, lighter UAVs to be designed with equivalent or improved performance. The resulting reduction in the size of many electronic components lessened the weight and power requirements for UAS systems, allowing longer flight durations and increased payloads. These technologies were critical in improving the performance of UAVs during various mission types [21].

At the same time, big strides in propulsion were being made. Fuel-efficient engines and propulsion systems that allowed longer-range flights for much longer durations started to enable long-distance UAV operations. While propulsion has attracted a large application in UAVs, it did not just extend the range but enabled them to work under adverse and angelic conditions where humans meet physical constraints. The developments in material science were also crucial to enable the enhancement of these UAVs during that period. Composites and advanced alloys are used in UAVs to enhance structural integrity and lifespan, making them more reliable when used cheek-by-jowl.

GPS (Global Positioning System) was a game changer and made autonomous navigation of UAVs realistic. GPS could communicate UAVs with the aircraft flying by itself along preplanned flight paths or change in

CHAPTER 11   THE ESSENTIAL ROLE OF CYBERSECURITY IN UAV SWARM OPERATIONS

real time to meet mission requirements. These operational enhancements not only boosted accuracy in mission planning but also allowed the capability of effectuating complex missions in diverse locations throughout the world. The evolution of sensor technology also provided UAVs with their intelligence, surveillance, and reconnaissance (ISR) capabilities. Equipped with high-resolution cameras and infrared sensors, among other environmental detection and analysis equipment. These sensors made it possible for UAVs to acquire imagery and data, enabling the completion or reconnaissance involving even specialized other operational tasks such as monitoring critical infrastructure (bridges or dams) disaster response operations that have been performed during missions in an unprecedented high-quality manner [22].

Modern UAVs

The emergence of AI, machine learning, and autonomous systems has been a renaissance for the potential functions of unmanned aerial vehicles (UAVs) in the 21st century. Adding these recent developments to the equation raised UAV capacity into an entirely new domain and enabled it to perform complex, highly automated missions with increased productiveness and precision [23]. The most important changes in contemporary UAVs, for example, include advanced onboard computers embedded with AI and machine learning solutions. These systems allow UAVs to input huge amounts of data and instantly analyze complicated environmental states while imposing performance autonomously throughout a mission. AI-enabled UAVs, using machine learning algorithms and deep neural network models can shape a dynamic environment utilizing its advanced properties so that it can adapt itself if situations change. With AI capabilities embedded in any such unmanned aerial vehicle or flying object, they not only optimize flight path planning but also perform tasks autonomously with minimal human intervention as part of enhancing operational efficiency.

CHAPTER 11   THE ESSENTIAL ROLE OF CYBERSECURITY IN UAV SWARM OPERATIONS

Another key factor contributing to modern UAV capabilities is improvements in battery technology. UAVs have seen flight time and range maximized by advancements in energy storage technologies. With the recent advances, still there is space to improve without extra parametric complexities that can affect the sustained flight of traditional UAS reasonably, and it could be employed for long-duration surveillance missions or disaster response operations and other research requiring a very long exposure time [24]. In addition, the widespread use of lightweight composite materials has brought a revolution in UAV design and performance. There has been reduced energy consumption in lightweight UAVs and an increase in the agility of these machines which allows them to get better maneuverability, operations speed, and flexibility for different kinds of environments. This triad of AI-powered onboard systems, advanced battery technology, and lightweight materials has turned modern UAVs into versatile multipurpose platforms for use in military applications, as well as civilian or commercial worlds. These breakthroughs have broadened UAV applications, making them beneficial not only for armed services ISR missions but also in disaster response, infrastructure inspection, and precision agriculture [23, 24].

Emergence of UAV Swarms

At the forefront of recent UAV technology innovation is the appearance of UAV swarms, transitioning from single-unit operations toward a collaborative collective swarm unit paradigm. Compared with traditional single UAVs, a swarm is constituted by several UAVs working as a whole to complete various tasks collectively [24, 25]. The two main benefits of a single UAV operation are that the concept of swarm introduces its kind of redundancy, and more importantly, it reduces or eliminates single points of failure. For starters, swarms provide better coverage and redundancy.

CHAPTER 11 THE ESSENTIAL ROLE OF CYBERSECURITY IN UAV SWARM OPERATIONS

Swarm operations leverage UAV swarms for larger geo-subject areas, increased data collection, and complex tasks, making them ideal for long-term surveillance, reconnaissance, and search and rescue.

Every UAV of a swarm flies autonomously but exchanges information with other members in real time. This decentralized concept of control allows swarms to self-organize in real time and achieve environmental adaptation or mission goals based on collective decision-making [26]. This coordination and behavior of the UAV swarms are regulated by sophisticated algorithms that mimic biological systems such as bird flocking behavior and insect swarming patterns. These algorithms enable effective and robust swarm operation through path planning, collision avoidance, and task allocation within the swarm members. Distributed UAV swarms achieve robustness, scalability, and adaptability across a wide range of operational scenarios by mimicking biological principles. The cyber-physical system being developed will also extend the future capabilities of swarm intelligence further down to more advanced communication protocol and use AI for even higher-order decision-making capability in a swarm. As these technologies develop, UAV swarms have the potential to revolutionize the fields of aerial operations by providing new levels of capabilities and efficiencies in a variety of applications [24–26].

CHAPTER 11 THE ESSENTIAL ROLE OF CYBERSECURITY IN UAV SWARM OPERATIONS

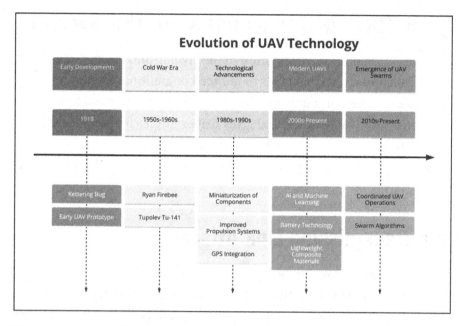

Figure 11-2. *Evolution of UAV Technology*

Figure 11-2 shows the evolution of UAV technology, highlighting the advancements and improvements over time. This figure discusses the development path from basic remote-piloted vehicles to high-end autonomous systems that can accomplish advanced tasks. Such progression has been facilitated by developments in the field of battery technology, light materials, sensors, and computing, originally conceived for military applications in the early 20th century and widely used since then; more recently, it is largely associated with some degree of autonomy, measuring requirements exceeding a human's capabilities. Early examples began during the First World War with technology advanced enough to complete things out-of-sight over long distances without direct human involvement. The advent of UAV swarms, powered by AI and machine learning, is revolutionizing autonomous decision-making in the realm of new battery and material technologies.

CHAPTER 11   THE ESSENTIAL ROLE OF CYBERSECURITY IN UAV SWARM OPERATIONS

Cyber Threats and Attacks on UAV Swarms

UAV swarms become more common in the civilian and military realms, and information stolen that leads to reduced operational capacity is vulnerable to cyber targeting. This section explains the various forms of cyber threats for UAV swarms and gives real examples that represent the effects of these attacks on their work in practice.

The text outlines the types of cyber threats and attacks that are specific to UAV swarms.

Data Transmission Link Attacks

Among all the objectives of the UAV swarm operations, the primary goal is the interruption of the communication link between the UAVs to their GCS or between UAVs. These links can be either jammed or eavesdropped, thus asserting that their communication is prone to interferences.

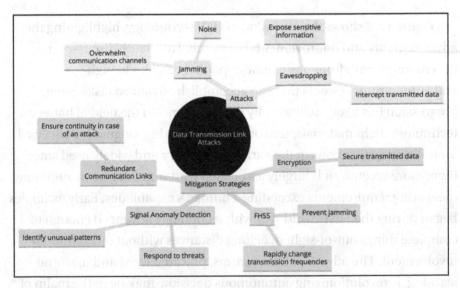

Figure 11-3. Data Transmission Link Attack

CHAPTER 11 THE ESSENTIAL ROLE OF CYBERSECURITY IN UAV SWARM OPERATIONS

In jamming attacks, the quality of the communication channels is diminished by the addition of noise, hence denying the UAVs the ability to either receive or transmit data. Interception of the transmitted data may enable one to access that important information. To combat these threats, transmitted data can be encrypted to improve the security of the data and FHSS to prevent jamming as it frequently changes the transmission frequencies and signals that are out of the ordinary identification systems to highlight abnormal communication signals. Also, a backup data link ensures operation in case the network is attacked since it functions as a backup pointer. All these strategies contribute largely to the protection of UAVs' operations against the attack of data transmission links [27]. Figure 11-3 illustrates a data transmission link attack. It shows how data transmission can be intercepted or disrupted by malicious entities.

GPS Spoofing

GPS spoofing remains a major threat to unmanned aerial vehicle swarm functionality since it alters GPS signals to provide the UAV with wrong coordinates. It is a technique whereby fake GPS signals are relayed to UAVs and create an illusion of the position of the UAVs so that they change their path. It might lead to several problems that range from accidents, entry into a restricted zone, or even ground interference with the set mission.

CHAPTER 11 THE ESSENTIAL ROLE OF CYBERSECURITY IN UAV SWARM OPERATIONS

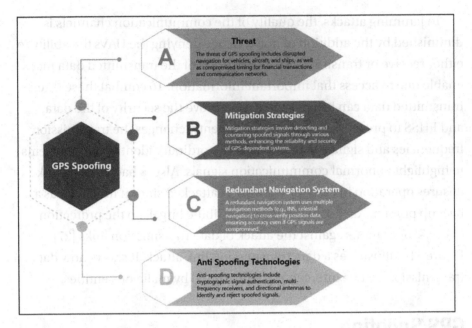

Figure 11-4. GPS spoofing

Spoofing attacks take advantage of the criticality of dependency of UAVs to GPS to correctly navigate, hence making them sensitive to cheating. To prevent GPS spoofing, cryptographic methods and antispoofing technologies are used to verify original GPS signals, alongside other navigation systems like inertial navigation or visual odometry [28]. These measures are very essential for the safety of UAV swarm operations from the jolting effects of GPS spoofing attacks. Figure 11-4 shows the GPS spoofing attack, demonstrating how false GPS signals can mislead UAV navigation.

CHAPTER 11   THE ESSENTIAL ROLE OF CYBERSECURITY IN UAV SWARM OPERATIONS

Authentication Attacks

Identification and authorization as a critical vulnerability have a severe impact on the security of UAV swarm missions, which is why they can compromise the methods for authenticating users to UAV systems. Such attacks include password guessing and token cracking where hackers try to guess passwords or tokens used to authenticate the UAVs and man-in-the-middle attacks where the attackers interrupt or intercept the signals between the UAV and the control station. Alas, the presented scheme of the AA interface and the overall system shows that by using known bugs and weaknesses in the authentication protocols, the attackers will have access to the UAV swarms. Once inside, they can perform arbitrary commands, modify data, or interrupt the control synchronism among UAVs that possibly may endanger the mission confidentiality, integrity, and availability [29].

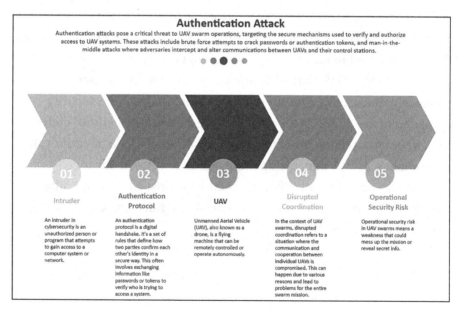

Figure 11-5. Authentication Attack

CHAPTER 11   THE ESSENTIAL ROLE OF CYBERSECURITY IN UAV SWARM OPERATIONS

Preventing authentication attacks entails using strong authentication measures such as utilizing encrypted authentication, multiple-factor authentication, and regularly scanning for any malicious activity to conduct an authentication attack on the users' accounts. That is why such measures are necessary to protect UAV swarm activities from unauthorized data access and maintain confidentiality of the critical missions' information. Figure 11-5 shows an authentication attack, highlighting vulnerabilities in verifying user identities.

Malware and Software Exploits

Malware and software vulnerabilities are some of the most shocking threats to UAVs threatening their coordinative and system integrity and data protection. Malware can be installed through a takedown of ground control stations, software updates, or tampered software which affects the UAVs' operations, and/or access sensitive information while the system can also be hijacked by malicious actors. A virus infects a hole within the UAV software system for further unauthorized control or to crash the UAV system and create DoS situations that cease normal functioning [30, 31]. To deal with these risks, one should apply strict information security management according to the plan that prescribes applications' coded build, frequent updates, antivirus and firewall services, data monitoring for anomalies, etc. By addressing these vulnerabilities in advance, UAV operators can enhance their protection against viruses and other malicious software attacks, ensuring crucial functions and confidential data are appropriately shielded from hacking.

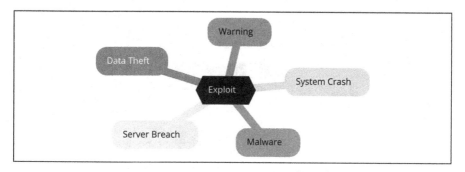

Figure 11-6. *Malware and Software Exploits*

Figure 11-6 illustrates software exploits and malware, demonstrating how malicious software can compromise UAV systems.

Physical Attacks on Hardware

These attacks focus on physical interference with the actual physical components of UAVs; these interventions may encompass the inclusion of unwanted parts, introducing data transfer chips, or compromising the essential physical components to cause system failures [32]. Although such attacks on a UAV often demand direct physical access to the UAV, their result can be disastrous and may cause a range of critical mission failures, losses, or even safety threats. Measures of protection against physical attack entail locking physical access points, seals that show signs of physical interference, and periodic physical checks for any interferences. Also, the incorporation of stringent cybersecurity measures and surveillance mechanisms enhances the identification of physical security threats, which, when addressed, maintains the efficiency and security of UAVs in different contexts of operation.

CHAPTER 11 THE ESSENTIAL ROLE OF CYBERSECURITY IN UAV SWARM OPERATIONS

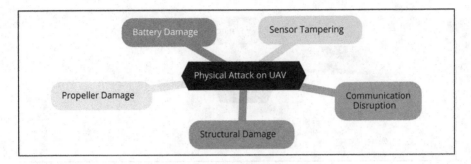

Figure 11-7. *Physical Attack on Hardware*

Figure 11-7 describes the physical attack on hardware, highlighting the risks of direct physical tampering with UAV components.

In this section, we carried on describing the fact of increased vulnerability of UAV swarms, which are actively incorporated in the civilian and military sectors, to various types of cyber threats able to negatively influence the performance of the programs and systems. These threats are related to data transmission link attacks, GPS spoofing, break-ins of authentication, viruses, software cracks, H/W tampering, and similar issues. Transmission of data can be observed or intercepted whereas GPS can be trapped to relay wrong locations to UAVs. Authentication attacks may lead to vulnerability and breach on UAV access, and malware or software attacks may lead to a decrease in UAV's performance. The physical attack is performed to physically change or damage the limbs and nodes of UAV. There are certain solutions required to minimize the above threats by encrypting, strong authenticating, and securing source code which is also trying to avoid imitations and physical security that are used for securing UAV swarm functions.

CHAPTER 11   THE ESSENTIAL ROLE OF CYBERSECURITY IN UAV SWARM OPERATIONS

Main Challenges and Issues in Cybersecurity for UAV Swarm Operations

Communication Security

It is essential to have secure and reliable communication among the UAVs in a swarm as well as between different vehicles of swarms and ground control stations [33]. Wireless communication protocols are normally used by UAVs which can be intercepted, jammed, or spoofed. Intercept may rival typically placed intelligence secrecy at risk (SCADA), by spying on transmitted data. Unauthorized access to identity data can result in a breach of sensitive information and security organization mission privacy. A jamming attack is simply a matter of bombarding the frequencies used by UAVs for communication with noise, rendering them unable to receive commands or GPS data. Spoof attacks deceive UAVs by broadcasting false signals, causing the scene to be mistakenly identified as a specific location or instruction device error which can fail a mission or security breach.

Data Protection and Privacy

UAVs bring about the collection and transmission of sensitive data through their onboard sensors which include imagery and telemetry [34]. Keeping this data secure from unauthorized access and tampering is vital for ensuring operational integrity and meeting privacy compliance. Encrypting data both in transit and at rest is paramount to protect from sniffing as well as servicing that guarantees the confidentiality of information. Yet, secure encryption and standardization of the data is another major challenge as well, across heterogeneous UAV swarm platforms. Additionally, the incorporation of implementable security standards must take into account agile service areas and local regulations within which UAVs will be performing duties.

CHAPTER 11   THE ESSENTIAL ROLE OF CYBERSECURITY IN UAV SWARM OPERATIONS

Vulnerability Management

Moreover, UAVs are subject to such exploitation in the forms of software vulnerabilities and hardware tampering which can deteriorate their operational security [35]. UAV systems may have vulnerabilities that can be exploited by an attacker to gain unauthorized access, disrupt operations, or even hijack control. Your software needs to be regularly updated, and you should carry out vulnerability assessment and penetration testing regularly to find a flaw before the attacker can capitalize on it. Still, maintaining updated information on multiple UAV swarm platforms with different levels of operational readiness spread over time and space creates many logistical challenges. Additionally, the physical securing of UAV hardware and assurance that key components have not been tampered with are continuous challenges.

Autonomy and Decision-Making

Autonomous systems are susceptible to attacks where malign actors could abuse vulnerabilities built into the system, making it possible for these malefactors to influence mission objectives and redirect flights or disrupt the behavior of some other autonomous machine in a closed-loop [36]. It is crucial to guarantee the robustness of decision-making algorithms and protect them against adversarial attacks and errors. In addition, managing the trade-off between autonomy and human oversight as well as web intervention is crucial to reduce the risks of having autonomous systems induce decisions in such dynamic and unpredictable scenarios.

Integration and Interoperability

While bonding was considered internally, at least the implementation requires heterogeneous systems which leads us to a first significant challenge: UAV swarms are typically organized from different manufacturers and heterogeneous platforms, each with its own cybersecurity needs and strengths [37]. It is also difficult to maintain unified cybersecurity measures across vastly different UAV swarm components that will be required for seamless integration and interoperability. The answer to this problem can be standardizing of communication protocols, data formats, and security interfaces across platforms but that needs collaboration between all stakeholders. Another continuing challenge is the compatibility and effectiveness of cybersecurity solutions in new UAV technologies and operational environments.

Resilience Against Advanced Threats

UAV swarm cybersecurity evolves quickly due to the emergence of cyber threats such as AI-driven attacks [38], zero-day exploits, and sophisticated malware. These new technologies and emerging vulnerabilities in UAV systems have the potential to expose organizations to advanced threats that can compromise operational resilience and mission objectives. It is necessary to create adaptive defenses that can prevent, identify, and react to advanced cyber threats in real time. Continuous monitoring, threat intelligence sharing, as well as strategic applications defense are necessary for resilience improvement and operational continuity.

Regulatory and Ethical Considerations

Regulatory and ethical guidance is essential to operate UAV swarms responsibly [39]. Drone operations have to follow national and global rules related to airspace use, data privacy, and cybersecurity. Further, legal and ethical implications of data sovereignty, liability for cyber incidents, and public trust require an ongoing dialogue with regulators. Transparency in UAV operations and ethical implications for cybersecurity is essential to uphold public trust and maintain regulatory compliance.

CHAPTER 11   THE ESSENTIAL ROLE OF CYBERSECURITY IN UAV SWARM OPERATIONS

Human Factors

Human operators and personnel are key in the UAV swarm operations, which will impact the cybersecurity practices and response capacity [40]. Improving cybersecurity awareness, training, and human-centric security measures are key to addressing problems that arise because of human mistakes, insider threats, or operational vulnerabilities. With dynamic operational environments, timeliness in threat detection and incident response is highly dependent on ensuring effective collaboration between human operators and automated cybersecurity systems.

Overcoming these challenges will take stakeholders in academia, industry, government, and regulatory bodies working together. Further research, innovation, and investment in the domain of UAV swarm cybersecurity are required to be able to put together wide-ranging solutions for a secure deployment framework that must enable safe, secured, and effective deployment paradigms over the air with an all-encompassing operational envelope.

Figure 11-8 shows the UAV swarm cybersecurity challenges and issues. In this figure, some of the most prominent threats affecting the swarm of UAVs have been discussed: how can we safeguard these UAV swarms from being intercepted, jammed, and spoofed; safeguarding data by employing the right encryption measures; and how to protect UAV swarms against any form of susceptibility to software and hardware. Thus, the security of the fully autonomous decision of the person is important to protect it from being influenced and used in a particular manner, namely, in a wrong way. Malfunction syntropy is described as issues arising out of different types of UAV platforms while future security threats are slimy in nature encompassing issues such as the deployment of artificial intelligence in crafting the attack. The outlined issues presuppose the need for cooperation between scholars carrying out research in the field, the industry, the government, and regulators to develop efficient and safe computational platforms for UAV swarm applications.

CHAPTER 11   THE ESSENTIAL ROLE OF CYBERSECURITY IN UAV SWARM OPERATIONS

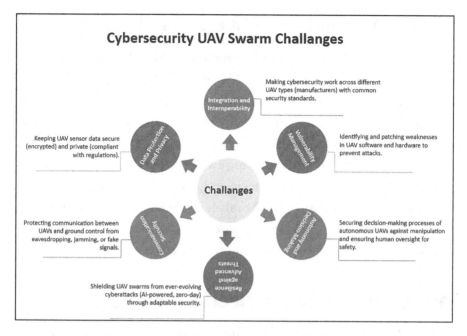

Figure 11-8. Cybersecurity UAV Swarm Challenges

Future Trends and Research Directions in UAV Swarm Cybersecurity

With the further development of technologies associated with UAV swarms, the sphere of cybersecurity will seem to have new opportunities and threats. Hence, other research efforts and relevant organizational and technical trials are needed to predict further threats and, on the contrary, to discover the utility of UAV swarms for certain tasks. Here are key future trends and research directions in UAV swarm cybersecurity.

Advanced Threat Detection and Response

In the future, the research in the cybersecurity field will be oriented toward creating new detection methods that will be effective for UAV swarms [41]. This entails the use of artificial intelligence (AI) and machine learning (ML) in identifying and eliminating effective cyber threats in real time. Automated anomaly detection systems implemented by artificial intelligence can be useful in improving awareness of behavioral patterns that are deviant and would signal a cyberattack so that preventive and immediate countermeasures can be taken. If applied to UAV swarms, these studies are intended to enhance the system's defense against emerging threats such as zero-day vulnerabilities targeted by AI attacks.

Secure Autonomy and Decision-Making

Another crucial research area will address the security and dependability of auto-maneuvering by UAV swarms [42]. Future work will include the development of rigorous cybersecurity to ensure that adversarial manipulation and mistakes do not compromise decision-making algorithms. This work includes advancing the robustness of AI decision systems using formal testing, validation, and verification processes. The call will also look to explore hybrid autonomy models that combine human oversight and intervention where necessary with the largely automated operation for reasons of risk in dynamic, uncertain environments.

Blockchain Technology for Secure Communication

Blockchain could improve security and transparency of communication among the nodes in UAV swarms [43]. Investigators will examine the potential for utilizing blockchain-based solutions in developing secure, decentralized communication networks between UAVs and ground control stations. In addition to the operation of UAV swarms, blockchain also can make sure data stored is irremovable and that any command or transaction shall be recognized legally. Based on our extensive evaluation and learning, we suggest several future research directions about optimizing blockchain protocols for low-latency, high-throughput UAV comms as well as addressing scalability challenges in decentralized USAs.

Resilient Edge Computing and Cyber-physical Systems

The future of UAV swarm cybersecurity will rely heavily on edge computing [45] and cyber-physical systems (CPS); both are the tools to increase resilience and performance in these network types [43–45]. Compound eye will develop secure edge computing architectures for processing and taking real-time decisions close to the network edge, which minimizes latency and dependency on centralized cloud infrastructure. This requires a secure integration of CPS components, such as UAVs and their sensors/actuators to ensure data integrity, system reliability, and the resilience against physical/cyber threats. Lightweight security protocols and encryption mechanisms could potentially be designed to secure the communication architecture in swarm UAV operations in an edge computing environment which would support the establishment of a strong link layer.

Regulatory Frameworks and Ethical Guidelines

The design of principled regulatory frameworks and ethics in UAV swarm cybersecurity is crucial for addressing legal, privacy, and ethical concerns [46]. Their future work will push for uniform cybersecurity guidelines in addition to compliance with the international rules governing UAV operations, airspace control, and data security. This will involve dealing with issues of legal liability, data sovereignty, and ethical considerations when dedicated UAV swarms are deployed. Industry, policymakers, and cybersecurity experts need to come together if the governance frameworks that allow for responsible UAV swarm operations including security among others are developed.

Human-Centric Security Solutions

To help address human mistakes and mitigate insider threats in UAV swarm operations [42, 43], the development of security solutions that are designed for people (human-centric) will be key, as well as more widespread cybersecurity training programs. Future directions of study encompass user-friendly human–machine interfaces (HMI) and situation awareness tools, together with interactive training modules designed for UAV operators or unit personnel. To illustrate these points, integrating human factor research concepts into the design and implementation of cybersecurity will provide improved incident response capabilities for cyber defenders in dynamic operational environments.

Interdisciplinary Collaboration and Innovation

Future directions in UAV swarm cybersecurity research will highlight interdisciplinary efforts and innovation among academia, industry, and government [45-47]. This will contribute to efforts to bridge gaps that exist between cybersecurity, aerospace engineering, AI/ML, and communications as well as regulations. Joint programs will spearhead the creation of secure UAV swarm architectures, robust communication protocols, and flexible cybersecurity frameworks. This interdisciplinary approach intends to assist in handling comprehensive cybersecurity problems and at the same time supports innovation, promoting faster deployment of UAV swarms within safe operation ranges across multipurpose operational scenarios.

CHAPTER 11 THE ESSENTIAL ROLE OF CYBERSECURITY IN UAV SWARM OPERATIONS

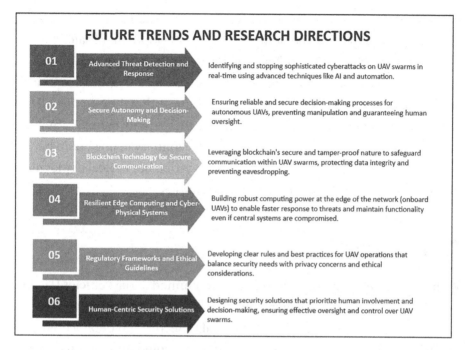

Figure 11-9. Future Trends and Research Directions

In consideration of these future trends and the directions of research needed for them, stakeholders may develop mature UAV swarm cybersecurity solutions to secure the safe operation of autonomous aerial technologies in civilian, commercial, and military environments.

Figure 11-9 illustrates future trends and research directions in UAV swarm cybersecurity. This figure explores the security of UAV swarms and lists possible future research topics regarding their cybersecurity. This is why advanced threat detection with AI and machine learning is developed to quickly spot well-planned cyberattacks that could happen in real time. Ensuring the security of autonomous decision-making processes is essential, requiring not only robust frameworks and models but also hybrid autonomy approaches that integrate human oversight. Blockchain offers secure communication over decentralized networks and

CHAPTER 11   THE ESSENTIAL ROLE OF CYBERSECURITY IN UAV SWARM OPERATIONS

resilient edge computing architectures, reducing reliance on centralized infrastructure with no single point of failure. Development of safety- and privacy-respecting regulatory frameworks for full-scale applications in compliance with ethical constraints is required. You either need to train your personnel or make those interfaces so simple that they do not fall victim of humans and insider threats, hence the security courses. An industry, academia, and government interdisciplinary team should be formed that will work together on the security-rich but ethical deployment of UAV swarms for different operational scenarios.

Conclusion

With the advancement in technology of unmanned aerial vehicles (UAV), security issues become a bigger concern with these developments. Securing swarm deployment: It is essential to secure and deploy uniquely identifiable civilian, commercial, or military-oriented unmanned aerial vehicle swarms. The study aims to enhance advanced threat detection technologies through A-I and raise awareness and understanding of blockchain in communication security. The study explores the integration of cyber and physical systems in the development of AI-based secure systems for autonomous device communication. Edge computing solutions and human-centric cybersecurity approaches will standardize UAV swarms' cybersecurity, addressing deficits and promoting sensible employment in various contexts, thereby enhancing the overall security of these systems. At the end of this process and by injecting their expertise in defense into the trial, stakeholders could secure a lifeline for safe, reliable, or ethical deployments with UAV swarms over the coming years.

References

[1] Kelner, J. M., Burzynski, W., & Stecz, W. (2024). Modeling UAV swarm flight trajectories using Rapidly-exploring Random Tree algorithm. *Journal of King Saud University-Computer and Information Sciences*, 36(1), 101909.

[2] Khalil, H., Rahman, S. U., Ullah, I., Khan, I., Alghadhban, A. J., Al-Adhaileh, M. H., ... & ElAffendi, M. (2022). A UAV-Swarm-Communication Model Using a Machine-Learning Approach for Search-and-Rescue Applications. *Drones*, 6(12), 372.

[3] Mazhar, T., Irfan, H. M., Khan, S., Haq, I., Ullah, I., Iqbal, M., & Hamam, H. (2023). Analysis of Cyber Security Attacks and Its Solutions for the Smart Grid using Machine Learning and Blockchain Methods. *Future Internet, 15(2), 83*.

[4] Zhang, X., & Liu, J. (2023). Research on UAV Swarm Network Modeling and Resilience Assessment Methods. *Sensors*, 24(1), 11.

[5] Chen, Z., Yan, J., Ma, B., Shi, K., Yu, Q., & Yuan, W. (2023). A Survey on Open-Source Simulation Platforms for Multi-Copter UAV Swarms. *Robotics*, 12(2), 53.

[6] Horyna, J., Baca, T., Walter, V., Albani, D., Hert, D., Ferrante, E., & Saska, M. (2023). Decentralized swarms of unmanned aerial vehicles for search and rescue operations without explicit communication. *Autonomous Robots*, 47(1), 77-93.

[7] Su, X., Ullah, I., Wang, M., & Choi, C. (2021). Blockchain-Based System and Methods for Sensitive Data Transactions. *IEEE Consumer Electronics Magazine*.

[8] Mazhar, T., Talpur, D. B., Shloul, T. A., Ghadi, Y. Y., Haq, I., Ullah, I., ... & Hamam, H. (2023). Analysis of IoT Security Challenges and Its Solutions Using Artificial Intelligence. *Brain Sciences*, 13(4), 683.

[9] Ray, A. (2023). Machine Learning Based Spectrum Fingerprinting of Drones for Defensive Cyber Operations *(Master's thesis, Harvard University)*.

[10] Han, K. (2023). Employing automotive security to improve the security of unmanned aerial vehicles. *Frontiers in Communications and Networks*, 4, 1122231.

[11] Mohsan, S. A. H., Othman, N. Q. H., Li, Y., Alsharif, M. H., & Khan, M. A. (2023). Unmanned aerial vehicles (UAVs): practical aspects, applications, open challenges, security issues, and future trends. *Intelligent Service Robotics*, 16(1), 109-137.

[12] Kashyap, A., Chakravarthy, A., Subbarao, K., Casbeer, D., Weintraub, I. E., & Hencey, B. (2024). Modeling and Detection of Cyber-Attacks in UAV Swarms using a 2D-LWR Model and Gaussian Processes. *In AIAA SCITECH 2024 Forum* (p. 0529).

[13] Li, J., Yue, Q., Huang, Z., Xie, X., & Yang, Q. (2024). Vulnerability Analysis of UAV Swarm Network with Emergency Tasks. *Electronics*, 13(11), 2005.

[14] Hjelle, J. D., & Omli-Moe, L. E. (2023). Cybersecurity Threats to the Internet of Drones in Critical Infrastructure: An Analysis of Risks and Mitigation Strategies *(Master's thesis, NTNU)*.

[15] Wang, Z., Li, Y., Wu, S., Zhou, Y., Yang, L., Xu, Y., ... & Pan, Q. (2023). A survey on cybersecurity attacks and defenses for unmanned aerial systems. *Journal of Systems Architecture*, 138, 102870.

[16] Niyonsaba, S., Konate, K., &Soidridine, M. M. (2023). A Survey on Cybersecurity in Unmanned Aerial Vehicles: Cyberattacks, Defense Techniques and Future Research Directions. *International Journal of Computer Networks and Applications*, 10(5), 688-701.

[17] Arnolnt, S. (2023). A study of cybersecurity threats in UAVs and threat model approaches.

[18] Omolara, A. E., Alawida, M., & Abiodun, O. I. (2023). Drone cybersecurity issues, solutions, trend insights and future perspectives: a survey. *Neural computing and applications*, 35(31), 23063-23101.

[19] Sharma, A. (2024). An analytical view on Unmanned Aircraft Systems. *Computer and Telecommunication Engineering*, 2(2), 2620.

[20] Okpaleke, F. N. (2023). Drones and US Grand Strategy in the Contemporary World. *Palgrave Macmillan Limited*.

[21] Orgeira-Crespo, P., & García-Luis, U. (2024). Brief Introduction to Unmanned Aerial Systems. In Applying Drones to Current Societal and Industrial Challenges (pp. 1-22). Cham: *Springer Nature Switzerland*.

[22] Zahid, N. (2023). Design and Optimization of Unmanned Aerial Vehicles for Planetary Exploration. *Liberal Journal of Language and Literature Review*, 1(01), 114-122.

[23] Telli, K., Kraa, O., Himeur, Y., Ouamane, A., Boumehraz, M., Atalla, S., & Mansoor, W. (2023). A Comprehensive Review of Recent Research Trends on Unmanned Aerial Vehicles (uavs). *Systems*, 11(8), 400.

[24] Folorunsho, S. O., & Norris, W. R. (2024). Redefining Aerial Innovation: Autonomous Tethered Drones as a Solution to Battery Life and Data Latency Challenges. *arXiv preprint arXiv*:2403.07922.

[25] Ren, M., Wang, B., & Yang, H. (2023, October). Key Technologies for Heterogeneous Unmanned Aerial Vehicle Swarm Cross-Domain Collaborative Operations. *In China Conference on Command and Control (pp. 116-123). Singapore: Springer Nature Singapore.*

[26] Javed, S., Hassan, A., Ahmad, R., Ahmed, W., Ahmed, R., Saadat, A., &Guizani, M. (2024). State-of-the-Art and Future Research Challenges in UAV Swarms. *IEEE Internet of Things Journal.*

[27] Sharma, J., & Mehra, P. S. (2023). Secure communication in IOT-based UAV networks: A systematic survey. *Internet of Things*, 100883.

[28] Hadi, H. J., Cao, Y., Nisa, K. U., Jamil, A. M., & Ni, Q. (2023). A comprehensive survey on security, privacy issues and emerging defence technologies for UAVs. *Journal of Network and Computer Applications*, 213, 103607.

[29] Oracevic, A., & Salman, A. (2024, May). Unmanned Aerial Vehicles in Peril: Investigating and Addressing Cyber Threats to UAVs. *In 2024 International Conference on Smart Applications, Communications and Networking (SmartNets)* (pp. 1-7). IEEE.

[30] Hadi, H. J., Cao, Y., Nisa, K. U., Jamil, A. M., & Ni, Q. (2023). A comprehensive survey on security, privacy issues and emerging defence technologies for UAVs. *Journal of Network and Computer Applications*, 213, 103607.

[31] Mohsan, S. A. H., Othman, N. Q. H., Li, Y., Alsharif, M. H., & Khan, M. A. (2023). Unmanned aerial vehicles (UAVs): practical aspects, applications, open challenges, security issues, and future trends. *Intelligent Service Robotics*, 16(1), 109-137.

[32] Olsson, E., Funk, P., &Sohlberg, R. (2023, June). Using a Drone Swarm/Team for Safety, Security and Protection Against Unauthorized Drones. *In International congress and workshop on industrial AI (pp. 263-277). Cham: Springer Nature Switzerland.*

[33] Shafik, W., Matinkhah, S. M., &Shokoor, F. (2023). Cybersecurity in Unmanned Aerial Vehicles: A Review. *International Journal on Smart Sensing and Intelligent Systems*, 16(1).

[34] Olsson, E., Funk, P., &Sohlberg, R. (2023, June). Using a Drone Swarm/Team for Safety, Security and Protection Against Unauthorized Drones. *In International congress and workshop on industrial AI (pp. 263-277). Cham: Springer Nature Switzerland.*

[35] Hafeez, S., Khan, A. R., Al-Quraan, M. M., Mohjazi, L., Zoha, A., Imran, M. A., & Sun, Y. (2023). Blockchain-Assisted UAV Communication Systems: A Comprehensive Survey. *IEEE Open Journal of Vehicular Technology*, 4, 558-580.

[36] Sun, Q., Li, H., Zhong, Y., Ren, K., & Zhang, Y. (2024). Deep reinforcement learning-based resilience enhancement strategy of unmanned weapon system-of-systems under inevitable interferences. *Reliability Engineering & System Safety*, 242, 109749.

[37] Betti Sorbelli, F. (2024). UAV-Based Delivery Systems: A Systematic Review, Current Trends, and Research Challenges. *Journal on Autonomous Transportation Systems*, 1(3), 1-40.

[38] Shandilya, S. K., Datta, A., Kartik, Y., & Nagar, A. (2024). Advancing Security and Resilience. In Digital Resilience: Navigating Disruption and Safeguarding Data Privacy (pp. 459-529). Cham: *Springer Nature Switzerland*.

[39] Mykytyn, P., Brzozowski, M., Dyka, Z., & Langendoerfer, P. (2024). A Survey on Sensor-and Communication-Based Issues of Autonomous UAVs. *CMES-Computer Modeling in Engineering & Sciences*, 138(2).

[40] Blakcori, N., Stathakis, L. I., Koutsoukos, L. D., & Kirilov, L. K. (2024). The Evolving UAS Threat: Lessons from the Russian-Ukrainian War Since 2022 on Future Air Defence Challenges and Requirements. *NATO, Integrated Air and Missile Defence Center of Excellence*.

[41] Shah, S. F. A., Mazhar, T., Al Shloul, T., Shahzad, T., Hu, Y. C., Mallek, F., & Hamam, H. (2024). Applications, challenges, and solutions of unmanned aerial vehicles in smart city using blockchain. *PeerJ Computer Science*, 10, e1776.

[42] Bakar, R. A., Paolucci, F., Cugini, F., Castoldi, P., & Olmos, J. J. V. (2024). DPUAUT: Secure Authentication Protocol with SmartNiC Integration for Trustworthy Communications in Intelligent Swarm Systems. *IEEE Access.*

[43] Kaushik, K., Khan, A., Kumari, A., Sharma, I., & Dubey, R. (2024). Ethical Considerations in AI-Based Cybersecurity. In Next-Generation Cybersecurity: AI, ML, and Blockchain (pp. 437-470). *Singapore: Springer Nature Singapore.*

[44] Ullah I, Khan IU, Ouaissa M, Ouaissa M, El Hajjami S, editors. Future Communication Systems Using Artificial Intelligence, Internet of Things and Data Science. CRC Press; 2024 Jun 14.

[45] Mallick, P. K. (2024). Artificial intelligence, national security and the future of warfare. In Artificial Intelligence, Ethics and the Future of Warfare (pp. 30-70). *Routledge India.*

[46] Abid, S. G., Rabbani, M., & Sarker, A. Comparative Analysis of Threat Detection Techniques in Drone Networks.

[47] Ali SH, Ali SA, Ullah I, Khan I, Ghadi YY, Tao Y, Khan MA, Khongorzul D. Optimizing network lifespan through energy harvesting in low-power lossy wireless networks. International Journal of Data Science and Analytics. 2023 Nov 6:1-5.

CHAPTER 12

Cloud Computing for UAV Swarms

Faisal Rehman[1,2], Shanza Gul[1], Saman Aziz[1],
Muhammad Hamza Sajjad[1]

[1]Department of Statistics & Data Science, University of Mianwali, Mianwali, Pakistan

[2]Department of Robotics & Artificial Intelligence, National University of Sciences & Technology, NUST, Islamabad, Pakistan

Correspondence: Faisal Rehman[1,2] (faisalrehman0003@gmail.com)

Introduction

The concept of coordination among multiple swarms of UAVs was proven to be useful in applying to a range of domains and was said to be effective. In precision agriculture [1], UAV swarms help to perform comprehensive crop analysis and manage applicable action. They afford total surveillance and data collection in surveillance [2] and remote sensing [3] in coverage. As it is mentioned in [4] and [5], for various search and rescue missions which are essentially multistep tasks, UAV swarms work effectively in identification and aiding of the targeted subjects. They also play a central role in disaster operations by emplacing and enabling swift evaluation and intervention [6] and in creating communication networks where these could possibly be disrupted. Furthermore, aerial performers have implemented UAV swarms within light shows, which reveal how these drones can accurately work as one. The aspect of using several UAVs in a coordinated manner leads to increased efficiency and functionality of tasks that require their use as well as their completion in shorter time and with minimal wastage of resources.

Firstly, UAV swarm systems are already interdisciplinary by definition; hence, someone needs to be knowledgeable in engineering, robotics systems, and computational methods. In this field, the application of simulation studies as well as the experimental setup have to be successful in contexts that require careful balancing of both methods to obtain accurate and reliable outcomes. But when the UAVs are used in realistic applications, there are several obstacles that may show up. Nonetheless, these sorts of systems are defined as being highly vulnerable to a number of disruptions no matter how well they had been planned out. Wind, rain, and temperature fluctuations may affect the performance of drones, signal jamming is also an issue, and small spaces can create navigational challenges. This might also be a factor of physical materials used in construction of UAVs since they are known to have an influence on the UAV durability as well as the operating capacity of main UAV functions.

Additionally, UAV operations are often regulated by strict legislation, which plays an extra role of complexity when it comes to field operations. Another significant factor is that unpredictable factors must be observed as the UAVs move in swarms; this includes conditions such as weather changes, appearance of obstacles, and mechanical breakdowns. For delivering stable, safe, and highly dependable swarming operations in such intricately complex environments, there are multiple areas which still need extensive development. Effective and reliable methods of operation are required for the UAVs to be able to work in tandem; effective failure recovery which could involve a failed unit not bringing down the entire system requires sensitive strategies [7]. Further, there is a need to design adaptation strategies to adapt the swarms to the differing conditions in the environment.

It is noteworthy that in the current literature, there are many examples that indicate that most of the time, researchers use specific protective mechanisms in various fields and applications rather than a global approach to the organization's system. This tendency is mainly because the development time for creating systems, which will be resilient the world over, can be very time-consuming. For instance, application areas could be on improving fault tolerance in search and rescue situations, or directing particular attention toward integrating an efficient communication scheme in agricultural monitoring rather than coming up with a general goal.

However, these challenges are still formidable, yet research in UAV swarm remains to be exponentially expansive due to the extensive arrays of application such as disaster response, surveillance, and agriculture. A growing research focus has been seen in this area as depicted in the figure below obtained from Dimensions.ai, which demonstrated the rising trend in number of published papers on UAV swarm over the recent past decade [8]. This growth simply mirrors the awakened understanding of the functions of UAV swarms and their incorporation in diverse fields. As has been presented in Figure 12-1, UAV swarm research domain reported a sound and ascending trajectory, which signifies the continuous attempts to tackle the existing drawbacks and harness the entire capability of UAV swarms in real-world applications.

The current state of drone swarm development is evident in the analysis of the existing model based on the state of art and results from simulated environment and the physical model developed based on real end-to-end setup that has certain advantages and limitations as follows: actual physical environments enable the testing of swarm behaviors under conditions where outside disturbances can be easily simulated, and simulations offer a cheap way of experimenting with and fine-tuning algorithms based on realistic assumptions at low cost. The field of UAV includes aerodynamics, robotics, computer science, and cybersecurity to guarantee the all-round development on the use of UAV swarm technology.

This multidisciplinary approach is essential for addressing the complex challenges and optimizing the performance of UAV swarms in diverse applications.

CHAPTER 12 CLOUD COMPUTING FOR UAV SWARMS

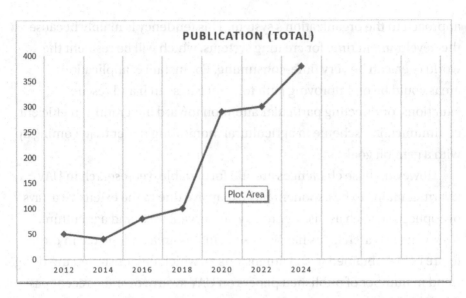

Figure 12-1. *Using Dimensions.ai, an ongoing evaluation conducted on unmanned aerial vehicles swarms from 2012 to 2024*

Complex, multifarious difficulties are unavoidably involved in building algorithmic and physical structures for the control, collaboration, travel, and task distribution in UAV swarms. Generally, suggested approaches focus on strong control, learning through neural networks, and optimization techniques. These approaches aim to enhance the efficiency and effectiveness of UAV swarms in executing tasks, managing dynamic environments, and overcoming challenges related to communication, stability, and adaptability. By integrating these advanced techniques, researchers strive to create sophisticated and resilient UAV swarm systems capable of performing a wide range of applications with high precision and reliability.

Based on the structure of our chapter, we find it appropriate to arrange our sections as follows. This is briefly described in the introduction where several aspects of UAV swarming are outlined and the potential of this technology in multiple fields, such as agriculture, security and public safety, emergencies, and recreation is discussed. Concerning the limitations and repercussions of using UAV swarms in real-world environments, the authors mentioned that it can cause disruptions in the environment where UAV swarms are applied, signal interference, and regulatory limitations. The following sections discuss the various related work in communicating protocols, means of fault tolerance, and suitable algorithms to address these challenges. We then discuss the open issues regarding the simulation and experimentation, also emphasizing the interdisciplinary approach for UAV swarm. Last but not least, we introduced the limitations of the research and recommend further research to improve the broader application of UAV swarm technologies.

Atmosphere of Swarm Studies

This state of affairs may look rather counterintuitive at first making it appear that UAV operations, being as practical and physical as a field can get, would not lend itself to a strong simulative and modeling bias in its research. But it is important to note that several remarkable trends exist which motivate this culture. Real-world environments cannot be taken lightly, and this makes simulation environments highly suitable for experimentation as they do not include the actual physical and monetary risks of UAV testing. Work of this nature underlines the delicate tension identified by Archer between operations that enable demands of actuality and conceptual analysis confined to the context of simulation.

Several Factors Explain Why Simulation Is Prevalent in UAV Research

The first is the capability of UAVs to fly and maneuver in three dimensions. An UAV can cover a three-dimensional space and can move in any direction. Simulation studies are crucial in UAV research to capture

CHAPTER 12 CLOUD COMPUTING FOR UAV SWARMS

nonlinearity and nonstationarity issues in motion, ensuring accurate and reliable understanding of UAV dynamics.

- **Controlled environment**
  - **Risk Mitigation:** In addition, simulations make it possible for the researchers to practice the behaviors and algorithms of the UAV swarm while avoiding any possible dangerous implications on the UAVs and the personnel engaged in the test. It is especially important in situations where an organization is introducing innovation or having yet untested ways of working, because the risk of failure is heightened.
  - **Safe Exploration of Extreme Conditions:** The concept of experimental or test UAVs may be extended to unrealistic situations (such as high wind, rough terrain, etc.) that we would not attempt to recreate in the real world.
- **Cost-effectiveness**
  - **Financial Savings:** Performing experiments in the context of simulators is significantly cheaper than carrying them out in real-life conditions. Real UAVs especially when scaled at a swarm level are a major investment given the costs required to purchase new systems let alone accounting for cases where the systems will require repairs or general maintenance than logistics of using the actual systems.

CHAPTER 12   CLOUD COMPUTING FOR UAV SWARMS

- **Resource Allocation:** This makes spending on actionable resources more effective because through simulations, it is possible to accomplish many more tests and constructions at the same time with the same amount of money.

- **Thorough and wavelike preliminary check**

  - **Repeatability:** Executed properly, simulations provide the opportunity to conduct experiments multiple times with the same stimuli, which is critical to scientific conduct. It is comparatively easy to perform excitation functions systematically, to introduce purposefully disturbances, and to document their impact, something that is barely possible when using physical experiments.

  - **Parameter Exploration:** It allows more extensive and exhaustive analysis of all the possible variables and conditions within a simulated environment ensuring that sophisticated procedures can be reviewed and adjusted in detail before they are sent to real live testing.

- **This section discusses regulatory constraints, as well as other aspects that might present significant logistical difficulties.**

  - **Regulatory Compliance:** The matters concerning UAVs are strictly regulated, and the existence of different legal systems around the world that allow the UAV operation increases the importance of formal legal regulation. It is often relatively difficult to gain the permission and always should follow these regulations and standards.

- **Safety Concerns:** While conducting actual test flights with real UAVs and real persons or other objects in close proximity, the most important issues are safety of the UAVs as well as people and objects in the surrounding area, which often makes the experiments confined and less frequent.

- **Logistical Planning:** Deploying a UAV swarm entails a careful configuration of the swarm as well concerning the area or environment within which the flying robots are deployed, in addition to resources that would support the swarm. Diligent scenario modeling is however often bracketed by this complication, hence making initial simulation studies even more feasible.

Preliminary Development and Validation

- **Feasibility Assessment:** This explains the use of simulations as an initial evaluation tool in determining the viability of such ideas and solutions. It enables researchers to eliminate frequent unique problems that may occur in elaborate and costlier techniques of hardware testing.

- **Concept Refinement:** Due to this sense-making process of the model, algorithm, and strategies, the methodology provides an iteration and optimization of the solution methods in "as real as possible" simulation before implementation of the solutions in real environment, hence achieving time and cost optimization.

CHAPTER 12 CLOUD COMPUTING FOR UAV SWARMS

Moreover, the testbed structures, research planning procedures, tutorials, and new structures of simulation [8] contribute to this further by offering better equipment to the investigators. Table 12-1 provides a summary of the tutorials, simulators, and programming frameworks available for UAV and swarm [9] programming. Other resource includes the Gazebo ROS package, the Webots simulator system, and the MATLAB UAV toolkit. Perhaps more information on them can be obtained, such as the specific comparison of the features in comprehensive reviews and examination of how they perform on various platforms [10–12].

However, it is worth recognizing that it is impossible to completely eliminate the drawbacks that relate to the use of simulations when analyzing R and D processes because such simulations generally simplify these or other procedures. One of these situations is related to the availability of real environment and actual technologies during the investigations in case of using the concept of simulations which influences the factor of generalization of the outcomes. Additionally, one could presuppose that simulation environment limitations hinder obtaining exact representations of physical UAV devices in terms of power consumption, actuator reply delay, and sensor accuracy. However, since these algorithms and methods are designed and tuned in simulation environments, their behavior when implemented in actual systems and networks may differ. There are different factors that indicators and simulations could overlook or underplay, such as changes in wind velocity, temperature difference, and behaviors not incorporated into the model. This possible discrepancy that we are facing may lead to other scenarios where results from simulation exercise could be less real and concordant to actual performance.

CHAPTER 12 CLOUD COMPUTING FOR UAV SWARMS

The strengths and weaknesses of simulation experiences in UAV swarm research are primarily attributed to aspects such as modeling, scope, and purpose

- **Application in business environments**
  - **Lack of Real-World Context:** Many simulations can partially leave out the important features inherent in real-life situations that allow for low generalization of the findings.
  - **Mismatch with Physical Attributes:** Simulations could potentially not capture the actual property and spatial limitations of the UAV systems and devices under consideration, which could impede the portability of premiums and strategies.
- **Performance discrepancy**
  - **Differences in Performance:** An important analyte is the behavior of the algorithms and methods developed and tested in simulations which may differ when applied to real systems.
  - **Missed Variables:** Often, simulations are incomplete or do not encompass all the variables within the social environment which might lead to differences in performance when simulated and actually implemented.

CHAPTER 12   CLOUD COMPUTING FOR UAV SWARMS

- **Reliability concerns**
    - **Less Reliable Results:** Some of the limitations found in studies include potential limitations of results from the simulation studies which may not represent real-world results and hence can pose a problem to the reliability of the simulation types of research in UAV swarm applications.

Table 12-1. A summary of the work done on testbeds, tutorials, and virtual platforms which can be utilized in UAV swarm creation

Reference	Year issued	Education label
[20]	2011	Free machines training engine in module form: MORSE
[21]	2013	CoppeliaSim (formerly V-REP): A flexible and expandable framework for robot modeling
[22]	2014	Using V-REP and ROS to set up an experiment for UAV vision-based control: An airborne visual examination case study
[24]	2017	AVENS: An innovative flying wireless network model that generates code automatically for unmanned aerial vehicles
[25]	2017	AirSim: High-resolution mechanical and visual modeling for self-driving cars
[15]	2018	FlyNetSim: A free to use, ns-3, and Ard pilot-based synchronized UAV network emulator
[26]	2019	Using the UB-ANC simulator to simulate swarms of unmanned aerial vehicles
[23]	2021	IoTactileSim: An online platform for interactive industrial IoT applications

(continued)

Table 12-1. (*continued*)

Reference	Year issued	Education label
[17]	2021	FlockAI: An ML-driven drone application evaluation suite
[28]	2022	An introduction to and overview of flight control cosimulation with flight simulation programmed and Simulink
[18]	2022	Drones go, three, two, one! A laboratory to launch unmanned aerial vehicle swarm intelligence for dispersed sensing
[29]	2022	GrADyS-SIM: An online of aerial things modeling platform using unmet++/INET
[27]	2022	Multidimensional image analysis and UAV flight control platform powered by MATLAB and the web
[3]	2023	NewSim2023: An enhanced training module that enables practicing of UAV swarm formation and leadership

In the rapidly advancing field of UAV swarms, the interplay between simulation-based research and hardware-centric validation is crucial. This dynamic ensures that theoretical advancements and innovative concepts developed within academic and research settings are effectively translated into practical, real-world applications. The ultimate goal is to achieve systems that are not only efficient and reliable but also safe for deployment across diverse sectors.

CHAPTER 12   CLOUD COMPUTING FOR UAV SWARMS

The Importance of Combining Simulation and Hardware Validation

- **Complementary approaches**

  - **Simulations as a Foundation:** UAV simulations are advantageous in that they offer a controlled environment, as well as reduced costs for testing the allocation and management of UAV swarm algorithms and strategies. It permits the researchers to test a number of different situations and variations of parameters systematically that are vital for the primary model conception verification and optimization.

  - **Hardware for Real-World Validation:** Even then, simulations are very helpful, but they are not a proper substitute for live networks. Hardware testing is inevitable for measuring how effective UAV swarms perform when exposed to practical operations and when handling physical constraints.

- **Ensuring practical applicability**

  - **Bridging Theory and Practice:** Hardware-in-the-loop testing is key to closing the gap between theoretical research and real implementation. This approach is important because it helps to ensure that the development for UAV swarm is feasible and effective in the actual settings.

- **Realistic Testing:** The hardware-centric validation familiarizes UAVs with the additional stability and unpredictability of the nature like unfavorable lightings which are quite hard to mimic during simulations.

Having Publications for Analyzing Research Trends

To understand how researchers approach UAV swarm studies and present their findings, examining a curated list of publications reveals significant trends:

- **Research methodologies**
    - **Simulation-Only Studies:** A significant number of publications are inclined on the simulation with goals being confined to improvements of algorithms within the simulation environment. Many of these studies simply demonstrate that a new idea is conceptually appealing and could work as a foundation for a more exhaustive investigation.
    - **Hybrid Approaches:** Many papers use simulations together with HW experimentation from many papers that are available; a large number of papers include simulations with HW experimentation. As with other types of papers dealing with simulation, these papers generally back up their simulation findings through practical trials of UAVs in real or at least semireal settings. The proposed techniques can be more systematically validated with this method and ascertain if it is fit for use in real-life settings.

CHAPTER 12 CLOUD COMPUTING FOR UAV SWARMS

- **Categorization by research questions**

  - **Algorithm Development:** Some papers focus on the optimization and enhancement of swarm control algorithms, the interactions, and synchronization of the swarm in different contexts and environmental adaption.

  - **Application-Specific Studies:** Other works concentrate on utilizing UAV swarm technology in certain contexts including farming, emergencies, or security. Such publications typically provide summaries of the simulation as well as the hardware testing phases to prove the relevance of their offered solutions.

  - **Communication and Coordination:** The behavior also concerns about effective communication and synchronization for coordinative tasks needed for swarm functionality, about how these can be improved and tested with respect to mathematics and reality either by simulations or constructing the hardware.

- **Perspectives and insights**

  - **Challenges and Solutions:** Real-world scenarios may consist of environmental changes, signal disturbances, and regulatory issues, and the implementation of these in publications is learned from. It is invaluable in learning how such issues can be alleviated by using sophisticated mathematics and the right layout of the system.

- **Future Directions:** Several papers provide suggestions for additional research, highlighting the fact that hardware testing should be carried out more comprehensively in the future and that creating ever more complex and realistic simulations should be the goal of the research.

Data Collection and Questions

This viewpoint does not intend to present an analysis of books as an evaluation, but some components of categories used in data search and evaluation do help in obtaining the understanding of the presumed tendencies.

Web of Science (WoS), which is considered one of the best sources for getting the bibliometric info necessary for such work, was selected for this purpose [13]. To ensure reliability for the first source, another transfer of information input was made through Google Scholar's database.

The nonrefereed manuscripts were, however, rejected. The nonrefereed papers were, however, excluded. To account for the heterogeneity of records and other papers that might not be ranked in WoS, another database was utilized.

To avoid the shortcomings prevalent in earlier structure mapping approaches, a thorough process of dataset refinement was used to determine the state of UAV swarm research. As such, this process was crucial in that any subsequent dataset derived would be timely as well as inclusive of the various works in the emergent and dynamic field. Several crucial processes were performed to prepare the dataset which would be appropriate for the analysis, encompassing important tendencies and development of UAV swarm analysis.

CHAPTER 12 CLOUD COMPUTING FOR UAV SWARMS

Steps in Refining the Dataset for Analysis

- **Exclusion of irrelevant and redundant entries**
  - **Removing Duplicates:** Many of the articles may have been retrieved from various sources and databases within the same context or with different versions of the same research article. These duplicates were however detected and excluded in the study to ensure that they do not influence the results of the study.
  - **Excluding Theses, Reports, and Extended Abstracts:** Articles not published in research journals were also omitted as well as Ph.D. theses, technical reports, and extended abstracts to capture articles contemporary in nature yet more impacting the current research development, either as research articles or conference proceedings. This leads to the reduction of the volume within the dataset and the success for standardized and directly comparable research outputs.
- **Inclusion of proceedings**
  - **Broader Range of Contributions:** The source was not limited to peer-reviewed journal articles, and to cover a broad view of the research going on in this area, proceedings of conferences were also included. The conference proceedings include presentations of original studies that present ideas and findings inaccessible in technologies of UAV swarm systems.

- **The proposed approach can then be supplemented with a manually conducted review for quality and relevance.**
  - **Evaluating Publisher and Author Affiliations:** In addition, because all articles were the focus of blog posts or web content, each piece was manually reviewed by the author to determine the quality and reliability of the publication. The reputation was taken into account as the quality and the context of the research; the affiliation country of the first author was also taken into account.
  - **Assessing Presence of Hardware Tests:** Articles were assessed regarding if they portrayed the utilization of equipment tests. This differentiation is important because it establishes exactly how the research results were put into practice and tested, between theoretical and purely modeling- and simulation-type research and research that employed some form of experimentation on the real world.
  - **Mentions of Future Hardware Experiments:** The review was also specific if the articles showed plans for future tests on the device (hardware). This aspect gives a clue of the direction of the research and the preparedness of the proposed procedures for practice.
  - **Type of Swarm Implementation:** The proposed swarm implementation type can be classified in each study. This includes decisions about whether the research was done on homogeneous/heterogeneous swarms, the level of autonomy a

swarm exhibited, as well as what specific tasks the swarms were executing – all of which are crucial to grasp the range and variability of application of swarm UAVs.

Creating an Analysis-Ready Dataset

The above-said steps were, therefore, considerate and critical in order to filter the dataset so that the end result matches the existing state of the art on UAV swarm research. The procedure of selecting and removing the content enhanced the work qualitatively which helped to make the dataset more suitable and useful for further analysis. The provided approach of analyzing multiple sources enables for a comprehensive review of the development of UAV swarm research with respect to utility and realism.

Outcome and Insights

- **Balanced representation**
  - A careful selection of the data offers the opportunity to analyze the current state of the field inclusively comparing simulation-based research and research that involves the usage of hardware, regarding the translation of theoretical findings into practical applications [19].

- **Nature of research and practice**
  - Connected with the previous point, the articles included in the dataset that contain hardware tests for the proposed algorithms or describe plans for future experiments pinpoint the sole focus of the researchers on the practical applicability of the developed approaches, as well as the efforts undertaken to assess the performance of the designed systems on real-world scenarios.

CHAPTER 12   CLOUD COMPUTING FOR UAV SWARMS

- **Geographical and institutional factors for diverse agency**
  - The investigation of authors according to affiliation demonstrates the geographical distribution of research activity which shows us the areas with high activity in UAV swarm development and how UAV swarm area is enhanced with international collaboration.
- **Innovation trajectories**
  - Another positive note about the review is that it encompasses different forms of swarm implementations to show the general direction that the UAV swarm research is towing, ranging from collaborative work to execute tasks to development of enhanced autonomous behaviors and communication techniques.

The data was collected and analyzed for the patterns and then asking four questions to get more insights about the current state of affairs when it comes to simulation with respect to the swarm development when physical tests are done or not. Summarizing these analytical research questions, as well as conclusion, we have Table 12-2.

Simulation Trends in Comparison to Hardware-Based Experiments

This section presents the results of an analysis and presentation of the methods used by field researchers across several parameters. Any concepts that are defined are defined appropriately. Based on the content of each

CHAPTER 12   CLOUD COMPUTING FOR UAV SWARMS

reviewed article, a key phrase associated with UAV swarm development was allocated. In cases where a keyword was not explicitly stated in the document, one was assigned.

The dataset included 1813 articles in total. Initially, the year of publication was used to categorize these articles. The number of publications reviewed in a given year is shown in Figure 12-3.

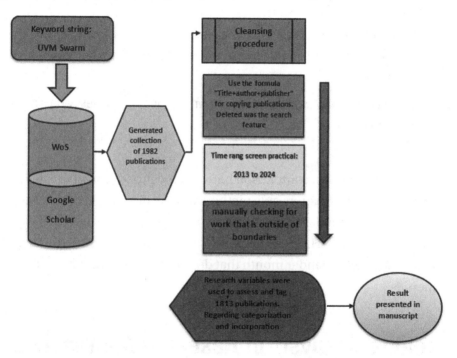

Figure 12-2. The curating process of the databases. Unmanned aerial device (UAV) and Web of Science (WoS)

CHAPTER 12   CLOUD COMPUTING FOR UAV SWARMS

Table 12-2. *Analysis tasks to assess the well-chosen dataset and look for patterns*

Q1	Did the study use physical experiments in addition to simulations to verify the effectiveness of the suggested methodology?
Q2	Did the authors explain why hardware testing weren't tried if any weren't included?
Q3	Did the researchers include enough information or a plan for carrying out hardware tests such that other people might duplicate their work using hardware?
Q4	Were there any new patterns in the way writers modeled studies pertaining to certain regions?

Figure 12-4 shows a comparison between the proportion of articles that adopted the simulation-only method and the proportion of articles with hardware implementations for the number of papers analyzed annually. The figure's Hardware-N% indicates the percentage of trainings evaluated for each calendar month that do not require computer hardware properties.

Regional Analysis of Research Approaches

In further dissecting the overall quantity of relevant projects released by area, a segregation based on the utilization of hardware was conducted. Figure 12-5 delineates the distribution of articles presenting hardware use (Hardware-Y) versus those that do not (Hardware-N). This breakdown offers insights into the prevalence of different research methodologies across regions, shedding light on the adoption of hardware-based and simulation-only approaches in UAV swarm studies globally. Moreover, Figure 12-6 provides a visual representation of the research questions

CHAPTER 12   CLOUD COMPUTING FOR UAV SWARMS

that the information trend evaluation and curation answer, aiding in the comprehension of research trends and patterns identified through data visualization techniques. Understanding the distribution of research approaches and visualizing analysis outcomes are crucial steps in unraveling the intricacies of UAV swarm research on a global scale.

The amount of work that researchers can replicate in the real world may be limited by cost, availability, and trade restrictions when obtaining gear and peripheral supporting technology in real life. This is in contrast to the ease with which these items can be simulated using simulation platforms and models. Second in importance are rules governing airplanes, airspace, and self-driving vehicles at the local, state, and federal levels. Lastly, as a number of recent works suggest using drones to travel over terrain, in hostile combat scenarios, and over extended periods of time, it makes more sense to test and report simulation results than to try to recreate scenarios.

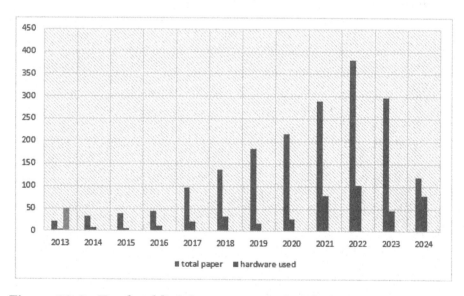

Figure 12-3. *Total publications contrasted with the annual total of papers utilizing hardware solutions*

CHAPTER 12 CLOUD COMPUTING FOR UAV SWARMS

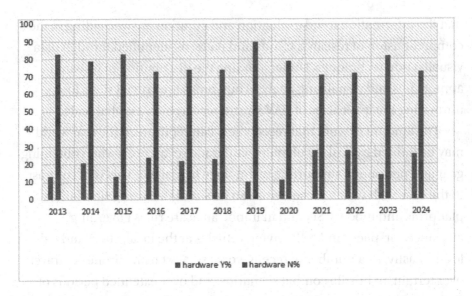

Figure 12-4. *An annual comparison, expressed as a proportion, between the number of articles that used hardware and those that didn't*

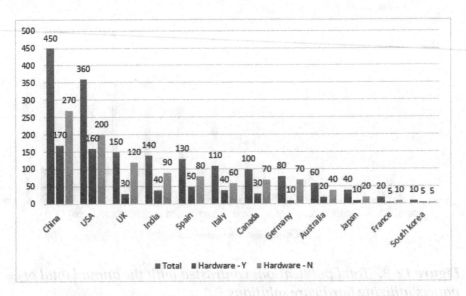

Figure 12-5. *A proportional analysis by region of the total number of articles, hardware that is present, and hardware that is not*

466

CHAPTER 12 CLOUD COMPUTING FOR UAV SWARMS

In the realm of swarm robotics, the scale of implementation in terms of the number of agents directly influences the complexity and feasibility of hardware experiments. This observation is vividly depicted where studies are categorized based on the number of agents employed [15]. The trend highlighted in this figure indicates that researchers predominantly favor hardware experiments for smaller swarm sizes, while larger swarms tend to be less frequently implemented in physical hardware setups [16].

Table 12-4 provides a comprehensive summary of insights derived from the analysis of the curated data, supported by figures and tables within the study. This summary highlights several critical aspects of UAV swarm research, particularly regarding the balance between hardware and simulation-based experimentation.

Key Insights from Table 12-4

- **Preference for simulations in large swarm studies**

    - **Scalability:** As indicated in the smaller swarms are typically used in hardware implementations due to the challenges associated with scaling up. On the other hand, with simulations, researchers are able to recommend and try out much larger swarms, whereby the limitation to implement this will be on factors such as costs, space, and the available resources.

    - **Flexibility and Risk Mitigation:** Computer simulations also allow the researcher a level of freedom in testing various input conditions and vary them within the scope of what could be largely expected and rarely can be observed in most hardware testing – edge cases and failure conditions.

CHAPTER 12 CLOUD COMPUTING FOR UAV SWARMS

- **Simulation as a validated method for performance evaluation**
  - **Trend Toward Acceptance:** An increase in the recognition of the model result as an authentic way of validating the efficiency of the UAV swarm is experienced. This trend can be seen from the relative recent increase in literary works that use simulation as a tool to model as well as validate swarm behaviors and algorithms.
  - **Supporting Evidence:** In order to justify simulations, Grench and Mort presented their simulation outcomes to correspond to experimental results from minisized hardware environments. Some authors confirm their simulation investigations by developing miniature mechanical robots, indicating reliability of similarly sized simulations for the giant swarms.
  - **Rapid Prototyping and Testing:** As an area-level tactic of progressive disclosure, simulations are excellent at quick prototyping and early exploration of novel algorithms and ideas. This increases the efficiency of research as it can be carried out on short, relevant timescales, affording the opportunity to make changes and reassess rapidly, which is not feasible when waiting for hardware to be distributed.

- **Comprehensive Analysis:** While hardware prototypes can be tested in both normal and abnormal cases, simulations give further specifics about the system and its behaviors under numerous conditions, which can help the analysts and designers better understand the swarm behaviors that can be hard to investigate in other ways.

Challenges in Bridging Simulation and Real-World Implementation

- **Simulation Fidelity:** Even though simulations have many benefits, their accuracy is reliant on how realistic the models are that are employed. When moving to real-world hardware, choices and simplifications made during simulation may cause disparities.

- **Real-World Variables:** While moving up from computer to hardware, factors like sensor noise, hardware defects, and ambient variability are typically challenging to recreate successfully in simulations.

CHAPTER 12 CLOUD COMPUTING FOR UAV SWARMS

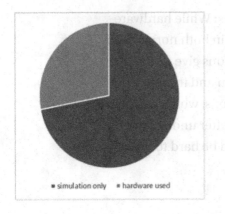
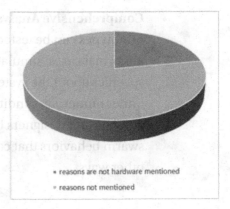
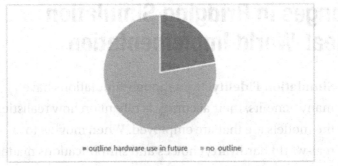

Figure 12-6. *Pie charts displaying data from quarters 1, 2, and 3 show trends*

CHAPTER 12   CLOUD COMPUTING FOR UAV SWARMS

Table 12-3. *Details about the number of pieces that include enough information to permit the use of the devices in the future*

Year	Hardware not used	The amount of research that suggested conducting hardware experiments
2013	18	0
2014	26	0
2015	33	2
2016	33	7
2017	74	14
2018	105	18
2019	165	11
2020	190	31
2021	210	23
2022	279	19
2023	251	21

Bearing this complexity in mind, comprehending the way scholars harness hardware while studying UAV swarms is important to the broader appreciation of a contested and emerging domain. This evolution is even more influenced by the advancement and availability of more complex systems used in simulations. These platforms help the researchers to verify the efficiency of the proposed techniques effectively and at times without the problems that may arise when developing practical hardware solutions. Table 12-3 considers the same research theme numerically, in terms of how many articles it contains with the phrase "hardware testing" or "testing of hardware" and how many are arguing for some form of hardware experiments in the period from 2013 through 2024. For the studies conducted in 2013, there was no specific experiment designed for testing the compiled hardware, while for the experiments conducted in 2023, there were 21 tests that were designed for this purpose. However, as evidenced by the current analysis of 2021 (210 articles) and 2022 (279 articles), most of the research exclusively used simulations as the primary means of data collection due to costs, risks, and numerous opportunities to alter parameters compared to actual situations. Another essential area is the manifestation of PC interfaces for operating a higher-level hardware emulation of conditions including disturbances and breakdowns [17].

The Role of Hardware in UAV Swarm Research

- **Essential for real-world validation**
  - **Practical Testing:** Critical for checking the correctness of UAV swarm principles and efficiency under real-world scenarios. Hardware experiments are necessary as they reveal the overheard realities of practical limitations and practical working difficulties which are hard to model while simulating.
  - **Complex Interactions:** Certain interactions, such as those involving physical contact or detailed environmental responses, can only be accurately tested using real hardware.
- **Challenges in hardware implementation**
  - **Resource Intensive:** Deploying and managing large-scale swarms in hardware is resource-intensive. It requires significant investment in terms of finances, infrastructure, and technical support.

CHAPTER 12   CLOUD COMPUTING FOR UAV SWARMS

- **Scalability Issues:** To highlight, Figure 12-7 shows that, often, the number of agents in hardware experiments is small due to these restrictions. Swarm algorithms are often targeted to be used on very large populations; however, this often limits overall experimentation by restricting the number of possible participants than can be used in experiments to those that can be effectively utilized in the simulations and theoretical modeling.

The Rise of Simulation Platforms

- **Increased accessibility and sophistication**

  - **Advanced Simulation Tools:** Recent simulations have proven to be very efficient and capable, providing a great amount of fidelity to actual UAV swarm characteristics. These applications provide the facility to test and optimize the swarm algorithms and the strategies at different levels without involving any type of physical system [18].

  - **Cost-Effective Exploration:** It offers a cheap setup to use in lieu of the hardware experiments given that they can be used to experiment various conceptions before an actual hardware can be used.

- **Validation and hybrid approaches**

  - **Complementary Use:** A times, researchers utilize simulations and some mini- and comparatively small-scaled hardware experiments. They build up the first level of confidence and get a clear picture as to what kind of hardware they are dealing with, through simulations, in a step-by-step process these results are then validated and again and again by hardware experiments.

- **Reducing Risk:** By screening and likely modifying the likely problems before field testing the hardware, this combined methodology lowers the risks and expenses that might be related to hardware testing.

Future Research Directions

- **Expanding bibliometric analysis**
  - **Broader Scope:** To advance the concerns presented in this paper, it is crucial to perform a more extensive bibliometric analysis of research conducted on aerial swarms, including a broader selection of databases and a longer timeframe. This would be more informative about the trends as well as the difficulties and innovations being experienced in the field.
  - **Diverse Sources:** A wider range of stocks and a decade range would yield a broader picture and valuable information about new technologies and methodologies in the field.
- **Longitudinal studies**
  - **Temporal Trends:** Such an analysis can help identify the trends in how the balance between physical implementation and simulation has occurred and how the new developments are influencing the future of the field of UAV swarm studies.

CHAPTER 12 CLOUD COMPUTING FOR UAV SWARMS

- **Impact of Innovations:** It is possible to evaluate the positive influence of the newly developed simulation platforms and the improvement of the hardware technologies at each stage in the context of the given field to determine the essence of this innovations' contribution to the development of simulation.

As for the future development of UAV swarm research, one of the biggest problems is the inconsistency and variety of scenarios that are quite realistic, which take place in UAVs. Despite advanced simulation environments for UAV swarm testing, they lack emulation of complex dynamic conditions. AI advancements have revolutionized sectors like IoT, energy, quantum computing, and image and signal analysis [30–32].

Real-World Disruptions and Simulation Limitations

- **Weather conditions**
  - **Unpredictable Environmental Factors:** Weather effects and climate conditions like rain, snow, dust, as well as hail disrupt the consistency of weather conditions in affects, which are almost impossible to model in a real-life setting. All of these weather conditions have been shown to impact UAVs in several ways including changes in flight characteristics and flying environment, obscuring visibility, and damaging the UAV physically [14]. Some of these characteristics could be accurately simulated; however, the complexity and flexibility of these conditions cannot be addressed by most simulators.

- **Core Disturbance Classification:** It is most often based on a fundamental categorization of disturbances, for instance, wind models in the case of wind turbines or the effects of temperature. However, the real-world weather is not solely a single variable changing over time but requires not only the change and systematization but also combination of the factors occurring in nature.

- **Material and component durability**

  - **Material Stress Testing:** The aeronautical components of the UAVs including the organic materials, composite materials, and metals used face stress and wear out due to environmental factors. Mimicking and testing materials' aging impact on UAV performance in long-endurance missions is complex due to actual practice revealing decisive areas and failures hard to predict and model.

  - **Component Failures:** A UAV system has various complex subsystems where each of them may have a unique way of failure. Wrong functioning of sensors, actuators, and communication modules can occur due to the manufacturing defects, deteriorating effects of environmental conditions, and excessive usage. Current simulations may include modeling of one or another component failure to a certain extent; nevertheless, the simulation of multiple components' failures and their interaction remains highly questionable.

CHAPTER 12 CLOUD COMPUTING FOR UAV SWARMS

- **Battery performance and degradation**

    - **Deteriorating Battery Efficiency:** As energy density in UAV batteries decreases on average over time, both flight time and sensor time are affected. While normal simulations can incorporate performing batteries or basic deterioration regimes in batteries, actual reduction of battery capacity over periods of time or in different environments that batteries are subjected to during dynamic real-world uses can be more complex and require more advanced modeling techniques [29]. Capacity fading due to daily routines like charging-discharging cycles, high/low temperature, and the use of high power can be attributed to the failure mode behavior of the battery.

    - **Impact on Mission Capabilities:** In this way, reduced battery efficiency has a rather practical impact on the UAV's functioning and characteristics, for instance, flight endurance and sensor capacity. This loss of capability can have major consequences on operating plans and mission performances which are noteworthy in case of highly synchronized swarms.

- **Bridging the gap between simulations and real-world conditions**

    However, there are still some shortcomings associated with the use of simulators as they try to closely mimic the real-life scenarios. In this connection, continuing efforts to enhance simulation technologies are likely to narrow the gap between virtual and actual environments. Ongoing efforts focus on designing better models so that simulation results are more detailed and realistic, and the effects of different features of the environment and detailed analysis of behavior of components of the system are included.

- **Enhanced environmental modeling**
  - **Dynamic Weather Simulation:** Efforts are ongoing to enhance actual weather models that will be capable of providing detailed interactions between UAVs and other related factors in the best real-life models. This includes mimicking how weather patterns develop and how UAVs are required to market themselves to new conditions.
  - **Incorporating Real-World Data:** The use of actual data helps to increase the realism of the models, as long as the simulations are constructed correctly. It involves retaining the basic tenets of testing but incorporates the history of weather and real-time updates to make the test conditions more realistic.
- **Advanced material and failure models**
  - **Material Fatigue and Wear:** Sophisticated modeling tools are available to predict the materials necessary to understand fatigue of the material and the wear out process reducing the stress all through the exposure time in a given environment. This is useful in estimating the durability, and expected routine maintenance period, of UAVs.

- **Comprehensive Failure Analysis:** Software's also becoming more complex where it is now possible to simulate not only the individual components but also their interconnections when a single component fails in the context of a system and whether this failure will cause other components in the system to fail also. The outcomes of this research are designed to align to the following recognized research questions: this action plan is done holistically to analyze how individual failures affect the overall swarm performance.

- **Realistic battery and energy management**
  - **Detailed Battery Models:** More sophisticated battery models are being developed to simulate the complex behaviors associated with battery degradation and energy management under varying operational conditions. These models can help predict how battery performance will change over time and under different usage scenarios.
  - **Energy Efficiency Strategies:** Research is also focused on optimizing energy usage within UAV swarms, developing strategies to manage power consumption and extend operational life, even as batteries degrade.

CHAPTER 12 CLOUD COMPUTING FOR UAV SWARMS

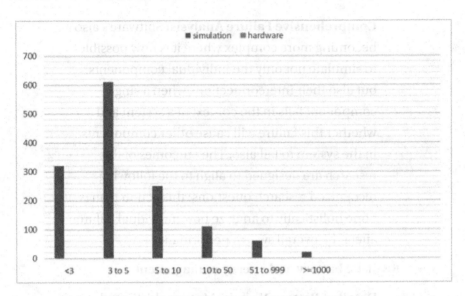

Figure 12-7. *The number of simulation-based implementations (as opposed to hardware validations) broken down by the quantity of agents employed*

Table 12-4. *A synopsis of the analysis that was done and the backing information that was shown in Figures*

Supportive information	Analysis
Figure 12-5	Of the publications in the dataset, 25% were from China, 21% from the United States, and 8% were from the UK. Of these, **61%**, **54%**, and **78%** of the work don't employ computer hardware to test presentations
Figure 12-6	It is not necessary for authors to aggressively defend the decision to restrict performance evaluation to simulation alone

(*continued*)

Table 12-4. (*continued*)

Supportive information	Analysis
Figures 12-3 and 12-4	The number of papers released has increased annually since 2013, but the number of articles suggesting the usage of hardware in the future has not increased. This suggests that further performance validations using modeling-only simulation techniques have been approved
Figure 12-7	The number of units that the suggested technique calls for is a key consideration when choosing between hardware validation and simulation

Tall levels are all important presumptions in modeling simulations. Although more efforts are being made at the model level as of late, some question the benefits of this strategy. The advantage of this simulation-only approach to UAV swarm research is that it can impact improvements in simulator software features and systems. It needs to be observed if this simulation-based validation strategy will just alter how the current research community perceives and acknowledges swarm performance outcomes.

Conclusion

In conclusion, cloud computing integration with UAV swarms offers a revolutionist perspective of numerous applications across various domains. Applying UAV swarms in target applications such as precision agriculture, aerial surveillance, disaster response, or entertainment has emerged as highly effective and efficient in recent years, yet their application in real-life scenarios still raises numerous issues. Some of the challenges include changes in the environment and the ability to interfere with the signals, stringent regulatory, and policy standards among others which require the enhancement of the need for complex communication protocols, reliable built-in-self-repair mechanisms, and learning capability of algorithms. It could be noted that focusing on the given paper allowed identifying the mentioned obstacles in the process of avoiding larger-scale discrepancies between simulation studies and real-world experiments, which proves the necessity of the multidisciplinary approach to the presented task. As UAV swarm technology matures, further developments, in both simulation capabilities of UAV swarms as well as hardware testing of swarming capabilities, will continue to play a major role in fully unlocking the potential of those systems. Subsequent studies should aim at improving the realism of simulations, especially on the newly proposed aspects and the dependability of the hardware to enable swarms of UAVs to perform efficiently and safely in new and open environments.

References

[1] P. Radoglou-Grammatikis, P. Sarigiannidis, T. Lagkas, and I. Moscholios, "A compilation of UAV applications for precision agriculture," *Computer Networks*, vol. 172, p. 107148, 2020.

[2] X. Zhang, W. Zhao, C. Liu, and J. Li, "Distributed Multi-Target Search and Surveillance Mission Planning for Unmanned Aerial Vehicles in Uncertain Environments," *Drones*, vol. 7, no. 6, p. 355, 2023.

[3] J. Detka, H. Coyle, M. Gomez, and G. S. Gilbert, "A Drone-Powered Deep Learning Methodology for High Precision Remote Sensing in California's Coastal Shrubs," *Drones*, vol. 7, no. 7, p. 421, 2023.

[4] Riaz, N., Shah, S. I. A., Rehman, F., & Gilani, S. O. (2020). An Intelligent Approach to Detect Actuator Signal Errors Based on Remnant Filter. In Intelligent Technologies and Applications: Second International Conference, INTAP 2019, Bahawalpur, Pakistan, November 6-8, 2019, Revised Selected Papers 2 (pp. 675-683). Springer Singapore.

[5] Phadke, A., and Medrano, F.A. 2024. Increasing Operational Resiliency of UAV Swarms: An Agent-Focused Search and Rescue Framework. Aerosp.Res. Commun. 1.doi:10.3389/arc.2023.12420.

[6] Xiong, T., Liu, F., Liu, H., Ge, J., Li, H., Ding, K., and Li, Q. 2023. Multi-Drone Optimal Mission Assignment and 3D Path Planning for Disaster Rescue. Drones, 7.doi:10.3390/drones7060394.

[7] Alzahrani, O. S. Oubbati, A. Barnawi, M. Atiquzzaman, and D. Alghazzawi, "UAV assistance paradigm: State-of-the-art in applications and challenges," *Journal of Network and Computer Applications*, vol. 166, p. 102706, 2020.

[8] Phadke, A., Medrano, F.A., Sekharan, C.N., and Chu, T. 2023b. Designing UAV Swarm Experiments: A Simulator Selection and Experiment Design Process. Sensors, 23: 7359. doi:10.3390/s23177359.

[9] Pitonakova, L., Giuliani, M., Pipe, A., and Winfield, A. 2018. Feature and Performance Comparison of the V-REP, Gazebo and ARGoS Robot Simulators. Springer International Publishing, Cham. pp. 357-368.

[10] Tselegkaridis, S., and Sapounidis, T. 2021. Simulators in Educational Robotics: A Review. Educ. Sci. 11.doi:10.3390/educsci11010011.

[11] Calderón-Arce, C., Brenes-Torres, J.C., and Solis-Ortega, R. 2022. Swarm Robotics: Simulators, Platforms and Applications Review. Computation, 10.doi:10.3390/computation10060080.

[12] Riaz, N., Shah, S. I. A., Rehman, F., Gilani, S. O., & Udin, E. (2020). A Novel 2-D Current Signal-Based Residual Learning With Optimized Softmax to Identify Faults in Ball Screw Actuators. IEEE Access, 8, 115299-115313.

[13] Iqbal, U., Riaz, M.Z.B., Zhao, J., Barthelemy, J., and Perez, P. 2023. Drones for Flood Monitoring, Mapping and Detection: A Bibliometric Review. Drones, 7.doi:10.3390/drones7010032.

[14] Baidya, S., Shaikh, Z., and Levorato, M. 2018. FlyNetSim: An Open Source Synchronized UAV Network Simulator based on ns-3 and Ardupilot. In Proceedings of the 21st ACM International Conference on Modeling, Analysis and Simulation of Wireless and Mobile Systems. Montréal, Québec, Canada.

[15] Humayoun, M., Sharif, H., Rehman, F., Shaukat, S., Ullah, M., Maqsood, H., ... & Chandio, A. H. (2023, March). From Cloud Down to Things: An Overview of Machine Learning in Internet of Things. In 2023 4th International Conference on Computing, Mathematics and Engineering Technologies (iCoMET) (pp. 1-5). IEEE.

[16] Trihinas, D., Agathocleous, M., Avogian, K., and Katakis, I. 2021. FlockAI: A Testing Suite for ML-Driven Drone Applications. Future Internet, 13. doi:10.3390/fi13120317.

[17] Qin,C., Candan, F., Mihaylova, L.S., and Pournaras, E. 2022. 3, 2, 1, Drones Go! A Testbed to Take Off UAV Swarm Intelligence for Distributed Sensing. arXiv preprint arXiv:2208.05914.

[18] Riaz, N., Shah, S. I. A., Rehman, F., & Gilani, S. O. (2020). An Approach to Measure Functional Parameters for Ball-Screw Drives. In Intelligent Technologies and Applications: Second International Conference, INTAP 2019, Bahawalpur, Pakistan, November 6–8, 2019, Revised Selected Papers 2 (pp. 398-408). Springer Singapore.

[19] Echeverria, G., Lassabe, N., Degroote, A., and Lemaignan, S.E. 2011. Modular open robots simulation engine: MORSE. In International Conference on Robotics and Automation. IEEE, Shanghai, China.

[20] Rohmer, E., Singh, S.P.N., and Freese, M. 2013. CoppeliaSim (formerly V REP): a versatile and scalable robot simulation framework. In IEEE/RSJ International Conference on Intelligent Robots and Systems. IEEE, Tokyo, Japan.

[21] Olivares-Mendez, M.A., Kannan, S., and Voos, H. 2014. Setting up a testbed for UAV vision based control using V-REP & ROS: A case study on aerial visual inspection. In 2014 International Conference on Unmanned Aircraft Systems (ICUAS).

[22] Islam, M.Z., Ali, R., Shahzad, Haider, A., and Kim, H. 2021. IoTactileSim: A Virtual Testbed for Tactile Industrial Internet of Things Services. Sensors (Basel), 21.doi:10.3390/s21248363.

[23] Marconato, E.A., Rodrigues, M., Pires, R.M., Pigatto, D.F., Filho, L.C.Q., Pinto, A.S.R., and Branco, K.R.L.J.C. 2017. AVENS——A Novel Flying Ad Hoc Network Simulator with Automatic Code Generation for Unmanned Aircraft System.

[24] Shah, S., Dey, D., Lovett, C., and Kapoor, A. 2017. AirSim: High-Fidelity Visual and Physical Simulation for Autonomous Vehicles. doi:10.48550/arXiv.1705.05065.

[25] Modares, J., Mastronarde, N., and Dantu, K. 2019. Simulating unmanned aerial vehicle swarms with the UB-ANC Emulator. Int. J. Micro Air Veh. 11.doi:10.1177/1756829319837668.

[26] Aliane, N., Muñoz, C.Q.G., and Sanchez-Soriano, J. 2022. Web and MATLAB-Based Platform for UAV Flight Management and Multispectral Image Processing. Sensors (Basel), 22.doi:10.3390/s22114243.

[27] Horri, N., and Pietraszko, M. 2022. A Tutorial and Review on Flight Control Co-Simulation Using Matlab/Simulink and Flight Simulators. Automation, 3: 486–510. doi:10.3390/automation3030025.

[28] Lamenza, T., Paulon, M., Perricone, B., Olivieri, B., and Endler, M. 2022. GrADyS-SIM——A OMNET++/INET simulation framework for Internet of flying things.

[29] Riaz, N., Shah, S. I. A., Rehman, F., & Khan, M. J. (2021). An Intelligent Hybrid Scheme for Identification of Faults in Industrial Ball Screw Linear Motion Systems. IEEE Access, 9, 35136-35150.

[30] H. Sharif, F. Rehman and A. Rida, "Deep Learning: Convolutional Neural Networks for Medical Image Analysis - A Quick Review," 2022 2nd International Conference on Digital Futures and Transformative Technologies (ICoDT2), Rawalpindi, Pakistan, 2022, pp. 1-4, doi: 10.1109/ICoDT255437.2022.9787469.

[31] A. Ashfaq, M. Kamran, F. Rehman, N. Sarfaraz, H. U. Ilyas and H. H. Riaz, "Role of Artificial Intelligence in Renewable Energy and its Scope in Future," 2022 5th International Conference on Energy Conservation and Efficiency (ICECE), Lahore, Pakistan, 2022, pp. 1-6, doi: 10.1109/ICECE54634.2022.9758957.

[32] I. Manan, F. Rehman, H. Sharif, N. Riaz, M. Atif and M. Aqeel, "Quantum Computing and Machine Learning Algorithms - A Review," 2022 3rd International Conference on Innovations in Computer Science & Software Engineering (ICONICS), Karachi, Pakistan, 2022, pp. 1-6, doi: 10.1109/ICONICS56716.2022.10100452.

CHAPTER 13

Enhancing UAV Swarm Security with Blockchain: Solutions for Secure Communication, Data Integrity, and Autonomous Operations

Inam Ullah Khan[1], Taj Rahman[1], Asim Zeb[2,] *, Inam Ullah[3], Inayat Khan[4], M. A. Al-Khasawneh[5,6,7]

[1]Department of Computer Science, Qurtuba University of Science & Information Technology, Peshawar, Pakistan.
[2]Department of Computer Science, Abbottabad University of Science and Technology, Abbottabad 22010, Pakistan
[3]Department of Computer Engineering, Gachon University, Seongnam 13120, Republic of Korea
[4]Department of Computer Science, UET Mardan, Pakistan
[5]School of Computing, Skyline University College, University City Sharjah, 1797, Sharjah, UAE
[6]Applied Science Research Center. Applied Science Private University, Amman, Jordan
[7]Jadara University Research Center, Jadara University, Irbid, Jordan

[1]inam1software@gmail.com, [1]tajuom@gmail.com, [2]asimzeb1@gmail.com, [3]inam.fragrance@gmail.com, [4]inayatkhan@uetmardan.edu.pk, [5,6,7]mahmoudalkhasawneh@outlook.com

Corresponding Author: Asim Zeb (asimzeb1@gmail.com)

CHAPTER 13   ENHANCING UAV SWARM SECURITY WITH BLOCKCHAIN: SOLUTIONS FOR SECURE COMMUNICATION, DATA INTEGRITY, AND AUTONOMOUS OPERATIONS

Introduction

Blockchain technology is expanding beyond cryptocurrencies and is being used in various smart-tech products, including drones and UAVs. Securing and ensuring the reliability of these systems is crucial for smart cities that frequently use them [1]. Blockchain technology, such as decentralization, immutability, and transparency by nature of its contextual properties, is substantiated to be the very ideal solution in solving these intricate security-related concerns concerning UAV swarms.

UAV swarms are employed in various domains such as surveillance (e.g., security), disaster management, agricultural monitoring, logistics, and environmental monitoring among others [2]. Aiden Ko, founder of Point One Technologies, describes from his work in swarm intelligence algorithms, "Swarms need to communicate with each other so that they can have real-time data exchange and make collaborative decisions on how things should move together". At the same time, however, more interconnected also means more exposure to security issues like unauthorized entry and data manipulations with cyberattacks. These vulnerabilities are frequently inadequately addressed by traditional centralized control systems because of their reliance on single points of failure and limited scalability.

Blockchain offers a secure solution to security issues by utilizing the distributed nature of swarm systems and ensuring no single point is compromised by failure [3]. Any UAV in the swarm can serve as a node within a blockchain network, helping to verify transactions and keep data integrity. The distributed architecture allows more robust, secure, and scalable swarms.

Ever since technology caught up with the concept of drones, it has been one of those things that have revolutionized many segments like military, agriculture logistics, and surveillance. UAV swarms, on the other hand, where numerous autonomous drones can work together simultaneously, is a more practical approach in terms of efficiency and scalability as well as redundancy. Nevertheless, the additional complexity and interconnectivity of UAV swarms result in severe security concerns [4]. This comes in handy below water when the swarm wants to communicate wirelessly but has other applications as well including at-scale cyber defense, telecom resiliency, and distributed cloud storage.

Blockchain technology was designed initially to secure the foundation of diseases such as Bitcoin but has developed into an exciting option for protection around a range of use cases. These properties along with its decentralized, immutable, and transparent nature make it a potential candidate to handle the security challenges of UAV swarms [1]. The use of blockchain enables a safe and secure framework for UAV swarm operations only by allowing each communication between drones to be tamperproof and make trustful collaborative decisions.

Role of UAV Swarm Security

To enable its deployment and functioning across a variety of sectors, security provision for unmanned aerial vehicle (UAV) swarms is critical. UAV swarms are efficient, coverage-rich, and redundant drone fleets that prioritize security to prevent malicious activities, ensure trustworthy operations, and safeguard sensitive data [5].

CHAPTER 13   ENHANCING UAV SWARM SECURITY WITH BLOCKCHAIN: SOLUTIONS FOR
SECURE COMMUNICATION, DATA INTEGRITY, AND AUTONOMOUS OPERATIONS

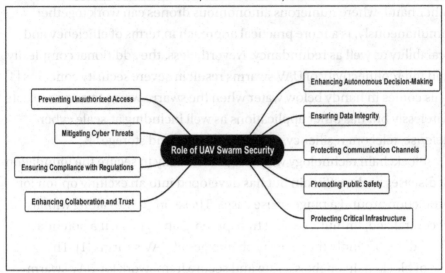

Figure 13-1. Role of UAV swarm security

UAV swarms rely on secure communication for coordinating their behavior and sharing information from different entities that may suffer eavesdropping, jamming, or spoofing attacks. UAV systems, secure communication frameworks, and encryption protocols ensure data exchanged between UAVs and control stations [6] from unauthorized access or tampering by adversaries during operations to prevent any adversary attack on the flight system. Since all the decisions in UAV swarm operations are based on data collected by drones, maintaining accurate information through this entire process is key (line 9). Blockchain provides a ledger that is decentralized and immutable making sure the data stored on it is correct.

Once again, recording data on a blockchain makes it nearly impossible to change or retrofit any mistakes into the information collected and renders them too traceable – all while ensuring that every piece of data is trustworthy. A certain level of operational security is a prerequisite for safe autonomous decision-making capabilities in UAV swarms. Autonomous

CHAPTER 13   ENHANCING UAV SWARM SECURITY WITH BLOCKCHAIN: SOLUTIONS FOR
                SECURE COMMUNICATION, DATA INTEGRITY, AND AUTONOMOUS OPERATIONS

drones use algorithms and artificial intelligence to process data while making real-time [1] decisions that inform them during their missions. However, it is very important to secure these algorithms and prevent malicious modification successfully as their integrity directly impacts the reliability/security of autonomous operations. Blockchain ensures that software updates and decision-making algorithms are verified as integrity.

Facility intrusion in UAVs and their control system results in severe consequences such as hijacking and misuse of drones. Blockchain-powered identity management ensures unauthorized access to UAVs, utilizing user data storage, verification, and processing to verify identities for secure platform interaction. Cyber threats [7] encompassed malware attacks, denial of service (DoS), and zero-day attack languages as well as UAV swarms and electronic warfare vectors. The use of tools like IDSs and frequent security audits can help detect these threats as well, though performing such operations in a regimen will require advanced cybersecurity measures. With a blockchain record of all transactions and interactions within the UAV network, cybersecurity could be greatly improved by making any cyberattacks detected at once.

In accordance with the provisions of regulatory frameworks and agreements, it is necessary to conduct activities honestly using UAV swarms. The regulations for privacy, keeping operations safe, and preventing malicious activities all require that a particular security measure is in place. With a blockchain-based solution, it can easily be proven that the flight happened according to regulation and data was processed securely even as unmanned aerial vehicles leverage this feature. In multistakeholder environments, such as disaster response and humanitarian aid settings, many organizations may come together to cooperate in deploying swarms of UAVs. The need to secure these collaborative endeavors is paramount for the confidence of all stakeholders. Utilizing blockchain ensures that data sharing is

CHAPTER 13   ENHANCING UAV SWARM SECURITY WITH BLOCKCHAIN: SOLUTIONS FOR
             SECURE COMMUNICATION, DATA INTEGRITY, AND AUTONOMOUS OPERATIONS

secured and transparent to all involved parties who can directly check the authenticity of shared information. This results in making UAV swarm operations more effective due to better collaboration and trust. Monitoring and protecting critical infrastructure such as power grids, pipelines, or a transportation network provides one key task for swarms of UAVs. Security of these UAV operations is important to ensure that they don't get interrupted and ruined. The UAV swarms and the critical infrastructure they operate around require robust security standards, including encryption and blockchain-based data integrity [8]. The safety of UAV swarms is a prerequisite for safer urban applications with UAM and in dense environments. This way, secure operations can guarantee that incidents like collisions, unauthorized surveillance, and sabotage are not possible. Adhering to strict security practices and using blockchain for data transparency and integrity are vital in preserving the trust of the general public, ensuring that UAV swarms can be deployed safely.

Blockchain Application

UAV swarm operations have always been a hot topic in the blockchain community and are renowned for their decentralization and immunity to tampering with historical transaction data properties of blockchain technology [9]. There are several points in UAV swarms where blockchain can be useful to enhance security, coordination, and data integrity.

Secure Communication

On the one hand, it is necessary for multiple UAVs to communicate with each other securely in a swarm of drones in order not to interfere with malicious actors. Blockchain can serve as a decentralized and encrypted communication layer, where every message or command is kept in a

CHAPTER 13   ENHANCING UAV SWARM SECURITY WITH BLOCKCHAIN: SOLUTIONS FOR
            SECURE COMMUNICATION, DATA INTEGRITY, AND AUTONOMOUS OPERATIONS

record pry-inviolable ledger. Smart Contract is a self-executing contract that automates and enforces secure communication protocols in swarms by directly writing the agreement into lines of code [10].

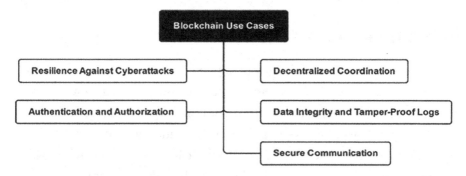

Figure 13-2. *Blockchain application*

Data Integrity and Tamperproof Logs

The UAV swarm is a data asset, using blockchain's immutable ledger to track every byte within a given information asset, aiming for maximum deployment and minimal casualties. All telemetry data, operational status, and mission advancement of each drone in the swarm are logged onto the blockchain. This results in an immutable record that is indisputable and can be audited anytime. When a security breach or system failure occurs, operators can pivot to the blockchain as an authoritative data source for tracing back and identifying why it happened.

Decentralized Coordination

Optimally, all drones operate in a coordinated manner, and for this, UAV swarms require some kind of robust coordination mechanisms. Traditional central control systems are susceptible to single points of failure and even remote cyber blockchain allows for decentralized coordination [11] so that every drone can operate on its own while restricted by the general mission goals kept in the blockchain. To achieve agreement between drones on the state of the swarm, consensus algorithms like proof Of stake (PoS) or Byzantine fault tolerance (BFT) can be used.

Authentication and Authorization

The drones' security is significantly enhanced by their ability to be vetted by authorized members. This is where blockchain comes in as it allows secure authentication to perform transactions with the help of cryptographic keys and identity management systems. These digital identities are recorded on the blockchain for each drone – from where its credentials can be easily verified. If needed, access control policies can be defined by smart contracts that enforce what drones may and are able to do within the swarm.

Resilience Against Cyberattacks

Fail safes due to the blockchain can assist a specialist prevent cyberattacks by decentralizing the control and data storage of UAV swarms. In a single control [12] point system, if the central controller is lost/ınıtiated, all swarm operation fails. However, in a blockchain, the general functionality and integrity of the swarm will not get completely destroyed because even if some drones are faulty, as long as part from all records for misbehaves is distributed by lights itself. This is what makes blockchain so secure and best to combat multiple cybersecurity threats.

Importance of Blockchain in UAV Swarm Security

Blockchain technology serves an invaluable role in the secure operation of UAV swarms by alleviating these core issues concerning communication, coordination, data integrity, and attack resilience. With the increasing

CHAPTER 13 ENHANCING UAV SWARM SECURITY WITH BLOCKCHAIN: SOLUTIONS FOR SECURE COMMUNICATION, DATA INTEGRITY, AND AUTONOMOUS OPERATIONS

adoption of UAV swarms and more extensive applications, it becomes ever more imperative to guarantee their safety and reliability. An integral role of blockchain in UAV swarm security [13] is data integrity during their operation. UAV swarms generate huge repositories of data from telemetry and sensor readings down to mission logs. This data needs to be trustworthy or else the system will not have enough information for the organizations to make sound decisions and process activities that ensure operational efficiency. Transaction data is incredibly secure and cannot be tampered due blockchain's immutable ledger, and no one can make unauthorized changes. This practice is critical in situations where the trust of data leads to a direct mission result, for example, military or emergency response operations.

Decentralized control and coordination then is another huge net positive for blockchain. Legacy-centralized control systems for UAV swarms can be a single point of failure, and they provide an attractive target for cyberattacks. It allows a decentralized way of control and coordination, so each drone acts independently while tuning to some common goals or principles (like the drones form what swarms to do). It carries less risk of catastrophic system collapse, and it increases resilience against attacks on the swarm as a whole. Consensus algorithms maintain the agreement of all drones in what is called a swarm state, leading to synchronized and reliable operations.

In addition, blockchain has the characteristics of secure and transparent communication necessary for UAV swarm operation [14]. It provides a secure and transparent channel for messages; all the communications sent from or to this blockchain are encrypted and recorded in an immutable ledger. This makes it easy to detect and trace any effort to intercept or modify communications. Trust between operators and stakeholders can only be truly established through transparency of communication, which is further enabled in cases where the node will also record actions taken by each drone.

It also makes a lot easier the elaborate authentication and authorization process by way of making use of cryptography keys and digital identities. This is a key aspect of security where all drones in the swarm should be authenticated. Every drone will have its identity registered on the blockchain, and smart contracts can be used to implement access control policies. The strategy restricts the attempt of illegal drones to enter the swarm and also makes sure that every drone adheres to its set limits for action.

Furthermore, blockchain development improves cyberattack immunity. Its swarms of UAVs are vulnerable to cyber threat exploits such as spoofing, jamming, or hacking. Blockchain improves the robustness of UAV swarms to this type of intrusion by distributing data and control everywhere, instead. As a result of the distributed nature of blockchain, even if some drones get hacked or are compromised to cause interference between each other, integrity and functionality overall survive [15]. By distributing the logic, it becomes much harder for an attacker to take down the entire swarm.

In multistakeholder environments where UAV swarms can be operated between different organizations or entities, trust as well as accountability is further supported. The transparency and immutability of blockchain provide drones with a reliable approach for recording their actions. This creates trust with all stakeholders, as anyone can verify how the swarm functions and what it produces. The accountability is also increased since any violation or an unauthorized transaction can be traced back to the owner.

Last but not least, blockchain helps in compliance and auditing. Compliance with regulatory standards and auditing requirements is a necessity for the successful deployment of UAV swarms in different industries [16]. Their activities in the swarm are all recorded immutably and transparently on blockchain, allowing for easier demonstration of regulatory compliance, auditing, etc. The blockchain provides regulators and auditors with access to verify that the swarm operations are compliant with a preagreed set of standards, thus alleviating operators from administrative complexity yet ensuring regulation at all levels.

CHAPTER 13   ENHANCING UAV SWARM SECURITY WITH BLOCKCHAIN: SOLUTIONS FOR SECURE COMMUNICATION, DATA INTEGRITY, AND AUTONOMOUS OPERATIONS

Evolution of UAV Technology

The history of unmanned aerial vehicle (UAV) technology has evolved into a number of increasingly characteristic, functional effects and applications. Originally focused on the military, UAVs have made their way into many commercial and civilian sectors, redefining agriculture, logistics delivery processes, or surveillance industries [17]. The following chapter provides a chronological overview of the main stages by showcasing specific breakthroughs and technological leaps in the development of drone technologies.

Early Developments and Military Applications

UAV (unmanned aerial vehicle) technology is deeply rooted in history starting from the early 20th century specially driven by military needs [18]. It all began with the creation of unmanned aerial vehicles, and on this front is quite long-range travel in terms of advancements made.

The Kettering Bug, a prototype of a functional UAV, was tested by the US Army in 1918 as a cruise missile during WWI. Although not a primary military force, it demonstrated autonomous flying potential [19].

In WWII, the advancements in UAV technology continued when target drones were developed, such as Radioplane OQ-2. Most of these drones were targets/MQMs used for training SAM crews and testing missile systems. For instance, the Radioplane OQ-2 [20] was effectively a mobile target and gunners could practice shooting at it; this model is an example of practical use UAV technology. This employment of UAVs greatly improved training impact and safety. It also underscored the expanded uses of drones beyond their usual articulations in live warfare to include training and testing.

During the Cold War period, under Project Buffalo House of Sight, the forerunner to modern UAVs was initially developed. The most famous example is the Ryan Firebee from the 1950s. It was a versatile UAV, performing multiple mission profiles up to and including reconnaissance as the Ryan Firebee. This was a major milestone in using UAVs as these were now being employed not only for training and target practice but also for intelligence and surveillance missions. In a highly publicized mission, the Ryan Firebee was flown in to dangerously allow over enemy territory whenever it proved that an unmanned drone may be employed relatively effectively as gunship and gathering intelligence [21] against the surrendered without risking human life (or invasion by hostile powers).

While these innovations were spurred by military applications, they set the course for present-day UAV systems. They highlighted the applications of UAVs as delivering cargo, participating in military training and reconnaissance activities among its hubq (parent domain) and intelligence-gathering missions. This brief history of UAV technology shows how it has developed from elementary pilotless aircraft into a sophisticated system capable of carrying out an extensive range of roles in both civilian [22] and military applications.

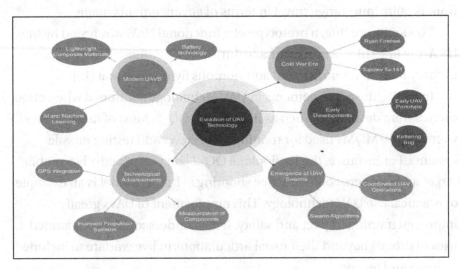

Figure 13-3. Evolution of UAV technology

CHAPTER 13   ENHANCING UAV SWARM SECURITY WITH BLOCKCHAIN: SOLUTIONS FOR
SECURE COMMUNICATION, DATA INTEGRITY, AND AUTONOMOUS OPERATIONS

Technological Advancements in the Late 20th Century

Mid- to late 20th century, there were major technological advancements during the latter half of the 20th century that allowed UAV technology to advance. Various advances of note were increases in materials, microprocessors [23] (which allow the minute and efficient instrumentation controls), and sensors for heat and radiometry that at once ramp up some electronics but shrink others by affecting their packaging real estate.

Materials science has developed materials for the construction of UAVs lighter and stronger, increasing their durability. Better avionics (the electronic systems used on aircraft) were instrumental [24] in the development of UAV capabilities. This resulted in improved control, stability, and automation of UAVs allowing them to be more reliable and easier to operate.

Based on engines, propulsion systems also incorporated mass improvements to prolong the range/endurance of UAVs with more efficient and serviced models. This made them better suited for more missions from long-duration surveys to high-speed reconnaissance.

One of the biggest game-changing inventions to come out in the 1980s and early 1990s was GPS (Global Positioning System). Powered by GPS, UAV navigation became a landmark revolution for the ability to be positioned exactly and pilot itself [25]. This development effectively enabled unlimited training and demonstration flights with a high level of overall accuracy, being especially valuable for conducting UAV flight maneuvers in the air or following prespecified vehicle paths.

The capabilities of UAVs were expanded with the introduction of satellite communication systems. Granted, these systems allowed for data transfer and remote control from a distance at short delays, but this had the effect of greatly enhancing the range and operational capabilities of

UAVs. Satellite communication ensures that UAVs [26] can penetrate into remote or hostile territories and at the same time always stay in touch with operators.

Put together, these material and technology leaps in avionics and propulsion units as a whole plus GPS + sat comms completely changed UAVs from simple pilotless aircraft into very advanced systems, showing the extreme usefulness of UAV technology [27] and paving the way for modern-day military and civilian uses.

Emergence of Multirole UAVs

The late 20th century and early 21st century saw the development of more versatile multirole UAVs that could perform a variety of missions. It was during this timeframe that UAVs were increasingly used for various roles in military operations.

It is authored here: one of the direct examples that is available today involves drones such as the Predator drone which was first introduced in the 1990s. Originally conceived for intelligence gathering, the Predator was later armed with Hellfire missiles to launch strike missions. This dual function revealed the adaptability of the Predator and was one of the major turning points in practical application for UAVs, where a single aircraft could perform both reconnaissance and attack roles.

The next big thing of this time was the Global Hawk: a UAV designed for high-altitude, long-endurance reconnaissance missions. Capable of flying at altitudes above 60,000 feet and for more than 30 hours at a time, the Global Hawk gave CBP extensive surveillance capabilities over wide swaths of territory. This ability to collect imagery in detail, as well as other intelligence data, made the Predator a vital tool for military operations famously around (but not only) the Middle East. Weapons platforms like Predator and Global Hawk are multirole UAVs used in warfare, providing tactical utility for ISR operations, ranging from reconnaissance to close combat and intelligence gathering. This flexibility demonstrated the

CHAPTER 13 ENHANCING UAV SWARM SECURITY WITH BLOCKCHAIN: SOLUTIONS FOR
SECURE COMMUNICATION, DATA INTEGRITY, AND AUTONOMOUS OPERATIONS

increasing dependence on UAV technology in contemporaneous military deployment and emphasized the improvements that had been made since then with regard to both drone design as well as capabilities in general. The appearance of multirole UAVs as a new kind marked a big event in unmanned aerial systems, in view that they may be used for numerous missions, including shut surveillance reconnaissance [28] roles conveying assault weapons and significantly evolved ISTAR (intelligence, surveillance, target acquisition, and reconnaissance). This evolution has cleared the path for more sophisticated UAVs with increased capability to handle a broad of range mission requirements in both military and civil applications.

Development of UAV Swarms

One of the most recent and important developments in UAV technology is that drones can now fly together. A UAV swarm is a team of connected small drones or minidrones that are able to operate together. The swarm features multiple components, including the number and density of individuals. Recognizing the by-product has its own virtues in efficiency and scale that can be translated to more complex tasks flown by a swarm of UAVs as opposed to an individual one.

It could in principle be used for a number of different types of applications to drone swarms. Large areas can be covered in search and rescue missions making up for the chances of finding survivors. Multiple UAVs can offer wide-area surveillance and real-time information from various viewpoints, which is important in large-area surveillance. Swarm configurations can strike in unison to overpower defenses and score targeted hits when needed.

This has been possible due to the recent advances in artificial intelligence (AI) and machine learning (ML), which have enabled UAV swarms for proper functioning and efficiency. The AI algorithms allow drones to make decisions autonomously [29] and coordinate in real time, without the need for human oversight. This makes them superior performing humans because machine learning techniques increase budget performance, clearer understanding of real-time environmental adaptability in a dynamic environment, and are faster to optimize swarm as they learn from their previous experience, thereby increasing the effectiveness of drones all over time.

UAV swarms work quite autonomously and in real time; with the help of AI/ML, they can change their behavior according to new conditions. This is an important feature for more complex missions that require a system able to make quick adjustments in response to rapidly changing conditions.

The UAV swarms are a generous step in the evolution of UAV technology toward collective and autonomous systems capable of dealing with different and complex tasks. This presents new options in both the military and commercial world that demonstrates how AI/ML is transforming UAV capabilities.

Integration of Blockchain and Advance Technology

The adoption of this technology has proven to be a huge step in the drone sector where the UAV (unmanned aerial vehicle) industry has better capabilities than it would have when operated traditionally. Blockchain equips a stable and decentralized platform for enhancing data integrity, communication security as well as cooperation among UAVs.

Blockchain integration enhances UAV operation security by ensuring trustful data and coordinating multiple drones for mission objectives, a crucial aspect of future UAV swarms.

Blockchain can be used to establish irrefutable records of UAV behavioral patterns. All transactions or data updates within the blockchain are securely encrypted and can give full traceability of mission-critical information. Accountability and legal safety in connection with these missions are further increased for both military and civil UAV operations.

The improvement of UAV capabilities is by no means limited to blockchain; advances in sensor technology, computer vision, and on-device ML inferencing have enabled further innovation in this space. Furthermore, sensors assist drones to gather data and process it quicker which allows them to navigate complex environments with greater ease and also carry out tasks with higher precision. By means of computer vision algorithms, UAVs can have the ability to recognize and track objects far more advanced than before. The system automatically detects and executes actions on interesting autonomous events based on the objects found.

In addition to this, edge computing augments UAV autonomy by processing and decision-making at the location of the drone in real time, thus reducing reliance on central systems and minimizing latency, which is something vital for military tasks like rapid intervention, emergency response, and dynamic surveillance missions.

Main Challenges and Issues

While UAV swarms present some promising opportunities, a number of issues and challenges remain to be resolved in order for them to operate safely, efficiently, and reliably. These technical, regulatory, and security aspects [30] form a myriad of issues when considering the application-specific integration of UAV swarms.

Technical Challenges

Scalability

The challenge is the management of many UAVs, a swarm. This is a large number due to the complexity of coordinating and communicating actions across what will god knows how many drones. It's also a big challenge to cope with these systems at scale without performance degradation.

CHAPTER 13   ENHANCING UAV SWARM SECURITY WITH BLOCKCHAIN: SOLUTIONS FOR
             SECURE COMMUNICATION, DATA INTEGRITY, AND AUTONOMOUS OPERATIONS

Communication and Connectivity

Communication is key to success in most aspects of life, and this principle holds true for the high-level functions that need to happen during a swarm operation. Telephone and radio networks can require significant amounts of power to maintain stable communication [31] links in highly variable or even antagonistic environments. The same can cause trouble interrupting the coordination and sharing of data in a swarm due to interference signal loss, bandwidth limitations, etc.

Energy Efficiency and Endurance

Most UAVs suffer from a battery endurance restriction, affecting their performance time. Similarly, energy efficiency is one of the most critical considerations for extended missions (where it might not be practical to recharge or refuel) These include developing energy-efficient propulsion systems and perfecting the flight of these aircraft.

Autonomous Decision-Making

Obtaining real autonomy in UAV swarms means dealing with heavy decision-making processes. In theory, drones must be able to understand their surroundings and then execute decisions in "real time" according to their changing conditions without the intervention of human beings. Trying to make these autonomous systems so reliable and sturdy is a big challenge.

Security Challenges

Cybersecurity Threats

UAV swarms are easy targets for hacking, spoofing, and jamming, which may easily disrupt operations, steal or access confidential information, or even hijack the UAV. It is necessary to have strong cybersecurity to prevent such attacks.

Data Privacy

Devices gather data from cameras and sensors mounted to the UAV. However, they also have stirred fears of overreach because of the possibilities for mass surveillance, raising worries that this would violate individual privacy rights. The key issue is definitely creating universal requirements and technologies that let you store personal data safely but still operate solutions.

Physical Security

UAV physical security while protecting data in UAVs is important. Drones could be physically attacked (shot down, captured) mitigation, e.g., having a self-destruct function for drones that are only under threat of loss and not engaging in anthropology data collection missions or the like.

Regulatory and Ethical Issues

Regulatory Compliance

There are a lot of regulatory issues, and these can be even more complicated for UAV operations which differ greatly by region. It is hard to follow all the rules and regulations in different countries at the national as well as international level. This in turn leads to a whole host of rules on everything from no-fly zones to airspace management and safety requirements. These regulations are essential if we want to enable the safe and legal use of UAVs. Second, regulations should also be standardized across all jurisdictions which will allow a seamless global operation, particularly for commercial and cross-border use cases.

CHAPTER 13 ENHANCING UAV SWARM SECURITY WITH BLOCKCHAIN: SOLUTIONS FOR SECURE COMMUNICATION, DATA INTEGRITY, AND AUTONOMOUS OPERATIONS

Ethical Considerations

The massive employment of UAV swarms, especially in military and surveillance-related scenarios, poses a number of ethical issues. One of the primary issues is then the danger of unsafe exploitation (e.g., during the conflict, or unintentional entry into areas with proper security measures). Another considered problem includes what entities may be held accountable for acts that are wholly autonomous: the more autonomy and available UAV has, this becomes increasingly unclear whose fault its actions will turn out to become. The hugely vast scope assigned to mass serial UAV monitoring forms a further big issue in discussions about dystopias and utopias because well-populated implementation pertaining to checking purposes might quickly degenerate unauthorized governmental "drones" while the lack thereof could just as surely lead police abdicating his own responsibilities. This indicates that the ethical implications need regulation and accountability.

Public Perception and Acceptance

The rapid proliferation of UAVs across each segment has raised public concerns about safety, privacy, and security. The first practical implementation of drone swarms is still arguably some way off, but trust is considered to be key in making the step into our lives happen successfully. Plain speaking: We need transparency as to what type of drones is and can be used everywhere, along with the checks in place that ensure these UAVs must not go astray. Engaging with communities and addressing concerns about UAVs can help bridge the gap between public safety jitters and improve service delivery enhancements.

Environmental Impact

Noise from drones, especially in city areas the sound of several drones all running at the same time, can be disruptive to both communities and wildlife. Quieter propulsion solutions and noise suppression measures are key to reducing this impact.

Drone flights can also have an impact on wildlife, especially in protected environments. The impact that UAVs may have on different species and ecosystems is important to grasp. With this in place, it will be necessary to work with implementation measures that minimize the disturbance and incompatible use of habitats, for example, by restricting UAV flight over critical areas or at certain times – likely remote from sensitive sites – as well as implementing design methods emending protected elements (i.e., environmental wise constructing your tasks out corridors avoiding potential habitat).

Reliability and Safety

As we have seen earlier, the reliability of UAV swarms is vital for safety. Failures of the system, such as from hardware errors, software bugs, or environmental conditions, can directly cause accidents and mission failures. Reliability is increased by writing good tests, sharing infrastructure that can be more easily maintained, and ensuring there are effective fail safe mechanisms to fall back on when things go wrong [32]. Frequent inspections, extensive testing programs, and the inclusion of redundant systems can help reduce failure risks and let UAV swarms work as expected in all conditions.

Another important safety issue is that of preventing collisions between drones within the swarm and with other objects. When collisions are near inevitable, advanced sense-and-avoid technologies allow aircraft to identify and respond in real time. One major requirement for reliable

CHAPTER 13   ENHANCING UAV SWARM SECURITY WITH BLOCKCHAIN: SOLUTIONS FOR
             SECURE COMMUNICATION, DATA INTEGRITY, AND AUTONOMOUS OPERATIONS

operation in such congested and dynamic environments is these supporting technologies together with well-defined communication and coordination protocols. They ensure the integrity of the entire UAV swarm, preventing potential damage or human injury in the process.

Future Trend and Research Direction in UAV Swarm Security

The field of UAV swarm security is a nascent one, but it continues to grow due to mounting advancements in technology and demand for secure, effective, and autonomous drone operations. As UAV swarms continue to be more integrated across different industries, a number of trends are forming for future work and research required to overcome current challenges as well as improve security.

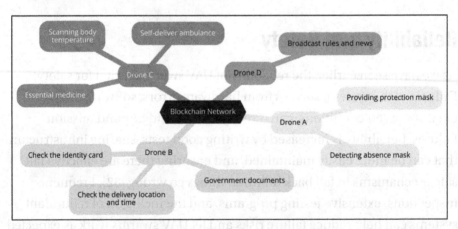

Figure 13-4. Future trend and research direction in UAV swarm security

Integration of Advanced AI and Machine Learning

The future of UAV swarm security is AI- and ML-based techniques. Advanced AI algorithms that allow drones to learn from their environment and respond dynamically and autonomously are currently such areas of research.

In reinforcement learning methods, its deliberative approach enhances the decision-making functionality of UAVs in situations that are not predictable or too complicated [33]. The idea is that drones can experiment with ways to efficiently navigate and complete tasks on their own.

Improved swarming performance and security can be realized by better understanding how drones share information, as well as learn together. This requires the design of algorithms that allow for better interaction and communication within a swarm, so they can work in concert to achieve shared goals. Real-time anomaly detection anomalies in drone behavior or communication can be detected using AI/ML, allowing security threats potential to be identified from the threat. This includes following patterns and abnormalities that could mean an attack or system failure with immediate proactive actions to ensure risks can be neutralized.

Blockchain and Distributed Ledger Technologies

The application of blockchain technology significantly increases decentralized security and reliability within UAV swarms. We speculate that future research on Oc2 will probably converge in the following directions:

- **Blockchain Scalability:** Creating scalable blockchain solutions needed to handle the high numbers of transactions that will be created by large UAV swarms at a rate required for real-time performance.

- **Scalability of Blockchain:** The goal is to develop robust consensus mechanisms and network architectures for unmanned aerial vehicle fleets, ensuring they can handle the high throughputs required for UAV operations.

- **Integration with IoT:** This includes a secure and synchronized network [34] of drones and ground stations, by implementing blockchain with IOT technologies. By using the blockchain, managing IoT data with a common ledger that is decentralized and tamperproof helps to assure secure communication as well as coordination between UAVs other devices by maintaining its integrity.

- **Smart Contracts:** Broader use of smart contracts to automate portions of UAV operations. In the context of project aims, smart contracts can enable automatic mission planning, coordination between drones, and secure data sharing protocols. This is an area where using UAV smart contracts will have to ensure a high level of security and compliance for broad coverage.

Quantum Computing in Cryptography

Quantum computing could solve this issue by providing the computational power needed for UAVs to use a form of encryption and decryption that would be impossible using classical means. The important directions of research are as follows:

Quantum-Resistant Cryptography: For creating cryptographic algorithms that can withstand quantum attacks. More importantly, these algorithms are necessary for the endurance of communication and data in UAVs against potential eavesdropping from quantum computers.

Key Research Areas: Developing postquantum cryptographic protocols that are secure against quantum computing capabilities.

> **Quantum Key Distribution (QKD):** The QKD protocols are proposed to be generated to ensure secure key exchange among the drones. QKD is based on principles of quantum mechanics and allows the distribution of an encryption key theoretically unbreakable. This effectively prevents any attempts for eavesdropping or spying on the communication between UAVs and ground stations.

Improved Communication Channels

Figure 13-2: Reliable and secure communication for UAV swarm applications [35] Using 5G and next-generation high-speed, low-latency connectivity Developing mesh networking for robust communication in challenging environments; Creating adaptive communication systems that adapt to changing conditions - future research areas [36].

Robust Cybersecurity measures

It is vital to protect the UAV swarm from cyber threats. The study focuses on developing specialized Intrusion Detection Systems (IDS) using Blockchain Technology for real-time threat detection and secure identity management, thereby ensuring applications can withstand cyberattacks [37, 38].

Ethical and Regulatory Frameworks

Using UAV swarms in such a way raises ethical and regulatory issues. Thus, follow-on research is expected to focus on creating transparent, ethical AI for autonomous decision-making and global standards for UAV swarm operations that account for security, privacy, and safety concerns, harmonizing regulations globally to ease the adoption of these systems across borders while at the same time ensuring compliance [39, 40].

Impacts on the Environment and Society
Environmental

The ecological and welfare effects of UAV swarms must be understood if such systems are to ever operate indoors. Future pathways that range from the development of sustainable propulsion solutions [41, 42] to minimize noise and emissions, transparent communication, and community involvement are key for greater social acceptance.

Conclusion

In this chapter, the future of UAV swarm technology pertains to new era drone operations in various verticals, offering unparalleled security, operational efficiency, and autonomy. With the rising tide of AI, blockchain, quantum computing, and communication protocol advancements are areas where new capabilities for UAV swarms can be applied as well. Of course, long before we can shrink the microgrid control space to a single abstraction in our head (if it ever happens), this raises all sorts of messy issues here: scalability concerns that influence cybersecurity risks and regulatory compliance vs. societal acceptance.

Adding AI ML-type methods would increase decision-making and coordination between the UAV swarms to reach their optimum mission. Professionals can build more resilient communication capabilities

CHAPTER 13   ENHANCING UAV SWARM SECURITY WITH BLOCKCHAIN: SOLUTIONS FOR SECURE COMMUNICATION, DATA INTEGRITY, AND AUTONOMOUS OPERATIONS

grounded in accurate data through blockchain technology courses and alignment of decentralization power by highly competitive operational muscle to defend against cyber threats. Meanwhile, quantum computing is poised to revolutionize encryption and secure key distribution which will further shore up [40] UAV swarm security too. Real-time data exchange between drones, advanced IDS implementation, and proactive design and resilient architectures are critical for cybersecurity against malicious attack endeavors. Effective 5G networks require communication protocols for accurate information sharing, while ethical considerations and regulatory frameworks are crucial for responsible deployment of UAV swarm technology [41]. Sustainable UAV swarm integration will also have to consider addressing the above, as well as techniques in producing eco-friendly technologies for environmental mitigation benefits and other transparent communication/engagement with public outreach. Future UAV swarms could revolutionize military operations, commercial services, disaster response, and fall skies, with advanced research and development paving the way for safer, efficient applications in our interconnected global society.

References

[1]   Gupta, R., Nair, A., Tanwar, S., & Kumar, N. (2021). Blockchain-assisted secure UAV communication in 6G environment: Architecture, opportunities, and challenges. *IET communications*, 15(10), 1352-1367.

[2]   Koulianos, A., & Litke, A. (2023). Blockchain Technology for Secure Communication and Formation Control in Smart Drone Swarms. *Future Internet*, 15(10), 344.

[3] Alsamhi, S. H., Shvetsov, A. V., Shvetsova, S. V., Hawbani, A., Guizani, M., Alhartomi, M. A., & Ma, O. (2022). Blockchain-Empowered Security and Energy Efficiency of Drone Swarm Consensus for Environment Exploration. *IEEE Transactions on Green Communications and Networking*, 7(1), 328-338.

[4] Hafeez, S., Khan, A. R., Al-Quraan, M. M., Mohjazi, L., Zoha, A., Imran, M. A., & Sun, Y. (2023). Blockchain-Assisted UAV Communication Systems: A Comprehensive Survey. *IEEE Open Journal of Vehicular Technology, 4,* 558-580.

[5] Mehta, P., Gupta, R., & Tanwar, S. (2020). Blockchain envisioned UAV networks: Challenges, solutions, and comparisons. *Computer Communications,* 151, 518-538.

[6] Xiao, W., Li, M., Alzahrani, B., Alotaibi, R., Barnawi, A., & Ai, Q. (2021). A Blockchain-Based Secure Crowd Monitoring System Using UAV Swarm. *IEEE Network,* 35(1), 108-115.

[7] Khanh, T. D., Komarov, I., Iureva, R., & Chuprov, S. (2020, December). TRA: Effective Authentication Mechanism for Swarms of Unmanned Aerial Vehicles. *In 2020 IEEE symposium series on computational intelligence (SSCI) (pp. 1852-1858). IEEE.*

[8] Alladi, T., Bansal, G., Chamola, V., & Guizani, M. (2020). SecAuthUAV: A Novel Authentication Scheme for UAV-Ground Station and UAV-UAV Communication. *IEEE Transactions on Vehicular Technology, 69(12),* 15068-15077.

[9] Zhang, L., Xu, J., Obaidat, M. S., Li, X., & Vijayakumar, P. (2022). A PUF-based lightweight authentication and key agreement protocol for smart UAV networks. *IET Communications*, 16(10), 1142-1159.

[10] Lei, Y., Zeng, L., Li, Y. X., Wang, M. X., & Qin, H. (2021). A Lightweight Authentication Protocol for UAV Networks Based on Security and Computational Resource Optimization. *IEEE Access*, 9, 53769-53785.

[11] Wu, Y., Dai, H. N., Wang, H., & Choo, K. K. R. (2021). Blockchain-Based Privacy Preservation for 5G-Enabled Drone Communications. *IEEE Network,* 35(1), 50-56.

[12] Cosar, M. (2022). Cyber Attacks on Unmanned Aerial Vehicles and Cyber Security Measures. *The Eurasia Proceedings of Science Technology Engineering and Mathematics*, 21, 258-265.

[13] Khalil, H., Rahman, S. U., Ullah, I., Khan, I., Alghadhban, A. J., Al-Adhaileh, M. H., ... & ElAffendi, M. (2022). A UAV-Swarm-Communication Model Using a Machine-Learning Approach for Search-and-Rescue Applications. *Drones*, 6(12), 372.

[14] Guo, Y., & Liang, C. (2016). Blockchain application and outlook in the banking industry. *Financial innovation*, 2, 1-12.

[15] Zheng, Z., Xie, S., Dai, H. N., Chen, X., & Wang, H. (2018). Blockchain challenges and opportunities: a survey. *International journal of web and grid services*, 14(4), 352-375.

[16] Tasatanattakool, P., & Techapanupreeda, C. (2018, January). Blockchain: Challenges and applications. *In 2018 international conference on Information Networking (ICOIN)* (pp. 473-475). IEEE.

[17] Ullah, I., Qian, S., Deng, Z., & Lee, J. H. (2021). Extended Kalman Filter-based localization algorithm by edge computing in Wireless Sensor Networks. *Digital Communications and Networks*, 7(2), 187-195.

[18] Ahmad, I., Rahman, T., Zeb, A., Khan, I., Ullah, I., Hamam, H., & Cheikhrouhou, O. (2021). Analysis of Security Attacks and Taxonomy in Underwater Wireless Sensor Networks. *Wireless Communications and Mobile Computing*, 2021(1), 1444024.

[19] Yu, H., Hu, Q., Yang, Z., & Liu, H. (2021). Efficient Continuous Big Data Integrity Checking for Decentralized Storage. *IEEE Transactions on Network Science and Engineering*, 8(2), 1658-1673.

[20] Li, Y., Yu, Y., Chen, R., Du, X., & Guizani, M. (2020). IntegrityChain: Provable Data Possession for Decentralized Storage. *IEEE Journal on Selected Areas in Communications*, 38(6), 1205-1217.

[21] Zhang, Q., Zhang, Z., Cui, J., Zhong, H., Li, Y., Gu, C., & He, D. (2023). Efficient Blockchain-Based Data Integrity Auditing for Multi-Copy in Decentralized Storage. *IEEE Transactions on Parallel and Distributed Systems*.

[22] Kumar, A., Bhushan, B., Shristi, S., Chaganti, R., & Soufiene, B. O. (2023). Blockchain-based decentralized management of IoT devices for preserving data integrity. In Blockchain Technology Solutions for the Security of IoT-Based Healthcare Systems (pp. 263-286*). Academic Press.*

[23] Wang, H., & Zhang, J. (2019). Blockchain Based Data Integrity Verification for Large-Scale IoT Data. *IEEE Access*, 7, 164996-165006.

[24] Jensen, I. J., Selvaraj, D. F., & Ranganathan, P. (2019, June). Blockchain Technology for Networked Swarms of Unmanned Aerial Vehicles (UAVs). In 2019 IEEE 20th International Symposium on" A World of Wireless, *Mobile and Multimedia Networks"(WoWMoM)* (pp. 1-7). IEEE.

[25] Qureshi, M. S., Khan, I. U., & Kim, K. (2023, December). Securing the Smart Grid: A Comprehensive Analysis of Recent Cyber Attacks. *In 2023 5th International Conference on Electrical, Control and Instrumentation Engineering (ICECIE) (pp. 1-6). IEEE.*

[26] Wang, J., Liu, Y., Niu, S., & Song, H. (2021). Lightweight blockchain assisted secure routing of swarm UAS networking. *Computer Communications*, 165, 131-140.

[27] Mazhar, T., Irfan, H. M., Haq, I., Ullah, I., Ashraf, M., Shloul, T. A., ... & Elkamchouchi, D. H. (2023). Analysis of Challenges and Solutions of IoT in Smart Grids Using AI and Machine Learning Techniques: *A Review. Electronics*, 12(1), 242.

[28] Hafeez, S., Khan, A. R., Al-Quraan, M. M., Mohjazi, L., Zoha, A., Imran, M. A., & Sun, Y. (2023). Blockchain-Assisted UAV Communication Systems: A Comprehensive Survey. *IEEE Open Journal of Vehicular Technology, 4*, 558-580.

[29] Sullivan, J. M. (2006). Evolution or revolution? The rise of UAVs. *IEEE Technology and Society Magazine, 25(3)*, 43-49.

[30] Fan, B., Li, Y., Zhang, R., & Fu, Q. (2020). Review on the Technological Development and Application of UAV Systems. *Chinese Journal of Electronics*, 29(2), 199-207.

[31] Khan, I. U., Khan, Z. A., Ahmad, M., Khan, A. H., Muahmmad, F., Imran, A., ... & Hamid, M. K. (2023, May). Machine Learning Techniques for Permission-based Malware Detection in Android Applications. *In 2023 9th International Conference on Information Technology Trends (ITT) (pp. 7-13). IEEE.*

[32] Mohsan, S. A. H., Othman, N. Q. H., Li, Y., Alsharif, M. H., & Khan, M. A. (2023). Unmanned aerial vehicles (UAVs): practical aspects, applications, open challenges, security issues, and future trends. *Intelligent Service Robotics*, 16(1), 109-137.

[33] Zhou, Y., Rao, B., & Wang, W. (2020). UAV Swarm Intelligence: Recent Advances and Future Trends. *Ieee Access*, 8, 183856-183878.

[34] Ullah, I., Noor, A., Nazir, S., Ali, F., Ghadi, Y. Y., & Aslam, N. (2024). Protecting IoT devices from security attacks using effective decision-making strategy of appropriate features. *The Journal of Supercomputing*, 80(5), 5870-5899.

[35] Javed, S., Hassan, A., Ahmad, R., Ahmed, W., Ahmed, R., Saadat, A., & Guizani, M. (2024). State-of-the-Art and Future Research Challenges in UAV Swarms. *IEEE Internet of Things Journal.*

[36] Ullah I, Adhikari D, Khan H, Anwar MS, Ahmad S, Bai X. Mobile robot localization: Current challenges and future prospective. Computer Science Review. 2024 Aug 1;53:100651.

[37] Nawaz, H., Ali, H. M., & Laghari, A. A. (2021). UAV Communication Networks Issues: A Review. Archives of Computational Methods in Engineering, 28(3), 1349-1369.

[38] Ullah, I., Khan, I. U., Ouaissa, M., Ouaissa, M., & El Hajjami, S. (Eds.). (2024). Future Communication Systems Using Artificial Intelligence, Internet of Things and Data Science. *CRC Press.*

[39] Ullah, I., Shen, Y., Su, X., Esposito, C., & Choi, C. (2019). A Localization Based on Unscented Kalman Filter and Particle Filter Localization Algorithms. *IEEE Access*, 8, 2233-2246.

[40] Ullah I, Adhikari D, Ali F, Ali A, Khan H, Sharafian A, Kesavan SM, Bai X. Revolutionizing E-Commerce With Consumer-Driven Energy-Efficient WSNs: A Multi-Characteristics Approach. IEEE Transactions on Consumer Electronics. 2024 Jun 10.

[41] Khan, F. M., Muhammad Shoaib Akhtar, Inam Ullah Khan, Zeeshan Ali Haider, & Noor Hassan Khan. (2024). Clinical Prediction of Female Infertility Through Advanced Machine Learning Techniques. *International Journal of Innovations in Science & Technology*, 6(2), 943–960.

[42] Mazhar, T., Irfan, H. M., Khan, S., Haq, I., Ullah, I., Iqbal, M., & Hamam, H. (2023). Analysis of Cyber Security Attacks and Its Solutions for the Smart Grid Using Machine Learning and Blockchain Methods. *Future Internet*, 15(2), 83.

CHAPTER 14

Fundamentals and Applications of Unmanned Aerial Vehicle Swarms

Zahid Ullah Khan[1], Faheem Khan[2*], Rahim Khan[1], Sohaib Bin Altaf Khattak[3], Moustafa M. Nasralla[3], M. A. Al-Khasawneh[4,5,6], Inam Ullah[7]

[1]College of Information and Communication Engineering, Harbin Engineering University, Harbin 150001, China.
engr.zahidkhan09@hrbeu.edu.cn, rahim@hrbeu.edu.cn

[2]Department of Electrical Engineering, University of Engineering and Technology Peshawar Main Campus, KPK Pakistan.
faheemkhan@uetpeshawar.edu.pk

[3]Smart Systems Engineering Laboratory, College of Engineering, Prince Sultan University, Riyadh, Saudi Arabia.
skhattak@psu.edu.sa, mnasralla@psu.edu.sa

[4]School of Computing, Skyline University College, University City Sharjah, 1797, Sharjah, UAE

[5]Applied Science Research Center. Applied Science Private University, Amman, Jordan;

[6]Jadara University Research Center, Jadara University, Irbid, Jordan; mahmoudalkhasawneh@outlook.com

[7]Department of Computer Engineering, Gachon University, Seongnam 13120, Republic of Korea; inam.fragrance@gmail.com

Corresponding Author: Faheem Khan (faheemkhan@uetpeshawar.edu.pk)

CHAPTER 14   FUNDAMENTALS AND APPLICATIONS OF UNMANNED AERIAL VEHICLE SWARMS

Introduction

UAVs, sometimes known as drones, have swiftly advanced from their military origins to a wide range of civilian uses, including disaster relief, agriculture, logistics, and environmental monitoring. One of the most intriguing advancements in UAV technology is the construction of UAV swarms. When several drones work together, their combined capabilities and efficiency surpass those of individual UAVs. This phenomenon is known as a "swarm" [1]. This chapter aims to provide readers with a comprehensive understanding of the concepts behind UAV swarms. It highlights significant subjects such as the many types of UAVs, the components they need, communication networks, sensor technologies, and the opportunities and challenges they provide. By looking at these elements, we want to provide readers with the basic knowledge needed to comprehend the complexities and potential of UAV swarm technology. First, we categorize many UAVs, from tiny drones meant for small-scale activities to larger drones capable of carrying large payloads. Understanding the range of UAV types is crucial since each one performs a certain role inside a swarm, which helps accomplish the objective. Next, we look at the basic components of UAVs. These components include payloads, power supply, navigational aids, and propulsion systems. The advancements in these technologies have significantly improved UAVs' performance, reliability, and flexibility [2]. A thorough understanding of these components is necessary to understand how UAVs operate individually and as a swarm. The UAV swarm's ability to communicate depends on it. Effective networks and communication protocols are essential for the real-time coordination of several UAVs' operations [3]. This section will cover the many communication methods and techniques that enable UAVs to complete complex tasks effectively, collaborate to solve issues, and exchange information. Sensor technologies also make it feasible for UAV swarms. UAVs equipped with state-of-the-art sensors

CHAPTER 14   FUNDAMENTALS AND APPLICATIONS OF UNMANNED AERIAL VEHICLE SWARMS

can perceive and interact with their environment, enhancing situational awareness and mission effectiveness [4, 5]. We'll examine the many types of sensors used by UAVs and their applications in various scenarios. While UAV swarms have great potential, they also face several challenges. A few of them are managing energy, preventing accidents, and maintaining excellent communication in changing environments. Swarm operations and deployment of UAVs need to navigate these challenges effectively.

Additionally, UAV swarms offer a multitude of options across a wide range of industries. In agriculture, they can optimize resource utilization and track the health of crops [6]. In environmental monitoring, they can track animals and assess ecological changes [6]. In disaster response, UAV swarms can quickly assess affected areas and aid in search and rescue operations. In the military, they offer strategic advantages in reconnaissance and surveillance [7, 8]. Considering these options, we must also consider the safety, legal, and ethical implications of UAV swarm use. Finally, we shift our focus to the future and discuss fresh approaches and research directions that might further enhance the capabilities and applications of UAV swarms. By the end of this chapter, readers will have a solid understanding of the concepts behind UAV swarms, empowering them to understand and contribute to this rapidly evolving field. This chapter lays the groundwork for a thorough investigation of the fascinating subject of UAV swarms while offering theoretical and practical insights. With the knowledge gained here, navigating the future of UAV technology will be simpler for anybody using it for personal, professional, or academic reasons.

Basic Concepts of UAV Swarms

In this section, we briefly introduce the basic concept of UAV swarms which are as follows.

CHAPTER 14   FUNDAMENTALS AND APPLICATIONS OF UNMANNED AERIAL VEHICLE SWARMS

Comprehending UAVs

UAVs, or drones, are flying machines that function without a human pilot. They may be operated remotely by a person, or they can operate independently using dynamic automation systems or preprogrammed flight plans [9]. Since their beginnings, UAVs have evolved tremendously. UAVs have come a long way from simple remote-controlled aircraft to extremely intelligent, sophisticated robots that can execute intricate tasks. The design and functioning of UAVs, which usually include of parts including airframes, propulsion systems, avionics, sensors, and communication modules, are what give them their adaptability [10, 11]. These components enable the UAV to navigate, stabilize, and carry out missions. It is important to comprehend individual UAV capabilities in the context of UAV swarms [12]. A swarm of UAVs can communicate and cooperate due to the sensors and communication technology installed on each drone. The collective behavior of the UAVs is akin to the natural swarming behavior of insects and birds, which allows the UAVs to do tasks more robustly and effectively than a single UAV [13]. Swarm intelligence is based on essential components such as adaptive behavior, decentralized decision-making, and real-time data exchange. Utilizing these concepts, UAV swarms may achieve increased robustness, redundancy, and coverage, making them appropriate for various uses, from military operations to environmental monitoring [14, 15]. Understanding the fundamental ideas and parts of UAVs is crucial to appreciating the intricacies and possibilities of UAV swarm technology. Figure 14-1 shows the use cases for UAVs.

CHAPTER 14   FUNDAMENTALS AND APPLICATIONS OF UNMANNED AERIAL VEHICLE SWARMS

Figure 14-1. *Use cases for UAVs [4]*

Natural Swarm Behavior: An Inspiration

The swarming behavior of insects, fish, and birds in nature provides valuable inspiration for the design of UAV swarms. This natural phenomenon is typified by the enormous number of people moving in unison, leading to highly adaptive and effective group behaviors. For instance, without obvious central supervision, fish schools and bird flocks may swiftly alter course together, while insect swarms have exceptional capacities for nest-building and foraging due to decentralized decision-making mechanisms [16]. Simple local restrictions control these spontaneous swarming behaviors, such as keeping a specific distance from neighbors, moving in the group's direction, and avoiding obstacles. Applying these ideas to UAV swarms entails imitating the local interaction rules and decentralized control systems seen in nature [17]. Basic algorithms and local information guide each drone in a UAV swarm to determine how it should behave concerning its neighbors. With the help of this strategy, the swarm may effectively and adaptably carry out complex tasks, including environmental monitoring, search and rescue operations, and area coverage. Scalable, robust, and adaptable robotic systems may be achieved using a solid framework that is derived from the natural inspiration of UAV swarming [18, 19]. Engineers may create UAV swarms with dynamic reconfiguration, resilience against individual failures, and increased operational efficiency by modeling the adaptive tactics of natural swarms. This makes UAV swarms very effective instruments for a variety of applications.

Fundamental Features of UAV Swarms: Scalability, Autonomy, and Cooperation

The fundamental qualities of autonomous, cooperative, and scalable UAV swarms serve as the cornerstone of their operational efficacy. In UAV swarms, autonomy is the capacity of individual drones to decide for themselves and carry out missions without assistance from humans. Because of sophisticated algorithms and artificial intelligence, each autonomous UAV can navigate, adjust to changes in its surroundings, and carry out specific tasks on its own [20]. This kind of independence lessens the need for continual human supervision and makes task execution more effective, particularly in dangerous or complicated circumstances. Another essential quality is collaboration, in which UAVs cooperate to accomplish a shared objective. This cooperative behavior, which includes constant communication and data sharing among the UAVs, is modeled after natural swarming [18, 21]. By means of this interaction, the swarm may improve its overall performance and resilience by coordinating activities, allocating tasks, and working together to solve issues. Collaborative UAVs, as opposed to solitary UAVs, can cover broad regions more wholly and rapidly during a search and rescue mission. Scalability is the swarm's capacity to continue

CHAPTER 14 FUNDAMENTALS AND APPLICATIONS OF UNMANNED AERIAL VEHICLE SWARMS

operating effectively and efficiently as the quantity of UAVs grows. This feature guarantees that the swarm may grow or shrink to the demands of the mission without sacrificing its functioning [22, 23]. Applications needing varied coverage regions or adjusting to various operating sizes must be scalable. UAV swarms are invaluable in a variety of disciplines, including environmental monitoring, disaster management, and military operations, because of their capacity to accomplish complicated tasks with great efficiency, adaptability, and resilience when combined with three critical characteristics: autonomy, cooperation, and scalability.

Classification of UAVs

UAVs have undergone substantial diversification to accommodate a wide range of operating requirements and uses. Size, range, durability, and applicability are only a few of the parameters used to classify this diversity. To choose the best drone for a particular task and ensure maximum performance and efficiency, it is crucial to understand the various kinds of UAVs. Here, we look at the main categories of UAVs.

Based on Size and Weight

The size and weight of UAVs could be used to classify them, which significantly affects their uses and capabilities [24]. These categories include big, heavy UAVs for industrial and military applications and tiny, light drones for commercial and recreational purposes. It is essential to comprehend these categories to choose the right UAV for a given mission and set of operational needs [25]. This section will examine the various UAV kinds according to their weight and size, emphasizing their unique characteristics and applications. Table 14-1 shows a review of UAV swarms that compares them according to weight and size, illustrating the variations in their features, applications, and capacities.

CHAPTER 14 FUNDAMENTALS AND APPLICATIONS OF UNMANNED AERIAL VEHICLE SWARMS

Table 14-1. Review of UAV swarms that compares them according to weight and size, illustrating the variations in their features, applications, and capacities

Classification	Micro-UAV swarms	Mini-UAV swarms	Small UAV swarms	Medium UAV swarms	Large UAV swarms
Weight	250 g or less	2 kg to 250 g	2 to 20 kg	20 to 150 kg	150 kg or more
Size	Fits in hand and is very small	Fits in hand and is very small	Moderate and required little takeoff space	Larger, requiring a specific launch pad	Tremendous, requiring runways
Principal use cases	Indoor navigation	Surveillance and environmental observation	Delivery services, search and rescue, and agriculture	Mapping, military, and sophisticated surveillance	Missions with extended duration and cargo transportation
Time of flight	Short (up to half an hour)	Moderate (for up to an hour)	Prolonged (for several hours)	Extremely prolonged (hours or even a day)	Extended (for a minimum of 48 hours)
Capacity of Payload	Minimal (small cameras, sensors)	Light (from microscopic equipment and cameras)	Moderate (sensors, cameras, and compact packaging)	Large (larger sensors, more substantial payloads)	Heavy (many sensors, heavy load)

530

Range	Short (less than 1 km)	Moderate (up to 5 km)	Large (20 km or more)	Extended (to 100 km)	Extended (more than 100 km)
Cost	Low (cost-effective, consumer-level)	Moderate (cost-effective, expert-caliber)	High (industrial, professional quality)	High (military-grade, sophisticated)	Extremely high (specialized, made-to-order)
Control system	Simplified, often using a smartphone	More advanced, intermediate	Sophisticated, needs a separate controller	Very sophisticated, specialized systems	Complex, very specialized systems

Nano- and Micro-UAVs

The most miniature class of drones are called nano- and micro-UAVs, usually weighing less than 2 kg. These small UAVs are designed for specific uses, and their small size offers major benefits [26]. Nano- and micro-UAVs are mostly used for interior surveillance since they can move stealthily across small areas and provide real-time data and video without calling attention to themselves. They are also perfect for environmental monitoring because of their lightweight construction, which enables them to collect information on temperature, air quality, and other factors in places that are hard to reach with bulkier equipment [27]. Additionally, these UAVs can maneuver in confined locations and provide vital situational awareness in reconnaissance missions, especially in urban or heavily wooded areas. Because of their tiny size and agility, they can fly under thick canopies, through windows, and into short passages that would be inaccessible to bigger drones [28, 29]. The special qualities that nano- and micro-UAVs provide are priceless for jobs needing accuracy, covertness, and entry into forbidden regions.

Mini-UAVs

Mini-UAVs, which range in weight from 2 to 25 kg, are essential in many commercial applications because of their ability to combine portability with capabilities. Mini-UAVs are widely used in agriculture to monitor crop health, determine irrigation demands, and control pests [30, 31]. With their sophisticated sensors, they can take specific insights that help farmers maximize productivity and resource efficiency. They can also record multispectral data and high-resolution images. Mini-UAVs provide a more secure and effective substitute for physical inspections and scaffolding when inspecting infrastructure. They can detailedly examine wind turbines, electricity lines, bridges, and other structures [32]. Their usefulness extends to short-range surveillance, providing real-time security monitoring at building sites, gatherings, or border patrols. Because of their small size and ability to carry a variety of payloads, such as cameras and LiDAR, tiny UAVs are useful tools for a variety of sectors. They can also be swiftly deployed.

Medium UAVs

With a maximum weight of 600 kg, medium UAVs are essential in a variety of applications because of their improved capabilities. These UAVs, mostly used in tactical military operations, provide strategic benefits on the battlefield by collecting real-time data and conducting sophisticated reconnaissance. Their large cargo capacity allows for highly advanced sensor arrays, which provide accurate and thorough long-range monitoring. For long-range surveillance operations, this capacity is essential for maintaining continuous observation of large regions [33]. Furthermore, medium UAVs play a crucial role in the delivery of freight, particularly in difficult terrain or conflict areas where traditional modes of transportation are dangerous or unfeasible. These UAVs' capacity to carry heavy loads makes it possible for them to deliver necessities such as medical assistance, equipment, and supplies in a timely and effective manner [34]. All things considered, medium UAVs are invaluable instruments in contemporary tactical, surveillance, and logistical operations due to their sophisticated sensors, long flight times, and strong payload capacities.

Large UAVs

Approximately 600 kg in weight, large UAVs are designed for strategic military applications and are essential to combat intelligence, surveillance, and reconnaissance (ISR) missions [35]. These UAVs can perform complex and dangerous missions because they are outfitted with state-of-the-art avionics and sensor systems. Their innovative equipment enables thorough intelligence collection, guaranteeing all-encompassing situational awareness and instantaneous data transfer to command centers. Large UAVs are used in surveillance and reconnaissance operations to continuously scan vast regions and accurately identify and track any threats [36, 37]. When engaged in warfare, they may be equipped with a variety of weapons to carry out precise attacks with little danger to human soldiers. Their operational efficacy is increased by their capacity to carry out extended duration missions, which enables them to provide constant coverage and assistance in dynamic and high-threat settings. Large UAVs, with their unique combination of endurance, cutting-edge technology, and agility to satisfy demanding mission requirements, are all-around invaluable assets in contemporary military strategy.

Based on Range and Endurance

This section provides the different range and endurance-based UAVs which are as follows.

Very Low-Altitude (VLA) UAVs

UAVs with VLA capability may function up to 100 m above the ground, making them ideal for close-quarters surveillance and inspection duties. These UAVs work exceptionally well in situations requiring in-depth, overhead surveillance [38]. They can take high-resolution pictures and movies due to their low-altitude flying capacity, which also provides precise and accurate data for various applications. This feature is especially helpful in fields where precise visual information is essential, such as public safety, infrastructure inspection, and agriculture. VLA UAVs are usually used in restricted locations because of their limited operating range, which allows them to effectively cover the required land without requiring a lot of travel [39]. This makes them the perfect option for jobs

CHAPTER 14   FUNDAMENTALS AND APPLICATIONS OF UNMANNED AERIAL VEHICLE SWARMS

like checking the health of crops, examining electricity lines, or carrying out search and rescue missions in tight areas. By reducing the possibility of interaction with human-crewed aircraft, their low-altitude operation also improves safety and complies with regulations.

Low-Altitude Short Endurance (LASE) UAVs

UAVs with the designation LASE are designed to fly for many hours at altitudes between 100 and 500 m. LASE UAVs are well suited for various applications that call for prolonged observation over limited distances because of their operating range and endurance [40, 41]. These UAVs are used in agriculture to monitor crop health, determine irrigation requirements, and find insect infestations. By doing so, they provide farmers with crucial information that helps them maximize yields and resource use. LASE UAVs are also helpful for monitoring wildlife since they let scientists keep an eye on animal habits, migration patterns, and habitats without affecting the surrounding ecosystem. Their ability to hover at intermediate altitudes provides enough detail to collect reliable data while providing an expansive view [42]. LASE UAVs are used in short-range reconnaissance for tasks including security patrols, environmental monitoring, and search and rescue operations that call for rapid deployment and dependable, real-time observation. They are valuable instruments in these many domains because of their durability and adaptability.

CHAPTER 14   FUNDAMENTALS AND APPLICATIONS OF UNMANNED AERIAL VEHICLE SWARMS

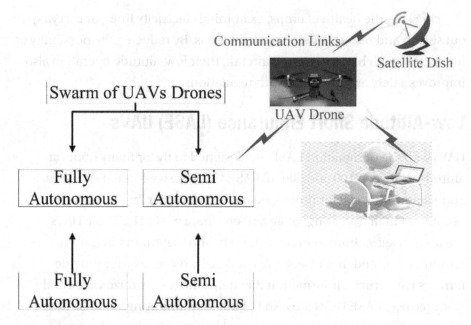

Figure 14-2. Swarm classification of UAV drones

Low-Altitude Long Endurance (LALE) UAVs

UAVs operating at altitudes comparable to low-altitude short endurance (LASE) UAVs, usually between 100 and 500 m, are known as LALE UAVs. On the other hand, LALE UAVs stand out due to their noticeably longer flight times, often more than 12 hours [43]. Their capacity to observe continuously for extended periods of time makes them perfect for activities involving continuous monitoring. LALE UAVs are useful for long-term environmental monitoring because they can monitor animal migrations, continually monitor ecosystems, and evaluate vegetation or water body changes. This long-term data collection is essential to comprehending environmental patterns and wise conservation choices [44]. LALE UAVs provide border patrol continuous surveillance capabilities, improving security by keeping an eye on extensive regions for long periods of time. They may identify and monitor any threats, illegal crossings, and smuggling operations, giving border security officers up-to-date information. Because LALE UAVs have longer endurance, they can cover larger areas and need fewer redeployments, which makes them very effective for long-term monitoring and surveillance activities.

Medium-Altitude Long Endurance (MALE) UAVs

UAVs classified as MALE are very successful for a variety of challenging applications since they can operate at altitudes ranging from 5,000 to 15,000 m and have a large flight endurance [45]. Their suitability for military intelligence, surveillance, and reconnaissance (ISR) activities stems from their capacity to sustain operations for prolonged periods of time at high altitudes. MALE UAVs provide these functions by supplying vital real-time data, such as signals intelligence and imaging, to aid in strategic planning and decision-making [46]. MALE UAVs are also perfect for long-range reconnaissance tasks. Their great endurance eliminates the need for regular refueling or redeployment, enabling them to cover large regions, provide thorough information, and monitor activities for extended periods. This capacity is critical for danger assessment, movement detection and tracking, and situational awareness. MALE UAVs are essential for monitoring large oceanic areas in maritime patrol [41]. They are able to keep an eye on fishing operations, identify illicit activities like piracy and smuggling, and maintain marine safety via constant observation. MALE UAVs are essential for preserving security and situational awareness over vast and distant regions because of their high altitude and extended endurance.

Based on Application

The detailed overview of application-based UAVs are briefly introduced in this section and are as follows.

Commercial UAVs

UAVs for commercial use are adaptable instruments used in a wide range of sectors, such as real estate, agriculture, film, and infrastructure inspection. These UAVs are outfitted with customized sensors and cameras designed to gather information for certain commercial uses, improving productivity and precision in their domains [47, 48]. Commercial UAVs are employed in agriculture to do precision agricultural activities, including insect identification, irrigation control, and crop monitoring. They provide farmers access to real-time data and high-resolution pictures, which helps them maximize resource usage and boost yields. UAVs are used in real estate to take beautiful aerial photos and videos, showcasing homes from unusual perspectives. This offers real estate brokers a thorough understanding of the property to assist them in designing effective marketing materials. UAVs are useful in filmmaking because they allow for dynamic aerial views that give movies a more cinematic feel. They make it feasible for filmmakers to record scenes from previously difficult or impossible angles [49]. Another crucial use for UAVs is infrastructure assessment, where they examine things such as wind turbines, electricity lines, and bridges. They increase productivity and safety by securely accessing difficult-to-reach places, spotting possible problems, and lowering the need for human inspections.

Military UAVs

UAVs in the military are vital for tactical and strategic missions, such as target acquisition, combat support, and intelligence, surveillance, and reconnaissance (ISR). These UAVs, outfitted with cutting-edge technology, are designed to operate efficiently in challenging conditions and accurately carry out intricate tasks. Military unmanned aerial vehicles (ISR) collect real-time intelligence, giving commanders vital information on enemy movements, topography, and other crucial aspects [50, 51].

CHAPTER 14   FUNDAMENTALS AND APPLICATIONS OF UNMANNED AERIAL VEHICLE SWARMS

Strategic decision-making is informed, and situational awareness is improved through this knowledge. UAVs may provide direct action for combat support by carrying out precise attacks on targets with little danger to human people. By spotting and eliminating threats, they also help ground troops by raising the success rates of missions. UAVs make target acquisition, which involves locating and designating targets for other weapon systems to use in combat, guaranteeing precision and efficacy [52]. They can operate in various challenging environments and yet retain their operational dominance because of their sophisticated sensors and communication systems. In contemporary warfare, military UAVs are indispensable due to their unparalleled capability in combat, observation, and strategic operations.

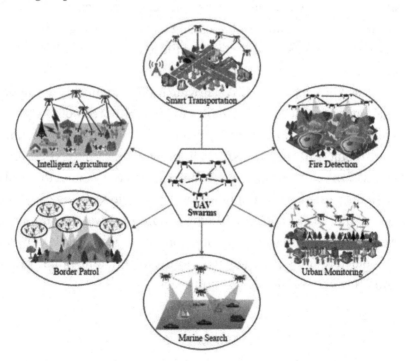

Figure 14-3. *UAV swarm networking and cooperation applications in both military and civilian sectors [3]*

CHAPTER 14 FUNDAMENTALS AND APPLICATIONS OF UNMANNED AERIAL VEHICLE SWARMS

Civil UAVs

Government and charitable organizations employ civil UAVs as indispensable instruments for a range of vital duties, such as environmental monitoring, law enforcement, search and rescue, and disaster response. They are crucial in disaster management and public safety operations due to their diversity and sophisticated capabilities. Regarding disaster response, civil UAVs provide quick assessments of impacted regions and real-time pictures and data to help coordinate relief operations [53]. They greatly improve the effectiveness of response operations by assisting in determining the level of damage, locating survivors, and allocating resources in a priority manner. UAVs are useful for search and rescue missions because they can swiftly cover huge regions and get access to hazardous or challenging terrain. By using infrared cameras and other sensors, they can find those who have vanished and provide essential supplies, increasing the likelihood that rescue operations will be successful. UAVs provide traffic monitoring, crime scene investigation, and airborne surveillance to police enforcement [54]. By giving police access to up-to-date information, controlling crowds, and assisting with tactical operations, they improve public safety. UAVs are used in environmental monitoring to follow animals, evaluate the health of ecosystems, and measure pollution levels. Their long-flying capabilities and high-resolution sensors provide continuous observation, supporting conservation and environmental protection initiatives. Civil UAVs perform a critical role in refining emergency response, augmenting public safety, and promoting environmentally sound practices.

CHAPTER 14 FUNDAMENTALS AND APPLICATIONS OF UNMANNED AERIAL VEHICLE SWARMS

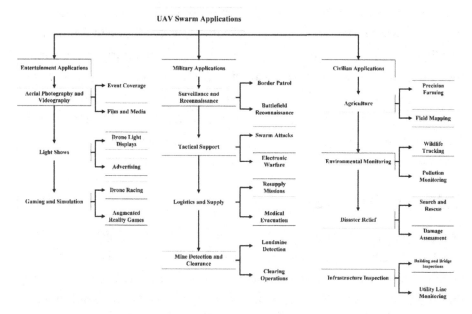

Figure 14-4. Applications of UAV swarms in different fields

Based on Design and Configuration

Design and configuration-based UAVs are briefly explained here in this section. The detailed explanations are as follows.

Fixed-Wing UAVs

UAVs with fixed wings mimic conventional aircraft with a fixed-wing structure that generates lift. They are perfect for long-range missions because of their design, allowing them to travel great distances quickly. Fixed-wing UAVs can remain in the air for longer periods because of their aerodynamic efficiency, enabling them to carry out activities like mapping, environmental monitoring, and surveillance over large regions [55, 56]. These UAVs need specialized infrastructure for takeoff and landing, usually runways or catapult launch systems. Fixed-wing UAVs, with their high efficiency over long distances, are ideal for agriculture and disaster response due to their ability to glide and soar, saving energy and extending flight time. They are well suited for use in agriculture, where they can monitor vast fields, or in disaster response, where they can scan enormous regions for damage assessment because of their high efficiency over long distances. Furthermore, fixed-wing UAVs are used in missions where long-range and endurance are essential, such as border patrol, maritime surveillance, and scientific research. Fixed-wing UAVs are useful for tasks needing broad coverage and continuous flying capabilities.

Rotary-Wing UAVs

Vertical takeoff and landing (VTOL) is a feature of rotary-wing UAVs, which include helicopters and multimotor devices like quadcopters and helicopters. Because they don't need runways to climb or descend vertically, they are very flexible and useful in a variety of settings. Their extraordinary mobility and capacity to hover in one spot are key features for activities that need precise and secure placement. Because of this, rotary-wing UAVs are perfect for close-quarters work and comprehensive inspections, such as checking wind turbines, power lines, and bridges [57]. Furthermore, when fixed-wing UAVs cannot fly, their VTOL characteristics allow them to operate in tight locations and challenging terrain. In search and rescue missions, rotary-wing UAVs are very widely used since they enable rescuers to reach difficult-to-reach locations and provide visual inspections in real time. They are essential for applications in photography and filmmaking because of their hovering and complicated maneuvering abilities, which allow for steady and dynamic pictures. Rotary-wing UAVs are indispensable instruments for various close-range, intricate, and dynamic jobs.

Hybrid UAVs

The advantages of both fixed-wing and rotary-wing UAVs are combined in hybrid UAVs, which provide the best of both worlds: the long-range, energy-efficient flying of fixed-wing UAVs and the VTOL capabilities of rotary-wing

CHAPTER 14   FUNDAMENTALS AND APPLICATIONS OF UNMANNED AERIAL VEHICLE SWARMS

UAVs. Because of this combination, hybrid UAVs may be used for a variety of tasks and are quite flexible. Without the need for runways, hybrid UAVs can fly and land in tight places, covering vast distances with efficiency. Applications include infrastructure inspection, which calls for accurate, up close inspection, and agricultural monitoring, which necessitates surveying broad areas, which benefit significantly from this dual capacity [58]. It is essential to comprehend the many kinds of UAVs, including hybrids, to choose the appropriate platform for a given application and guarantee successful deployment in a range of situations. These airborne systems will get even more potent as UAV technology develops, extending their applications in environmental monitoring, disaster response, and logistics and further assimilating into contemporary society.

Table 14-2. *A description of the multiple types of UAVs, emphasizing their features and applications*

UAV types	Weight and size	Endurance and range	Applications	Configuration and design
Micro-UAVs	Extremely small and light (less than 1 kg)	Limited endurance (minutes), limited range (a few kilometers)	Indoor operations, investigation, and spying	Unchangeable or rotary wings
Mini-UAVs	Having a small size and 1 to 10 kg weight	Medium endurance (1–2 hours) and medium range (10–30 km)	Law enforcement, hobbies, and environmental monitoring	Having Unchangeable, hybrid, or rotary wings

(*continued*)

CHAPTER 14　FUNDAMENTALS AND APPLICATIONS OF UNMANNED AERIAL VEHICLE SWARMS

Table 14-2. (*continued*)

UAV types	Weight and size	Endurance and range	Applications	Configuration and design
Tactical UAVs	Moderately sized and having 10–50 kg weight	Increased duration (4–6 hours) and range (up to 150 km)	Border enforcement, disaster relief, and military ISR	Hybrid or fixed wing
MALE UAVs	Large (between 100 and 500 kg)	Long endurance (24–48 hours) and range (hundreds of kilometers)	Long-range reconnaissance, marine patrol, and military ISR	Unchangeable wings
HALE UAVs	Very big (500+ kg)	Very long duration (several days) and range (thousands of kilometers)	High-altitude mapping, atmospheric research, and strategic military activities	Unchangeable wings
Commercial UAVs	Fluctuates (1–50 kg)	Changes in distance (10–150 km), duration (1–8 hours)	Filmmaking, real estate, agriculture, and infrastructure inspection	Having Unchangeable, hybrid, or rotary wings
Civil UAVs	Fluctuates (1–50 kg)	Changes in distance (10–150 km), duration (1–8 hours)	Operations involving search and assistance, law enforcement, monitoring the environment, and response to disasters	Having Unchangeable, hybrid, or rotary wings

(*continued*)

CHAPTER 14 FUNDAMENTALS AND APPLICATIONS OF UNMANNED AERIAL VEHICLE SWARMS

Table 14-2. (*continued*)

UAV types	Weight and size	Endurance and range	Applications	Configuration and design
Recreational UAVs	Small in size and 0.5 to 5 kg in weight	Short duration (30–60 minutes), limited range (a few kilometers)	Interest in photography and videography	Rotary wings
Unchangeable-wing UAVs	Fluctuates	Long duration (several hours) and long range (hundreds of kilometers)	Mapping, long-range missions, and environmental surveillance	Unchangeable wings
Rotary-wing UAVs	Fluctuates	Short duration (10 to 30 km), limited range (1 to 2 hours)	Close-quarters work, close examinations, and photographs	Hexacopters, quadcopters, and other unchangeable wings
Hybrid UAVs	Fluctuates	Long duration (several hours) and long range (hundreds of kilometers)	Missions that are versatile and require both long-range and VTOL capabilities	Combination of rotary and unchangeable wings

Critical UAV Components

UAVs are complex systems made up of many essential parts, each of which plays a distinct role in guaranteeing the general functionality and dependability of the drone. Anyone engaged in UAVs' design, operation, or maintenance must thoroughly understand these components. We examine the main parts of a UAV below.

CHAPTER 14   FUNDAMENTALS AND APPLICATIONS OF UNMANNED AERIAL VEHICLE SWARMS

Airframe

The airframe represents the essential framework of a UAV, serving as its main source of support and housing all other components, including the propulsion system, sensors, and payload. This essential component must be robust and lightweight to withstand the many stresses that flying entails, including collisions, vibrations, and aerodynamic forces. Airframes are often constructed from advanced materials, including carbon fiber, aluminum, or composites, to achieve the optimal strength-to-weight ratio [59]. UAVs that must be robust without sacrificing performance employ carbon fiber because of its exceptional strength-to-weight ratio. Aluminum is also widely used since it is lightweight and corrosion-resistant. Composites use many materials to enhance certain properties, offering tailored solutions for specific UAV applications [60]. The intended use of the UAV, weight, and required strength are some factors that affect the material choices. For high-performance UAVs employed in military or long-range missions, for instance, advanced composites may be given priority over simpler models meant for recreational use, which might use less costly, lighter materials. The effectiveness and operational efficiency of the UAV greatly depend on the design and material selection of the aircraft. Figure 14-5 illustrates the concept of airframe swarms in UAVs.

CHAPTER 14   FUNDAMENTALS AND APPLICATIONS OF UNMANNED AERIAL VEHICLE SWARMS

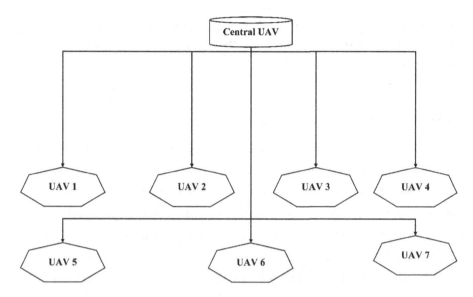

Figure 14-5. *The concept of airframe swarms in UAVs*

Propulsion System

The propulsion system of a UAV is crucial for generating the power needed to lift and maneuver the aircraft. Normally, this system is made up of many crucial components that work together to enable efficient flight. Electric motors, often found in small- and medium-sized UAVs, provide the necessary power because of their effective and transportable design. Larger UAVs often employ internal combustion engines to provide the extra power required for more challenging tasks. Propellers are necessary because they convert the spinning motion of the engines into thrust [61]. The UAV's carefully crafted size and design, tuned to enhance performance for specific flight scenarios, allow it to achieve the necessary speed and agility.

Moreover, the speed of the motors must be controlled using electronic speed controllers, or ESCs. They provide precise control over propulsion and maneuverability, enabling responsive and fluent flight. Together, these components provide the UAV with power while balancing power, efficiency, and control to meet the particular needs of its intended application.

Power Source

The power source is required to provide the energy required for the UAV to operate and successfully complete its intended duties. Among the most popular power sources are batteries, fuel cells, and internal combustion engines; each is suitable for a certain kind of UAV and its intended use. Due to its high energy density and low weight, lithium-polymer (LiPo) batteries are often used in small- to medium-sized UAVs to provide significant power without adding extra weight. This capability is beneficial for UAVs that have to balance flight time against performance. Fuel cells are an additional choice; they have longer flight times than batteries [62]. They are often used in complex UAVs with extended operating periods because they provide a consistent and stable power supply for endurance missions. Larger UAVs are often equipped with internal combustion engines that run on gasoline or diesel. These engines provide significant power output and extended endurance, making them suitable for heavy-duty applications requiring longer flight times and greater cargo capacity. Each power source determines the UAV's functioning and suitability for a mission.

Flight Control System

The UAV's brain is the flight control system necessary for accurate command execution and stable flying. The key element of this system is the flight controller, a central processing unit (CPU) that controls the flight control software. It guarantees accurate and responsive navigation by analyzing data from several sensors and transmitting control signals to the motors and actuators. Gyroscopes and accelerometers are crucial components of inertial measurement units (IMUs) that support flight controls [63]. These sensors provide the flight controller with vital information about the UAV's orientation and motion, enabling it to maintain stability and make the required corrections, even under challenging flying conditions. The GPS module is also essential to the flight control system since it provides real-time location data that facilitates navigation and waypoint-based missions. This makes it possible for the UAV to follow preset paths and arrive at specified places with accuracy [64]. Together, these components form a comprehensive flight control system that ensures the UAV can reliably and efficiently complete challenging missions.

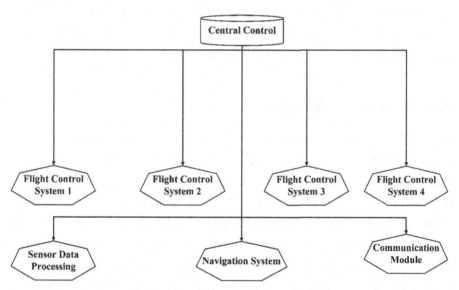

Figure 14-6. Controlling the flying system in a swarm of UAVs

Navigation and Communication Systems

Communication and navigation systems are necessary to operate UAVs remotely and transmit data to the operator. These systems' key components are the radio transmitter and receiver, which allow the UAV and the ground control station (GCS) to stay in continuous communication. This link allows operators to interact, give orders, and receive information that provides effective administration and monitoring throughout the flight. The telemetry system plays a vital role by giving the operator access to real-time flight data, including altitude, speed, and battery status. This data is essential for monitoring the UAV's performance and making informed decisions during flight, which raises efficiency and safety [65, 66]. According to the autopilot system, the UAV may also fly autonomously in line with preprogrammed flight plans. With this ability, the UAV can do complex tasks without constant human assistance, following precise paths and working independently. These navigation and communication technologies work together to ensure that the UAV remains under control, operates as intended, and provides the operator with pertinent information while in flight.

Autonomous Systems

Modern UAVs are more effective and adaptive because they can function independently according to updated technology. One crucial part is the obstacle detection and avoidance system, which uses a range of sensors to identify and steer clear of obstacles. These sensors, which include LiDAR, radar, and cameras, give real-time data that allows the UAV to alter its flight path and avoid collisions, ensuring safe and reliable operation in difficult scenarios. Furthermore, enhancing the UAV's decision-making ability requires the use of artificial intelligence (AI) techniques. These algorithms enable the UAV to perform complex target recognition and route planning tasks. For instance, AI may use visual data processing to identify certain items or traits, allowing the UAV to follow targets or scan areas with remarkable precision [67, 68]. Algorithms for AI-driven route planning enhance flight pathways to complete missions, even in dynamic and unforeseen situations successfully. Combining these autonomous systems gives sophisticated UAVs more independence and power, allowing them to do difficult jobs with minimal help from people and quickly adapt to changing environments. Understanding these fundamental components is necessary to comprehend how UAVs operate and how they may be modified for various applications. Each component is necessary for the UAV's overall functioning and performance, which ensures dependable operation and efficient mission completion.

UAVs with Communication Systems

UAVs need communication systems to transmit data to and from other UAVs, sensors, and GCS. These technologies provide real-time information and control capabilities, ensuring the UAV works efficiently and safely. We examine the several facets and varieties of UAV communication systems below.

Types of Communication Systems

Radio Frequency (RF) Communication

The most used technique for UAV communication is RF transmission, which sends data via frequencies usually in the 900 MHz, 2.4 GHz, or 5.8 GHz range. This technique is essential for data transfer and control signal transmission to provide a dependable connection between the UAV and the GCS. A transmitter at the GCS and a receiver on the UAV make up the communication arrangement [69, 70]. The UAV's motions and functions are guided by control instructions sent by the GCS transmitter. These orders are accepted by the UAV's receiver, giving the operator complete control over the vehicle's movements. Furthermore, telemetry systems that send back real-time flight data to the GCS are essential to RF communication. This data allows the operator to track the UAV's operation and make wise choices while on a mission. It contains vital information,

CHAPTER 14 FUNDAMENTALS AND APPLICATIONS OF UNMANNED AERIAL VEHICLE SWARMS

including altitude, speed, battery condition, and GPS locations. These RF communication parts work together to provide efficient and ongoing communication between the UAV and its operator, improving operational security and efficiency.

Satellite Communication (SATCOM)

SATCOM is essential for long-range UAV operations, especially those that operate beyond the line of sight (BLOS). This communication approach is essential for military and surveillance UAVs operating in isolated or difficult-to-reach locations where regular radio frequency connection is unfeasible. Near-global coverage offered by SATCOM systems guarantees that UAVs can always stay connected, no matter where they are. One of SATCOM's main benefits is its broad reach, which enables UAVs to function well in areas lacking radio frequency infrastructure [71]. This capacity is essential for operations when UAVs fly in remote areas or cover large distances. Even though SATCOM provides extensive coverage, a disadvantage of this technology is its increased latency. The large distances that signals must travel to and from satellites are the source of the increased latency, which may lead to communication delays. Despite this lag, SATCOM is still vital in keeping long-range UAVs connected and operating reliably, allowing them to carry out intricate tasks in far-off places under constant supervision and control.

Cellular Communication

The high bandwidth and low latency capabilities of cellular networks, such as 4G and 5G, are making them more and more useful for UAV communication. This approach is appropriate for metropolitan situations with strong and broad cellular network coverage. Cellular networks are suited for applications that need quick and efficient data relay, including surveillance, mapping, and real-time monitoring, because of their high-speed data transmission capabilities, allowing for the transfer of large datasets and real-time video streaming. One of the main advantages of cellular communication is the capacity to send large volumes of data and stream high-definition video in real time, improving situational awareness and decision-making [72]. However, the dependability and accessibility of cellular networks within the operational area greatly influence this communication method's effectiveness. Under solid cellular infrastructure conditions, UAVs can sustain stable and reliable communication relationships. On the other hand, in regions with inadequate cellular coverage, the dependability of the connection might be jeopardized, which could affect the operational effectiveness of the UAV. However, this drawback is that cellular networks provide a viable option for UAV communication, particularly in crowded, well-connected metropolitan environments.

Communication Protocols

TCP/IP (Transmission Control Protocol/ Internet Protocol)

Transmission Control Protocol/Internet Protocol, or TCP/IP, is a core set of communication protocols that is widely used for data transmission over the Internet and is being increasingly modified for use in UAV communication systems. TCP/IP is a widely used and very flexible protocol that can be combined with several communication technologies, such as cellular networks, SATCOM, and RF, offering flexibility in creating communication linkages [73, 74]. One of the critical benefits of TCP/IP is its dependability since it incorporates error-checking systems and permits the retransmission of lost packets, ensuring that data arrives precisely and undamaged [75]. This dependability is critical for applications requiring accurate data interchange, such as real-time control, telemetry, and high-definition video streaming.

User Datagram Protocol (UDP)

For applications, including live video streaming, where speed is more important than dependability, UDP is recommended. Because UDP does not have built-in error checking or retransmission facilities for dropped packets, it cannot ensure packet delivery as TCP/IP does. Due to this trade-off, UDP can transmit data more quickly and effectively for real-time data transfer. UDP is often used in UAV communication systems to provide real-time sensor data and live video feeds. UDP's low latency ensures that data gets to its destination fast, which is important for real-time surveillance and remote piloting applications, which call for speedy responses [76]. The constant flow of information guarantees that the overall quality of the live broadcast is maintained, even if some data packets may be lost during transmission. For certain UAV applications, where fast data transmission is more important than assured dependability, UDP is a necessary protocol since it prioritizes speed and minimizes delays, allowing the UAV to execute tasks that rely on real-time data.

Communication Security

UAV communication systems security is paramount since unauthorized access may lead to serious repercussions such as loss of control, data breaches, and other harmful actions. Several security measures have been implemented to safeguard the confidentiality and integrity of UAV operations and lessen these threats. An essential security precaution is encryption, which encrypts data flows to guard against manipulation and listening. Encrypting communications between the UAV and the GCS prevents unwanted parties from intercepting sensitive data, including video feeds, telemetry data, and control orders. Another crucial security measure is authentication, ensuring only those with permission can access and operate the UAV. This entails confirming the legitimacy of users trying to connect with the UAV, usually using cryptographic keys or secure login credentials. Authentication prevents unauthorized people from controlling the UAV or accessing its information. Furthermore, defense against spoofing and jamming is essential [77, 78]. By interfering with the UAV's communication signals, jamming might result in the device losing contact with the GCS. Sending fake signals to the UAV to trick it or seize control is known as spoofing. Advanced technologies are used to identify and foil such efforts to address these dangers. Spread spectrum technology and frequency hopping are examples of antijamming solutions that make it more difficult for attackers to tamper with the transmission. Verifying the legitimacy of incoming signals to ensure they come from reliable sources may be part of spoofing defense [79]. Integrating security mechanisms such as encryption, authentication, and jamming and spoofing protection may fortify UAV communication systems against unwanted access and guarantee UAVs' safe and secure functioning in diverse applications.

Swarm Communication

Effective communication between many drones in a UAV swarm is crucial for coordinated actions, which enable the swarm to operate as a cohesive entity. Swarm communication is made possible by many essential components that let drones coordinate their movements and share data. The foundation of swarm operations is inter-UAV communication, which allows drones to share sensor data, locations, and velocities [1, 80]. Due to this real-time data exchange, the swarm can carry out intricate tasks, including formation control, coordinated flight, and cooperative mapping. To ensure that the whole group can react dynamically to changes in the environment or mission goals, each drone in the swarm may modify its activities depending on the information received from its peers. Mesh networks are essential for improving the resilience and adaptability of swarm communication. Every UAV functions as a node in a mesh network, capable of communicating with many other nodes in the network. Because of its decentralized design, the network can automatically adjust to preserve communication channels even if some drones malfunction or lose connectivity. Due to its redundancy and durability, mesh networks allow drones to communicate continuously, even in the event of disturbances [81]. Using mesh networks and inter-UAV communication, UAV swarms may attain high degrees of coordination and flexibility. These communication methods allow the swarm to function more effectively and reliably while performing challenging tasks. Swarm communication's capabilities will grow as UAV technology progresses, creating new opportunities for coordinated drone operations across several industries, including military and agriculture.

UAV-Based Sensor Technologies

UAVs depend on sensor technology to collect, process, and analyze data from various situations. With the help of these sensors, UAVs may carry out various activities, from simple navigation to intricate data collecting and environmental monitoring. The various sensor technologies often found in UAVs are examined below.

Camera Sensors

UAVs are becoming increasingly outfitted with sophisticated imaging systems, each designed for a particular use case. RGB, multispectral, and hyperspectral cameras are the most notable due to their unique features and wide range of applications in different industries. The most common camera in UAVs is the RGB camera, which detects red, green, and blue light wavelengths to produce high-resolution color photos and films [82]. Applications like real-time surveillance, videography, and aerial photography need this capacity. RGB cameras are used in aerial photography and videography to provide vivid, detailed images essential for real estate marketing, media production, and tourist promotions.

These cameras provide high-quality images that aid in creating captivating material for various advertising and record-keeping uses [83]. RGB cameras provide live visual feeds for real-time surveillance, which makes it possible to monitor and analyze ground conditions immediately. This is especially helpful in security operations, disaster relief, and traffic management, where timely information is essential. RGB cameras also assist in mapping and inspection duties by providing precise and thorough visual recordings of urban settings, landscapes, and infrastructure. This helps with fault discovery, change monitoring, and maintenance assistance [84].

Multispectral cameras, using wavelengths outside the visual range, enhance imaging and are crucial for agricultural surveillance by analyzing plant health and stress in various spectral bands. Multispectral imaging makes it possible to see things invisible to the unaided eye, including crop and vegetation health. These cameras use reflectance patterns to spot problems, such as insect infestations, water stress, and fertilizer shortages, allowing farmers to make timely corrections. Multispectral cameras make it possible to create intricate maps of regions that need attention in precision agriculture, which improves crop yields and makes better use of available resources. Multispectral imaging-based vegetation analysis is essential for environmental monitoring as it tracks changes in forest health, monitors wetlands, and evaluates the effects of environmental policy. Decision-making procedures related to land management and environmental protection are informed by this data, which also helps conservation efforts. Providing comprehensive spectrum information for every pixel in a picture, hyperspectral cameras record an even greater variety of wavelengths than multispectral cameras [77]. This makes it possible to identify and analyze materials using their spectral characteristics precisely. Hyperspectral imaging is potent in agricultural research, environmental monitoring, and mineral prospecting.

Hyperspectral cameras are beneficial in mineral exploration because they can distinguish between minerals and materials according to their distinct spectral properties. This is especially useful in identifying and mapping mineral deposits, directing exploration efforts, and determining the feasibility of mining operations. Hyperspectral cameras' rich spectral data allow for more thorough and accurate analysis, supporting cutting-edge research and development across various industries. When these cutting-edge imaging technologies are combined, UAVs may carry out a variety of activities with greater accuracy and efficiency [85]. Hyperspectral cameras provide in-depth material analysis for specific applications, multispectral cameras improve agricultural and environmental management, and RGB cameras are perfect for visual documentation and real-time monitoring. The integration of these imaging systems will allow UAV technology to go even further, providing more advanced instruments for a wider range of sectors and research areas.

Thermal Imaging Sensors

UAVs can record and observe temperature differences in the surroundings using thermal imaging sensors, sophisticated devices that detect infrared radiation produced by objects. These sensors are very useful in many applications where heat signatures are important, improving UAV capabilities in a variety of disciplines. Search and rescue activities are one of the main uses for thermal imaging sensors in situations where it is difficult to find surviving or missing people. For example, in heavily forested, hilly, or disaster-affected areas, thermal sensors may identify body heat emissions from people. This feature greatly increases the likelihood of promptly locating and rescuing individuals who may be out of sight or buried behind rubble. Thermal imaging is very useful for search and rescue teams since it can see through smoke, fog, and darkness, giving them vital information that might save lives. Another crucial area where thermal imaging sensors on UAVs have a significant influence is firefighting. The hottest fire areas are revealed by these sensors' real-time thermal data, which also shows the direction and intensity of the flames. This information is helpful to firefighters. This knowledge makes it possible to use resources more strategically, which improves fire control and extinguishment [86].

Thermal imaging may also be used to watch hotspots after a fire has been put out to ensure it doesn't start again. This device improves firefighter safety by allowing firefighters to evaluate the situation remotely before venturing into potentially dangerous regions. In infrastructure inspection, thermal imaging sensors are used to find abnormalities and possible problems in structures such as electricity lines, buildings, bridges, and other structures.

For example, thermal imaging may detect building heat loss regions, indicating inadequate insulation or structural issues. Thermal sensors in electrical infrastructure may detect overheating parts, which may indicate upcoming malfunctions or the need for repair. Early detection of these problems using thermal imaging helps to avoid expensive repairs and guarantees the dependability and safety of vital infrastructure. Large regions may be swiftly and effectively inspected by UAVs fitted with heat sensors, eliminating the need for manual inspections and lowering the danger to human inspectors. Another significant use of thermal imaging sensors in UAVs is wildlife monitoring. Researchers and conservationists use thermal sensors to track and monitor wildlife, particularly nocturnal species or creatures that are hard to detect during the day. Thermal imaging offers a noninvasive method of studying animal behavior, population sizes, and habitat use without upsetting the animals. This technology is beneficial for monitoring endangered species because it gathers information that might guide conservation plans and activities [87].

Furthermore, by detecting human heat signatures in protected regions, thermal imaging may aid in detecting poachers and improve antipoaching activities. Thermal imaging sensors are a useful and versatile technology that may be used to improve UAV functioning in a variety of applications. By recording and evaluating temperature fluctuations, these sensors provide significant insights that are not readily apparent. With these capabilities, UAVs may be used for tasks including search and rescue, firefighting, infrastructure inspection, and wildlife monitoring that need more sophisticated solutions than can be achieved with standard visual imaging. Incorporating thermal imaging sensors into UAVs is expected to advance in sophistication as the technology develops, increasing the range of applications and enhancing operational efficiency. This current research is expected to satisfy the needs of contemporary difficulties by using thermal imaging to improve the safety, efficacy, and efficiency of UAV operations across various areas.

Light Detection and Ranging (LiDAR) Sensors

LiDAR sensors are cutting-edge instruments that estimate one object's distance from another using laser pulses, producing very precise and detailed three-dimensional maps of the surrounding area. Because of its accuracy and capacity to pierce foliage, this technology has become increasingly essential to UAV operations, offering useful data for various uses. Topographic mapping is one of the main applications of LiDAR sensors. Conventional techniques for mapping topography may be labor- and time-intensive, and they often call for thorough ground surveys. LiDAR, on the other hand, can quickly produce high-resolution maps of vast regions, accurately capturing the landscape's elevations and contours. This capacity is crucial for geological research, environmental monitoring, and land use planning [82]. Because LiDAR can produce precise digital elevation models (DEMs), scientists and researchers may more confidently plan building projects, evaluate flood hazards, and examine topographical characteristics. Because LiDAR can see through the forest canopy and get precise data on the underlying topography and plant structure, it is

CHAPTER 14 FUNDAMENTALS AND APPLICATIONS OF UNMANNED AERIAL VEHICLE SWARMS

very useful in forestry management. With this technology, accurate three-dimensional (3D) models of forests can be created, which is essential for inventorying tree species, calculating the biomass of forests, and tracking changes over time. LiDAR data may be used by forest managers to monitor the impacts of environmental changes, such as insect infestations or climate change, and to plan sustainable harvesting strategies. LiDAR supports educated decision-making that strikes a balance between resource use and ecological protection by providing precise and comprehensive data.

LiDAR technology is also very beneficial for urban planning. Accurate and current geographical data is essential for managing resources, planning infrastructure, and guaranteeing sustainable development as cities expand and change. With great accuracy, LiDAR can create comprehensive 3D representations of urban settings, including roads, buildings, plants, and other characteristics. Architects and urban planners use this data to create disaster-resilient plans, evaluate the effects of new developments, and build infrastructure projects. LiDAR, for example, may assist in locating locations vulnerable to landslides or floods, enabling the installation of mitigating measures before the start of construction. LiDAR-equipped UAVs are an effective tool for evaluating the state of important infrastructure, including pipelines, power lines, bridges, and dams [88]. By taking exact measurements and monitoring even the smallest changes in the structure over time, these sensors may spot any problems that conventional inspection techniques would miss. By flying over and around buildings, UAVs fitted with LiDAR technology may produce intricate 3D models that show fractures, deformations, and other wear-and-tear indicators. Using this data, maintenance staff may carry out preventative maintenance and prioritize repairs, ultimately improving the infrastructure's lifetime and safety [89]. LiDAR sensors are essential in these applications because they can provide precise and in-depth 3D maps. They are able to collect data that would be difficult or impossible to get with other techniques because of their high degree of accuracy and capacity to pass through foliage and other impediments. LiDAR sensor integration is expected to become much more common as UAV technology develops, increasing the range of applications and enhancing performance.

Radar Sensors

UAVs employ radar sensors, a key piece of technology, to detect objects and use radio waves to calculate their distance, speed, and other properties. Radar sensors, despite being hampered by external conditions like clouds, fog, and rain, are crucial in UAV applications for obstacle detection, ensuring reliable operation. UAVs must be able to recognize and avoid obstacles to operate safely in a variety of contexts. By continually emitting radio waves that bounce off of things, radar sensors enable the UAV to detect and map impediments in its path [90]. The UAV may instantly alter its flight course, thanks to this real-time data, preventing crashes into buildings, other obstacles, or the landscape. Accurate and timely obstacle detection is particularly critical in crowded or complicated situations, including metropolitan areas or thick woodlands, where optical sensors may have difficulty.

Radar sensors are an essential component that improves UAV navigation capabilities. Accurate navigation is crucial for UAVs to carry out their tasks, such as surveillance, delivering cargo, or mapping terrain. Radar sensors make more exact location and movement tracking possible, which offer accurate speed and distance readings. Integrating this data with GPS and other navigational systems may increase the accuracy and dependability of the UAV's navigation. Radar sensors provide an extra layer of navigational help when GPS signals are poor or nonexistent, including beneath thick canopies or in urban canyons, ensuring that the UAV can

CHAPTER 14   FUNDAMENTALS AND APPLICATIONS OF UNMANNED AERIAL VEHICLE SWARMS

continue functioning as intended. One essential feature that radar sensors in UAVs provide is collision avoidance. The chance of mid-air collisions with human aircraft or other UAVs rises as UAVs are used in more sectors. When determining the speed and trajectory of other flying objects, radar sensors are essential [91]. According to the knowledge mentioned earlier, the UAV can anticipate possible collisions and conduct evasive maneuvers to avoid them. The radar sensors' all-weather capabilities provide an additional degree of safety, which guarantees that the collision avoidance system will continue to function in inclement weather. This is especially crucial for UAVs that operate in airport vicinity or shared airspace, where keeping a safe distance from other aircraft is essential. Besides these primary applications, radar sensors are used in UAVs for other complex tasks.

For example, they may be used for surface monitoring and terrain mapping, where precise distance measurements are required to produce intricate maps or track changes over time. Radar sensors may assist in the location of survivors during search and rescue operations by using synthetic aperture radar (SAR) technology to identify heat signatures or detect movement. Radar sensors are very reliable and rugged, making them ideal for various UAV activities [92]. The safety and effectiveness of UAV operations are greatly increased by their capacity to operate in all-weather situations and to provide precise data on object attributes, speed, and distance. Advanced radar systems will be integrated into UAVs to significantly increase their capabilities and enable more complicated and autonomous missions as UAV technology develops.

Ultrasonic Sensors

Unmanned aerial vehicles, or UAVs, employ ultrasonic sensors as a fundamental technology to measure objects' distances from them using sound waves. These sensors emit high-frequency sound waves, which reflect off of things and back to the sensor. The sensor can precisely measure the distance to an item by measuring the time it takes for sound waves to return. Numerous important applications of this technology, including obstacle avoidance, proximity sensing, and landing aid, are commonly used in UAVs. Obstacle avoidance is one of the main uses of ultrasonic sensors in UAVs. For UAVs to operate safely and effectively in a variety of contexts, the capacity to recognize and avoid obstacles is essential. By using ultrasonic sensors to measure the distance to objects in real time, the UAV may modify its flight route and prevent accidents [93]. This is especially crucial in situations where visual sensors could be less reliable, including dimly lit areas or densely vegetated areas. Ultrasonic sensors ensure the UAV can fly securely by continually scanning the surroundings and spotting any dangers that might cause accidents. Another essential function made possible by ultrasonic sensors is proximity detection. Keeping a safe distance from obstacles and other UAVs requires proximity sensing, which measures the presence and distance of objects within a certain range. Because ultrasonic sensors can determine distance with accuracy and dependability, the UAV can carry out a variety of activities while keeping a safe buffer zone. This is especially helpful in cases when numerous UAVs are operating in close vicinity or in complicated surroundings like metropolitan regions with plenty of barriers. In addition to ensuring smooth and coordinated flight operations, proximity detection helps avoid collisions. Another essential use of ultrasonic sensors in UAVs is landing aid. Precise and safe landings depend on accurate distance measures, especially in hazardous situations or on uneven ground [89]. The UAV may modify its descent speed and angle for a smooth landing by using real-time data from ultrasonic sensors, which can estimate the distance to the ground. This feature is particularly crucial for autonomous UAVs that have to land without human assistance. The overall safety and dependability of the operation are improved by ultrasonic sensors, which guarantee accurate and regulated landings and assist avoid damage to the UAV and its supplies. Moreover, ultrasonic sensors serve these main purposes and enhance other cutting-edge UAV features. For example, they can be used in terrain following, where the UAV needs to maintain a constant altitude relative to the ground. Ultrasonic sensors provide continuous altitude measurements, allowing the UAV to adjust its flight path to follow the terrain's contours accurately. This is particularly useful in agricultural applications, where maintaining a consistent altitude is crucial for crop spraying or monitoring tasks. The reliability and precision of ultrasonic sensors make them an invaluable component of UAV systems. Their ability to provide accurate distance measurements in various conditions ensures that UAVs can navigate safely and perform their tasks effectively. As UAV technology continues to advance, integrating ultrasonic sensors will further enhance their capabilities, enabling more complex and autonomous operations.

Acoustic Sensors

Specialized instruments known as acoustic sensors can detect sound waves. This makes them useful in many different industries, particularly when combined with UAVs. These sensors can detect sound and process, yielding useful information for acoustic surveillance, structure health assessments, and animal monitoring [94, 95]. They are a vital tool when sound-based information is required due to their ability to detect faint acoustic signals. Monitoring wildlife is one of the main uses of sound sensors in UAVs. Acoustic sensors are especially helpful for following animals that are hard to see with the naked eye and for researching animal behavior [96-98]. For example, certain noises made by birds, bats, and marine mammals may be tracked to collect information on an animal's existence, population, and behavior.

Researchers may explore vast and inaccessible regions by using UAVs fitted with acoustic sensors. This allows them to gather audio data that offers valuable insights about the health and dynamics of animal populations. This noninvasive approach is perfect for ecological research and conservation initiatives since it permits ongoing observation without upsetting the animals. Effective habitat management and the protection of endangered species may both be aided by the data gathered. Acoustic sensors on UAVs are also essential for structural health assessments [99, 100]. These sensors can pick up noises and vibrations from buildings, bridges, and industrial sites, often signs of a structure's structural stability. Examples of distinct acoustic signatures include the emergence of fractures, corrosion, or other types of damage. By flying near these buildings, UAVs fitted with acoustic sensors may record sound waves and vibrations that indicate possible problems. This skill makes it possible to identify structural issues early on, allowing for prompt maintenance and repairs that avert disastrous failures. Acoustic sensors are an economical and effective means of keeping an eye on the structural integrity of vital infrastructure, contributing to its durability and safety. Another important use for these sensors is acoustic surveillance.

Acoustic sensors may be employed in security and defense scenarios to identify and examine ambient noises like footfall, gunfire, or vehicle movements [101, 102]. These sensors allow UAVs to cover broad regions while delivering real-time audio surveillance, improving situational awareness. This feature is especially helpful when visual monitoring may be restricted, including at night or in places with a lot of greenery. Acoustic surveillance may be used in conjunction with other sensor types, including thermal and video cameras, to provide a complete picture of the surroundings and any dangers. Acoustic sensors may also be useful for monitoring the surroundings. They may be used, for instance, to track the amount of noise pollution in cities, gathering information on the volume and origins of noise. Since excessive noise may have serious negative consequences on people's health and well-being, this knowledge is important for both public health and urban planning. Wide-area noise assessments may be effectively carried out using UAVs and acoustic sensors to find hotspots and guide mitigation techniques [103]. The usefulness and adaptability of UAVs in these diverse applications are improved by including sound sensors. These sensors are crucial to sophisticated UAV systems because they provide vital and distinct data that cannot be gathered from optical or other sensors. The capabilities of acoustic sensors are expected to grow as UAV technology progresses, providing new opportunities for data gathering and analysis in a variety of situations.

Control Algorithms and Swarm Intelligence

Swarm Intelligence Fundamentals: Concepts and Methods

The field of swarm intelligence, which draws inspiration from the collective behavior of social animals such as ants, bees, and birds, is concerned with ideas and techniques that enable decentralized and self-organizing systems. Simple rules and local interactions between individual agents are the foundation of swarm intelligence, allowing for the development of complex and adaptive collective behavior [104]. Stigma, negative feedback, and positive feedback are the three main concepts. Positive feedback encourages productive activity, while negative input discourages damaging ones, creating a balance that guides the swarm. Ant pheromone trails are an example of stigmergy, an indirect communication method where people utilize their environment to influence others [105]. These trails guide colony members to food sources. Other methods, including particle swarm optimization (PSO), ant colony optimization (ACO), and genetic algorithms (GA), are based on the same concepts. PSO mimics the social behavior of flocks of birds or schools of fish, where individual particles adjust their placements to find the optimal solutions based on their individual and collective experiences [106]. ACO uses artificial pheromones to simulate pathfinding, a behavior inspired by ant foraging, to tackle challenging optimization problems. GAs, which are modeled after natural selection, produce solutions via a sequence of steps that include crossing, mutation, and selection. These techniques enable the development of robust, scalable, and adaptable systems that find application in a range of domains, such as robotics, artificial intelligence, and optimization. Researchers and engineers may design durable, adaptable, and effective systems that are evocative of the remarkable abilities of natural swarms by using the notions of swarm intelligence [107].

Basic Algorithms: Genetic, Ant Colony, and Particle Swarm Optimization

Common swarm intelligence methods, such as PSO, ACO, and genetic methods (GA), use collective behaviors seen in nature to solve complex problems. PSO draws influence from the social dynamics seen in fish schools and avian groups. In this method, potential solutions are represented by a swarm of particles that move over the problem space and adjust their placements based on individual and collective experiences to find the optimal solution. Each particle changes its location and velocity based on the best placements of itself and its neighbors, promoting convergence toward the ideal solution. The ACO model explains ant foraging, where ants communicate through pheromone trails, guiding each other to find food sources [108]. In ACO, artificial ants employ probabilistically reinforced pheromone trails to construct optimality solutions. As ants find better solutions, the connected routes get stronger until the swarm settles on an ideal or almost optimal solution. This approach is particularly effective for combinatorial optimization problems such as the traveling salesman problem. Genetic algorithms, or GAs, are primarily inspired by natural selection and genetic evolution [109]. In GA, chromosomes are a population of potential solutions that develop throughout the course of several generations. Genetic operators that produce novel offspring solutions include selection, crossover, and mutation. The healthiest individuals are selected for reproduction,

CHAPTER 14 FUNDAMENTALS AND APPLICATIONS OF UNMANNED AERIAL VEHICLE SWARMS

ensuring the positive traits are passed forward. Over time, the population tends to gravitate toward optimum solutions. These algorithms use the ideas of swarm intelligence and natural evolution to give robust, flexible, and efficient solutions to a range of optimization problems, making them valuable tools in robotics, engineering, and artificial intelligence.

Mechanisms for Real-Time Control and Adaptation

UAV swarms' capacity to make rapid judgments and modifications based on real-time data is crucial to their ability to operate successfully in dynamic and unpredictable situations. This skill ensures the swarm can respond swiftly to obstacles and changes during missions. The integration of cutting-edge sensors and communication systems that continually collect and exchange data about the UAVs' internal conditions and external environment is a crucial part of this capacity. Sophisticated algorithms that regulate the swarm's behavior are used to analyze this data, enabling fast adjustments to job distribution, formation, and flight trajectories [12, 110]. UAV swarms need to learn from their experiences and surroundings in order to improve performance in the future. Reinforcement learning and other machine learning methods are widely employed to enable UAVs to adapt their tactics based on feedback from their actions. For example, a UAV swarm engaged in search and rescue operations may dynamically adapt its search patterns to account for topographical obstacles or signs of human habitation. Decentralized control strategies are also necessary to preserve the robustness and resilience of the swarm [110]. Each UAV operates in a semiautonomous way, using local laws to coordinate with other units and maintain swarm unity. This decentralized approach increases the system's overall reliability by allowing the swarm to operate even if certain UAVs break or are disrupted. Thus, UAV swarms have the flexibility and agility to complete challenging tasks in various operational scenarios, thanks to real-time control and adaptation systems.

CHAPTER 14   FUNDAMENTALS AND APPLICATIONS OF UNMANNED AERIAL VEHICLE SWARMS

Challenges and Opportunities

UAV swarms are a fast-developing sector that offers much potential for further study and development. As technology advances, several essential areas require more research to improve UAV capabilities, effectiveness, and applications.

Advanced Sensor Integration

More sophisticated sensor integration into UAV systems should be the main emphasis of future research. This entails creating small, light, and energy-efficient sensors that can operate in a variety of environments and provide high-resolution data. Advances in multisensor fusion, which integrates data from many sensor types (such as LiDAR, thermal, and hyperspectral), may enhance situation awareness and more precise data interpretation.

Autonomous Navigation and AI

Prospective research areas include autonomous navigation using machine learning (ML) and AI. The operational capabilities of UAVs may be significantly improved by creating more complicated algorithms that enable them to explore complex surroundings independently, avoid obstacles, and make judgments in real time. The lifetime and dependability of UAVs may also be increased by research into AI-driven predictive maintenance and self-diagnosis systems.

Swarm Technology

Swarm technology has a wealth of research options as it allows numerous UAVs to collaborate to execute tasks. Future research might investigate cooperative algorithms, decentralized control systems, and optimal communication protocols to increase the effectiveness and resilience of UAV swarms. Swarm technology advancements, including environmental surveillance, agricultural monitoring, and search and rescue, may significantly impact applications.

Robust Communication Systems

UAVs' safe and effective operation depends on developing more dependable and secure communication technologies. Enhancing long-range communication capabilities, incorporating 5G technology, and creating robust networks that can resist interference and cyberattacks should be the main areas of future study. Investigating mesh networks and satellite communication may also provide reliable options for UAV operations in rugged and isolated locations.

Regulatory Frameworks and Standards

Comprehensive regulatory frameworks and standards are becoming increasingly necessary as UAV use grows to guarantee safety and interoperability. Developing rules for UAV certification, airspace management, and integration with manned aircraft should be the goal of this field's research. Business, academia, and regulatory agencies must work together to solve these issues and create a unified worldwide framework.

Application-Specific Innovations

Research on improvements that are tailored to a certain application has several prospects. For example, UAVs in agriculture may be fitted with cutting-edge sensors to maximize resource use and monitor crop health. UAVs may be created for disaster management in order to help rescue operations, distribute supplies, and evaluate damage. Each application sector has distinct possibilities and difficulties for UAV technological improvements. The UAV industry may continue to develop, providing new capabilities and solutions across a variety of industries by concentrating on these future research objectives. Unlocking the full potential of UAVs and tackling the obstacles related to their broad deployment will require sustained research and development investment.

Conclusion

UAV swarms are a quickly developing technology that have crucial applications across multiple fields. The basic ideas of UAV swarms are covered in this chapter, along with the many kinds of UAVs, their parts, communication systems, and sensor technologies. Through UAV classification and analysis of their propulsion, power, navigation, and payload systems, we illustrate the technical developments that improve UAV performance and dependability. Robust communication systems and sophisticated sensors that allow for real-time data exchange and situational awareness are essential for efficient UAV swarm operations. We spoke about energy management, collision avoidance, and communication issues in dynamic situations, along with the algorithms and protocols that enable coordinated operations. For environmental monitoring, agricultural, disaster relief, and military uses, UAV swarms provide scalable, flexible, and effective solutions. The transforming power of case studies is displayed. To guarantee responsible usage, complete frameworks are necessary, and safety, ethics, and legal considerations must be taken into account while deploying UAV swarms. Future developments in AI, machine learning, and autonomous systems will improve UAV swarm capabilities even further. With a firm grasp of UAV swarms provided by this chapter, readers will be ready to make contributions to this quickly developing subject. UAV swarms have the potential to transform several industries with their unparalleled capabilities and efficiency as technology develops.

References

[1] Majee, A.; Saha, R.; Roy, S.; Mandal, S.; Chatterjee, S. Swarm UAVs Communication. *arXiv preprint arXiv:2405.00024* **2024**.

[2] Bhat, G.; Dudhedia, M.; Panchal, R.; Shirke, Y.; Angane, N.; Khonde, S.; Khedkar, S.; Pansare, J.; Bere, S.; Wahul, R. Autonomous Drones and their Influence on Standardization of Rules and Regulations for Operating-A Brief Overview. *Results in Control and Optimization* **2024**, 100401.

[3] Cao, P.; Lei, L.; Cai, S.; Shen, G.; Liu, X.; Wang, X.; Zhang, L.; Zhou, L.; Guizani, M. Computational Intelligence Algorithms for UAV Swarm Networking and Collaboration: A Comprehensive Survey and Future Directions. *IEEE Communications Surveys & Tutorials* **2024**.

[4] Javed, S.; Hassan, A.; Ahmad, R.; Ahmed, W.; Ahmed, R.; Saadat, A.; Guizani, M. State-of-the-Art and Future Research Challenges in UAV Swarms. *IEEE Internet of Things Journal* **2024**.

[5] Shikhola, T. An Introduction to Swarm Intelligence in Communication. In *Applications of Computational Intelligence Techniques in Communications*; CRC Press: 2024; pp. 1-10.

[6] Yablokova, A.; Kovalev, D.; Kovalev, I.; Podoplelova, V.; Astanakulov, K. Environmental safety problems of swarm use of UAVs in precision agriculture. In Proceedings of the E3S Web of Conferences, 2024; p. 04018.

[7] Fedorovych, O.; Kritskiy, D.; Malieiev, L.; Rybka, K.; Rybka, A. Military logistics planning models for enemy targets attack by a swarm of combat drones. *Radioelectronic and Computer Systems* **2024**, *2024*, 207-216.

[8] Khalil, H.; Rahman, S.U.; Ullah, I.; Khan, I.; Alghadhban, A.J.; Al-Adhaileh, M.H.; Ali, G.; ElAffendi, M. A UAV-Swarm-Communication Model Using a Machine-Learning Approach for Search-and-Rescue Applications. *Drones* **2022**, *6*, 372.

[9] Yue, P.; Xin, J.; Zhang, Y.; Lu, Y.; Shan, M. Semantic-Driven Autonomous Visual Navigation for Unmanned Aerial Vehicles. *IEEE Transactions on Industrial Electronics* **2024**.

[10] Hassan, M. Integration of UAVs in society: Opportunities and challenges from public viewpoint on UAVs in Civilian Airspace. **2024**.

[11] Ullah, I.; Noor, A.; Nazir, S.; Ali, F.; Ghadi, Y.Y.; Aslam, N. Protecting IoT devices from security attacks using effective decision-making strategy of appropriate features. *The Journal of Supercomputing* **2024**, *80*, 5870-5899.

[12] Jia, Z.; You, J.; Dong, C.; Wu, Q.; Zhou, F.; Niyato, D.; Han, Z. Cooperative Cognitive Dynamic System in UAV Swarms: Reconfigurable Mechanism and Framework. *arXiv preprint arXiv:2405.11281* **2024**.

[13] Phadke, A.; Medrano, F.A.; Sekharan, C.N.; Chu, T. An analysis of trends in UAV swarm implementations in current research: simulation versus hardware. *Drone Systems and Applications* **2024**, *12*, 1-10.

[14] Puente-Castro, A.; Rivero, D.; Pedrosa, E.; Pereira, A.; Lau, N.; Fernandez-Blanco, E. Q-Learning based system for Path Planning with Unmanned Aerial Vehicles swarms in obstacle environments. *Expert Systems with Applications* **2024**, *235*, 121240.

[15] Men, J.; Zhao, C. An advanced cooperative multi-hive drone swarm system for global dynamic multi-source information awareness. *Journal of Industrial Information Integration* **2024**, *40*, 100608.

[16] Yin, S.; Wang, X.; Luo, L.; Pan, N.; Zhao, D.; Zhang, X. Collaborative strategy research of target tracking based on natural intelligence by UAV swarm. *Proceedings of the Institution of Mechanical Engineers, Part G: Journal of Aerospace Engineering* **2024**, 09544100241233313.

[17] Phadke, A.; Medrano, F.A.; Chu, T.; Sekharan, C.N.; Starek, M.J. Modeling Wind and Obstacle Disturbances for Effective Performance Observations and Analysis of Resilience in UAV Swarms. *Aerospace* **2024**, *11*, 237.

[18] Wang, Z.; Li, J.; Li, J.; Liu, C. A decentralized decision-making algorithm of UAV swarm with information fusion strategy. *Expert Systems with Applications* **2024**, *237*, 121444.

[19] Zhang, X.; Wang, Y.; Ding, W.; Wang, Q.; Zhang, Z.; Jia, J. Bio-Inspired Fission–Fusion Control and Planning of Unmanned Aerial Vehicles Swarm Systems via Reinforcement Learning. *Applied Sciences* **2024**, *14*, 1192.

[20] KEERTHANA, B. DESIGN AND DEVELOPMENT OF COMMUNICATION LINK FOR COLLABORATIVE MULTI UAV SYSTEM. **2024**.

[21] Kong, L.; Wang, L.; Cao, Z.; Wang, X. Resilience evaluation of UAV swarm considering resource supplementation. *Reliability Engineering & System Safety* **2024**, *241*, 109673.

[22] Cetinsaya, B.; Reiners, D.; Cruz-Neira, C. From PID to swarms: A decade of advancements in drone control and path planning-A systematic review (2013–2023). *Swarm and Evolutionary Computation* **2024**, *89*, 101626.

[23] Li, J.; Yu, D.; Ma, W.; Liu, J.J.; Liu, Y.-J. Cooperative Control of Air–Ground Swarms Under DoS Attacks via Cloud–Fog Computing. *IEEE Transactions on Network Science and Engineering* **2024**.

[24] Cabuk, U.C.; Tosun, M.; Dagdeviren, O.; Ozturk, Y. Modeling Energy Consumption of Small Drones for Swarm Missions. *IEEE Transactions on Intelligent Transportation Systems* **2024**.

[25] Fan, X.; Li, H.; Chen, Y.; Dong, D. UAV Swarm Search Path Planning Method Based on Probability of Containment. *Drones* **2024**, *8*, 132.

[26] Srihari, P. DoA Estimation for Micro and Nano UAV Targets using AWR2243 Cascaded Imaging Radar.

[27] Zulkarnain, S.A.; Zulkifli, S.; Ali, A.M. Identification and Analysis of Micro-Doppler Signature of a Bird Versus Micro-UAV. *Journal of Advanced Research in Micro and Nano Engineering* **2024**, *16*, 102-113.

[28] Valente, L.; Nadalini, A.; Veeran, A.H.C.; Sinigaglia, M.; Sá, B.; Wistoff, N.; Tortorella, Y.; Benatti, S.; Psiakis, R.; Kulmala, A. A Heterogeneous RISC-V Based SoC for Secure Nano-UAV Navigation. *IEEE Transactions on Circuits and Systems I: Regular Papers* **2024**.

[29] Friess, C.; Niculescu, V.; Polonelli, T.; Magno, M.; Benini, L. Fully Onboard SLAM for Distributed Mapping With a Swarm of Nano-Drones. *IEEE Internet of Things Journal* **2024**.

[30] Piancastelli, L.; Leon-Cardenas, C.; Pezzuti, E.; Sali, M. Cost effectiveness and feasibility considerations on the design of mini-UAVs for balloon takedown. Part 3: reliability and availability. *Cogent Engineering* **2024**, *11*, 2297515.

[31] Nory, H.; Yildiz, A. High-Speed Permanent Magnet Synchronous Motor Design for Mini Unmanned Aerial Vehicle. In Proceedings of the 2024 Third International Conference on Power, Control and Computing Technologies (ICPC2T), 2024; pp. 49-54.

[32] Çoban, S. Stochastic redesign of mini UAV wing for maximizing autonomous flight performance. *Aircraft Engineering and Aerospace Technology* **2024**, *96*, 113-120.

[33] Sen, C.; Singh, P.; Gupta, K.; Jain, A.K.; Jain, A.; Jain, A. UAV Based YOLOV-8 Optimization Technique to Detect the Small Size and High Speed Drone in Different Light Conditions. In Proceedings of the 2024 2nd International Conference on Disruptive Technologies (ICDT), 2024; pp. 1057-1061.

[34] Wu, K.; Chin, K.-W.; Soh, S. Multi-UAVs Network Design Algorithms for Computed Rate Maximization. *IEEE Transactions on Mobile Computing* **2024**.

[35] Liu, H.; Wu, K.; Huang, K.; Cheng, G.; Wang, R.; Liu, G. Optimization of large-scale UAV cluster confrontation game based on integrated evolution strategy. *Cluster Computing* **2024**, *27*, 515-529.

[36] Kuru, K.; Ansell, D.; Jones, D.; Watkinson, B.; Pinder, J.M.; Hill, J.A.; Muzzall, E.; Tinker-Mill, C.; Stevens, K.; Gardner, A. IoTFaUAV: Intelligent remote monitoring of livestock in large farms using autonomous unmanned aerial vehicles with vision-based sensors. *Biosystems Engineering* **2024**.

[37] Efrem, R.; Coutu, A.; Saeedi, S. Sensor Integration and Performance Optimizations for Mineral Exploration using Large-scale Hybrid Multirotor UAVs. *arXiv preprint arXiv:2402.11810* **2024**.

[38] Ostler, J.; Dudek, A.; Bertoncini, J.; Russ, M.; Stütz, P. MissionLab: A Next Generation Mission Technology Research Platform Based on a Very Light Aircraft. In Proceedings of the AIAA Scitech 2024 Forum, 2024; p. 0309.

[39] Dun, Y.; Li, L.; Wencan, B.; Weiyong, T. Conceptual design and configurations selection of S/H-UAVs based on Q-rung dual hesitant double layer FQFD. *Chinese Journal of Aeronautics* **2024**.

[40] Chi, Z.; Chun, Z.; Ruyi, K.; Shiyong, X.; Jiao, Z. An Overview of Countermeasures Against Low-altitude, Slow-speed Small UAVs. In Proceedings of the Journal of Physics: Conference Series, 2023; p. 012164.

[41] Neveling, L.; Götz, J.; Zoghlami, S.; Dominik, S.; Babetto, L.; Stumpf, E. *Conceptual Design of a High-Altitude, Long-Endurance Aerial Vehicle System for Rapid Network Deployment*; Deutsche Gesellschaft für Luft-und Raumfahrt-Lilienthal-Oberth eV: 2023.

[42] Hare, D.; Asgedom, N.G.; Jordan Jr, M.S.; Steedley, J.A. *Small High Endurance Aircraft Technology (sHEAT)-Application of Solid-State Hydrogen Storage to Extend Flight Time of sUAS*; Savannah River National Laboratory (SRNL), Aiken, SC (United States): 2023.

[43] Cao, X.; Liu, L.; Ge, J.; Yang, D. Conceptual design of long-endurance small solar-powered unmanned aerial vehicle with multiple tilts and hovers. *Proceedings of the Institution of Mechanical Engineers, Part G: Journal of Aerospace Engineering* **2023**, *237*, 3185-3201.

[44] Barros, J.; Henriques, J.; Reis, J.; Rosado, D.P. Unmanned Aerial Systems: A Systematic.

[45] Pramana, D. Cost-Benefit Analysis of Medium Altitude Long Endurance Unmanned Aerial Vehicle (Male UAV) Operator Training. Acquisition Research Program, 2024.

[46] Nugroho, R.C.; Sakti, M.A.K.; Mukti, S.; Fuadi, A.P.; Wahdiyat, A.I.; Suprianto, A.; Hayoto, V.; Samanhudi, R.D.; Wahidin, A.; Permana, A.D. Dataset for Experimental Measurement of Medium Altitude Long Endurance Unmanned Aerial Vehicle Propulsion System Performance Under Static Condition.

[47] Coletti, C.; Williams, K. Real-Time Control Interface for Research and Development Using Commercial-Off-The-Shelf (COTS) Mobile Robotics and UAVs. In Proceedings of the AIAA SCITECH 2024 Forum, 2024; p. 0235.

[48] Thantharate, P.; Thantharate, A.; Kulkarni, A. GREENSKY: A fair energy-aware optimization model for UAVs in next-generation wireless networks. *Green Energy and Intelligent Transportation* **2024**, *3*, 100130.

[49] Chen, C.-L.; Deng, Y.-Y.; Zhu, S.; Tsaur, W.-J.; Weng, W. An IoT and blockchain based logistics application of UAV. *Multimedia Tools and Applications* **2024**, *83*, 655-684.

[50] Zeng, B.; Gao, S.; Xu, Y.; Zhang, Z.; Li, F.; Wang, C. Detection of Military Targets on Ground and Sea by UAVs with Low-Altitude Oblique Perspective. *Remote Sensing* **2024**, *16*, 1288.

[51] Alexan, W.; Aly, L.; Korayem, Y.; Gabr, M.; El-Damak, D.; Fathy, A.; Mansour, H.A. Secure Communication of Military Reconnaissance Images Over UAV-Assisted Relay Networks. *IEEE Access* **2024**.

[52] Pandey, G. Military Advancement of Chinese Unmanned Aerial Vehicles: Security Implications for India. *RESEARCH REVIEW International Journal of Multidisciplinary* **2024**, *9*, 222-228.

[53] AL-Dosari, K.; Hunaiti, Z.; Balachandran, W. Civilian UAV Deployment Framework in Qatar. *Drones* **2023**, *7*, 46.

[54] Hadi, H.J.; Cao, Y.; Nisa, K.U.; Jamil, A.M.; Ni, Q. A comprehensive survey on security, privacy issues and emerging defence technologies for UAVs. *Journal of Network and Computer Applications* **2023**, *213*, 103607.

[55] Zhang, Y.; Li, S.; Wang, S.; Wang, X.; Duan, H. Distributed bearing-based formation maneuver control of fixed-wing UAVs by finite-time orientation estimation. *Aerospace Science and Technology* **2023**, *136*, 108241.

[56] Zheng, J.; Zhao, S.; Wang, X.; Tang, J.; Zuo, Z. Prescribed-Time Maneuvering Target Closing for Multiple Fixed-Wing UAVs. *IEEE Transactions on Vehicular Technology* **2024**.

[57] Elamassie, M.; Uysal, M. Performance Characterization of Rotary Wing UAV-Mounted FSO Links in the Presence of Pointing Errors. In Proceedings of the IEEE Wireless Communications and Networking Conference (WCNC), 2024.

[58] Tian, W.; Liu, L.; Zhang, X.; Shao, J.; Ge, J. A coordinated optimization method of energy management and trajectory optimization for hybrid electric UAVs with PV/Fuel Cell/Battery. *International Journal of Hydrogen Energy* **2024**, *50*, 1110-1121.

[59] Šančić, T.; Brčić, M.; Kotarski, D.; Łukaszewicz, A. Experimental Characterization of Composite-Printed Materials for the Production of Multirotor UAV Airframe Parts. *Materials* **2023**, *16*, 5060.

[60] Thuc, N.X. Structural design of the light UAV airframe. *Science and Technology Development Journal* **2023**, *26*, 3069-3073.

[61] Kulanthipiyan, S.; Rajendran, P.; Raja, V. Multi-Perspective Investigations Based Design Framework of an Electric Propulsion System for Small Electric Unmanned Aerial Vehicles. *Drones* **2023**, *7*, 184.

[62] Saravanakumar, Y.N.; Sultan, M.T.H.; Shahar, F.S.; Giernacki, W.; Łukaszewicz, A.; Nowakowski, M.; Holovatyy, A.; Stępień, S. Power Sources for Unmanned Aerial Vehicles: A State-of-the Art. *Applied Sciences* **2023**, *13*, 11932.

[63] Mostafa, S.A.; Mustapha, A.; Gunasekaran, S.S.; Ahmad, M.S.; Mohammed, M.A.; Parwekar, P.; Kadry, S. An agent architecture for autonomous UAV flight control in object classification and recognition missions. *Soft Computing* **2023**, 1-14.

[64] Kanokmedhakul, Y.; Bureerat, S.; Panagant, N.; Radpukdee, T.; Pholdee, N.; Yildiz, A.R. Metaheuristic-assisted complex H-infinity flight control tuning for the Hawkeye unmanned aerial vehicle: A comparative study. *Expert Systems with Applications* **2024**, 123428.

[65] Politi, E.; Purucker, P.; Larsen, M.; Reis, R.J.D.; Rajan, R.T.; Penna, S.D.; Boer, J.-F.; Rodosthenous, P.; Dimitrakopoulos, G.; Varlamis, I. Enabling Technologies for the Navigation and Communication of UAS Operating in the Context of BVLOS. *Electronics* **2024**, *13*, 340.

[66] Ullah, I.; Adhikari, D.; Ali, F.; Ali, A.; Khan, H.; Sharafian, A.; Kesavan, S.M.; Bai, X. Revolutionizing E-Commerce With Consumer-Driven Energy-Efficient WSNs: A Multi-Characteristics Approach. *IEEE Transactions on Consumer Electronics* **2024**.

[67] Chai, R.; Guo, Y.; Zuo, Z.; Chen, K.; Shin, H.-S.; Tsourdos, A. Cooperative motion planning and control for aerial-ground autonomous systems: Methods and applications. *Progress in Aerospace Sciences* **2024**, *146*, 101005.

[68] Yasir, M.; Ullah, I.; Choi, C. Depthwise channel attention network (DWCAN): An efficient and lightweight model for single image super-resolution and metaverse gaming. *Expert Systems* **2024**, *41*, e13516.

[69] Ozyurt, A.B.; Tinnirello, I.; Popoola, W.O. LiFi-enabled UAV swarm networks. *Applied optics* **2024**, *63*, 1471-1480.

[70] Ahmad, I.; Zhu, M.; Liu, Z.; Shabaz, M.; Ullah, I.; Tong, M.C.F.; Sambas, A.; Men, L.; Chen, Y.; Chen, S. Multi-Feature Fusion Based Convolutional Neural Networks for EEG Epileptic Seizure Prediction in Consumer Internet of Things. *IEEE Transactions on Consumer Electronics* **2024**.

[71] Hamdi, M.M.; Rashid, S.A.; Nafea, A.A. Resource Allocation and Edge Computing for Dual Hop Communication in Satellite Assisted UAVs enabled VANETs. *Iraqi Journal For Computer Science and Mathematics* **2024**, *5*, 108-127.

[72] Nomikos, N.; Giannopoulos, A.; Kalafatelis, A.; Özduran, V.; Trakadas, P.; Karagiannidis, G.K. Improving Connectivity in 6G Maritime Communication Networks With UAV Swarms. *IEEE Access* **2024**.

[73] Zhou, X.; Yang, L.; MA, L.; He, H. Towards secure and resilient unmanned aerial vehicles swarm network based on blockchain. *IET Blockchain* **2023**.

[74] Shah, S.M.; Sun, Z.; Zaman, K.; Hussain, A.; Ullah, I.; Ghadi, Y.Y.; Khan, M.A.; Nasimov, R. Advancements in Neighboring-Based Energy-Efficient Routing Protocol (NBEER) for Underwater Wireless Sensor Networks. *Sensors* **2023**, *23*, 6025.

[75] Ahmad, I.; Rahman, T.; Zeb, A.; Khan, I.; Ullah, I.; Hamam, H.; Cheikhrouhou, O. Analysis of Security Attacks and Taxonomy in Underwater Wireless Sensor Networks. *Wireless Communications and Mobile Computing* **2021**, *2021*, 1444024.

[76] Bakirci, M. A Novel Swarm Unmanned Aerial Vehicle System: Incorporating Autonomous Flight, Real-Time Object Detection, and Coordinated Intelligence for Enhanced Performance. *Traitement du Signal* **2023**, *40*.

[77] Javaid, S.; Saeed, N.; Qadir, Z.; Fahim, H.; He, B.; Song, H.; Bilal, M. Communication and control in collaborative UAVs: Recent advances and future trends. *IEEE Transactions on Intelligent Transportation Systems* **2023**.

[78] Khan, R.; Yang, Q.; Ullah, I.; Rehman, A.U.; Tufail, A.B.; Noor, A.; Rehman, A.; Cengiz, K. 3D convolutional neural networks based automatic modulation classification in the presence of channel noise. *IET Communications* **2022**, *16*, 497-509.

[79] Rotta, R.; Mykytyn, P. Secure Multi-hop Telemetry Broadcasts for UAV Swarm Communication. *arXiv preprint arXiv:2401.11915* **2024**.

[80] Horyna, J.; Baca, T.; Walter, V.; Albani, D.; Hert, D.; Ferrante, E.; Saska, M. Decentralized swarms of unmanned aerial vehicles for search and rescue operations without explicit communication. *Autonomous Robots* **2023**, *47*, 77-93.

[81] Ming, R.; Jiang, R.; Luo, H.; Lai, T.; Guo, E.; Zhou, Z. Comparative Analysis of Different UAV Swarm Control Methods on Unmanned Farms. *Agronomy* **2023**, *13*, 2499.

[82] Li, Z.; Jiang, C.; Gu, X.; Xu, Y.; Cui, J. Collaborative positioning for swarms: A brief survey of vision, LiDAR and wireless sensors based methods. *Defence Technology* **2024**, *33*, 475-493.

[83] Arshad, T.; Zhang, J.; Ullah, I. A hybrid convolution transformer for hyperspectral image classification. *European Journal of Remote Sensing* **2024**, 2330979.

[84] Budiyono, A.; Higashino, S.-I. A Review of the Latest Innovations in UAV Technology. *Journal of Instrumentation, Automation and Systems* **2023**, *10*, 7-16.

[85] Horyna, J.; Krátký, V.; Pritzl, V.; Báča, T.; Ferrante, E.; Saska, M. Fast Swarming of UAVs in GNSS-Denied Feature-Poor Environments Without Explicit Communication. *IEEE Robotics and Automation Letters* **2024**, *9*, 5284-5291.

[86] Angueira Irizarry, K. Autonomous UAV Swarms: Distributed Microservices, Heterogeneous Swarms, and Zoom Maneuvers. The Ohio State University, 2024.

[87] Milidonis, K.; Eliades, A.; Grigoriev, V.; Blanco, M. Unmanned Aerial Vehicles (UAVs) in the planning, operation and maintenance of concentrating solar thermal systems: A review. *Solar Energy* **2023**, *254*, 182-194.

[88] Oğuz, S.; Heinrich, M.K.; Allwright, M.; Zhu, W.; Wahby, M.; Garone, E.; Dorigo, M. An Open-Source UAV Platform for Swarm Robotics Research: Using Cooperative Sensor Fusion for Inter-Robot Tracking. *IEEE access* **2024**, *12*, 43378-43395.

[89] Chandran, N.K.; Sultan, M.T.H.; Łukaszewicz, A.; Shahar, F.S.; Holovatyy, A.; Giernacki, W. Review on Type of Sensors and Detection Method of Anti-Collision System of Unmanned Aerial Vehicle. *Sensors* **2023**, *23*, 6810.

[90] Conte, C.; Verini Supplizi, S.; de Alteriis, G.; Mele, A.; Rufino, G.; Accardo, D. Using Drone Swarms as a Countermeasure of Radar Detection. *Journal of Aerospace Information Systems* **2023**, *20*, 70-80.

[91] Khawaja, W.; Yaqoob, Q.; Guvenc, I. RL-Based Detection, Tracking, and Classification of Malicious UAV Swarms through Airborne Cognitive Multibeam Multifunction Phased Array Radar. *Drones* **2023**, *7*, 470.

[92] Hanif, A.; Doroslovački, M. Operating a Battery-Limited Drone Swarm in 6G Network by Joint Power Transfer and Radar Imaging. *IEEE Transactions on Aerospace and Electronic Systems* **2024**.

[93] Phadke, A.; Medrano, F.A.; Sekharan, C.N.; Chu, T. Designing UAV Swarm Experiments: A Simulator Selection and Experiment Design Process. *Sensors* **2023**, *23*, 7359.

[94] Khan, Z.U.; Gang, Q.; Muhammad, A.; Muzzammil, M.; Khan, S.U.; Affendi, M.E.; Ali, G.; Ullah, I.; Khan, J. A Comprehensive Survey of Energy-Efficient MAC and Routing Protocols for Underwater Wireless Sensor Networks. *Electronics* **2022**, *11*, 3015.

[95] Gang, Q.; Muhammad, A.; Khan, Z.U.; Khan, M.S.; Ahmed, F.; Ahmad, J. Machine Learning-Based Prediction of Node Localization Accuracy in IIoT-Based MI-UWSNs and Design of a TD Coil for Omnidirectional Communication. *Sustainability* **2022**, *14*, 9683.

[96] Liu, L.; Sun, B.; Li, J.; Ma, R.; Li, G.; Zhang, L. Time-Frequency Analysis and Recognition for UAVs Based on Acoustic Signals Collected by Low-Frequency Acoustic-Electric Sensor. *IEEE Sensors Journal* **2024**.

[97] Aman, M.; Gang, Q.; Shang, Z.; Khan, Z.U.; Khan, M.S.; Ullah, I. Realization of RSSI Based, Three Major Components (Hx, Hy, Hz) of Magnetic Flux Created around the MI-TD Coil. In Proceedings of the 2023 IEEE International Conference on Electrical, Automation and Computer Engineering (ICEACE), 2023; pp. 1012-1017.

[98] Khan, S.; Ullah, I.; Ali, F.; Shafiq, M.; Ghadi, Y.Y.; Kim, T. Deep learning-based marine big data fusion for ocean environment monitoring: Towards shape optimization and salient objects detection. *Frontiers in Marine Science* **2023**, *9*, 1094915.

[99] Jiang, Y.; Renner, B.-C. Low-Cost Underwater Swarm Acoustic Localization: A Review. *IEEE access* **2024**.

[100] Khan, Z.U.; Aman, M.; Rahman, W.U.; Khan, F.; Jamil, T.; Hashim, R. Machine Learning-based Multi-path Reliable and Energy-efficient Routing Protocol for Underwater Wireless Sensor Networks. In Proceedings of the 2023 International Conference on Frontiers of Information Technology (FIT), 2023; pp. 316-321.

[101] ur Rahman, W.; Gang, Q.; Feng, Z.; Khan, Z.U.; Aman, M.; Bilal, M. A MACA-Based Energy-Efficient MAC Protocol Using Q-Learning Technique for Underwater Acoustic Sensor Network. In Proceedings of the 2023 IEEE 11th International Conference on Computer Science and Network Technology (ICCSNT), 2023; pp. 352-355.

[102] Mian, S.; Gang, Q.; Aman, M.; Khan, Z.U.; Khan, M.S.; Ullah, I. A Visible Light Geometric Multilateration Based Positioning System to Localize Nodes in an UWSNs. In Proceedings of the 2023 IEEE 11th International Conference on Computer Science and Network Technology (ICCSNT), 2023; pp. 392-396.

[103] Chen, Z.; Yan, J.; Ma, B.; Shi, K.; Yu, Q.; Yuan, W. A Survey on Open-Source Simulation Platforms for Multi-Copter UAV Swarms. *Robotics* **2023**, *12*, 53.

[104] Abualigah, L.; Falcone, D.; Forestiero, A. Swarm Intelligence to Face IoT Challenges. *Computational Intelligence and Neuroscience* **2023**, *2023*, 4254194.

[105] Su, J.; Ng, D.T.K.; Chu, S.K.W. Artificial Intelligence (AI) Literacy in Early Childhood Education: The Challenges and Opportunities. *Computers and Education: Artificial Intelligence* **2023**, *4*, 100124.

[106] Tiwari, S.; Kumar, A. Advances and bibliographic analysis of particle swarm optimization applications in electrical power system: concepts and variants. *Evolutionary Intelligence* **2023**, *16*, 23-47.

[107] Xiang, H.; Liu, X.; Song, X.; Zhou, W. UAV Path Planning Based on Enhanced PSO-GA. In Proceedings of the CAAI International Conference on Artificial Intelligence, 2023; pp. 271-282.

[108] Tang, J.; Duan, H.; Lao, S. Swarm intelligence algorithms for multiple unmanned aerial vehicles collaboration: a comprehensive review. *Artificial Intelligence Review* **2023**, *56*, 4295-4327.

[109] Aljalaud, F.; Kurdi, H.; Youcef-Toumi, K. Bio-Inspired Multi-UAV Path Planning Heuristics: A Review. *Mathematics* **2023**, *11*, 2356.

[110] Yu, D.; Li, J.; Wang, Z.; Li, X. An Overview of Swarm Coordinated Control. *IEEE Transactions on Artificial Intelligence* **2023**.

CHAPTER 15

Integration of IoT Devices with UAV Swarms

Muhammad Hamza Sajjad[1], Faisal Rehman[1,2], Muhammad Muneer[1]
[1]Department of Statistics & Data Science, University of Mianwali, Mianwali, Pakistan
[2]Department of Robotics & Artificial Intelligence, National University of Sciences & Technology, NUST, Islamabad, Pakistan
Correspondence: Faisal Rehman[1,2] (faisalrehman0003@gmail.com)

Introduction

The advancement in the technological front is an ever-growing field, and with the use of UAVs and IoTs, the fields have grown significantly. The operational capacity of several UAVs that operate in a coordinated and synchronized manner is known as UAV swarms; the technology is considered revolutionary in aerial engineering technology. These swarms are well-fitted with communication, identification, and data-assessing devices that make these swarms work automated. The incorporation of IoT devices in UAVs has introduced opportunities for the improvement of the swarms and the systems they are part of, particularly in smart cities [1].

Smart cities utilizing IoT platforms are urban infrastructures that allow managing available resources and using services with reduced expenses on services at the expense of using new technologies that would help enhance the quality of life of citizens. Such cities' inhabitants also stand to gain in the IoT device usage since UAV swarms can be used in surveillance, controlling of emergencies and disasters and planning of the infrastructure and environmental status. The concept of IoT-based UAV swarms allows the gathering of data and analysis of what is occurring within the urban environment, and this enables the decision-making to take place and create a timely response to various issues. This integration is essential because without it there can be no improvement of intelligent structures which are essential in developing and managing today's cities. Several smart IoT UAV swarm applications relevant to smart cities include surveillance and monitoring of particular areas. For instance, these UAVs can survey vast areas of land including expanded cityscapes of large metros and, at the same time, feed concurrent video coverage and climatic details to the city executive's technological infrastructure [2]. Thanks to this capability, the rate, protection of the people, and management of the traffic at mass events can be done without disruption. Besides, in cases of disasters, UAV swarms could gather information about the impact of the disaster, look for people who were trapped, and offer necessary goods and services. They are versatile in a crowded and low light environment, have the unique ability to efficiently and rapidly share information on what is happening in emergencies, and are invaluable. When integrated with IoT and UAVs, smart city applications are also most useful in the planning as well as the management of physical infrastructure within cities. Such systems can perform several functions like surveillance and inspection at different degrees of detail, monitoring of construction, and electrical power follow-up of structures like bridges, electric power transmission lines, etc. Enabled with rich data collection instruments for sensing, the IoT-UAV swarm flies to provide important information related to the construction of the infrastructures of urban city planning and engineering, making construction projects safer and smarter. Second, environmental control is another area, for air quality and pollution source identification, as individual UAVs with various sensors can be used to track them and

CHAPTER 15 INTEGRATION OF IOT DEVICES WITH UAV SWARMS

monitor wildlife and the quality of air in urban parks and beautification gardens. However, the connective integration of the IoT devices with the UAV swarms is not entirely without significant difficulty. Discrete challenges that entail technical aspects include guaranteeing connectivity, the capacity to handle data, and energy consumption on the UAVs are major challenges [3]. Another important issue is security since the gathered data by these systems are to be protected from illegitimate access or modification. However, certain limitations such as legal frameworks concerning privacy and airspace must be solved, and new ethical issues related to the usage of IoT in smart cities have to be further discussed for the common utilization of IoT-based UAV swarms. Quite a prospector, the bright future of IoT-enabled UAV swarms has been established and designed. Some of these are 5G, artificial intelligence systems, and edge computing which are among the sophisticated technologies that are expected to underpin and enhance such systems [4]. For instance, the introduction of 5G is expected to assist in enhancing the efficiency of control signals and video streams between the UAVs and GCS for the analysis of real-time data [5]. Increasing and consolidating AI techniques in identified areas can result in higher levels of self-organization that would enable the UAV swarms to execute larger levels of elaborate coordinated missions with relative autonomy. The use of edge computing implies that most of these computations will be performed near the origin of data which will reduce these time delays.

More specifically, this chapter is dedicated to discussing the possibilities of the integration of IoT devices with UAV swarms, opportunities, and hurdles of such establishment, and trends of further development. Specifically, it will cover the main multifaceted technologies and concepts that are at the core of UAV swarms and IoT integration. Moreover, it will also discuss the current methods and implementation of these systems and how future developments could expand the usability of such systems. Regarding the relationships between UAV swarms and IoT, it can be suggested that recognizing and valuing the opportunities of their interaction might contribute to the broader understanding of the key trends in smart city development among the stakeholders.

The content of the chapter is cross-sectional and includes a description of the basic principles of the IoT and UAV swarms, technology integration, application, and the key issues and approaches regarding the IoT and UAV swarm technology integration and application. It also examines what is likely to be the main characteristics or topics of the future trends and innovations that will play out on the IoT-enabled UAV swarm model. In this manner, this chapter endeavors to present useful suggestions for the researchers, policymakers, and practice professionals who are involved in the manufacturing of smart city solutions.

In this regard, then it shall be pertinent to approach these aspects in a bid to develop a better understanding of how IoT in conjunction with UAV technology can be realized to achieve a better response to the emergence of smart urban cities. This will also provide insights into the current state of affairs regarding the progress in the field of IoT and UAV swarm, threats that need to be tackled, as well as the possibilities of the future concerning IoT and UAV swarm integration.

Unmanned Aerial Vehicles (UAVS)

As the prevalence of technologies rises, uncrewed aerial vehicles (UAVs), also known as drones, have been widely used in many fields due to their flexibility and applicability. Unmanned aerial vehicles are aircraft systems that can fly without having an onboard pilot or even a copilot and can be controlled from a distance with the help of remote control equipment or can be fitted with automatic flight control systems and computers and can fly on their own [6]. They carry various sensor tools, microprocessors, and electronic tools that enable them to accomplish a countless number

CHAPTER 15 INTEGRATION OF IOT DEVICES WITH UAV SWARMS

of chores effectively and accurately [7]. UAV systems are complex entities that implement the functions by utilizing different subsystems integrated into one whole. This encompasses the body of the aircraft, engines, instruments, and communication and control devices and systems. These work together with other systems such as satellites, ground control stations, smartphones, and computers through the line of communication, whereby a human operator can control and manage the UAV from a comfortable distance. This helps UAVs to carry out encapsulated tasks under conditions that are hard or dangerous for human intervention, including disaster monitoring, lost people seeking and finding tasks, and environmental monitoring [8]. The transport advantages of UAVs make them quite suitable for logistics in general and especially in the Civilian Bionomic market where less expensive but functional UAVs have been applied in numerous ways. These include remote control searches, surveillance of disasters, surveillance of the environment, and the delivery of mail and medical supplies as well as parcels and other goods [9]. For instance, it has been used in cases that required aerial filming and photographing of a scene to determine the appropriate and immediate measures to provide in an emergency. However, many of the currently used UAVs are still controlled by humans using remote systems or remote controls. UAVs are available in several forms, and mediums, depending on the ability, speed, weight, and operations that are required for their use. There are difficulties inherent in piloting UAVs because of inherent problems such as human error and labor- and cost-intensive, extremely rigorous training. However, to this date, even as autopilot abilities have been incorporated in most UAVs, they are still framed by challenges that include short battery power, short self-endurance, and a problem in establishing accurate landings. Farm owners and managers could significantly benefit from the improved capabilities because it provides a more efficient way of operating UAVs [10]. With features of VTOL, these UAVs are capable of horizontal usage in addition to vertical usage as they can take off and land vertically. They have high speed, high rate, and quality for aerial hover, which is useful in, for example, observation missions or other jobs that demand vertical movement with limited airborne movement.

In conclusion, it can be pointed out that UAVs indeed provide a marked improvement to aerial systems and can find usage in almost all areas. Their capacity to work with precisions and initiate sophisticated operations independently makes them plausible in most technological and logistical undertakings today. The uptake of sophisticated communication interfaces, instruments, and control features also improves the flexibility and efficiency of UAVs, including their probable application range in consumer and business contexts [11].

Utilizing these new-age technologies, future automation in aerial movement through UAVs is set to fulfill exponential duties needed in functions including disaster management, environmental surveys, logistics, and so on.

UAV Swarm

This is defined as a set of UAVs or several UAVs that perform as a single system to achieve a specific mission or objective. This is generally accomplished with the help of algorithms that enable the UAVs to be controlled and also permit them to share data with every other UAV, thus mimicking the swarms of bees, birds, etc. These tasks may be achieved more so if the UAVs are to be used in formation as a swarm other than each UAV being used separately. The progress of UAV swarms can be divided into two types, which are semiautonomous UAVs and full autonomous UAVs, and the classification for the swarms can be done in single-layer swarms or multilayer swarms. Consequently, first-layer swarms lack a leader, yet at the same time, each UAV is their leader in a sense because each UAV of the first layer independently makes decisions. On the other hand, multilayered UAV swarms have multitier UAV swarm architectures for which different other layers are directed by some specially directed UAVs [12]. This makes it easier to conduct the swarm in a well-coordinated manner through a good number of hierarchy mechanisms as illustrated in the following topics. Each mobile node in every swarm possesses data collection and processing equipment; the CP is in the cloud or base station. This decentralized computing allows the large area search and many drones of the UAV swarms to cover vast areas at extremely high speeds and accuracy necessary in today's complex wars [13]. The self-organization and coordinated control within the swarm raise its efficiency within numerous missions, specialists in observing, and direction of the environment as well as during emergencies. Other works have offered other details regarding aspects like energy control, priority in handling tasks, and scheduling of the specific UAV swarm. For example, some earlier papers have touched on some of the aspects such as charging schemes for controlling multiple drones and the simulations in which some UAVs are given top priority for charging [14]. This approach enables great missions to be completed without much trouble because there will be no possibility of the helicopter having to land halfway through the mission due to battery discharge. Moreover, formation flying and hovering control of multiple UAVs is part of the system functionality of the instance for a mission for use in any that requires accurate positioning and coordination as is the case of quadcopter operations [15]. Another consideration is the signals that must be exchanged between UAV team members, including information related to aerial maneuverings. Novel approaches and methods have been proposed for problems of real-time cooperation for several UAV meetings over Barnes's recent networks [16]. These pauses help to make sure that the UAVs can maintain good interconnection and can also interconnect and exchange information as required.

CHAPTER 15   INTEGRATION OF IOT DEVICES WITH UAV SWARMS

Integration of IoT with UAV Swarms

Operating IoT devices as a part of UAV swarms brings the ability to add real-time information collection, processing, and sharing into the mix. These technologies can be integrated into the UAVs to collect environmental data and the condition of the systems on the UAVs and enhance decision-making on the UAVs. By integrating these components, it is possible to make the UAV swarms more independent within the application area and improve output, which can also be used for surveillance purposes, during disasters or to monitor infrastructure. In this article, the communication architecture of a UAV swarm system is depicted in Figure 15-1. The multiple UAVs operate in conjunction with satellites and a range of ground control HQs, portable and local. It is administered by a human to filter and coordinate its functions effectively. This architecture highlights how critical communication links are in supporting UAV systems swarm missions that cover as large an area as possible at the earliest time possible.

CHAPTER 15   INTEGRATION OF IOT DEVICES WITH UAV SWARMS

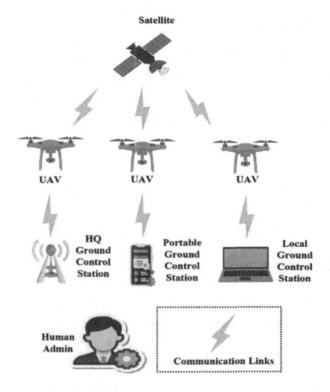

Figure 15-1. *Communication Architecture of a UAV System*

Technologies Enabling UAV Swarms
Communication Systems

Some of the critical issues that should be addressed include the need to create efficient ways of communication that will help in the execution of various operations involving UAV swarms. The 5G, ZigBee, and LoRa can also enable the UABV and the GCS to transmit data at high speed without any interference from each other [17]. Such systems facilitate real-time updating of the UAVs' operating status in reaction to an event, allowing them to coordinate their activities [18]. For example, 5G connections have low latency and high data traffic that are essential in swarming activities, which require conveying realistic data.

Sensor Technologies

Sensors are perhaps one of the most critical building blocks of a UAV, especially because they feed the UAV systems with the relevant data that enables the accomplishment of various diverse functions such as navigation, recognition, and avoidance of obstacles among others [19]. Some additional UAV sensors include a position sensor GPS, an orientation sensor IMU (inertial measurement unit), a visual sensor camera, and an environmental sensor which has temperature, humidity, air quality sensors, etc. These sensors allow the UAVs to complete several tasks independently and with a high level of precision [20].

Data Processing and AI

Besides, the particular elements of algorithms are essential for UAV swarms, as far as advanced data processing and AI integration are concerned. It also involves path planning, obstacle avoidance, target identification, and decision-making processes through the application of available AI algorithms. Hence, the large amounts of data collected by IoT sensors can be fully utilized by machine learning models to optimally adapt the swarm to changes in the environment [21]. For example, AI for path planning helps the UAVs be able to fly along the designed path, have a clear view of the environment, and have mechanisms to keep off from obstacles and hazards.

Power Management

In contrast to large-scale UAVs where energy management is generally centered on overall flight time, swarm UAVs need a better power management system to increase their durability. Some of the technologies are high-density batteries, solar power, and efficient motors which are used to make UAVs aerodynamically enable them to have better flight endurance [22]. Moreover, the telemetry module of UAVs can play an important role in controlling power consumption since they can control power smartly based on the workload and status of the UAVs [23]. This is because long-duration missions are possible and are important for the endurance of UAVs in various operations.

Edge Computing

At the core of edge computing, there is a concept that much of data processing should take place as closely as possible to the source of data to minimize latency and the amount of data transferred [24]. Toward UAV swarms, edge computing can facilitate the real-time data processing acquired by onboard sensors with the possibility of decision-making and execution [25]. This is especially the case when the response to changes in the environment is expected to be nearly instantaneous for safety reasons, as in avoiding obstacles and real-time surveillance. They have proposed a novel concept of integrating AI to process data at the periphery of UAV swarms for efficient handling of dynamic environments and achieving critical missions.

Technological Integration

Integrating IoT devices in UAV swarms affects the performance since the UAVs can now collect, process, and relay information to other devices in real time. This section moves deeper to look at some of the technologies and protocols that support the workings of IoT-based UAV swarms.

IoT Devices Used in UAV Swarms

Various IoT devices deployed in the UAV swarm are used for various roles as navigational aids, health monitoring systems, and as part of decision-making systems. Such components can range from several environment probes for detecting different environmental factors to GPS to help track

CHAPTER 15 INTEGRATION OF IOT DEVICES WITH UAV SWARMS

the location of the device and communication modules to ensure that data is relayed in real-time. Environmental conditions include temperature, humidity, air quality, and other important environmental factors which include the parameters of disaster management and infrastructure inspection.

- **Environmental Sensors:** This sensorial check information such as temperature, humidity, and air quality that is helpful in cases of environmental checking and incidences of disaster. For instance, temperature and smoke sensors installed on UAVs can assist in measuring the fire's behavior and evolution rates of a wildfire.

- **GPS Modules:** Global Positioning System (GPS) is used to give the exact position of the UAVs to help them in proper positioning and also help the individual UAVs to fly in formation in the swarm. This is essential if, for instance, all drones have to work together say in a surveillance mission of a large area or agricultural monitoring.

- **Communication Modules:** These are for real-time data communication between the flying UAVs and the ground control stations and also between different UAVs flying in the same environment [26]. Some of the features of effective communication modules are as follows. They provide precise and timely information for analysis and decision-making with the help of sensors.

CHAPTER 15 INTEGRATION OF IOT DEVICES WITH UAV SWARMS

Communication Protocols

Coordination and integration among several UAVs are most efficiently and successfully done by communication. Some of the features used to guarantee efficient and reliable data transfer between the UAVs and the ground control station include various communications technologies and protocols. Key technologies include:

- **5G:** This technology has a high data rate, low latency, and potential massive connectivity, which make it important in real-time data transmission and coordination on the UAVs [27]. The 5G networks ensure the delivery of data in large volumes at the same time which is important for instance in areas such as surveillance that require real-time video streams. The smoothness of 5G makes it possible to relay commands and various data in a very short time which enables efficient and synchronized forcible responses with multiple UAVs [28].

- **ZigBee and LoRa:** LOPAN provides low-powered communication over a large area, which makes it very ideal for IoT devices especially when it comes to power consumption. ZigBee is excellent for coverage solutions with less energy consumption for short-range applications, while LoRa has better coverage and excellent energy consumption for applications with extensive ranges [29]. These protocols are particularly useful for continuing communication in areas such as areas where there are little to no power resources like in the remote or rural regions.

- **Wi-Fi and LTE:** Typically employed for transmitting data over a short and mid-distance range Wi-Fi and LTE are critical for the transmission of data where high-speed access to the web is always available [30]. Wi-Fi is best used where clients need networking solutions for specific working spaces or a region interior space or where the range is limited such as outdoor covering a small area while LTE is more appropriate where clients require networking solutions that cover a wide area such as in urban area or where LTE networks are already established [31].

Moreover, there is a rigid formal organization structure where one can distinguish the usage of hierarchical networks and multiple levels of communication. In such structures, probably some of the UAVs may act as a hub or a master node where they facilitate communication to the swarm as well as ensure that data is effectively passed despite the environment's complexity or dynamism. This can be useful to have strong and stable connections, as this stage has crucial importance in UAVs swarm real-time applications.

Data Collection and Management

The incorporation of IoT devices and UAV swarms also facilitates the effective collection and managing large data systems and the use of sophisticated data processing strategies and agencies alongside AI technologies for real-time data analysis. Smart devices implemented in UAV systems perform various roles which include: acting as sensors to gather information about the environment, serving as a monitoring system on the status of UAVs in the system, and enhancing information management and decision-making procedures. These devices may involve the use of sensors for measurement of different conditions of the environment, GPS circuits for accurate determination of placing, and transmitter circuits for efficient relaying of data. Sensors include the temperature, humidity, air, quality, and other conditions that are important in disaster management and infrastructural assessment. Connectivity is very important for both aerial formation control and the ability of UAV swarms to perform their tasks. Techniques like full-duplex, as well as half-duplex and various communication protocols, cover for accurate and very fast exchange of data between the UAVs and the ground control stations.

Key technologies such as 5G, ZigBee, LoRa, Wi-Fi, and LTE provide high data rates, low latency, and high connection density for real-time data transfer and coordination. ZigBee and LoRa are used for LPWAN which is a low-power wide area network application that requires IoT Technology where the power is a considerable advantage. It can be seen that Wi-Fi and LTE are typically utilized for data communication and at higher data rates in short to medium-range applications [30]. Different levels of network affiliation and multiple tiers of interconnection schemes are implemented for effective management of the rate of communication and for guaranteed delivery of data which can be sent as the network pertains to contingencies that include the fluctuations that occur when changes in topographical hierarchy. Real-time data capture and ownership are made possible by integrating IoT devices with UAV swarms, improving the abilities of these systems. Data through IoT sensors from UAVs are then analyzed through machine learning algorithms that help in tasks such as navigation, avoiding obstacles, or even identifying objects of interest [32]. The data gathered in the system is managed closer to the source of data generation, and there is a minimum delay as well as usage of bandwidth, which is advantageous for time-sensitive applications such as collision avoidance systems and surveillance. As seen here, the combination of the two makes UAV swarms more capable and efficient in performing practical tasks.

CHAPTER 15 INTEGRATION OF IOT DEVICES WITH UAV SWARMS

Possible uses of IoT with flying drones include continuous tracking of environmental and infrastructural changes. Telemetry and environment conditions on IoT-based UAVs include gas concentration, temperature, humidity, and pollution levels, thus facilitating the evaluation of environmental conditions and potential dangers [33]. With UAVs at their disposal, roads, bridges, electrical, water and other supply lines, pipelines, and gas distribution systems can be surveyed for intelligence on maintenance or repair works. This is further complemented by cloud-based platforms and distributed computing infrastructures which also improve the abilities to manage data, reach analytical potential, and provide high levels of computation [34]. A synergistic combination of one or more edge computing nodes with cloud-support platforms provides the best computation ratio at the edge node and cloud layer to ensure UAV swarms are responsive to mission plans and environmental challenges. Overall, the analysis of combining IoT as well as the classification of UAVs establishes their effectiveness in improving modern technical and logistics operations.

Applications of UAV Swarms

UAV swarms that are operational with IoT keep numerous applications in association with different fields and contribute majorly to boosting the functionality of the management and responding systems of the urban environment. As can be seen in Figure 15-2, UAV swarms are highly adaptable; they can be effectively used in multiple fields and give their marks in various sectors. These applications include:

CHAPTER 15   INTEGRATION OF IOT DEVICES WITH UAV SWARMS

Figure 15-2. *Applications Areas of UAV Swarms*

Security and Monitoring

Unmanned aerial vehicles along with IoT devices nowadays are becoming popular for surveillance purposes in urban areas. These systems allow for real-time surveillance, and besides they are more secure since they cover areas that cannot be covered physically by fixed CCTV camera systems. For example, UAV models can have a radius of 2-3 km and offer the opportunity to take high-definition video for surveillance and decrease the crime rate in urban territories. This is because, unlike fixed surveillance systems, UAVs for surveillance have mobility and flexibility and can follow the subject of interest and survey large where necessary. Military intelligence is also served by UAVs as nations employ unmanned vehicles in surveillance activities such as antipoaching, identifying enemy movements, policing their national borders, and monitoring seaborne traffic over important water channels. High-quality, cost-effective, and flexible UAVs can be used for surveillance of certain areas or provide supervision to prevent certain unlawful activities in such areas [35].

Emergency Response and Disaster Management

In disaster scenarios, UAV swarms such as this could be advantageous as they would continually provide imagery support while being orchestrated with other teams. They're drones that are equipped with IoT sensors that can be used to assist with that assessment to locate victims and bring them food. For example, UAVs have been used for forest fires, aerial forest fire image processing, and forest fire alerts and as a way to prevent forest fires. That is, a UAV can be an asset, especially in disaster management because it can function in conditions that could be dangerous for traditional manned vehicles. Drones offer a safe entry point for data collection and rescue operations. Equipped with radars, sensors, and cameras, they aid in disaster assessment, evacuation, aerial surveys, alarms, and rapid response. A fleet of autonomous drones with fire suppressants could monitor impacted areas and alert potential danger [36].

CHAPTER 15 INTEGRATION OF IOT DEVICES WITH UAV SWARMS

Environmental Monitoring and Remote Sensing

The technology used in hobbyist drones has developed to cause high-quality videos and photos of distances, like coasts and islands, mountains, etc. UAV integrates with aerial and satellite remote sensing data and ground sensory data. The affordability, portability, and capacity for high temporal and spatial data collection make UAVs useful for observation. The functions of UAVs as remote sensing tools are enormous and versatile since they can facilitate several operations [37]. Some of these include disease diagnostics, water quality assessment, drought detection, oil and natural gas exploration, yield predictions, wildlife and wildlife habitat identification, assessment of geological hazards, hydrological simulations, topographic analysis, forest inventory, and land use monitoring [38]. The use of UAVs also plays an active role in crowd-sourcing or the creation of 3D view of the surroundings, making important contribution, such as in archaeology and cartography. Affordable drones could be easily update, which goes in line with the land planning budgets, while obsolete mapping tools could be avoided. UAVs applied to remote sensing offer critical insights and real-time data across various sectors, enhancing the precision and efficiency of environmental monitoring and resource management.

Rescue and Search

UAVs are very essential where there is a need to respond to disasters, help in rescue operations, and security services. This can be useful in economizing the deployment of manpower, funds, and time in that it avails real images of planned environments. This will enable SAR teams to identify and decide on where help is needed most. UAVs can help to increase the speed of SAR in unfortunate cases such as missing individuals, snow slides, fire incidences, and the infiltration of toxic gases. For example, it can follow up mountaineers who may be lost during their missions or search for people who may have gotten lost in the desert or dense forests.

CHAPTER 15   INTEGRATION OF IOT DEVICES WITH UAV SWARMS

UAVs aid in locating victims in steep or rugged areas or in conditions that are unfavorable for flights. That is, drones can transport crucial medical equipment and medication before the ambulance or a medical teams reach. It can contain medical kits, life jackets, food supplies for the needy, and support to the disaster-affected areas and other inaccessible places. From these sights, such drones can offer lifesaving items such as clothes, water, and items that the trapped people need before actual rescue teams can get to where they are stranded. Of particular importance is the improved operational capacity of SAR mission which makes it easier to provide aid to the victims in the shortest possible time [36].

Construction Monitoring and Infrastructure Inspection

The use of drones can offer effective and functional methods of as-built mapping, construction monitoring, and site inspection that are convenient, quick, and productive. The duty of overseeing the projects at the sites to the actual completion of the site construction work fosters the revelation of project progress. These UAVs can provide reports containing images and video clips and also 3D maps to any customer who may wish to be provided with the same, thus increasing the inspection of the infrastructure and constructions. There is a tendency to use UAVs for inspecting GSM towers, bridges, and other key structures with a lot of worry arising. This one indicates how, with the assistance of UAVs, one has the possibility of receiving a broad and repetitive check that does not mean sending one's employees to a dangerous zone. This reduces cost and in the same instance brings accuracy and coverage of the conducted inspections. Thus, constructors of the construction companies and managers of infrastructures can guarantee to practice the safety standards or even improve the effectiveness of managing such projects with UAVs [39].

CHAPTER 15 INTEGRATION OF IOT DEVICES WITH UAV SWARMS

Traffic Monitoring and Management

The road traffic monitoring domain is one of the fields where a relatively high number of activities related to the integration of UAVs has been observed. For the transportation sector, RTM has the capability to reach every sector of automation through UAV and is made up of rescue teams, road surveyors, traffic police, and field support teams [40]. Thus, highway traffic conditions can be quite informatively gathered or observed through reliable and intelligent UAVs in particular.

Comparing expensive microwave sensors, surveillance video cameras, and loop detectors, alternative cheaper UAVs can maneuver well in large road segments [39]. The UAV can have a clear vision of the road condition to be required. It can be controlled by local police to monitor on-road incidents or for police operations against criminal incidences associated with the highway, like carjackings. They can be used to count the vehicles in situations where the police may wish to apprehend the occupants of a certain car, in instances of armed robbery or carjacking, or any other criminal who may include any wrongdoer on the roads such as traffic offenders.

With UAVs, one can observe the speed of the vehicle, and accidents to help avoid traffic jams and provide input of traffic information to traffic management systems [33]. Some of the services that could be offered using UAVs for the monitoring of road traffic include the following: They may also be used in aerial photography and filming high-definition images and records of traffic flows that are so useful when it comes to monitoring traffic and traffic management in real time. This has the advantage of extending traffic management efficiency and response time if any emergency service is needed. In other ways, the UAVs could assist local police in policing the breakdown of order by observing traffic laws with a view to gathering evidence for the case.

CHAPTER 15   INTEGRATION OF IOT DEVICES WITH UAV SWARMS

Power Line Inspection and Maintenance

Fault detection and prevention for power lines are critical to ensure the availability of power supply. Conventional methods of inspecting power lines bring huge costs, cumbersome installation, and significant hazardous risks. As a result, for scholars and industry practitioners, the UAV-aided inspection of power distribution and transmission lines has been extremely appealing. They come fitted with digital cameras and take images of possible power corridor lines in a very convenient and efficient way for supporting the inspection work.

One of the capabilities of these UAVs is the ability to trace power pylons to identify issues such as damaged bolts, corrosion, rust, and lightning strikes. UAVs can also detect short circuiting of power lines which is mainly caused by harsh weather conditions, bushfires, and tree falls. Recent studies have installed UAVs to monitor the overhead power lines, showing that they can reduce the cost of examination operations and the risks associated with them compared to helicopter observation and foot patrol. There are different types of UAVs to be used based on specified inspection requirements. Fixed-wing UAVs are more applicable for large-area inspection and vegetation monitoring of power line corridors, mainly due to their higher speed and altitude. On the other hand, multirotor UAVs can fly close to the towers for detailed imaging [41]. Their high 3D maneuverability makes the multirotor UAVs eligible for operation in complicated environments that require close inspection of power lines.

However, autonomic task planning and piloting for multirotor UAVs within tight and complex operational environments is difficult. Future advancements in cooperative practices of UAV network mechanisms for data collection, sharing, and processing methods are the key to offering reliable, efficient, and faster services [42]. These technological advancements can be exploited in that the application of UAVs improves power line inspection work concerning safety and efficiency, thus being a critical component of the overall stability and reliability of the power transmission grid.

CHAPTER 15 INTEGRATION OF IOT DEVICES WITH UAV SWARMS

Urban Planning

The UAV swarms provide an urban planner with in-depth aerial views of city landscapes and data for designing and managing urban spaces. These UAVs can survey construction sites, monitor infrastructure development, and provide data to traffic management systems. The data thus gathered, with the IoT devices embedded within, will be of immense value to long-term urban planning and infrastructure management. This continuous data collection provides support for informed decision-making and ensures that the urbanization process is sustainable and meets the needs of the population.

UAVs for Automated Forest Restoration

Another application area on the rise is using UAVs for automated forest restoration (AFR). The tasks on which UAVs can be used to implement forest restoration are many, and they range from site surveys, restoration plan development, site infrastructure design, seed provision, site maintenance (including fertilizing and weeding), and biodiversity surveys postrestoration intervention. This includes UAVs, which can already undertake, among other uses, primary pre-restoration site surveys and monitoring of elements of biodiversity recovery [43]. These technologies would wish to monitor changes in climate, changes in ecosystem composition, and the functions of forests themselves that are, in turn, helpful in the inspection and ongoing management of restoration projects [44]. High-quality cameras fitted in UAVs get detailed data on forest ecosystems and help characterize and analyze forest areas. It is this capability that is so important in projects with an emphasis on forest restoration. That is, it allows accurate monitoring and the ability to assess the restoration process [45].

CHAPTER 15   INTEGRATION OF IOT DEVICES WITH UAV SWARMS

Challenges and Solutions

Several technical, cybersecurity, and regulatory challenges face the integration of IoT devices in UAV swarms for innovative city applications. Various challenges have to be addressed to attain a successful deployment and operation of these advanced systems.

Technical Challenges

The key technical concern of managing multiple UAVs is to guarantee that the communication links are fully functional and the data handling efficient. The ability to effectively communicate and coordinate is the de factor success of any operation with a UAV swarm. The 5G, ZigBee, and LoRa are some of the technologies that have been presented in this paper that is expected to enable secure, fast data link between the UAVs and their respective Ground Control Stations. However, the most complex issue is the ability to maintain a steady flow of communication, and this becomes even harder in such areas that consist of various structures and objects that can interfere with signals in urban settings [46].

Another important area that should not be overlooked is the data processing system. The huge amounts of data from applications like surveillance, disaster management, and environmental monitoring generate this need for real-time data processing. Data analysis is also another process that involves the use of several elements such as advanced data processing and other AI algorithms on edge processing. Edge computing, which centralizes the data processing at the edges closer to the data source, can ensure low latency and limit the use of bandwidth needed for data transfer, which allows for making an instant decision and take actions based on new data received [47]. At the same time, the introduction of any of these technologies into UAV swarms is fraught with compatibility at the computational level and energy consumption level, as well as taking into account the size of the equipment.

CHAPTER 15 INTEGRATION OF IOT DEVICES WITH UAV SWARMS

Cybersecurity Issues

One primary concern in UAV swarm operations is cybersecurity. UAVs have potential vulnerabilities to cyberattacks like data breaches, signal jamming, and even being hijacked. This means providing security of communication links and data integrity up to guaranteed levels will be required. In such a context, it will be necessary to encrypt the communication channels, ensure secure authentication protocols during the communication process, and provide frequent updating of software. It has been proposed that blockchain technology can make the communications of UAVs more secure by offering a decentralized, tamper-proof ledger in the storage of transaction records and data exchanges. Minimal effort has been made in the design considerations of small UAVs toward security and privacy concerns, which makes them susceptible to numerous cyberattacks [48]. UAVs' onboard wireless communication modules use open, unencrypted channels, exposing them to cyberattacks like eavesdropping, falsified data, and unauthorized reservation requests [49]. Attackers can easily interfere with drone communication because it is not encrypted. Drones are also highly susceptible to hacking, where attackers can gain control of UAVs to use them for illegal activities such as smuggling, invasion of privacy, and data theft [50]. It is also found that warfighting UAVs with secret information are pretty at risk from this type of attack, especially in military operational tasks. They also might spoof GPS signals to mislead the UAVs and, using false data insertion, add extra commands with additional commands to take complete control.

- **Denial of Service (DoS) and Distributed DoS Attacks:** Too many requests are flown into the UAV network to create network congestion, thus leading to severe challenges in availability. These attacks drain batteries, overload processing units, and flood communication links, all of which cause heavy interruptions in UAV operations. These have many sources targeting to send traffic and serve in rendering the UAV system unreachable.

- **Attacks on Ground Control System:** The GCS attack is hazardous because it allows attackers to steal data and send malicious commands to UAVs. Such acts are primarily performed through critical loggers, viruses, and malware. Compromised GCS will allow adversaries to have remote control over the UAVs, thus causing breaches in confidentiality by using social engineering or malware.

- **Eavesdropping and Trajectory Monitoring:** In several attacks, the trajectory of a UAV is monitored for the attacker to exploit the location information for malicious activities. Eavesdropping attacks exploit the absence of encryption, a fact that enables adversaries to get access to UAV data. It is essential to maintain the integrity of the UAV missions because, if not, malicious attacks can tamper with the original data by corruption for it to be worthless.

- **Malicious Hardware Attacks:** In a malicious hardware attack, a UAV's hardware is tampered with such that its normal behavior is altered, confidential data is stolen, or missions fail. Consider, for instance, the malicious attacks that can disrupt flight control and communication links that change mission parameters. The tampering of the onboard navigational sensors can misguide UAV trajectories.

Regulatory and Ethical Issues

Regulatory and ethical issues are significant problems. UAV operations have to meet some stringent regulations regarding airspace usage, privacy, and safety [51]. Such regulatory navigations are complex, especially when UAV swarms happen in different jurisdictions. Entities like the Federal Aviation Administration in the United States and their counterparts worldwide may have regulations on flight permissions, operational limits, and safety standards, among others. Some ethical issues may be related to privacy and limiting the intrusiveness of UAV surveillance. The public acceptance of UAV technology should dovetail with the transparent policies and practices driving these concerns. Engage communities and stakeholders with the benefits and safeguards of UAV applications.

Solutions

- **Improved Connectivity and Data Processing:** High-end communication technologies, such as 5G for fast data transmission and edge computing for real-time data processing, will significantly reduce the problems of connectivity and data processing. Specialized hardware and software developed for UAV swarm operations will enhance performance and efficiency [52].

- **Strong Cybersecurity:** Comprehensive cybersecurity, including encryption, secure authentication, and blockchain technology, will protect UAV swarms from cyber threats [53]. Regular security audits and updates will maintain the integrity and safety of UAV operations.

- **Regulatory Compliance and Ethical Practices:**
  International and local regulations need to be
  complied with for thoughtful monitoring and changes
  in regulation. Developing and adhering to ethics
  that respect privacy and safety will encourage trust.
  Put differently, it is possible to create an enabling
  environment for integrating UAV swarms in smart
  cities through a collaborative effort among regulatory
  agencies, policymakers, and the public.

Future Trends and Innovations

While various emerging technologies and cutting-edge research directions march rapidly toward integration with the IoT where swarming UAVs are concerned, this section discusses the most promising future trends and innovations [58]. Recent areas include 5G, AI, and potential applications in smart cities. Furthermore, the advancement of AI has revolutionized various sectors, including IoT, the energy sector, quantum computing, and the fields of image and signal analysis [59-61].

- **5G and Beyond Communications:** 5G technology
  will revolutionize UAV swarm operations by providing
  ultra-reliable, low-latency communication, massive
  connectivity, and enhanced data transmission rates.
  Based on the high data rates and low latency of 5G
  networks, this will allow the exchange of information in
  real time and coordination among UAVs for performing
  different tasks—more specifically, a collaborative
  task comprising environmental mapping and disaster
  response. Further research is directed toward
  integrating 5G with UAVs to improve the efficiency in
  communication and operation [54].

- **Artificial Intelligence and Machine Learning:** AI and machine learning are essential for implementation in the autonomous operation of UAV swarms. These enable the UAVs to conduct finer tasks such as obstacle avoidance, path planning, target recognition, and resource allocation. Advanced ML algorithms and improvements in onboard computational power will bring in the development of more intelligent and capable UAVs [55]. Deep learning techniques are also being applied toward enhancing the performance of UAVs in missions related to precision agriculture and urban monitoring.

- **Power Management and Energy Harvesting:** One of the most significant challenges remains mastering the game of UAV operations, that of energy management [56]. Future innovations are likely based on developing lightweight, high-capacity batteries, and new energy harvesting solutions to expand both flight time and operational range for UAVs. This concept of energy-efficient UAV swarms might be further enhanced through paradigms such as wireless power transfer (WPT) using solar panels.

- **Blockchain Technology:** Blockchain is an emergent technology that can be adopted to enhance the security and reliability of UAV operations [57]. Blockchain secures communication and data exchange between the UAVs by providing a decentralized and immutable ledger. This technology will further support different intelligent contracts in the autonomous management of missions, ensuring the UAVs are running safely and efficiently without human intervention.

CHAPTER 15 INTEGRATION OF IOT DEVICES WITH UAV SWARMS

- **Advanced Sensing and Navigation:** Improvements in sensor technology and navigation algorithms will continue to benefit future UAVs. Advanced GPS systems, along with more advanced IMUs and visual odometry, will support enhanced precision and reliability in navigation. This is important for applications focusing on exact location and movement, including infrastructure inspection and environmental monitoring.

- **Edge Computing:** The function of edge computing will be essential in the future for the swarming operations of UAVs. Reduce latency and conserve bandwidth as data is processed nearer to the source. Real-time decision-making and immediate action through data processing are crucial for low-latency-sensitive applications like search and rescue operations or real-time surveillance.

- **New Antenna Designs:** These will increase the communication capabilities of UAVs. Using small, aerodynamic antennas and directional antennas will increase data transmission rates and reliability. These will enable high-speed communication in tight operational conditions and ensure the preservation of solid and constant connectivity of the UAV swarm during tasks.

Conclusion

The integration of UAV swarms with IoT devices paves the way for a significant step change in smart city technologies, with a wide array of applications that improve urban management and safety. Principal points of the selected papers are providing valuable, real-time data through a swarm of UAVs fitted with IoT sensors for applications in surveillance, disaster management, urban planning, environmental monitoring, and precision agriculture. These UAV swarms shall include autonomy in their operations, communicate efficiently through advanced communication technologies like 5G, and use local processing of data through edge computing to reduce latency and bandwidth. The impact that this has on the future of smart cities is enormous. Consumer-friendly surveillance systems with UAV swarms will ensure the effectiveness of commands in detecting incidents for a quick response, hence improving public safety. UAV swarms can be highly effective means of damage assessment, locating survivors, and distributing necessary supplies during an emergency in disaster management. The more detailed aerial data that UAV swarms provide is helpful for urban planners, and environmental monitoring continues to benefit since data collection on air quality, water resources, and wildlife conservation is continuously growing. The conclusions are extended to further research and development, such as the need for an effective solution to robust connectivity and reliable data processing. Methods for superior cybersecurity have to, therefore, be put forward to ensure cybersecurity by the UAV swarms from a wide variety of cyber threats like data loss and signal jamming. Relevant regulations and ethical concerns would also be considered to ensure the whole system is respectful and safe toward privacy. Future innovative technologies are to research further advanced AI algorithms; AI will enable autonomous operations and power management solutions that are energy efficient and utilize blockchain technology for securing communication and data exchange.

References

[1] Giyenko and Y. Im Cho, "Intelligent UAV in smart cities using IoT," in 2016 16th International Conference on Control, Automation and Systems (ICCAS), IEEE, 2016, pp. 207–210.

[2] K. Kuru, "Planning the Future of Smart Cities with Swarms of Fully Autonomous Unmanned Aerial Vehicles Using a Novel Framework," IEEE Access, vol. 9, pp. 6571–6595, 2021.

[3] Copiaco et al., "An innovative deep anomaly detection of building energy consumption using energy time-series images," Eng Appl Artif Intell, vol. 119, p. 105775, 2023.

[4] K. Telli et al., "A Comprehensive Review of Recent Research Trends on Unmanned Aerial Vehicles (UAVs)," Systems, vol. 11, no. 8, p. 400, 2023.

[5] Sungheetha and R. Sharma, "Real Time Monitoring and Fire Detection Using Internet of Things and Cloud Based Drones," Journal of Soft Computing Paradigm (JSCP), vol. 2, no. 03, pp. 168–174, 2020.

[6] H. Ucgun, U. Yuzgec, and C. Bayilmis, "A review on applications of rotary-wing unmanned aerial vehicle charging stations," Int J Adv Robot Syst, vol. 18, no. 3, p. 17298814211015864, 2021.

[7] Nourmohammadi, M. Jafari, and T. O. Zander, "A Survey on Unmanned Aerial Vehicle Remote Control Using Brain–Computer Interface," IEEE Trans Hum Mach Syst, vol. 48, no. 4, pp. 337–348, 2018.

[8] S. Hayat, E. Yanmaz, and R. Muzaffar, "Survey on Unmanned Aerial Vehicle Networks for Civil Applications: A Communications Viewpoint," IEEE Communications Surveys & Tutorials, vol. 18, no. 4, pp. 2624–2661, 2016.

[9] Mairaj, A. I. Baba, and A. Y. Javaid, "Application specific drone simulators: Recent advances and challenges," Simul Model Pract Theory, vol. 94, pp. 100–117, 2019.

[10] Yinka-Banjo and O. Ajayi, "Sky-Farmers: Applications of Unmanned Aerial Vehicles (UAV) in Agriculture," Autonomous vehicles, pp. 107–128, 2019.

[11] M. R. Palattella et al., "Internet of Things in the 5G Era: Enablers, Architecture, and Business Models," IEEE journal on selected areas in communications, vol. 34, no. 3, pp. 510–527, 2016.

[12] F. Fabra et al., "MUSCOP: Mission-Based UAV Swarm Coordination Protocol," IEEE Access, vol. 8, pp. 72498–72511, 2020.

[13] Fotouhi, M. Ding, and M. Hassan, "Understanding autonomous drone maneuverability for Internet of Things applications," in 2017 IEEE 18th International Symposium on A World of Wireless, Mobile and Multimedia Networks (WoWMoM), IEEE, 2017, pp. 1–6.

[14] Riaz, N., Shah, S. I. A., Rehman, F., & Khan, M. J. (2021). An Intelligent Hybrid Scheme for Identification of Faults in Industrial Ball Screw Linear Motion Systems. IEEE Access, 9, 35136-35150.

[15] J. Pestana, J. L. Sanchez-Lopez, P. de la Puente, A. Carrio, and P. Campoy, "A Vision-Based Quadrotor Swarm for the Participation in the 2013 International Micro Air Vehicle Competition," in 2014 International Conference on Unmanned Aircraft Systems (ICUAS), IEEE, 2014, pp. 617–622.

[16] J. O. de Souza and M. Endler, "Coordinating movement within swarms of UAVs through mobile networks," in 2015 IEEE international conference on pervasive computing and communication workshops (PerCom Workshops), IEEE, 2015, pp. 154–159.

[17] S. Zhang, H. Zhang, L. Song, Z. Han, and H. V. Poor, "Sensing and Communication Tradeoff Design for AoI Minimization in a Cellular Internet of UAVs," in ICC 2020-2020 IEEE International Conference on Communications (ICC), IEEE, 2020, pp. 1–6.

[18] E. Vinogradov, F. Minucci, and S. Pollin, "Wireless Communication for Safe UAVs: From Long-Range Deconfliction to Short-Range Collision Avoidance," IEEE Vehicular Technology Magazine, vol. 15, no. 2, pp. 88–95, 2020.

[19] Tahir, J. Böling, M.-H. Haghbayan, H. T. Toivonen, and J. Plosila, "Swarms of Unmanned Aerial Vehicles—A Survey," J Ind Inf Integr, vol. 16, p. 100106, 2019.

[20] F. Koohifar, I. Guvenc, and M. L. Sichitiu, "Autonomous Tracking of Intermittent RF Source Using a UAV Swarm," IEEE Access, vol. 6, pp. 15884–15897, 2018.

[21] Humayoun, M., Sharif, H., Rehman, F., Shaukat, S., Ullah, M., Maqsood, H., ... & Chandio, A. H. (2023, March). From Cloud Down to Things: An Overview of Machine Learning in Internet of Things. In 2023 4th International Conference on Computing, Mathematics and Engineering Technologies (iCoMET) (pp. 1-5). IEEE.

[22] Lee, S. Kwon, P. Park, and K. Kim, "Active power management system for an unmanned aerial vehicle powered by solar cells, a fuel cell, and batteries," IEEE Trans Aerosp Electron Syst, vol. 50, no. 4, pp. 3167–3177, 2014.

[23] Q. Zhang, W. Fang, Q. Liu, J. Wu, P. Xia, and L. Yang, "Distributed Laser Charging: A Wireless Power Transfer Approach," IEEE Internet Things J, vol. 5, no. 5, pp. 3853–3864, 2018.

[24] M. A. Khan, B. A. Alvi, A. Safi, and I. U. Khan, "Drones for Good in Smart Cities: A Review," in Proc. Int. Conf. Elect., Electron., Comput., Commun., Mech. Comput. (EECCMC), 2018, pp. 1–6.

[25] Q. Zhang, J. Chen, L. Ji, Z. Feng, Z. Han, and Z. Chen, "Response Delay Optimization in Mobile Edge Computing Enabled UAV Swarm," IEEE Trans Veh Technol, vol. 69, no. 3, pp. 3280–3295, 2020.

[26] H. Nawaz and H. M. Ali, "Implementation of Cross Layer Design for Efficient Power and Routing in UAV Communication Networks," Stud. Informat. Control, vol. 29, no. 1, pp. 111–120, 2020.

[27] Li, Z. Fei, and Y. Zhang, "UAV Communications for 5G and Beyond: Recent Advances and Future Trends," IEEE Internet Things J, vol. 6, no. 2, pp. 2241–2263, 2018.

[28] N. H. Motlagh, T. Taleb, and O. Arouk, "Low-Altitude Unmanned Aerial Vehicles-Based Internet of Things Services: Comprehensive Survey and Future Perspectives," IEEE Internet Things J, vol. 3, no. 6, pp. 899–922, 2016.

[29] Zohourian et al., "IoT Zigbee device security: A comprehensive review," Internet of Things, p. 100791, 2023.

[30] J. M. Meredith, "Technical Specification Group Radio Access Network: Study on Enhanced LTE Support for Aerial Vehicles." 3GPP, 2015.

[31] H. F. ETSI, "European telecommunications standards institute," Human Factors (HF); User, 2005.

[32] Rehman, M. Muneer, M. H. Sajjad, and N. Riaz, "Data Science and Big Data Analytics," in Future Communication Systems Using Artificial Intelligence, Internet of Things and Data Science, CRC Press, pp. 92–109.

[33] M. Elloumi, R. Dhaou, B. Escrig, H. Idoudi, and L. A. Saidane, "Monitoring road traffic with a UAV-based system," in 2018 IEEE wireless communications and networking conference (WCNC), IEEE, 2018, pp. 1–6.

[34] T. Sands, "Virtual Sensing of Motion Using Pontryagin's Treatment of Hamiltonian systems," Sensors, vol. 21, no. 13, p. 4603, 2021.

[35] N. G. La Vigne, S. S. Lowry, J. A. Markman, and A. M. Dwyer, "Evaluating the Use of Public Surveillance Cameras for Crime Control and Prevention," Washington, DC: US Department of Justice, Office of Community Oriented Policing Services. Urban Institute, Justice Policy Center, pp. 1–152, 2011.

[36] S. A. H. Mohsan, M. A. Khan, F. Noor, I. Ullah, and M. H. Alsharif, "Towards the Unmanned Aerial Vehicles (UAVs): A Comprehensive Review," Drones, vol. 6, no. 6, p. 147, 2022.

[37] Riaz, N., Shah, S. I. A., Rehman, F., & Gilani, S. O. (2020). An Approach to Measure Functional Parameters for Ball-Screw Drives. In Intelligent Technologies and Applications: Second International Conference, INTAP 2019, Bahawalpur, Pakistan, November 6–8, 2019, Revised Selected Papers 2 (pp. 398-408). Springer Singapore.

[38] Omia et al., "Remote Sensing in Field Crop Monitoring: A Comprehensive Review of Sensor Systems, Data Analyses and Recent Advances," Remote Sens (Basel), vol. 15, no. 2, p. 354, 2023.

[39] H. Shakhatreh et al., "Unmanned Aerial Vehicles (UAVs): A survey on Civil Applications and Key Research Challenges," Ieee Access, vol. 7, pp. 48572–48634, 2019.

[40]   H. Menouar, I. Guvenc, K. Akkaya, A. S. Uluagac, A. Kadri, and A. Tuncer, "UAV-Enabled Intelligent Transportation Systems for the Smart City: Applications and Challenges," IEEE Communications Magazine, vol. 55, no. 3, pp. 22–28, 2017.

[41]   Riaz, N., Shah, S. I. A., Rehman, F., Gilani, S. O., & Udin, E. (2020). A Novel 2-D Current Signal-Based Residual Learning with Optimized Softmax to Identify Faults in Ball Screw Actuators. IEEE Access, 8, 115299-115313.

[42]   H. A. Foudeh, P. C.-K. Luk, and J. F. Whidborne, "An Advanced Unmanned Aerial Vehicle (UAV) Approach via Learning-Based Control for Overhead Power Line Monitoring: A Comprehensive Review," IEEE Access, vol. 9, pp. 130410–130433, 2021.

[43]   Tiansawat and S. Elliott, "Unmanned Aerial Vehicles for Automated Forest Restoration," Faculty of Science, Chiang Mai University, 2020.

[44]   R. A. De Almeida et al., "Monitoring restored tropical forest diversity and structure through UAV-borne hyperspectral and lidar fusion," Remote Sens Environ, vol. 264, p. 112582, 2021.

[45]   M. M. Moura, L. E. S. de Oliveira, C. R. Sanquetta, A. Bastos, M. Mohan, and A. P. D. Corte, "Towards Amazon Forest Restoration: Automatic Detection of Species from UAV Imagery," Remote Sens (Basel), vol. 13, no. 13, p. 2627, 2021.

[46]   Skorobogatov, C. Barrado, and E. Salamí, "Multiple UAV Systems: A Survey," Unmanned Systems, vol. 8, no. 02, pp. 149–169, 2020.

CHAPTER 15   INTEGRATION OF IOT DEVICES WITH UAV SWARMS

[47] Altan and R. Hacıoğlu, "Model predictive control of three-axis gimbal system mounted on UAV for real-time target tracking under external disturbances," Mech Syst Signal Process, vol. 138, p. 106548, 2020.

[48] D. Grieve et al., "The challenges posed by global broadacre crops in delivering smart agri-robotic solutions: A fundamental rethink is required," Glob Food Sec, vol. 23, pp. 116–124, 2019.

[49] E. Vattapparamban, I. Güvenç, A. I. Yurekli, K. Akkaya, and S. Uluağaç, "Drones for smart cities: Issues in cybersecurity, privacy, and public safety," in 2016 international wireless communications and mobile computing conference (IWCMC), IEEE, 2016, pp. 216–221.

[50] G. L. Krishna and R. R. Murphy, "A review on cybersecurity vulnerabilities for unmanned aerial vehicles," in 2017 IEEE international symposium on safety, security and rescue robotics (SSRR), IEEE, 2017, pp. 194–199.

[51] M. Mozaffari, W. Saad, M. Bennis, Y.-H. Nam, and M. Debbah, "A Tutorial on UAVs for Wireless Networks: Applications, Challenges, and Open Problems," IEEE communications surveys & tutorials, vol. 21, no. 3, pp. 2334–2360, 2019.

[52] Y. Che, Y. Lai, S. Luo, K. Wu, and L. Duan, "UAV-Aided Information and Energy Transmissions for Cognitive and Sustainable 5G Networks," IEEE Trans Wirel Commun, vol. 20, no. 3, pp. 1668–1683, 2020.

[53] Jiang, J. Yang, and H. Song, "Protecting Privacy from Aerial Photography: State of the Art, Opportunities, and Challenges," in IEEE INFOCOM 2020-IEEE Conference on Computer Communications Workshops (INFOCOM WKSHPS), IEEE, 2020, pp. 799–804.

[54] Dong et al., "UAVs as an Intelligent Service: Boosting Edge Intelligence for Air-Ground Integrated Networks," IEEE Netw, vol. 35, no. 4, pp. 167–175, 2021.

[55] X. Liu, M. Chen, Y. Liu, Y. Chen, S. Cui, and L. Hanzo, "Artificial Intelligence Aided Next-Generation Networks Relying on UAVs," IEEE Wirel Commun, vol. 28, no. 1, pp. 120–127, 2020.

[56] Lee, P. Park, C. Kim, S. Yang, and S. Ahn, "Power managements of a hybrid electric propulsion system for UAVs," Journal of mechanical science and technology, vol. 26, pp. 2291–2299, 2012.

[57] Qin, P. Li, J. Liu, and J. Liu, "BLOCKCHAIN-ENABLED CHARGING SCHEDULING FOR UNMANNED VEHICLES IN SMART CITIES," Journal of Internet Technology, vol. 22, no. 2, pp. 327–337, 2021.

[58] Riaz, N., Shah, S. I. A., Rehman, F., & Gilani, S. O. (2020). An Intelligent Approach to Detect Actuator Signal Errors Based on Remnant Filter. In Intelligent Technologies and Applications: Second International Conference, INTAP 2019, Bahawalpur, Pakistan, November 6–8, 2019, Revised Selected Papers 2 (pp. 675-683). Springer Singapore.

[59] H. Sharif, F. Rehman and A. Rida, "Deep Learning: Convolutional Neural Networks for Medical Image Analysis - A Quick Review," 2022 2nd International Conference on Digital Futures and Transformative Technologies (ICoDT2), Rawalpindi, Pakistan, 2022, pp. 1-4, doi: 10.1109/ICoDT255437.2022.9787469.

[60] A. Ashfaq, M. Kamran, F. Rehman, N. Sarfaraz, H. U. Ilyas and H. H. Riaz, "Role of Artificial Intelligence in Renewable Energy and its Scope in Future," 2022 5th International Conference on Energy Conservation and Efficiency (ICECE), Lahore, Pakistan, 2022, pp. 1-6, doi: 10.1109/ICECE54634.2022.9758957.

[61] I. Manan, F. Rehman, H. Sharif, N. Riaz, M. Atif and M. Aqeel, "Quantum Computing and Machine Learning Algorithms - A Review," 2022 3rd International Conference on Innovations in Computer Science & Software Engineering (ICONICS), Karachi, Pakistan, 2022, pp. 1-6, doi: 10.1109/ICONICS56716.2022.10100452.

CHAPTER 16

Navigating the Urban Landscape: A Comprehensive Review of IoT and UAV Integration in Smart Cities

Oroos Arshi[1] Aprajita Kashyap[2] Khadija Slimani[3] Inam Ullah Khan[4]
[1]School of Computer Science, University of Petroleum and Energy Studies, Dehradun, India, oroosarshi523@gmail.com
[2]School of Computer Science, University of Petroleum and Energy Studies, Dehradun, India, Kashyapaprajita2@gmail.com,
[3]Higher school of automatic electronic Computing, Paris, France, Khadija.slimani@esiea.fr
[4]Department of Computer Science, National University of Technology, Islamabad, Pakistan, Inamullahkhan05@iee.org

CHAPTER 16   NAVIGATING THE URBAN LANDSCAPE: A COMPREHENSIVE REVIEW OF IOT AND
             UAV INTEGRATION IN SMART CITIES

Introduction

The incorporation of Internet of Things (IoT) technology into smart city structures marks a big step forward in improving metropolitan efficiency and promoting innovation [1]. As our globe rapidly industrializes, there is an urgent need for creative techniques to manage money, infrastructure, and service delivery in densely populated locations [1, 2]. IoT integration into smart cities creates a sophisticated network of networked devices, such as powerful data processing platforms, and sensors, including actuators [2]. This networked infrastructure allows for the seamless gathering, transmission, and analysis of massive amounts of real-time data, affording city administrators with unprecedented insights into various aspects of urban life [3]. The use of IoT goes beyond data collecting to predictive analytics, which allows city managers to make educated decisions in areas such as energy management, public safety, and transportation efficiency. For example, IoT-enabled sensors may monitor energy consumption trends in buildings, allowing for more efficient use and lowering overall carbon footprints. Similarly, smart traffic management systems use real-time data from IoT devices to reduce congestion and shorten commuting times, ultimately improving urban mobility [3].

In parallel, the incorporation of UAVs (unmanned aerial vehicles) into smart city frameworks broadens the capabilities of IoT technology. UAVs play an important role in urban contexts, offering airborne surveillance, building inspection, and response services during emergencies. UAVs, outfitted with modern sensors and cameras, may monitor important infrastructure such as bridges and roads and detecting potential problems before they become emergencies. In catastrophe scenarios, UAVs aid in the fast evaluation of damage and the coordination of timely responses, hence increasing urban resilience and public safety [4]. UAVs play a crucial role in environmental monitoring in smart cities, detecting pollution sources, and supporting sustainability initiatives. By integrating UAV data with IoT analytics, city planners can mitigate environmental impacts and promote sustainable development. This approach enhances quality of life, improves planning, and enables cities to adapt to complex global challenges.

Related Works

Bellini et al. [4] review IoT-enabled smart city research literature, identifying trends and challenges in adopting IoT technologies for sustainable and efficient development. It surveys key technologies and reviews main smart city approaches and frameworks, categorized into eight domains. Rejeb et al. [5] analyzed the academic literature on IoT applications in smart cities, analyzing 1,802 articles from the Scopus database. Results show significant growth in IoT research, with major applications in smart buildings, transportation, healthcare, parking, and grids. The study uses VOS viewer software to build a keyword co-occurrence network and cluster relevant literature. Nagarajan et al. [6] propose an IoT-based Dynamic Food Supply Chain for Smart Cities, enhancing food quality, intelligent vehicle routing, and tracing contamination sources. The system uses smart sensor data collection and vehicle routing algorithms, outperforming existing approaches in performance metrics such as tracing accuracy and execution time. Whaiduzzaman et al. [7] explore smart cities, their concepts, characteristics, and applications, focusing on cloud and fog IoT ecosystems, machine learning approaches, and security and privacy. It provides a conceptual model of smart city mega-events and highlights the impact of emerging technologies on futuristic smart cities. Arshi and Chaudhary [8] survey IoT-based smart cities, discussing their potential, trends, amenity architecture, application areas, real-world involvement, and open challenges. It discusses implementation constraints and integration of IoT-based application areas, providing a useful panorama for future research and reference points. Ali et al. [9] review existing IoT middleware challenges and propose a novel solution, Generic Middleware for Smart

City Applications (GMSCA), combining big data and IoT for sustainable solutions. The middleware is implemented, tested, and evaluated through load balance and performance tests, proving its excellent functionality and usability. Almalki et al. [10] review smart city techniques, strategies, IoT capabilities, and challenges, discussing opportunities for future research in sustainable, eco-friendly, and smart city applications. Rai et al. [11] explore the use of IoT in various smart community applications, including transportation, water management, garbage management, lighting, parking, and infrastructure. It outlines the process of implementing IoT in these applications, its characteristics, and specific applications. The study also highlights the importance of IoT devices in smart city project implementations, including automated water collection methods.

Overview of Smart Cities

Smart cities offer a novel approach to urban living that enhances the efficiency, sustainability, and quality of life of residents through the integration of cutting-edge technologies [12]. These cities optimize waste management, public services, energy systems, transportation, and other aspects of urban infrastructure through the use of networked sensors, data analytics, and the Internet of Things [12]. By utilizing real-time data, smart cities can adjust swiftly to changing conditions, leading to improved resource utilization, reduced energy use, and better traffic flow. Smart technology integration also encourages citizen participation through digital channels, giving residents easy access to a range of services and participation in decision-making [12, 13]. Enhancing liability, resilience, and sustainability in urban areas to address the evolving issues facing contemporary society is the ultimate goal of smart cities. This section will cover the evolution of smart cities, their constituent parts, and the use of IoT in smart cities, providing a succinct summary.

CHAPTER 16   NAVIGATING THE URBAN LANDSCAPE: A COMPREHENSIVE REVIEW OF IOT AND UAV INTEGRATION IN SMART CITIES

Evolution of Smart Cities

As smart cities have developed, a wide range of applications covering various aspects of urban life have emerged. Smart parking options, real-time navigational apps, and sophisticated traffic management systems have all evolved in the transportation space, relieving gridlock and improving mobility [13]. To achieve sustainability goals, energy applications integrate energy-efficient technologies, sources of clean energy, and smart grids. Smart city technology has improved public safety by enabling IoT-enabled monitoring, response to emergency systems, and predictive policing models [14]. As smart cities have developed, digital platforms have also changed public engagement by facilitating community-driven projects, e-services, and participatory governance. Improved public health in healthcare is facilitated by videoconferencing and IoT-enabled monitoring equipment. Applications for environmental monitoring also assist in addressing problems with the quality of the air and water. Figure 16-1 illustrates the evolution of smart cities [13, 14].

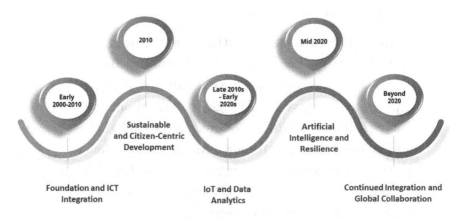

Figure 16-1. *The Evolution of Smart Cities*

- **The Early 2000s–2010s: Foundation and ICT Integration:** The progressive integration of ICT (information and communication technology) from the early 2000s to the 2010s made the creation of smart cities possible [15]. The concept of "smart cities" first gained popularity in the 2000s, demonstrating an innovative use of ICT to enhance government and municipal services. During this period, urban development revolutionized with experimentation with sensor networks, statistical data analysis, and technological infrastructure. [16]. These initiatives used technology in novel ways that enhanced public services, energy management, and commuting network efficiency [17]. These two generations collectively lay the framework for the development of smart cities, spreading the seeds for the increasingly sophisticated and linked urban ecosystems that are going to emerge throughout the years to come [17].

- **Mid-2010s: Sustainable and Citizen-Centric Development:** A crucial phase in the construction of smart cities emerged in the mid-2010s, defined by programs that gave equal weight to the environment as well as individuals [18]. Between 2014 and 2017, there was a clear shift toward environmentally friendly growth, with a focus on increasing energy efficiency, fixing environmental challenges, and aggressively incorporating renewable energy into urban infrastructure [19]. At the same time, the relevance of public engagement increased significantly from 2015 to 2018. As a method of building a more accessible and participatory model for municipal administration, smart city projects have emphasized the involvement of residents in decision-making as well as service delivery [20]. During this time, cities recognized that individuals, in addition to taking environmental responsibility, had a critical role to play in shaping whether smart cities were going to work [21].

- **Late 2010s–Early 2020s: IoT and Data Analytics:** When the Internet of Things, or IoT, first gained traction in the latter part of the decade and early 2020s, data analytics was vital to the creation of smart cities [22]. The Internet of Things gained traction in 2018 as cities embraced networked devices and systems, allowing for massive real-time data collection between 2018 and 2020 [23]. Governments may now draw significant inferences from the deluge of created data because data analytics are developing as a crucial instrument for making educated choices at the same time. This trend continued from 2020 to 2022, with a concentration on data-driven methods [24]. IoT technology deployment progressed, significantly enhancing operations by allocating personnel more efficiently, adjusting dynamically to changing requirements in urban situations, and utilizing the cloud [25].

- **Mid-2020s: Artificial Intelligence and Resilience:** By the end of the 2020s, robustness was becoming increasingly crucial, and artificial intelligence (AI) was being heavily incorporated into the development of smart cities [26]. As automation and forecasting algorithms evolved, artificial intelligence (AI) became increasingly essential to smart city systems in 2023 [26]. As a result of this relationship, cities are now better equipped to foresee trends, make smarter judgments, and automate operations to save money and time. From 2023 to 2025, resilience planning underwent a substantial shift [27]. It was widely accepted that cities were required to quickly build adaptable methods to deal with complex problems such as health crises, catastrophes caused by Mother Nature, and climate change.

- **Beyond 2020: Continued Integration and Global Collaboration:** The development of smart cities after 2020 will necessitate the persistent pursuit of continued integration and global collaboration [29]. Cities are ready to integrate emerging technologies into their infrastructure in the coming years, such as 5G, blockchain, and edge computing [30]. We anticipate that this connection will further enhance connectivity, boost output, and create new opportunities for innovation. At the same time, there appears to be a trend toward increased international collaboration among cities [31]. As more people become aware of the advantages of sharing best practices, standardizing norms, and exchanging technological ideas, collaboration is becoming more and more crucial. With cities working together to address common issues and propel urban living toward a connected, technologically advanced, and sustainable urban ecosystem, this collaborative effort advances a more inclusive and global approach to the development of smart cities [32].

CHAPTER 16 NAVIGATING THE URBAN LANDSCAPE: A COMPREHENSIVE REVIEW OF IOT AND UAV INTEGRATION IN SMART CITIES

Components of Smart Cities

The components of a smart city are numerous and crucial, as Figure 16-2 shows. Applications for smart cities typically involve four key elements: information collection, transmission/reception, storage, and analysis. The collection of data differs depending on the specific applications, which drives advancements in the development of sensors across multiple disciplines [32]. The second component is data transfer, which involves delivering information from data collection units to the cloud for processing and storage. This transmission is made possible by numerous local networks with different data transfer rates, citywide Wi-Fi networks, and 4G and 5G technologies [33]. The third phase, which includes cloud storage, arranges data using a variety of techniques to make it accessible in the fourth stage, data analysis. Identifying trends and conclusions in the data is the process of analyzing data to help guide decision-making [34]. In certain situations, basic analysis might be sufficient, but heterogeneous data collection, storage, and real-time processing made possible by cloud computing allow for more sophisticated decision-making. In the ecosystem of smart cities, machine learning, deep learning algorithms, and statistical techniques all play a part in complex, real-time data processing [35].

CHAPTER 16   NAVIGATING THE URBAN LANDSCAPE: A COMPREHENSIVE REVIEW OF IOT AND UAV INTEGRATION IN SMART CITIES

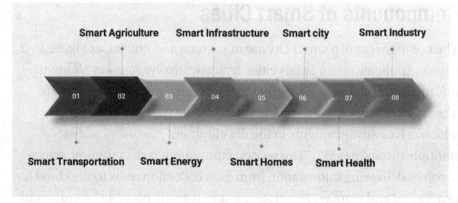

Figure 16-2. The Components of Smart Cities

Smart City Services

The services provided by smart cities include municipal duties including garbage management, water supply, environmental control and monitoring, and other things that keep a city's inhabitants alive. Water quality sensors can be installed to continuously report on the state of the water utilized in the city and find leaks [36]. One common feature of initiatives related to smart cities is waste management, which has been incorporated into many of the previously mentioned initiatives. Examples of these include Barcelona's chutes and bins with sensors and cloud connections that can use artificial intelligence (AI) to identify the most cost-effective route while also alerting the appropriate authorities when a bin needs to be emptied. Sensors can also be used to direct traffic to the closest open space [37] and track pollution levels in a city, including monitor environmental conditions to identify the sources of pollution.

Smart Health

Smart health refers to the use of ICT to improve healthcare quality and access. The growing population and rising costs of healthcare have drawn a lot of attention from researchers and medical experts to this topic [38]. The public's expanding demand cannot be met by the overburdened current health systems. As a result, the objective of smart health is to ensure that the greatest number of individuals has access to healthcare through the use of telemedicine services [38] and improved AI-powered diagnosis support for medical professionals [39]. The increasing prevalence of mobile phones and health trackers [39] has made it possible for researchers to analyze this data using cloud-based computing capabilities to improve healthcare decisions. These devices can record daily activity, identify abnormal movements using inertial sensors, and collect real-time data about the wellness of individuals (ECGs, temperature, body oxygen saturation, and other biosensors). As a result, overall costs and the burden on healthcare facilities are reduced [40].

Smart Industry

Global industries are constantly striving to become more efficient and economical, boosting output while cutting expenses. Industry 4.0 focuses on a networked factory, enabling seamless, interconnected, and collaborative work among its intermediary functionaries through the Internet of Things [41]. IoT applications in production and industrial processes and cyber-physical systems that combine humans and machines have produced several advantages for the sector, such as improved and quicker invention and manufacturing scheme optimization (resources and procedures), improved product quality, and increased plant safety employees. However, there are some barriers to IoT implementation in smart industries [42]. Cyber-physical systems with flexible configurations, quick deployment, and connection are required for use in Smart Industry IoT applications, as controlling a collection of different equipment and devices has particular challenges [43]. Industry 4.0 services are being developed and implemented more quickly because of the collaboration of IoT and AI. Sensors are included in machinery as well as other production processes; thus, using data from these sources allows AI techniques to be used for business intelligence activities, increase automation, and more. Academics have presented frameworks [44] for integrating AI with IoT for the Smart Industry.

Smart Transport

Traffic issues are common in many cities and include congestion, pollution, and challenges in planning and implementing cost-effective public transportation [45]. Information and communication technologies (ICT) are being employed widely as solutions because of their rapid evolution, especially in the area of vehicle-infrastructure-pedestrian connectivity [46]. They include vehicle to pedestrian (V2P), vehicle to infrastructure (V2I), and vehicle to vehicle (V2V) (P2I) among other communication modalities. These technologies are the cornerstone of smart transportation systems because of the pervasiveness of GPS devices in cars and cell phones among drivers [47]. GPS data is used in many inventive ways to track driver conduct and analyze traffic patterns in real time. Applications like Waze and Google Maps currently utilize this type of data to schedule public transport and plot routes in real time. Furthermore, parking systems with sensors help to solve today's transportation issues by guiding cars to the nearest available parking spots, hence promoting efficient urban mobility [49].

Smart Agriculture

Since it is becoming harder to produce food due to the rising global population and severe climate change, food security is a key component of the United Nations Sustainable Development Goals for 2030. One of the most important global issues today is ensuring sustainable agriculture and wise resource usage, particularly about water. To make decisions easier, prevent diseases and pests, and detect many features, smart agriculture is a critical alternative [50]. This involves placing sensors into fields and plants. Precision agriculture—a subset of smart agriculture—allows for exact care operations by utilizing sensors for targeted measurements. Precision farming is seen as essential to achieving food security in the future. With an emphasis on data-driven decision-making, disease diagnosis, and crop monitoring, artificial intelligence (AI) [50] is a key component of the Internet of Things (IoT) for agriculture. By tackling the intricate problems brought on by a growing population and a changing climate, this collaboration between AI and IoT in agriculture represents a determined push toward sustainable food production.

Smart Energy

Typical electrical systems consist of a primary generating source (usually a hydrological or fossil fuel-based power plant) that provides electricity in a one-way flow. For there to be a steady supply of power, the energy production plan employed with these networks needs the power provided by these sources to be significantly greater than the demand. While the substations provide feedback to control power generation, no consumer-provided data is incorporated [51]. Error detection and correction are also time-consuming tasks in these kinds of systems. Furthermore, modern consumers have access to both their primary utility supply and their generation as renewable energy technologies become more reasonably priced. ICT are used to develop smart grids, which enable distributed energy generation from consumer and utility sources, boost the observability across newly installed and existing grids, and provide self-repairing capabilities. Real-time power data delivery to utilities at different grid points, all the way via supply lines to the customer, is a feature of smart grids. Because they give real-time data on consumer demand, smart grids allow for greater electrical power management through prediction models developed from acquired consumption data, integrating many energy sources, and healing themselves [51] of the network to ensure a continuous supply of electricity.

Smart Homes

One of the most important components in the subject of smart cities is the smart home, which is recognized as vital to the everyday lives of urban dwellers. Sensing units strategically positioned throughout a person's house are incorporated into "smart homes," that serve as a vast data store on the residence and its occupants [53]. Among other technologies, these sensors include motion trackers, sensors for the environment, and power/energy usage monitors. By collecting and analyzing sensor data, smart homes enable a thorough understanding of household utilization of resources, conditions in the environment, and behavioral patterns. This wealth of data helps to accomplish the larger goal of optimizing citywide assets and amenities within the larger context of smart city development, in addition to increasing the efficiency of home administration [52].

Importance of Applications of UAVs in Smart City

UAV infrastructure in smart cities is a low-cost and efficient way to collect data, monitor it, conduct surveillance, and analyze infrastructure. They help to improve planning for cities, management of resources, responding to emergencies, public safety, and environmental sustainability, ultimately increasing the overall standard of life for city residents as seen in Figure 16-3 [54].

CHAPTER 16   NAVIGATING THE URBAN LANDSCAPE: A COMPREHENSIVE REVIEW OF IOT AND UAV INTEGRATION IN SMART CITIES

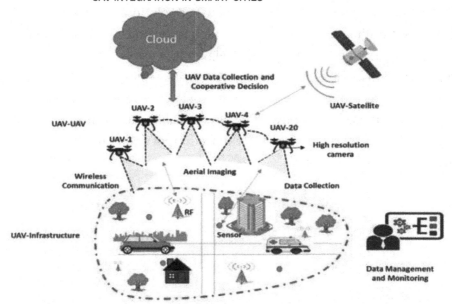

Figure 16-3. *UAV Infrastructure for Smart Cities [54]*

UAVs, equipped with sensors and cameras, collect data for urban planning, infrastructure management, environmental monitoring, and decision-making in smart cities. They can inspect critical infrastructure, assess disaster-affected areas, and provide situational awareness. In emergency situations, UAVs can detect fires and deliver supplies. They also monitor environmental parameters, aiding in environmental planning and conservation efforts. Delivery drones offer an environmentally friendly alternative for transporting goods, reducing traffic congestion and carbon emissions. Overall, UAVs are essential tools for smart cities to improve safety, reduce costs, and enhance logistics operations.

Advances in electronics including manufacturing technologies have allowed controllers, sensors, and CPUs to be miniaturized while still functioning. This advancement has resulted in more compact UAV models. The potential for reducing size is significant, and it offers numerous advantages. PwC's 2016 report "Clarity from Above" found that the target market value of drone-powered solutions exceeds USD 127 billion, suggesting that the drone revolution is causing huge disruptions across a wide range of industries [19]. Figure 16-4 shows a schematic representation of the estimated values for various significant sectors. UAVs, often known as drones, have a wide range of applications in a variety of fields due to their adaptability, efficiency, and ability to reach inaccessible locations. Here are some of the key applications of unmanned aerial vehicles in various sectors:

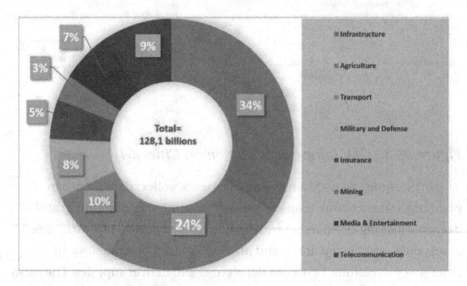

Figure 16-4. The Potential Value of UAV-Based Solutions in Important Industries for the Worldwide Market [54]

(a) **Drones Are Used in Precision Farming:** They may assess crop health, monitor livestock, and even assist with agricultural planting and spraying. They can provide detailed airborne imagery, enabling farmers to make informed decisions about their crops and animals.

(b) **Construction and Infrastructure:** Drones are used to survey land, examine constructions, and monitor construction progress. They can give high-resolution photos and videos that can be used to plan, monitor, and check construction sites.

(c) **Disaster Management:** Drones are frequently used in situations of disaster to conduct search and rescue missions, assess damage, and transport emergency supplies. They can reach regions that people find difficult or unsafe to visit.

(d) **Environmental Monitoring and Conservation:** Drones are utilized in wildlife monitoring, forest protection, and environmental research. They can monitor environmental changes, follow animal movements, and gather information on wildlife populations.

Characteristics of UAVs

Because of their unique capabilities and qualities, several types of UAVs are useful for a variety of applications. The function of a drone determines which one to use. Understanding the uses of UAVs in society requires a study of their structural types. A UAV's structure has a significant impact on its capabilities, performance, and adaptability for various jobs [56]. Figure 16-5 depicts the various designs of UAVs, which include fixed-wing, rotary-wing (e.g., helicopters), and hybrid forms. Each type has advantages and disadvantages that make it appropriate for certain purposes.

CHAPTER 16 NAVIGATING THE URBAN LANDSCAPE: A COMPREHENSIVE REVIEW OF IOT AND UAV INTEGRATION IN SMART CITIES

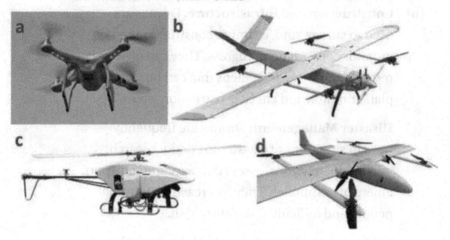

Figure 16-5. Types of UAV Rotors: (a) Rotary Wing (Multicopter), (b) Fixed Wing, (c) Rotary Wing, and (d) Fixed Wing [57]

Studying the structure of UAVs helps researchers and engineers understand their capabilities and limitations, enabling them to design and develop optimized UAVs for specific applications. This knowledge is crucial for advancing UAV technology and maximizing its benefits in sectors like agriculture, transportation, environmental monitoring, and disaster management. UAVs are typically operated by a remote pilot or autonomously using preprogrammed flight paths. Understanding the characteristics of UAVs helps make informed decisions about which type is best suited for a specific task, leading to more efficient use of UAV technology.

Internet of Things for Smart Cities

The Internet of Things (IoT), a game-changing technology that powers the widespread digitalization central to the idea of smart cities, is at the center of smart city projects. IoT refers to the pervasive Internet connectedness of gadgets, which allows them to easily send data to the cloud and maybe get commands for different tasks [54]. With over 75 billion devices expected

CHAPTER 16 NAVIGATING THE URBAN LANDSCAPE: A COMPREHENSIVE REVIEW OF IOT AND UAV INTEGRATION IN SMART CITIES

to be connected by 2025, this network of devices serves as the basis for a vast amount of application development [55]. IoT makes it easier to install sensors throughout urban areas to gather and send data to a central cloud in the context of smart cities. After that, this data is processed and mined to extract insightful patterns and information that are essential for well-informed policy formulation and decision-making within the dynamic framework of a smart city.

IoT Architectures for Smart Cities

Data transmission/reception, monitoring, and other functions are unified by the Internet of Things. Cloud services are utilized for data processing and storage in a generic Internet of Things architecture consisting of five layers that process information from the previous layers. The earlier layer is depicted in Figure 16-6. Additionally, it displays the three distinct architectures that are available for use by Internet-connected systems.

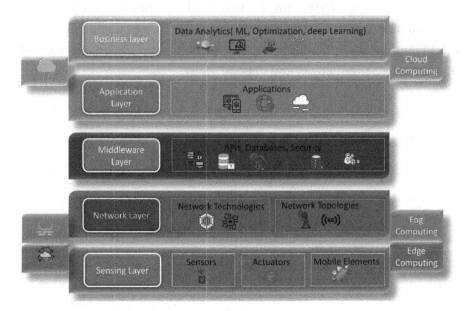

Figure 16-6. *IoT Architecture*

CHAPTER 16 NAVIGATING THE URBAN LANDSCAPE: A COMPREHENSIVE REVIEW OF IOT AND UAV INTEGRATION IN SMART CITIES

The sensors that collect data regarding tangible quantities of interest for any application, along with actuators that may act on physical things like RFID readers that scan RFID tags and other similar devices, make up the sensing layer, also known as the perception layer. The sensing layer receives data from the networking layer, which uses wireless network technologies like Wi-Fi, cellular internet, Zigbee, and Bluetooth, forwarded to the middleware layer [56]. While the application layer uses the data to give services to customers through database management services and various APIs, the middleware layer offers a general interface for the hardware of the sensing layer. The business layer, which is used to develop policies and plans that support system management, is linked to the application layer.

IoT architectures are typically categorized based on the types of tasks allocated to different IoT system components; this categorization is mostly driven by data processing responsibilities [57]. There are three architectures for IoT systems in the IoT framework stage where data processing could take place: cloud, fog, and edge models. Table 16-1 lists the attributes of the three IoT system levels. It is important to understand that the three IoT topologies discussed here are not adversarial to one another. Rather, the hierarchy aims to improve the system by providing the higher layer with only relevant data that boosts productivity and dependability.

Table 16-1. *Cloud, Fog, and Edge Computing Model Comparison*

Cloud computing model	Fog computing model	Edge computing model
Worldwide contextual awareness incorporating every facet of the application	The local sensing scenario is contextually sensitive to the fog layer	Usually, edge devices are simply aware of their state. Possible exchange technique, but restricted to the immediate neighborhood
Farthest from the edge, which causes decision-making to be sluggish and latency to be high	The fog layer may combine data from sensors and other devices and respond to it faster because it is the unit closest to the edge	Quickest decision-making feasible; yet, choices will be made by local states
Utilizes diverse data from many sensor devices	Makes use of diverse data, but only in a limited area	Typically lack access to several kinds of data
Elevated network connectivity	As data flow decreases, medium network cost	The least possible network cost
Possible privacy issue due to the possibility of sending raw data to the cloud	More privacy as opposed to cloud computing	Potential for even stronger privacy enforcement than the fog computing model
Least resilient since centralized decision-making	More resilient than the cloud computing paradigm	Greatest resource-related capabilities
Scalability is poor	Scalability outperforms cloud computing	Scalability is at its peak

CHAPTER 16 NAVIGATING THE URBAN LANDSCAPE: A COMPREHENSIVE REVIEW OF IOT AND
UAV INTEGRATION IN SMART CITIES

Cloud Computing Model

The basic architecture for Internet of Things systems is based on the notion that cloud computing should handle data processing from various IoT system components. Without the need for human involvement, cloud computing offers a dynamic distribution of shared resources such as processing power, storage, and other amenities through remote access. This approach is available on multiple platforms and provides hardware and software services for applications related to smart cities [58]. The central administration platform of the cloud offers the benefit of vast computational and storage capabilities for complicated operations like information mining and pattern extraction, along with the ability to observe, control, and distribute command actions depending on incoming data. Nevertheless, there are shortcomings with this cloud computing paradigm. More network traffic from sending all data to the cloud could result in higher costs and transmission overheads, particularly in scenarios where measurements are taken often. Data latency results from the separation of the processing and sensing layers and network stability issues are raised by the volume of data traffic growing in large-scale IoT systems, endangering dependable data transmission techniques.

Fog Computing Model

A different approach to resolving issues with the cloud computing model for the Internet of Things was fog computing, which addressed issues with information processing at the edge or sensing end of the network. By outsourcing processing tasks to local network devices, particularly routers and other network devices on the Internet of Things network layer, fog computing allows for a more decentralized division of labor when compared to classical cloud designs. By utilizing the enhanced computational capability of these network devices, fog computing makes it possible to do fundamental activities locally, such as data gathering, basic processing, and decision-making. This approach optimizes data transport by reducing the amount of data traveling to the cloud layer. Fog layers are required to support distributed decision-making processes and to create local networks with non-Internet technologies like RFID, radio frequency, Bluetooth, and Zigbee [59]. Decreased latency, data overhead, and transmission faults improve system resilience, lower IoT installation costs, and boost application efficiency by enabling faster decision-making in life-threatening scenarios. The fog layer's capacity to gather information from many devices and offer application-neutral Internet of Things systems makes it easier to create cloud-based policies and provides a deeper understanding of how systems behave. Notwithstanding these benefits, determining who is responsible for what in the cloud and fog layer necessitates taking the related expenses into account.

Edge Computing Model

Fog computing was first intended to improve data processing and reduce reliance on centralized cloud servers by bringing decision-making closer to the edge of the network. Recent years have seen the emergence of ever more powerful devices connected to "edge" nodes, bringing straightforward data analysis and decision-making closer to these devices. Edge computing, then, is the processing of data at the level of individual "things," such as sensors and other Internet of Thing's devices. A similar concept, as shown from various perspectives, such as [60], characterizes the edge computing layer as a link that connects the fog layer as well as the "things" or sensors, with edge nodes for computing serving as smaller-scale units of aggregation and decision-making concerning the broader focus on data integrity and connectivity. The primary goal of the fog and edge computing paradigms, which also aim to reduce costs, increase scalability, and fortify overall system resilience, is to decentralize the Internet of Things (IoT) [61]. In an intentional effort to leverage edge device capabilities to build more efficient, responsive, and cost-effective Internet of Things systems, deeper distributed decision-making methods are emerging.

Sensing Technologies for Smart Cities

Technologies for smart cities are centered around the senses. Through sensors, new ideas for smart cities can be generated with the information and data that are needed. Since sensors are so diverse and varied, using them in smart city initiatives is a perfect fit. A list of sensors used for the Internet of Things, as well as a comparison technique for IoT sensors, is supplied by the authors in [62].

For our investigation of the sensing technologies employed in smart cities, their work acts as a guide. Figure 16-7 shows how IoT sensors are categorized: surroundings, motion, electric, biological sensors, identification, being present hydraulic, and chemical sensors. Sensors are the key component of Internet of Things systems in smart cities that enable communication between the infrastructure of the smart city and its citizens and make it possible to develop new services. It is important to remember that many of the sensors have several uses as mentioned. Each application will also require measuring different physical quantities and will require the use of many different types of sensors. For example, ambient sensors for motion, electricity, location, identification, chemistry, and hydraulics are all available for use in houses with intelligent systems.

CHAPTER 16   NAVIGATING THE URBAN LANDSCAPE: A COMPREHENSIVE REVIEW OF IOT AND UAV INTEGRATION IN SMART CITIES

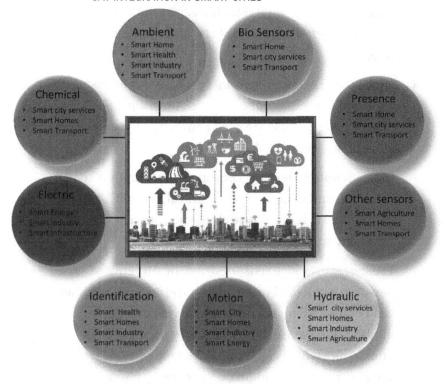

Figure 16-7. Sensing Technology in Smart Cities

- **Ambient Sensors:** A variety of sensors that track physical values that show elements of the environment including light intensity, humidity, temperature, and pressure [62]. Ambient sensors find numerous applications in smart cities, one of which is smart homes. In addition to controlling comfort levels, they are used for smart city services [63].

- **Chemical Sensors:** Chemical sensors analyze the chemical properties of materials. These sensors include smoke detectors, pH meters, water quality monitoring sensors [64], and sensors for measuring and detecting gases like carbon monoxide (CO) and carbon dioxide (CO_2) [64].

- **Hydraulic:** Hydraulic sensors are those that are used for liquid measurements, like level, flow, and leak detection [65]. These are used to measure the amount of liquid in storage tanks [66].

- **Identification:** Identifier sensors include things like RFID tags and near-field communication (NFC) equipment. Data exchange apps, smart transport, and smart city services all use these sensors [67].

- **Presence:** Presence sensors signal the presence of persons or objects. For security purposes, reed switches can be installed on windows and doors; passive infrared (PIR) sensors are commonly used to detect human movements; and inductive loop sensors, which use electromagnetic induction to detect presence in transit systems, are also available [68]. Ultrasonic sensors can also be used to estimate the distance between objects. Capacitive sensors are another type in this category that can be used to locate something [69].

- **Biosensors:** The health of a living thing can be assessed in several ways using biosensors [70]. Biosensors are used to help with healthcare monitoring in smart cities. Examples of these sorts of sensors are electrocardiograms (ECG), electromyograms (EMG), electrocardiograms (ECG), resistance to touch, heart rate, respiration sensors, pulmonary oximetry, blood pressure, and other sensors [71].

- **Electric Sensors:** In smart homes and smart grids, electric sensors are commonly used to monitor consumer and appliance power usage [72]. They make electrical power measuring possible. These include devices that sense voltage and current, respectively, such as voltage sensors and current transformers [73].

- **Motion Sensors:** Motion sensors are those that have the potential to detect motion. Among the inertial sensors used in motion detection systems are gyroscopes and accelerometers. These sensors are used for activity tracking in smart health applications and vibration detection in smart homes and businesses [74].

- **Other Sensors:** Applications for smart cities use a range of sensing modalities, including audio and visual data and other signal-measuring devices like Bluetooth and Wi-Fi signal strength. The acquired data typically goes through additional processing before it can be utilized to identify the target variable under investigation because the sensors of these modalities record raw signal information (visual, audio, signal strength, etc.).

Integration of IoT in Urban Infrastructure

The Internet of Things (IoT) integration into urban infrastructure is redefining how cities operate and interact with their surroundings. This represents a paradigm shift in city administration. The urban landscape needs to be extensively equipped with sensors, actuators, as well as smart devices to build an interconnected network that transcends traditional silos. Urban mobility is changing as a result of the Internet of Things applications [75]. These applications, which lessen traffic and boost overall transportation efficiency, include intelligent traffic management, connected automobiles, and flexible public transit systems. By utilizing renewable energy sources, facilitating real-time control along with monitoring, and improving energy consumption patterns, the implementation of smart grids revolutionizes the energy infrastructure. While preserving a secure urban environment, public safety is increased to previously unheard-of levels with the aid of rapid emergency response systems, IoT-enabled monitoring, and statistical analysis for authorities. Because they communicate when they are full, smart bins streamline garbage management while lowering operational costs and environmental impact. IoT-enabled air and water quality monitoring lessens the environmental effect of cities by enabling data-driven solutions for sustainable resource management and pollution control. The copious amounts of data generated by these interconnected technologies offer pragmatic perspectives to municipal authorities, facilitating adaptable decision-making to promptly address novel problems. The integration of the Internet of Things with urban infrastructure shows a commitment to creating intelligent, resilient, and sustainable cities that prioritize the well-being of their residents and employ technology to manage cities holistically.

The Internet of Things (IoT) is revolutionizing urban infrastructure by transforming transportation, waste management, lighting, and energy infrastructure. Smart parking solutions optimize parking space utilization, reduce traffic congestion, and improve mobility. Smart waste management systems use sensors to monitor fill levels, while smart grids enable real-time monitoring and management of renewable energy sources. IoT data empowers city officials to respond to emerging challenges and create resilient, sustainable, and citizen-centric urban environments [76].

Smart Infrastructure

Smart cities create an efficient and sustainable infrastructure for the country. An intelligent infrastructure makes earlier designs and constructions more prudent and cost-effective. Smart infrastructure installation is necessary for energy conservation and environmental system friendliness. To make systems smart in all domains—commercial, industrial, environmental, etc.—many IoT applications are integrated into smart infrastructure, as seen in Figure 16-1. All of the current applications, including intelligent waste management and intelligent parking, are configured using IoT-connected communication devices. It also enumerates all currently in-use applications for smart cities, together with the region they are intended for and the equipment they use.

Smart Parking

In today's world of rapid change, smart parking is essential. It turns out that smart parking is an IoT application that can significantly enhance everyone's quality of life. In this crowded environment, parking-related problems are lessened for individuals, thanks to Internet-connected smart parking technology [47]. Sensors linked to the Internet of Things are used to locate vacant parking spaces. Information collected at the parking lot is transmitted to a cloud server [48]. The user can then access the necessary instructions via smartphones or other communication devices. Smart parking systems frequently use sensors that assess proximity, distance, and ultrasonic waves. This system recognizes when a car is coming or going, shows where to park, and shows how far apart each automobile is. This system can be connected to any smartphone, using any application or web browser. This makes the parking process very easy and quick to complete [77]. This approach doesn't require an expensive or complicated infrastructure. MQTT, also known as Message Queuing Telemetry Transport, is a thin message translation mechanism that apps, servers, and devices use. The cloud server organizes data about every parking spot. Finally, the users can get the required direction through the communication devices connected to the cloud servers.

Smart Waste Management System

Waste materials are gathered, moved, processed, disposed of, managed, repurposed, and monitored, among other minor processes that make up the larger process of waste management. A smart waste management system may alleviate a lot of problems, like air pollution, trash buildup in the streets, infectious disease spread from mishandled garbage, inadequate waste collection, etc. [76]. Using level detection sensors, the driver of the garbage disposal vehicle can receive a smartphone alert indicating the level of the waste once it reaches a certain level set by the system management. This contributes significantly to the intelligent infrastructure of the country [77]. IoT technology must be used in several sectors for waste management. Sensing important data, like the quantity and type of waste, is the main job of the sensors used in smart waste management. This data is then sent either directly or indirectly to the system via the nearby communication devices that are linked to the cloud server and the smart city system. This method separates waste that is biodegradable from that that is not [76]. Our environment will remain clean as a result of this decreasing the possibility of trash cans overflowing [78]. It would be easier to organize different waste management operations, like figuring out when collection should occur and categorizing different kinds of garbage, with the help of this kind of network, which is made up of sensors and data collection. Thanks to IoT, there will be a database system to maintain a clean environment and a fully coordinated system with a far larger vision [78].

Smart Environment Monitoring

One of the main causes of concern right now is the environmental hazards we are facing. Pollution, water quality, and air quality are just a few of the critical components that may be monitored and managed with the help of IoT. Smart environment monitoring with IoT apps is seen in Figure 16-3. Temperature, humidity, and water sensors are used. To monitor with conventional means, it will take a long period. By evaluating gathered data and drawing conclusions, the instrumental approach makes use of instruments and software. Quicker, quicker, and more efficient is the instrumental approach [79]. This helps to avert serious pollution of our resources and the disasters that follow. For example, the water monitoring system not only reports on the availability and purity of the water but also on its source and the way it combines with potable water, among other things. Soil productivity is impacted by pollution. Pesticide misuse affects the environment. Each year, hundreds of people lose their lives as a consequence of using chemicals and pesticides. By using the sensors that are fixed to monitor the plant's nutrient and water needs, better agricultural practices can be implemented. A further strategy is to use IoT for animal protection [80].

Smart Grid

An electrical network and a communication network together are called a smart grid. A smart grid is a component of the framework for smart cities, allowing for the control of all monitoring systems in one grid, including smart environment monitoring, smart parking, smart traffic management, smart lighting, and smart roads [81]. The smart grid uses sensors, embedded systems, stations, substations, transformers, software, transmission networks, and transformers to do this. To respond quickly to users, two-way communication is established between the hardware and the user. A smart grid can supply electricity more efficiently and dependably than a standard power system by implementing bidirectional communication and power flow [82].

IoT in Healthcare Services

The patient experience, including facility administration, has undergone significant change as a result of the Internet of Things (IoT) incorporation into healthcare services. An important use for remote patient monitoring is enabling medical professionals to monitor patients' vital signs, medication compliance, and general health from a distance [83]. In the event of problems, this technology allows for early treatment and promotes patient autonomy. IoT is used by smart hospitals and other healthcare institutions to improve patient experiences, expedite procedures, and allocate resources optimally [84]. Smart beds and linked medical equipment are examples of IoT-enabled products that let healthcare workers work more efficiently and access data in real time. To protect patients and employees, IoT also supports public surveillance in healthcare settings. Healthcare facilities are safer and more secure because of the addition of environmental sensors, video surveillance, and access control systems to security measures. In general, the use of IoT in healthcare applications is revolutionizing patient outcomes, streamlining the delivery of healthcare, and creating a safer, more cutting-edge healthcare ecosystem [85].

Remote Patient Monitoring

Remote patient monitoring, or RPM, is an innovative treatment strategy that uses technology to provide care outside of traditional settings. It enables ongoing surveillance of patients' vital signs, signs and symptoms, and various other indicators of wellness outside the boundaries of clinics and hospitals, resulting in a paradigm change [86]. Healthcare professionals may now obtain real-time information, thanks to this novel approach, allowing them to analyze patients' problems from faraway locations and treat them as soon as possible. RPM's key benefit is that it can be tailored to suit a wide range of people, including the elderly, those with chronic illnesses, and those who require continuing medical care. RPM seamlessly incorporates into a patient's daily life by delivering an in-depth and precise understanding of their health through the use of several wearable devices and in-home monitoring systems. Continuous surveillance is a key component of RPM because this helps identify patterns and irregularities in health data over time. The final result of this is improved patient outcomes because it facilitates proactive treatment of chronic illnesses in addition to making early intervention easier [87]. Healthcare providers receive real-time information over secure routes to centralized systems, which in turn fosters a dynamic and adaptive healthcare ecosystem. When a patient's status changes drastically, RPM systems with built-in alarm mechanisms alert medical workers, which is why they are vital. Proactive therapy can be performed quickly, minimizing the incidence of problems and potentially avoiding readmissions to the hospital. The active engagement of patients in the process of healthcare promoted by RPM fosters a sense of responsibility along with responsibility [88]. Patients can obtain details about their health, consult with doctors of their choice, and benefit from educational tools to foster a collaborative and knowledgeable approach to healthcare. Remote patient monitoring (RPM) is an innovative healthcare technology that provides an emphasis on patient care and real-time monitoring [89]. Strong security and privacy protocols and collaboration with existing systems such as electronic health record (EHR) and regulatory compliance are also required. RPM is an important component in long-term illness treatment because it can improve healthcare for patients and well-being as technology advances. Figure 16-8 shows an overview of RPM in smart cities.

CHAPTER 16   NAVIGATING THE URBAN LANDSCAPE: A COMPREHENSIVE REVIEW OF IOT AND UAV INTEGRATION IN SMART CITIES

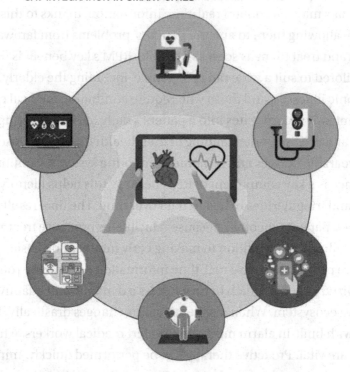

Figure 16-8. *An Overview of Remote Patient Monitoring*

The use of Internet of Things (IoT) technology for remote patient monitoring, or RPM, provides various benefits that contribute to better patient experiences, enhanced efficiency, and improved healthcare outcomes. Figure 16-9 depicts a few key advantages.

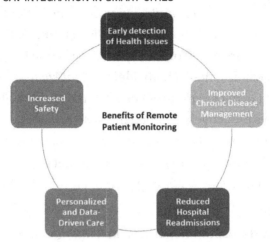

Figure 16-9. The Applications of Remote Patient Monitoring

(a) **Early Detection of Health Issues:** The Internet of Things (IoT) enables remote patient monitoring (RPM), which allows for continuous, real-time monitoring of vital signs and other health data. This capability makes it easy to recognize deviations from normal ranges and to take appropriate action as soon as possible. RPM via IoT plays a critical role in preventing the progression of illnesses, hence lowering the risk of complications and supporting more efficient and proactive healthcare management by identifying possible health issues at their early stages [90].

(b) **Improved Chronic Disease Management:** Those who suffer from long-term conditions like diabetes, hypertension, or heart issues might benefit greatly from routinely monitoring their health indicators. Devices with Internet of Things capabilities are necessary to give a complete picture of the patient's condition. By developing personalized treatment plans that are based on each patient's unique needs and reactions, medical providers can utilize this vast amount of data to more effectively manage patients' illnesses proactively. For patients with chronic illnesses, timely interventions and continuous monitoring made possible by the Internet of Things can enhance care quality, quality of life, and health outcomes [91].

(c) **Reduced Hospital Readmissions:** Maintaining patients under continuous remote monitoring is an important method for reducing hospital readmissions. By identifying possible health problems before they worsen, healthcare providers can enhance patient outcomes and reduce the need for follow-up hospital stays. Along with increasing people's overall health, this has a major positive impact on the healthcare system's finances. Remote monitoring effectively safeguards patient safety and optimizes medical resource execution, as indicated by an overall reduction in hospital readmissions [92].

(d) **Personalized and Data-Driven Care:** Because of the vast amounts of data collected by IoT-enabled devices, a new era of individualized integrated data-driven healthcare has come. Healthcare workers have access to a multitude of data, allowing them to examine minute details and find developments and patterns in medical data about patients [93]. This degree of understanding enables treatment programs to be tailored to each patient's unique needs and reactions. Healthcare professionals may enhance patient care and promote better health outcomes by utilizing a data-driven approach that allows for more focused treatments and a personalized, patient-centered healthcare experience [94].

(e) **Increased Safety:** Continuous monitoring, the cornerstone of remote patient monitoring via the Internet of Things, significantly enhances patient safety. Watchful surveillance allows for the prompt identification of any safety concerns, such as falls and abnormalities in vital signs; prompt communication is ensured by the integration of alerts and notifications, which notify carers or healthcare personnel in the event of an emergency. The quick reaction system underscores the significance of IoT-enabled monitoring in ensuring the well-being of patients in care by lowering the likelihood of unfavorable events and establishing a proactive approach to patient safety [95].

Smart Hospitals and Healthcare Facilities

The application of cutting-edge technologies by smart hospitals and healthcare facilities to enhance patient care, operational efficacy, and efficiency represents a paradigm shift in the delivery of healthcare services. These facilities use an extensive array of networked devices and technologies that comprise the Internet of Things (IoT) to provide a seamless and seamless healthcare environment. Smart hospitals may optimize resource allocation, enhance communication, and streamline operations by utilizing wearable technology, remote monitoring of patients, electronic health records (EHRs), and telehealth solutions. Automation frees up staff time to concentrate on patient care by reducing administrative duties. Healthcare professionals may make prompt, well-informed judgments with the help of real-time data analytics. By providing individualized care and encouraging patient involvement, smart technology integration also improves patient experiences. Examples of features found in smart hospitals are RFID tagging, smart beds, and state-of-the-art imaging technologies. These hospitals embrace innovation to support improved patient outcomes, operational efficiency, and a patient-centric model of care [96].

Public Healthcare Surveillance

A thorough and methodical approach to tracking, gathering, evaluating, interpreting, and sharing health-related data to prevent and control diseases within a population is known as public healthcare surveillance. This monitoring system plays a vital role in the infrastructure of public health by revealing information about community health, spotting new dangers, and directing evidence-based responses. Data for public health surveillance is collected from a range of sources, including laboratories and physician reports, electronic health records, as well as social media monitoring, using both modern and traditional technology. By reviewing the collected data, health professionals can discover patterns, trends, and prospective outbreaks and take quick action to prevent disease spread. Determining the risk variables related to chronic illnesses, monitoring vaccine coverage, and monitoring infectious disease outbreaks are just a few examples. As technology evolves, the incorporation of machine learning and artificial intelligence into surveillance systems improves their speed and accuracy. Monitoring public health is critical for assuring a proactive and responsive response to emerging health concerns at the population level, protecting community health, and guiding public health policies.

CHAPTER 16   NAVIGATING THE URBAN LANDSCAPE: A COMPREHENSIVE REVIEW OF IOT AND UAV INTEGRATION IN SMART CITIES

IoT Challenges for Smart City

The digitalization of every aspect of our lives is the promise of the Internet of Things. For shrewd cities, this digitalization trend means that more and more sensing nodes will be placed across every aspect of the city's operational mechanism. With this wide application reach, the invention and the subsequent installation of IoT technologies in smart cities present significant obstacles that must be taken into account. We address the difficulties that IoT systems face in this section deployment challenges that designers encounter with apps for smart cities. In this essay, we highlight the technical obstacles that come with using IoT in smart cities, which have been shown in Figure 16-10, and illustrate the several challenges that smart city IoT system adoption faces, including big data analytics, networking, security and privacy, and smart sensors.

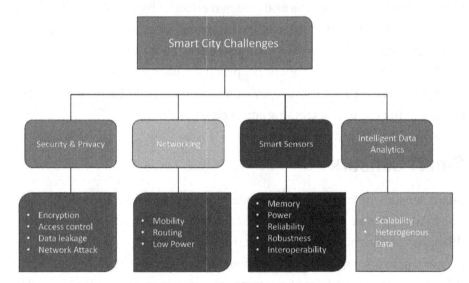

Figure 16-10. *Challenges Faced by Smart Cities*

Security and Privacy

Security and privacy are the main concerns of smart cities, which are places with online connections to vital city infrastructures. A disruption in the operation of city services may put lives and property in danger, in addition to causing discomfort to the local population. The current political scene, which is marked by cybercrime and warfare approaches in international relations, puts smart cities at greater risk of damaging attacks. When sensitive data is being transferred over a network, data encryption becomes crucial for its protection. Encouraging individuals to trust and engage with smart city initiatives is crucial to their success. Because smart cities use a lot of sensors, there are concerns that people's private behaviors could end up being made public. Companies utilizing the Internet of Things (IoT) network might improperly utilize citizen data for things like eavesdropping and targeted advertising. To solve these problems, solutions that protect the contextual integrity of measured tasks while anonymizing data collection are needed. These solutions need to carefully balance the needs for appropriate decision-making, security, and privacy. This paper thoroughly examines the subtleties of security and privacy in the context of creating smart cities.

Smart Sensors

Smart sensors are crucial for data collection in smart cities; however, due to their diverse origins, there are interoperability problems because different vendors have varying standards for measurements, data formats, networking protocols, and sensing processes. To get around this, free protocols and data formats that encourage communication and interoperability across devices made by various manufacturers must be created and used. Standard Internet of Things access point nodes offer an additional means of enhancing interoperability because they support several communication protocols [97]. Smart sensors do have reliability and robustness issues, though. IoT systems, which will serve as the foundation for smart cities in the future, must function accurately and consistently to provide customers with a positive experience. Smart grids with self-healing properties are two examples of essential services that ought to have decentralized systems to enhance resilience. Effective compression algorithms, low-power data transmission methods, and inventive memory and storage solutions are also needed because smart city sensors are often mobile and battery-operated. Innovations in battery technologies and the study of energy harvesting devices are needed to address power constraints and increase service availability in smart cities and IoT ecosystems. The continuous innovation is crucial in addressing the technical challenges inherent in the dynamic nature of smart city development [98].

Networking

The smooth interaction between sensing devices and the cloud is essential to the Internet of Things (IoT) effectiveness. The difficulty in developing strong networking solutions to guarantee these devices' constant connectivity comes with the emergence of new applications for smart cities. When it comes to meeting the unique requirements of smart city components—especially those with mobility and high data throughput requirements—current networking techniques are frequently less than ideal [99]. Many strategies, such as the definition of access points and local networks, have been put forth to provide a service of a reasonable caliber. In addition, overcoming the networking obstacle calls for the creation of effective and flexible routing protocols that can meet IoT needs, particularly for mobile devices. The diversified and dynamic nature of IoT devices in smart cities makes many existing protocols unable to serve them adequately, which has prompted researchers to look for creative ways to improve connectivity and communication in this changing environment.

Big Data Analytics

IoT devices generated 13.6 zettabytes of data in 2018 and are expected to grow to 79.4 zettabytes by 2025 [100]. To improve services in smart cities, new data analytics algorithms are needed to analyze varying data types, including structured and unstructured. Deep learning is an area of interest as it can leverage large amounts of data for various applications. Scalability is crucial for these algorithms, as they can be used across various applications. Concept drift is another concern, as data properties may change over time. Explainability is another crucial factor for smart city analytics, particularly in smart health. Hybrid deep learning classifiers and semantic web technologies are being explored for flood monitoring and COVID-19 care.

Conclusion

In conclusion, this paper provides a comprehensive overview of the role of IoT in urban environments. It explores the evolution of smart cities, IoT integration, and its impact on various aspects of urban infrastructure, including transportation and healthcare. The paper also addresses challenges related to security, privacy, and interoperability in implementing IoT in smart cities. The paper serves as a valuable resource for researchers, policymakers, and practitioners, highlighting the complex interactions between technology, infrastructure, and urban well-being. The scope also includes offering insightful analyses to scholars, decision-makers, and professionals engaged in the continuous creation and improvement of smart urban settings.

References

[1] Al-Turjman, Fadi, Hadi Zahmatkesh, and Ramiz Shahroze. "An overview of security and privacy in smart cities' IoT communications." *Transactions on Emerging Telecommunications Technologies* 33, no. 3 (2022): e3677.

[2] Cirillo, Flavio, David Gómez, Luis Diez, Ignacio Elicegui Maestro, Thomas Barrie Juel Gilbert, and Reza Akhavan. "Smart City IoT Services Creation Through Large-Scale Collaboration." *IEEE Internet of Things Journal* 7, no. 6 (2020): 5267-5275.

[3] Badii, Claudio, Pierfrancesco Bellini, Angelo Difino, and Paolo Nesi. "Smart City IoT Platform Respecting GDPR Privacy and Security Aspects." *IEEE Access* 8 (2020): 23601-23623.

[4] Bellini, Pierfrancesco, Paolo Nesi, and Gianni Pantaleo. "IoT-Enabled Smart Cities: A Review of Concepts, Frameworks, and Key Technologies." *Applied Sciences* 12, no. 3 (2022): 1607.

[5] Rejeb, Abderahman, Karim Rejeb, Steve Simske, Horst Treiblmaier, and Suhaiza Zailani. "The big picture on the internet of things and the smart city: a review of what we know and what we need to know." *Internet of Things* 19 (2022): 100565.

[6] Nagarajan, Senthil Murugan, Ganesh Gopal Deverajan, Puspita Chatterjee, Waleed Alnumay, and V. Muthukumaran. "Integration of IoT based routing process for food supply chain management in sustainable smart cities." *Sustainable Cities and Society* 76 (2022): 103448.

[7] Whaiduzzaman, Md, Alistair Barros, Moumita Chanda, Supti Barman, Tania Sultana, Md Sazzadur Rahman, Shanto Roy, and Colin Fidge. "A Review of Emerging Technologies for IoT-Based Smart Cities." *Sensors* 22, no. 23 (2022): 9271.

[8] Oroos Arshi and Aryan Chaudhary (2023) "Fortifying the Internet of Things: A Comprehensive Security Review", *EAI Endorsed Transactions on Internet of Things*, 9(4), p. e1. doi: 10.4108/eetiot.v9i4.3618.

[9] Ali, Zulfiqar, Azhar Mahmood, Shaheen Khatoon, Wajdi Alhakami, Syed Sajid Ullah, Jawaid Iqbal, and Saddam Hussain. "A Generic Internet of Things (IoT) Middleware for Smart City Applications." *Sustainability* 15, no. 1 (2023): 743.

[10] Almalki, Faris A., Saeed H. Alsamhi, Radhya Sahal, Jahan Hassan, Ammar Hawbani, N. S. Rajput, Abdu Saif, Jeff Morgan, and John Breslin. "Green IoT for Eco-Friendly and Sustainable Smart Cities: Future Directions and Opportunities." *Mobile Networks and Applications* 28, no. 1 (2023): 178-202.

[11] Rai, Hari Mohan, Aditya Pal, Sandeep Mishra, and Kaustubh Kumar Shukla. "Use of Internet of Things in the context of execution of smart city applications: a review." *Discover Internet of Things* 3, no. 1 (2023): 8.

[12] Khang, Alex, Shashi Kant Gupta, Sita Rani, and Dimitrios A. Karras, eds. Smart Cities: IoT Technologies, Big Data Solutions, Cloud Platforms, and Cybersecurity Techniques. CRC Press, 2023.

[13] Basford, Philip J., Florentin MJ Bulot, Mihaela Apetroaie-Cristea, Simon J. Cox, and Steven J. Ossont. "LoRaWAN for Smart City IoT Deployments: A Long Term Evaluation." *Sensors* 20, no. 3 (2020): 648.

[14] Jeong, Young-Sik, and Jong Hyuk Park. "IoT and Smart City Technology: Challenges, Opportunities, and Solutions." *Journal of Information Processing Systems* 15, no. 2 (2019): 233-238.

[15] Wu, Huaming, Ziru Zhang, Chang Guan, Katinka Wolter, and Minxian Xu. "Collaborate Edge and Cloud Computing With Distributed Deep Learning for Smart City Internet of Things." *IEEE Internet of Things Journal* 7, no. 9 (2020): 8099-8110.

[16] Lin, Chuan, Guangjie Han, Jiaxin Du, Tiantian Xu, Lei Shu, and Zhihan Lv. "Spatiotemporal Congestion-Aware Path Planning Toward Intelligent Transportation Systems in Software-Defined Smart City IoT." *IEEE Internet of Things Journal* 7, no. 9 (2020): 8012-8024.

[17] Li, Xiaoming, Hao Liu, Weixi Wang, Ye Zheng, Haibin Lv, and Zhihan Lv. "Big data analysis of the Internet of Things in the digital twins of smart city based on deep learning." *Future Generation Computer Systems* 128 (2022): 167-177.

[18] Chen, Zhong, C. B. Sivaparthipan, and BalaAnand Muthu. "IoT based smart and intelligent smart city energy optimization." *Sustainable Energy Technologies and Assessments* 49 (2022): 101724.

[19]   Hassan, Rondik J., Subhi RM Zeebaree, Siddeeq Y. Ameen, Shakir Fattah Kak, Mohammed AM Sadeeq, Zainab Salih Ageed, Adel AL-Zebari, and Azar Abid Salih. "State of Art Survey for IoT Effects on Smart City Technology: Challenges, Opportunities, and Solutions." *Asian Journal of Research in Computer Science* 8, no. 3 (2021): 32-48.

[20]   Zhang, Yang, Zehui Xiong, Dusit Niyato, Ping Wang, and Zhu Han. "Information Trading in Internet of Things for Smart Cities: A Market-Oriented Analysis." *IEEE Network* 34, no. 1 (2020): 122-129.

[21]   Alam, Tanweer. "Cloud-Based IoT Applications and Their Roles in Smart Cities." *Smart Cities* 4, no. 3 (2021): 1196-1219.

[22]   Ghazal, Taher M., Mohammad Kamrul Hasan, Muhammad Turki Alshurideh, Haitham M. Alzoubi, Munir Ahmad, Syed Shehryar Akbar, Barween Al Kurdi, and Iman A. Akour. "IoT for Smart Cities: Machine Learning Approaches in Smart Healthcare—A Review." *Future Internet* 13, no. 8 (2021): 218.

[23]   Almalki, Faris A., Saeed H. Alsamhi, Radhya Sahal, Jahan Hassan, Ammar Hawbani, N. S. Rajput, Abdu Saif, Jeff Morgan, and John Breslin. "Green IoT for Eco-Friendly and Sustainable Smart Cities: Future Directions and Opportunities." *Mobile Networks and Applications* 28, no. 1 (2023): 178-202.

[24]   Neelakandan, S. B. M. A. T. S. D. V. B. B. I., M. A. Berlin, Sandesh Tripathi, V. Brindha Devi, Indu Bhardwaj, and N. Arulkumar. "IoT-based traffic prediction and traffic signal control system for smart city." *Soft Computing* 25, no. 18 (2021): 12241-12248.

[25] Singh, Saurabh, Pradip Kumar Sharma, Byungun Yoon, Mohammad Shojafar, Gi Hwan Cho, and In-Ho Ra. "Convergence of blockchain and artificial intelligence in IoT network for the sustainable smart city." *Sustainable cities and society* 63 (2020): 102364.

[26] Santos, Rodrigo, Gabriel Eggly, Julián Gutierrez, and Carlos I. Chesñevar. "Extending the IoT-Stream Model with a Taxonomy for Sensors in Sustainable Smart Cities." *Sustainability* 15, no. 8 (2023): 6594.

[27] Hameed, Aroosa, John Violos, and Aris Leivadeas. "A Deep Learning Approach for IoT Traffic Multi-Classification in a Smart-City Scenario." *IEEE Access* 10 (2022): 21193-21210.

[28] Siddiqui, Shahbaz, Sufian Hameed, Syed Attique Shah, Abdul Kareem Khan, and Adel Aneiba. "Smart contract-based security architecture for collaborative services in municipal smart cities." *Journal of Systems Architecture* 135 (2023): 102802.

[29] Bedi, Pradeep, S. B. Goyal, Anand Singh Rajawat, Rabindra Nath Shaw, and Ankush Ghosh. "Application of AI/IoT for Smart Renewable Energy Management in Smart Cities." *AI and IoT for Smart City Applications* (2022): 115-138.

[30] Ibrahim, Amin S., Khaled Y. Youssef, Ahmed H. Eldeeb, Mohamed Abouelatta, and Hesham Kamel. "Adaptive aggregation based IoT traffic patterns for optimizing smart city network performance." *Alexandria Engineering Journal* 61, no. 12 (2022): 9553-9568.

[31] Singh, Jagendra, Mohammad Sajid, Suneet Kumar Gupta, and Raza Abbas Haidri. "Artificial Intelligence and Blockchain Technologies for Smart City." *Intelligent Green Technologies for Sustainable Smart Cities* (2022): 317-330.

[32] Khan, Jawad, Muhammad Amir Khan, N. Z. Jhanjhi, Mamoona Humayun, and Abdullah Alourani. "Smart-City-based Data Fusion Algorithm for Internet of Things." *Computers, Materials & Continua* 73, no. 2 (2022).

[33] Rajawat, Anand Singh, Pradeep Bedi, S. B. Goyal, Rabindra Nath Shaw, and Ankush Ghosh. "Reliability Analysis in Cyber-Physical System Using Deep Learning for Smart Cities Industrial IoT Network Node." *AI and IoT for Smart City Applications* (2022): 157-169.

[34] D'Ortona, Cristian, Daniele Tarchi, and Carla Raffaelli. "Open-Source MQTT-Based End-to-End IoT System for Smart City Scenarios." *Future Internet* 14, no. 2 (2022): 57.

[35] Rao, P. Muralidhara, and B. D. Deebak. "Security and privacy issues in smart cities/industries: technologies, applications, and challenges." *Journal of Ambient Intelligence and Humanized Computing* (2022): 1-37.

[36] Tekouabou, Stéphane Cédric Koumetio, Walid Cherif, and Hassan Silkan. "Improving parking availability prediction in smart cities with IoT and ensemble-based model." *Journal of King Saud University-Computer and Information Sciences* 34, no. 3 (2022): 687-697.

[37] Ghazal, Taher M., Mohammad Kamrul Hasan, Haitham M. Alzoubi, Muhammad Alshurideh, Munir Ahmad, and Syed Shehryar Akbar. "Internet of Things Connected Wireless Sensor Networks for Smart Cities." In *The Effect of Information Technology on Business and Marketing Intelligence Systems*, pp. 1953-1968. Cham: Springer International Publishing, 2023.

[38] Setiawan, Roy, Maria Manuel Vianny Devadass, Regin Rajan, Dilip Kumar Sharma, Ngangbam Phalguni Singh, K. Amarendra, Rama Koteswara Rao Ganga, Ramkumar Raja Manoharan, V. Subramaniyaswamy, and Sudhakar Sengan. "IoT Based Virtual E-Learning System for Sustainable Development of Smart Cities." *Journal of Grid Computing* 20, no. 3 (2022): 24.

[39] Ahmed, Imran, Yulan Zhang, Gwanggil Jeon, Wenmin Lin, Mohammad R. Khosravi, and Lianyong Qi. "A blockchain- and artificial intelligence-enabled smart IoT framework for a sustainable city." *International Journal of Intelligent Systems* 37, no. 9 (2022): 6493-6507.

[40] Fadda, Mauro, Matteo Anedda, Roberto Girau, Giovanni Pau, and Daniele D. Giusto. "A Social Internet of Things Smart City Solution for Traffic and Pollution Monitoring in Cagliari." *IEEE Internet of Things Journal* 10, no. 3 (2022): 2373-2390.

[41] Singh, Shailendra Pratap, Wattana Viriyasitavat, Sapna Juneja, Hani Alshahrani, Asadullah Shaikh, Gaurav Dhiman, Aman Singh, and Amandeep Kaur. "Dual adaption based evolutionary algorithm for optimized the smart healthcare communication service of the Internet of Things in smart city." *Physical Communication* 55 (2022): 101893.

[42] Alsubaei, Faisal S., Fahd N. Al-Wesabi, and Anwer Mustafa Hilal. "Deep Learning-Based Small Object Detection and Classification Model for Garbage Waste Management in Smart Cities and IoT Environment." *Applied Sciences* 12, no. 5 (2022): 2281.

[43] Dogan, Onur, and Omer Faruk Gurcan. "Applications of Big Data and Green IoT-Enabling Technologies for Smart Cities." In *Research anthology on big data analytics, architectures, and applications*, pp. 1090-1109. IGI Global, 2022.

[44] Dashkevych, Oleg, and Boris A. Portnov. "Criteria for Smart City Identification: A Systematic Literature Review." *Sustainability* 14, no. 8 (2022): 4448.

[45] Twahirwa, Evariste, James Rwigema, and Raja Datta. "Design and Deployment of Vehicular Internet of Things for Smart City Applications." *Sustainability* 14, no. 1 (2022): 176.

[46] Kumar, Priyan Malarvizhi, Bharat Rawal, and Jiechao Gao. "Blockchain-enabled Privacy Preserving of IoT Data for Sustainable Smart Cities using Machine Learning." In *2022 14th International Conference on COMmunication Systems & NETworkS (COMSNETS)*, pp. 1-6. IEEE, 2022.

[47] He, Xing, Qian Ai, Jingbo Wang, Fei Tao, Bo Pan, Robert Qiu, and Bo Yang. "Situation Awareness of Energy Internet of Thing in Smart City Based on Digital Twin: From Digitization to Informatization." *IEEE Internet of Things Journal* (2022).

[48] Jee, Anand, and Shankar Prakriya. "Performance of Energy and Spectrally Efficient AF Relay-Aided Incremental CDRT NOMA Based IoT Network with Imperfect SIC for Smart Cities." *IEEE Internet of Things Journal* (2022).

[49] Kuo, Yong-Hong, Janny MY Leung, and Yimo Yan. "Public transport for smart cities: Recent innovations and future challenges." *European Journal of Operational Research* 306, no. 3 (2023): 1001-1026.

[50] Sharma, Rohit, and Rajeev Arya. "UAV based long range environment monitoring system with Industry 5.0 perspectives for smart city infrastructure." *Computers & Industrial Engineering* 168 (2022): 108066.

[51] Menon, Varun G., Sunil Jacob, Saira Joseph, Paramjit Sehdev, Mohammad R. Khosravi, and Fadi Al-Turjman. "An IoT-enabled intelligent automobile system for smart cities." *Internet of Things* 18 (2022): 100213.

[52] Bhardwaj, Kartik Krishna, Siddhant Banyal, Deepak Kumar Sharma, and Waleed Al-Numay. "Internet of things based smart city design using fog computing and fuzzy logic." *Sustainable Cities and Society* 79 (2022): 103712.

[53] Arshi, O., Mondal, S. Advancements in sensors and actuators technologies for smart cities: a comprehensive review. *Smart Constr. Sustain. Cities* 1, 18 (2023). https://doi.org/10.1007/s44268-023-00022-2

[54] Aditya, Amara, Shahina Anwarul, Rohit Tanwar, and Sri Krishna Vamsi Koneru. "An IoT assisted Intelligent Parking System (IPS) for Smart Cities." *Procedia Computer Science* 218 (2023): 1045-1054.

[55] Chourabi H, Nam T, Walker S, Gil-Garcia JR, Mellouli S, Nahon K, Pardo TA, Scholl HJ (2012) Understanding Smart Cities: An Integrative Framework. In: 2012 45th Hawaii international conference on system sciences, IEEE, pp 2289-2297

[56] Allam Z, Newman P (2018) Redefining the Smart City: Culture, Metabolism and Governance. Smart Cities 1(1):4-25

[57] Neirotti P, De Marco A, Cagliano AC, Mangano G, Scorrano F (2014) Current trends in Smart City initiatives: Some stylised facts. Cities 38:25-36

[58] Nam T, Pardo TA (2011) Conceptualizing smart city with dimensions of technology, people, and institutions. In: Proceedings of the 12th annual international digital government research conference: digital government innovation in challenging times, pp 282-291

[59] Ullah F, Sepasgozar SM, Wang C (2018) A Systematic Review of Smart Real Estate Technology: Drivers of, and Barriers to, the Use of Digital Disruptive Technologies and Online Platforms. Sustainability 10(9):3142

[60] Felli F, Liu C, Ullah F, Sepasgozar S (2018) Implementation of 360 videos and mobile laser measurement technologies for immersive visualisation of real estate & properties. In: Proceedings of the 42nd AUBEA Conference

[61] Ullah F, Sepasgozar S (2019) A study of information technology adoption for real-estate management: A system dynamic model. Innov Prod Constr Transf Constr Through Emerg Technol pp 469-484

[62] Ullah F, Samad Sepasgozar P, Ali TH (2019) Real estate Stakeholders Technology Acceptance Model (RESTAM): User-Focused Big9 Disruptive Technologies for Smart Real Estate Management. In: Proceedings of the 2nd International Conference on Sustainable Development in Civil Engineering (ICSDC 2019), Jamshoro, Pakistan. pp 25-27

[63] Ullah F, Sepasgozar SM (2020) Key Factors Influencing Purchase or Rent Decisions in Smart Real Estate Investments: A System Dynamics Approach Using Online Forum Thread Data. Sustainability 12(11):4382

[64] Nowicka K (2014) Smart City Logistics on Cloud Computing Model. Procedia-Social Behav. Sci. 151(Supplement C):266-281

[65] Zygiaris S (2013) Smart City Reference Model: Assisting Planners to Conceptualize the Building of Smart City Innovation Ecosystems. J. Knowl Econ 4(2):217-231

[66] Sepasgozar SM, Hawken S, Sargolzaei S, Foroozanfa M (2019) Implementing citizen centric technology in developing smart cities: A model for predicting the acceptance of urban technologies. Technol Forecast Soc Change 142:105-116

[67] Munawar HS, Qayyum S, Ullah F, Sepasgozar S (2020) Big Data and Its Applications in Smart Real Estate and the Disaster Management Life Cycle: A Systematic Analysis. Big Data Cogn Comput 4(2):4

[68] Narayanan A, Bonneau J, Felten E, Miller A, Goldfeder S (2016) Bitcoin and Cryptocurrency Technologies: A Comprehensive Introduction. Princeton University Press, New Jersey

[69] Ciaian P, Rajcaniova M (2018) Virtual relationships: Short-and long-run evidence from BitCoin and altcoin markets. J Int Financ Markets Inst Money 52:173–195

[70] Quest M (2018) Cryptocurrency 101: Your Guide to Understanding How to Trade Bitcoin, Altcoin, and Other Online Currencies.

[71] Dewan S, Singh L (2020) Use of blockchain in designing smart city. Smart and Sustainable Built Environment

[72] Seigneur J-M, Pusterla S (2020) Socquet-Clerc X Blockchain real estate relational value survey. In: Proceedings of the 35th annual ACM symposium on applied computing. pp 279-285

[73] Veuger J (2020) Dutch blockchain, real estate and land registration. J Prop Plan Environ Law

[74] Kanak A, Ugur N, Ergun S (2019) A Visionary Model on Blockchain-based Accountability for Secure and Collaborative Digital Twin Environments. In: 2019 IEEE international conference on systems, man and cybernetics (SMC), IEEE, pp 3512-3517

[75] Sun M, Zhang J (2020) Research on the application of block chain big data platform in the construction of new smart city for low carbon emission and green environment. Comput Commun 149:332–342

[76] Karamitsos I, Papadaki M, Al Barghuthi NB (2018) Design of the Blockchain Smart Contract: A Use Case for Real Estate. J Inform Security 9(3):177–190

[77] Leiding B, Memarmoshrefi P, Hogrefe D (2016) Self-managed and blockchain-based vehicular ad-hoc networks. In: Proceedings of the 2016 ACM international joint conference on pervasive and ubiquitous computing: adjunct. pp 137-140

[78] Liu S (2020) Blockchain - statistics & facts. Statista.com. https://www.statista.com/topics/5122/blockchain/. Accessed 12 October 2020

[79] Zīle K, Strazdiņa R (2018) Blockchain Use Cases and Their Feasibility. Applied Computer Systems 23(1):12–20

[80] Nakamoto S (2008) Bitcoin: A Peer-to-Peer Electronic Cash System.

[81] Yli-Huumo J, Ko D, Choi S, Park S, Smolander K (2016) Where Is Current Research on Blockchain Technology?—A Systematic Review. PloS one 11(10):e0163477

[82] Linoy S, Stakhanova N, Ray S (2020) De-anonymizing Ethereum blockchain smart contracts through code attribution. Int J Netw Manag :e2130

[83] He D, Zhang Y, Wang D, Choo K-KR (2018) Secure and Efficient Two-Party Signing Protocol for the Identity-Based Signature Scheme in the IEEE P1363 Standard for Public Key Cryptography. IEEE transactions on dependable and secure computing

[84] Ying B, Nayak A (2019) Lightweight remote user authentication protocol for multi-server 5G networks using self-certified public key cryptography. J Netw Comput Appl 131:66–74

[85] Casino F, Dasaklis TK, Patsakis C (2019) A systematic literature review of blockchain-based applications: Current status, classification and open issues. Telemat Inform 36:55–81

[86] Amani S, Bégel M, Bortin M, Staples M (2018) Towards verifying ethereum smart contract bytecode in Isabelle/HOL. In: Proceedings of the 7th ACM SIGPLAN international conference on certified programs and proofs, pp 66-77

[87] Falazi G, Hahn M, Breitenbücher U, Leymann F, Yussupov V (2019) Process-Based Composition of Permissioned and Permissionless Blockchain Smart Contracts. In: 2019 IEEE 23rd international enterprise distributed object computing conference (EDOC). IEEE, pp 77-87

[88] Watanabe H, Fujimura S, Nakadaira A, Miyazaki Y, Akutsu A, Kishigami J (2016) Blockchain contract: Securing a blockchain applied to smart contracts. In: 2016 IEEE international conference on consumer electronics (ICCE). IEEE, pp 467-468

[89] Wright C (2017) Serguieva A Sustainable blockchain-enabled services: Smart contracts. In: 2017 IEEE international conference on big data (Big Data). IEEE, pp 4255-4264

[90] Sepasgozar S, Karimi R, Farahzadi L, Moezzi F, Shirowzhan S, Ebrahimzadeh M, S, Hui F, Aye L, (2020) A Systematic Content Review of Artificial Intelligence and the Internet of Things Applications in Smart Home. Appl Sci 10(9):3074

[91] Ali Q, Thaheem MJ, Ullah F, Sepasgozar SM (2020) The Performance Gap in Energy-Efficient Office Buildings: How the Occupants Can Help? Energies 13(6):1480

[92] Veuger J (2018) Trust in a viable real estate economy with disruption and blockchain. Facilities

[93] Spielman A (2016) Blockchain: Digitally Rebuilding the Real Estate Industry. Massachusetts Institute of Technology

[94] Li M, Shen L, Huang GQ (2019) Blockchain-enabled workflow operating system for logistics resources sharing in E-commerce logistics real estate service. Comput Indus Eng 135:950–969

[95] School SB, Group A (2019) Direct and indirect investments in Proptech firms by real estate companies worldwide in 2019, by type of technology. Statista.com. https://www.statista.com/statistics/1128676/cre-real-estate-investment-direct-indirect-proptech-firm-global/. Accessed 12 October 2020

[96] Hoffmann T (2019) Smart Contracts and Void Declarations of Intent. In: International conference on advanced information systems engineering. Springer, pp 168-175

[97] Ma F, Fu Y, Ren M, Wang M, Jiang Y, Zhang K, Li H, Shi X (2019) EVM*: From Offline Detection to Online Reinforcement for Ethereum Virtual Machine. In: 2019 IEEE 26th international conference on software analysis, evolution and reengineering (SANER), IEEE, pp 554-558

[98] Kolluri A, Nikolic I, Sergey I, Hobor A, Saxena P (2019) Exploiting the laws of order in smart contracts. In: Proceedings of the 28th ACM SIGSOFT international symposium on software testing and analysis. pp 363-373

[99] Molina-Jimenez C, Sfyrakis I, Solaiman E, Ng I, Wong MW, Chun A, Crowcroft J (2018) Implementation of Smart Contracts Using Hybrid Architectures with On and Off–Blockchain Components. In: 2018 IEEE 8th international symposium on cloud and service computing (SC2). IEEE, pp 83-90

[100] Kapsoulis N, Psychas A, Palaiokrassas G, Marinakis A, Litke A, Varvarigou T (2020) Know Your Customer (KYC) Implementation with Smart Contracts on a Privacy-Oriented Decentralized Architecture. Future Internet 12(2):41

CHAPTER 17

Advancement in ML Techniques and Applications for UAV Swarm Management

Yasir khan[1,*], Sabitha Banu A[2], Junaid Yousaf[3]

[1]Department of Science and Technology & Information Technology (ST&IT), New Delhi, India Email: imyasir.308@gmail.com
[2]Department of Cybersecurity, PSGR Krishnammal College for Women, Coimbatore, India sabithabanu@psgrkcw.ac.in
[3]Ghulam Ishaq Khan Institute of Engineering Science and Technology (GIKI), Khyber Pakhtunkhwa, Pakistan junaidyousaf432@gmail.com

Introduction

Machine learning (ML) deals with the development of self-learning computer programs that learn gradually from experience and data usage. In other words, machine learning is the ability of computers to perform certain tasks on input data without having to be directly programmed by user input for that specific task. ML includes a broad category of techniques including unsupervised learning, supervised learning, reinforcement learning, and semisupervised learning. Reinforcement learning (RL) is one of the most active and developing areas of ML that deals with the examination and handling of environmental issues and activities [1]. There has been enhanced use of ML in diverse sectors including economics, maths, engineering, psychology, neurosciences, and energy. In particular, machine learning in healthcare is used to forecast disease outbreaks and individual patient treatment regimens and improve the diagnostic accuracy of medical imaging. For instance, Google launched DeepMind Health to work with doctors to build machine learning that is useful in diagnosing diseases early and enhancing patients' health. In addition, the finance industry has benefited greatly from machine learning for credit scoring, algorithm trading, and fraud detection. According to another global poll [2], 56% of executives admitted the utilization of AI and ML in their financial crime compliance. Figure 17-1 depicts a comparison that has been made on different domains in AI. AI enables the creation of programs that can learn and reason like any human. Similarly, there is use in construction of programs which can learn on their own without being programmed to do so. Finally, deep learning (DL) is a branch in ML in which deep neural networks attempt to learn from large databases. Figure 17-1 presents a comparison between AI, ML, and DL.

CHAPTER 17 ADVANCEMENT IN ML TECHNIQUES AND APPLICATIONS FOR UAV SWARM MANAGEMENT

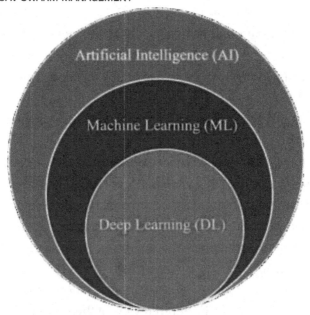

Figure 17-1. *Comparison of different AI domains*

Unmanned aerial vehicles (UAVs), also known as drones, are flying vehicles that do not require the services of an operator. UAVs are incorporated in a multitude of tasks and environments due to their mobility, flexibility, and the capability of varying their height [3]. The evolution of these contemporary modes of communication reflected by such features such as innovative wireless networks has further enriched the operability of unmanned aerial vehicle (UAV) systems. For example, as mentioned by Bithas et al. in [4], all these enhancements offer improved reliability, lower delay, and higher network throughput for UAVs. Thus, flexibility has increased in UAVs and can accomplish many other tasks today.

Machine learning has the potential of revolutionizing the use of drones and integration of solutions for problems that would have not been well dealt with using conventional methods. Another quite unique attribute of UAVs is their ability to effectively fly at low altitude. In this regard, UAVs can also be considered as user equipment (UEs) connected to a cellular network. This makes it possible to have numerous possible uses. They can use probe sensors for environmental checking. They are suitable for short distance air transports of items that need to be delivered quickly and with a lot of flexibility. UAVs similarly produce excellent results for real-time video streaming, which presents a live visual feed. Besides, they are ideal in performing surveillance activities that help in increasing security as well as monitoring. All these diverse functions point to the fact that UAV's are capable of serving multiple vital roles depending on the context. Also, UAVs can act as base stations (BS) and provide wireless communication on demand within specific coverage areas. In contrast to conventional BS, UAV BS can change the height, avoid buildings, and make LoS connections with ground users [5].

The rest of the book chapter is structured as follows: the "Literature Review" section provides an in-depth review of the existing literature on the role of ML in UAVs, detailing the objectives and methods used in each technique. In the "Classification of ML Techniques for UAVs" section, a holistic comparison is drawn between traditional and modern-day ML techniques used in UAVs, followed by the "Applications of Machine Learning for UAV Swarms" section which actually discusses how different techniques of ML can be applied in the field of UAVs. Moreover, the "Future Research Directions" section proposes certain areas which can be explored further through future research. Lastly, the "Conclusion" section concludes the chapter.

Literature Review

Self-sufficient systems are being developed in networks of UAVs through the utilization of ML algorithms that enable the network to perform efficiently irrespective of changes in environment conditions. Such UAV networks can alert each other, make decisions, learn from their past experiences, and adjust themselves autonomously without intervention. This makes the UAVs functional in executing activities like navigation, data gathering, and mission planning more efficiently and effectively. Sun et al. [6] describe historical development of ML in UAV communication systems and prospects for its use. Additionally, they consider how UAVs can be adopted in mobile networks as one and the same with airborne UE and BS while exposing the technology's potential in dealing with issues like increasing coverage or creating temporary hotspots. Table 17-1 provides a literature review on the applicability of ML methods in various UAV components.

Khan et al. [7] have developed an intrusion detection system (IDS) employing the Markov chain to calculate attack probability for the purpose of detecting flooding attacks in aerial networks that are used in UAVs. Further, this IDS employs the precise threshold for balancing false positive and false negative states with the application of the Markov binomial distribution and Markov chain stochastic models. Similarly, Khalil et al. [8] introduce a model for UAV swarm communication utilizing ML techniques for search and rescue operations. In relation to UAV communication, random forest regression has been used to model the received signal strength (RSS) and power loss. Likewise, Ullah et al. [9] explore advanced techniques to improve the accuracy of mobile robot localization. The authors leverage drones to enhance sensor fusion processes, combining various sensor data to achieve more precise positioning.

Table 17-1. Literature Review on the Role of ML in UAVs

Description	Objective	Methods	Reference
This article provides an overview of machine learning in aerial wireless networks	Enhancing UAVs in wireless networks using ML for optimal positioning, resource management, and interference control	Analyzed ML applications and discussed future research in UAV-aided wireless networks	[10]
The paper surveys ML methods for managing UAV flocks and addressing existing approaches and open issues	The paper discusses challenges in UAV flocks and surveys machine learning methods to address them	Various ML methods have been proposed in the literature to address these challenges	[11]
This paper utilized deep convolutional neural networks with batch normalization, ReLUs, and dropout for improved performance	This project uses the Blackbird Dataset to develop ML models for UAVs, emphasizing the need for real data in training and deployment	Unsupervised learning models partitioned flight tests, while DNN models enabled nonlinear dynamic inversion	[12]
This review discusses advanced sensors, communication technologies, computing platforms, and ML techniques in autonomous UAVs	UAVs are popular for their autonomous operation with sensors, software, and communication technologies	Study and presentation of key performance metrics and comparative study of various technologies	[13]

(*continued*)

Table 17-1. (continued)

Description	Objective	Methods	Reference
This research presents a machine learning approach combining NNs and LR for estimating air data parameters for a tiny fixed-wing UAV	In this paper, a machine learning approach to estimate air data parameters for a tiny fixed-wing UAV is proposed	Two ML algorithms based on NNs and LR are used to estimate air data parameters	[14]
This survey aims to offer a thorough overview of ML techniques applied in UAV operations and communications, highlighting growth opportunities and research gaps	We categorize current popular ML tools based on their applications to the four components and perform gap analyses	Conducted an analysis on different ML methods used in UAV swarms	[15]
This paper examines joint optimization issues in UAVs and explores the impact of AI, ML, DRL, MEC, and SDN on their operational efficiency	The authors developed a review to investigate the UAV joint optimization problems to enhance system efficiency	Developing a review to explore joint optimization challenges in UAVs. Classifying these challenges based on parameter count	[16]

(*continued*)

Table 17-1. (*continued*)

Description	Objective	Methods	Reference
The proposed design utilizes SVM as a machine learning algorithm for object detection and classification	A solar-powered UAV employs SVM for object detection and classification, utilizing IoT technology for real-time data collection and analysis	IoT technology is used to facilitate communication and coordination between the UAV and other devices	[17]
The paper outlines a preliminary system for UAV damage detection and classification using machine learning approaches	This paper introduces a preliminary UAV damage detection and classification system based on machine learning approaches, validated using data from a piezoelectric sensor network	The data from a piezoelectric sensor network undergoes analysis using independent component analysis and machine learning methods	[18]

Classification of ML Techniques for UAVs

Artificial intelligence (AI) is a vast arena of computer science which incorporates methods for letting machines learn from given inputs or experiences with the aim of enhancing their efficiency and to predict new observations. Recall that ML approaches are now deeply discussed and implemented in transport, finance, manufacturing, wireless communications, healthcare, retail, energy, and entertainment industries, among others. In general, ML is based on architecture of pattern recognition; the primary purpose of the ML is to learn about environment changes by identifying correlations between sets of data and utilizing effective previous activities [19]. In relation to problem solving and enhancement of operations in UAV communication, the following are other opportunities that could be pursue through ML. In the context of WN, ML offers high deterministic and intelligent WA for UAVs, contributing to the overall WN performance ranging from environment characteristics prediction. Also, it has been used as a fair tool for controlling different resources for communication with UAVs especially frequency resource allocation, spectrum control, and smart beam forming. Figure 17-2 showcases various ML techniques that can be applied to UAVs, showcasing their real-world applications.

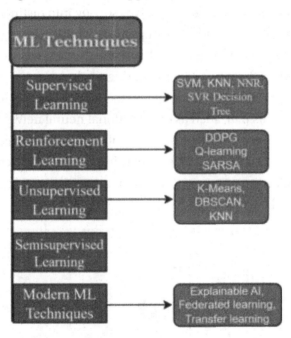

Figure 17-2. Categories of ML techniques

Supervised Learning

Supervised learning is a basic method in machine learning where models are trained using labeled data. This means the data includes both the inputs and the correct outputs. The goal is for the model to learn how to match the inputs to the outputs. Usually, the data is split into two parts: a training set, which the model uses to learn, and a test set, which is used to see how well the model makes predictions based on what it learned. For example, supervised learning can help identify different types of UAVs (drones), such as telling apart fixed-wing drones from rotary-wing drones, by training on data that clearly labels these categories. This method works well when you have a lot of data where both the features of the UAVs and their types are known. Supervised learning problems usually fall into two groups: regression and classification. Classifying UAVs, as mentioned, is a classification problem because it involves sorting into distinct categories. In classification problems, the model predicts which category an input belongs to. In regression problems, the model predicts a continuous outcome based on the input, like predicting a value that can range freely. Supervised learning can be combined with neural networks to create more versatile algorithms [20], such as convolutional neural networks (CNNs), recurrent neural networks (RNNs), and multilayer perceptron (MLP). This mix of supervised learning and deep learning has led to the creation of advanced applications. Figure 17-3 demonstrates the working mechanism of supervised ML algorithms.

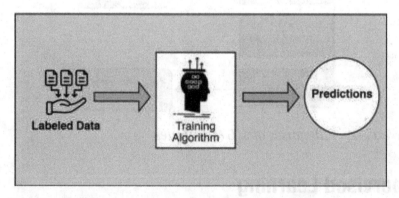

Figure 17-3. Working mechanism of supervised ML

CHAPTER 17 ADVANCEMENT IN ML TECHNIQUES AND APPLICATIONS FOR
UAV SWARM MANAGEMENT

Unsupervised Learning

Unsupervised learning does not presuppose the existence of an output vector that has been previously marked. This approach enables one deal with issues with little or nil information on what would be achieved. This is true even if an effect of a variable is determined; the organization of the data might still persist as a problem. Unsupervised learning is the process of partitioning the data variables and assigning them into categories with respect to the corresponding relationships and hierarchical structures that exist between them. We can then harness these patterns for decision-making and to predict actions. For instance, environment-based partitions in unsupervised learning help in identifying the UAV features to choose the correct UAV for certain conditions or in training UAVs to behave correctly in certain contexts as per the features of the environment. The main areas that unsupervised learning can solve include clustering of data, loss of several dimensions from the data, and generation of data. When it comes to UAV communication, many of these are commonly used applications that are widely used. The types are hard clustering in which data points are assigned to a single cluster, for instance, in K-means, and soft clustering that allows data points to belong to several clusters, for example, in the GMMs [21]. Figure 17-4 shows the typical functioning of unsupervised ML algorithms.

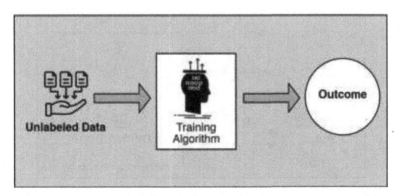

Figure 17-4. Functioning of unsupervised ML algorithm

Reinforcement Learning

Reinforcement learning (RL) is an ML methodology in which an agent is always in contact with the environment. An agent is an object that can perform an action or possesses certain characteristics of behavior. RL has an agent-based point of view of machine learning. The agent thus learns in order to attain its objective of performing a particular task optimally. Each attempt to finish the task is called an episode, which consists of a number of steps, observations, and rewards. RL can also involve an underlying model, which differentiates between two primary types of RL problems: there are two major approaches, which include the model-based and the model-free approaches. In model-based RL, a model is employed to predict how the environment will respond to the actions of the agent. The aim here is to forecast the subsequent states and actions given the current state and action known as hidden Markov models [22]. But in many cases, supervised learning can offer a better solution to these issues [23]. Model-free RL has become popular because it let the agent decide what action will give the most rewards without predicting the future state of environment. One of the most popular model-free approaches to RL is Q-learning, whereby the agent determines the action with the maximum Q-value, which reflects the maximum achievable reward [24]. Figure 17-5 represents a demonstration of reinforcement learning mechanism.

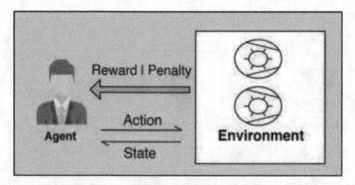

Figure 17-5. Reinforcement learning mechanism

CHAPTER 17 ADVANCEMENT IN ML TECHNIQUES AND APPLICATIONS FOR UAV SWARM MANAGEMENT

Semisupervised Learning

A unique kind of machine learning that falls between supervised and unsupervised learning models is called semisupervised learning. It is the technique where a model is trained using both a few labeled examples and a number of unlabeled examples. The aim is to create a function that will estimate the output variable based on the input variables, which is similar to supervised learning. The main difference is that while in this case, the algorithm is trained using both labeled and unlabeled data simultaneously. For example, text classification is used in categorizing a given text into one or more specific categories. In text classification, there is the use of semisupervised learning, where a set of labeled data can be used together with a large number of samples of texts that are unlabeled.

Role of Modern Machine Learning Techniques for UAV Swarms

Modern machine learning techniques and their application in control and coordination of UAV swarms are instrumental in improving the flow, synchronization, and learning capacities of the swarms. Functions like formation flight, negotiation of obstacles, and cooperative target chasing may be achieved by using all those machine learning methods for UAV swarms. UAVs can utilize advanced ML techniques like XAI, federated learning, and transfer learning to adapt to their environment, improve routes, and make immediate decisions. This does not only enhance their usage efficiency but also widens its uses starting from search and rescue operations, environmental concern probing, and military usages.

CHAPTER 17 ADVANCEMENT IN ML TECHNIQUES AND APPLICATIONS FOR UAV SWARM MANAGEMENT

Explainable Artificial Intelligence (XAI) in Drones

In most of the UAV operations, UAVs are programmed to operate and even make critical decisions independently with the help of advanced artificial intelligence systems and even communicate with other UAVs or ground control systems. Some of them are bound to fail, or act in a manner that is unpredictable, and this can have undesirable consequences – usually, such consequences should not be tolerated. Nevertheless, cognitive researchers have inadequate perception of the AI and ML processes to understand what the decision-making power is based upon or when the decisions made might contain mistakes. This instability in UAV reliability along with the possible threats complicates the observance of local legislation. XAI enhances the AI systems' ability to provide clear and understandable reasoning for their decisions and actions [25]. XAI contributes to the decisions and actions of UAVs being comprehensible to human operators in complex, dynamic, and multifaceted scenarios that may affect options. This increases confidence of the operators in the system and helps them to easily diagnose any problem [26]. This is an area of major improvement by XAI especially in drone navigation and collision avoidance due to the ability to provide basic explainability in autonomous decisions. The authors in [27] present a reactive controller by neural networks for small UAVs; this controller uses a low amount of the computational power to self-navigate in unknown outdoor terrains. The following is an example of this kind of controller, whose function is to solve a navigation problem formulated as a Markov decision process (MDP) through the use of deep reinforcement learning.

Federated Learning

A distributed machine learning method called federated learning (FL) enables simultaneous learning of the same model by many clients without the exchange of data between them. Popularized by Google in 2016 [28], FL works by applying ML algorithms in parallel across nodes while avoiding the centralization of training data. The features and architecture of FL are adequate for UAV communication, which is a client- and server-based system. FL addresses the need for data protection and limited on-board computation through its applicability to limited networks that cannot perform extensive on-board computations [29]. FL also enables UAVs to switch and work in tandem with each other. When a node is offline, the learning process can persist for a brief time, and UAVs can continue updates once they reconnect to the network. This ensures that the overall functionality of the system is not heavily compromised in the situations where a UAV requires recharging or experiences a sudden failure [30].

Transfer Learning

One of the invaluable characteristics of the neural networks is their ability to apply the learned function to particular input data for a different purpose, which is called transfer learning (TL). TL is a process of acquiring knowledge from one problem and using it in the solution of another related problem. It was established in the study [31] that layers closer to the input tend to learn features that are not specific to the task and the final output of the neural network. TL has been adopted in UAV technologies where the drones are able to use prior knowledge of pretrained models to enhance performance on new and diverse tasks. This method helps to reduce the reliance on large amounts of data and time spent on training, thus increasing the versatility of UAVs in various scenarios. TL is crucial in enabling UAVs to overcome obstructions, improve their interaction, and achieve real-world utility. As stated in [32], knowledge distillation is used to transfer knowledge from a teacher policy to a student agent in a different environment. First, a pretrained DDQN model is trained, but in doing this, emphasis is placed on various domains to enhance the transfer of knowledge from one environment to another. In [33], the authors investigate continuous learning in order to efficiently transfer knowledge from a pretrained stage to other contexts in order to reduce the amount of time required to retrain DRL models and improve the ability of such models to address constantly changing environments.

Applications of Machine Learning for UAV Swarms

Solutions based on ML for UAV swarms are changing many branches by optimization of its coordination, autonomy, and execution. Real time was referred in this context as to mean the ways by which ML algorithms allow UAV swarm to collectively perform coordinated choreographed flight, simultaneous real-time obstacle avoidance, and also on demand changes in mission type and planning. These applications are indispensable in environmental controlling as UAVs can move in large groups, gather data at the same time, and process it collectively. In security and surveillance, it is important that the usage of ML for UAV swarms is quite fast in offering protection against attacks and data falsification. Furthermore, in the fault diagnosis, UAV swarms with the integration of ML can identify the faults in UAVs which will help in early identification of damages. In general, the application of ML in UAV swarms paves the way to accomplish dynamic and intelligent operations in numerous fields. Figure 17-6 depicts the major areas of utilization of ML in UAV swarms.

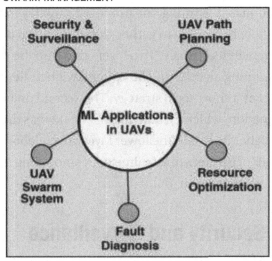

Figure 17-6. *Applications of ML in UAV systems*

UAV Path Planning Based on Machine Learning

The path planning is one of the most important topics of the guidance navigation and control (GNC) of UAV. This means that the aim of path planning is to find the end point of the best path that begins from the start point to the terminal point. Another crucial aspect of path planning is the avoidance of obstacles, meaning during the execution of the planned path, the UAV cannot run into the obstacles whether they are stationary or moving. But with the rapid advancements in machine learning (ML) and reinforcement learning (RL), UAV path planning techniques have entered a new era. ML techniques have been thoroughly researched and deployed in a vast number of applications in the autonomous agents and unmanned systems due to its capability of learning adaptation [34, 35].

The GNC system includes the path planning stage. In this paper, Cui and Wang [36] developed a multilayer path planning algorithm using RL technique for autonomous robots. Albeit being more complex than the conventional Q-learning, the proposed multilayer algorithm has a clear benefit of being able to gather global and local data which significantly enhances its efficacy. Two layers make up the suggested basic reinforcement learning algorithm. The upper layer handles local input and can be thought of as a short-term strategy. The second one interacts with the global information, which I would like to consider as the long-term approach. The higher layer and the lower layer are collaborating to find a collision-free path of the network. Online path smoothing is implemented using B-spline curve approach [36].

ML-Based Security and Surveillance in UAV Swarms

Challita et al. [37] proposed a machine learning scheme for sharing data among the UAV swarms during the attack. To overcome this problem, they propose the use of federated learning for UAV swarms which would go some way into alleviating the problems associated with the raw data transmission between UAVs, problems such as adversarial machine learning. Korki et al. [38] proposed the architecture of three prototypes for automating UAV to identify power lines in real time. These systems are designed to incorporate a sound and strong neural network (NN) and utilize efficient sensors, thereby capable of detecting different failures and defects on the power lines with considerable accuracy. Lodeiro-Santiago et al. [39] offered a new approach to identifying small boats using UAVs. In this system, UAVs send signals to a server where the image is processed using neural network, and for communication, LTE is used. To minimize the security risks of the proposed system, proper measures have been put in place to enhance the integrity of the data in the database.

CHAPTER 17 ADVANCEMENT IN ML TECHNIQUES AND APPLICATIONS FOR UAV SWARM MANAGEMENT

Resource and Network Optimization

Thus, while managing resources, planning networks, caching contents, and offering user associations in the UAV cellular networks, some of the important objectives to be considered are high throughput, low overhead, scalability in terms of the number of device support, and dynamic nature of the scenario. AI frameworks and ML have been utilized to enhance resource management systems by analyzing UAV integrity, network conditions, remaining battery power, and the surrounding environment. Park et al. in [40] wanted to predict success and failure rates of UAV networks through ML methods, linear regression, and SVM. The mobility of UAV connections makes their connections time varying, thus reducing success of transmission as the mode with wireless distance is taken into account. In this case, UAVs can use ML techniques to elicit real-world information about the likelihood of connecting with nodes that are in close proximity. This has been demonstrated in simulations indicating that if trained, logistic regression (LR) and support vector machines (SVM) can be trained with higher precision and within considerably shorter time due to the enhanced features of the latter.

Swarm UAV System

Swarm UAV systems have been used in the recent past because they are able to accomplish tasks that would be difficult or even impossible for one UAV to accomplish. These systems are typically composed of several UAVs that work together to fulfill a certain task or goal. It is for this reason that, as opposed to the use of a single UAV, there are so many advantages that come with the use of swarm UAV systems which include increased efficiency, flexibility, modularity, and reliability [41]. The swarm UAV base stations can be constructed with the capability to support different swarm formations and scale of the swarm, making it possible to support various scales of missions. It is especially useful in areas that would be hard to cover with a single UAV due to size constraints. Furthermore, swarm UAV systems can operate on a high level of sophistication to accomplish different tasks. The swarm UAV system primarily features two common organizational structures: swarm infrastructure-based architecture and FANET [42]. Infrastructure-based swarm architecture integrates ground-based infrastructures to enhance the ability of the swarm's interaction and cooperation. In this configuration, swarm UAVs are able to exchange information within themselves and the ground structures in order to smoothly function. The control infrastructure may usually consist of a base station, which is responsible for controlling and managing the UAVs. A FANET (Flying Ad-Hoc Network) is a network where UAVs are involved and the UAVs are fitted with routers for communication among them. In FANETs, UAVs act as other nodes through the sharing of information and operation among them. One of the major strengths of FANETs is their dynamic nature, which makes it possible to rapidly deploy FANETs, as well as easily modify their parameters based on current conditions in the field. Also, FANETs can be easily scaled as they can effectively manage a large number of nodes.

Fault Diagnosis in UAVs Using Machine Learning

Within a short time, various research studies have been conducted to address the application of machine learning methods in identifying fault in UAVs. These studies stress the need to verify the uniqueness of the acquired dataset, using signal processing or any other methods [43, 44]. Al-Haddad and Jaber [45] offered a systematic review on the application of machine learning approaches to fault diagnosis. Moreover, they argue that it is neural network methods that are used most often to diagnose faults in different components of UAVs, comprising 56%. Blades are one of the most vulnerable parts of the wind turbine, with more than a third of fault diagnosis studies conducted on them. The high usage rates of UAVs in these studies also suggest that drones are more frequently used and thus more likely to be destroyed [45].

Future Research Directions

Energy limitation is one of the most important research challenges that provide great improvements to UAV's longer endurance. Some of them concern progressive technologies for energy harvesting, storage, and transfer, for instance, wireless or laser energy transfer. Based on the findings of Fontanesi et al. [46], these technological developments are crucial for achieving longer flight times and hence improve the productivity of UAVs. Another essential facet, where machine learning system is central to an improved UAV power system, is assisting in the location of the charging stations and the proper organization of the charging activities. UAVs' versatility improves due to enhanced mission longevity and reduced idle time, enhancing their suitability for more reliable missions.

The path planning of the UAV has been mainly looked at in cases where the UAV is alone, and the environment is static. However, the future direction is to find methods for MT Blick multi-UAV trajectory planning in dynamic environment by means of online learning. This includes not only avoiding the obstacles, choosing the proper moment for their selection, and avoiding trajectory interferences between several UAVs. These processes require optimization techniques involving AI to increase the preciseness of the path planning. It will also be crucial for further expansion of potential capabilities and robustness of the UAV swarms operating in more dynamic and unpredictable scenarios.

NTN integration is the process of merging the UAV networks with the satellite networks to develop other large nonterrestrial networks which is a new development. These networks, together with the support of terrestrial and aerial components, provide higher communication capacity, more extensive coverage, and better networks' fault tolerance. Thus, the application of machine learning can become effective in the management and control of such complex networks where UAV swarms operate smoothly and efficiently within the proposed integrated structure [47]. Pursuing this known fact that amalgamation of approaches has the propensity to revolutionize the communication strategies especially in the LAMA areas or where the terrestrial infrastructure has been or is in the process of being degraded.

Conclusion

Thus, one can conclude that machine learning is being adopted and implemented in the various fields and is providing positive results. It is well known that they heavily contributed to the successes of so many domains. Nonetheless, it is necessary to find out that the application of machine learning is very important for UAV swarms. This paper presents a type of literature review study that seeks to map ML concerning enhancing various aspects of UAV swarms. First, several types of machine learning were described in order to stress that this branch of computer science is applicable to UAVs. Similarly, when explaining the enhancement of UAV swarms' general capability, efficiency, and capacity of work through the concept of ML, the detailed exposition was done. Besides that, the problems which can be encountered at the stage of applying machine learning into the UAV swarms are specified. Last but not the least, the last subtopic of this book chapter states several topics that requires more research and should be focused on. Machine learning also can influence greatly the UAV industry and its further enhancement of the UAVs in terms of the usage productivity and efficiency.

CHAPTER 17 ADVANCEMENT IN ML TECHNIQUES AND APPLICATIONS FOR UAV SWARM MANAGEMENT

References

[1] W. Shafik, S. M. Matinkhah and M. Ghasemzadeh,"Theoretical Understanding of Deep Learning in UAV Biomedical Engineering Technologies Analysis," *SNComputer Science*, vol. 1, no. 6, pp. 1–13, 2020.

[2] Security Newswire, "AI and machine learning have been added to financial crime programs", Url: https://www.securitymagazine.com/articles/99596-ai-and-machine-learning-have-been-added-to-financial-crime-programs, Accessed: 20 June, 2024.

[3] H. Ahmadi, K. Katzis, and M. Z. Shakir, "A Novel Airborne Self-Organising Architecture for 5G+ Networks," in 2017 IEEE 86th Vehicular Technology Conference (VTC-Fall). IEEE, 2017, pp. 1–5.

[4] P. S. Bithas, E. T. Michailidis, N. Nomikos, D. Vouyioukas, and A. G. Kanatas, "A Survey on Machine-Learning Techniques for UAV-Based Communications," Sensors, vol. 19, no. 23, p. 5170, 2019

[5] J. Won, D.-Y. Kim, Y.-I. Park, and J.-W. Lee, "A survey on UAV placement and trajectory optimization in communication networks: From the perspective of air-to-ground channel models," ICT Express, 2022.

[6] Sun, Chenrui, Gianluca Fontanesi, Berk Canberk, Amirhossein Mohajerzadeh, Symeon Chatzinotas, David Grace, and Hamed Ahmadi. "Advancing UAV Communications: A Comprehensive Survey of Cutting-Edge Machine Learning Techniques." *IEEE Open Journal of Vehicular Technology* (2024).

[7] Khan, Inam Ullah, Asrin Abdollahi, Ryan Alturki, Mohammad Dahman Alshehri, Mohammed Abdulaziz Ikram, Hasan J. Alyamani, and Shahzad Khan. "Intelligent Detection System Enabled Attack Probability Using Markov Chain in Aerial Networks." *Wireless Communications and Mobile Computing* 2021, no. 1 (2021): 1542657.

[8] Khalil, H., Rahman, S. U., Ullah, I., Khan, I., Alghadhban, A. J., Al-Adhaileh, M. H., & ElAffendi, M. (2022). A UAV-Swarm-Communication Model Using a Machine-Learning Approach for Search-and-Rescue Applications. *Drones*, 6(12), 372.

[9] Inam Ullah, D Adhikari, Habib K, Shabir A, Christian E, C Choi, "Optimizing Mobile Robot Localization: Drones-Enhanced Sensor Fusion with Innovative Wireless Communication," 7th International Workshop on DroneCom: Drone-Assisted Wireless Communications for 5G and Beyond, IEEE International Conference on Computer Communications (INFOCOM), May 20, 2024.

[10] Klaine, Paulo V., Richard D. Souza, Lei Zhang, and Muhammad Imran. "An Overview of Machine Learning Applied in Wireless UAV Networks." *Wiley 5G Ref: The Essential 5G Reference Online* (2019): 1-15.

[11] Azoulay, Rina, Yoram Haddad, and Shulamit Reches. "Machine Learning Methods for UAV Flocks Management-A Survey." *IEEE Access* 9 (2021): 139146-139175.

[12] McNamee, Patrick. "Machine Learning for Aerospace Applications using the Blackbird Dataset." (2021).

[13] Wilson, A. N., Abhinav Kumar, Ajit Jha, and Linga Reddy Cenkeramaddi. "Embedded sensors, communication technologies, computing platforms and machine learning for UAVs: A review." *IEEE Sensors Journal* 22, no. 3 (2021): 1807-1826.

[14] Borup, Kasper Trolle, Thor I. Fossen, and Tor Arne Johansen. "A Machine Learning Approach for Estimating Air Data Parameters." (2019).

[15] Kurunathan, Harrison, Hailong Huang, Kai Li, Wei Ni, and Ekram Hossain. "Machine Learning-Aided Operations and Communications of Unmanned Aerial Vehicles: A Contemporary Survey." *IEEE Communications Surveys & Tutorials* (2023).

[16] Ullah, Zaib, Fadi Al-Turjman, Uzair Moatasim, Leonardo Mostarda, and Roberto Gagliardi. "UAVs joint optimization problems and machine learning to improve the 5G and Beyond communication." *Computer Networks* 182 (2020): 107478.

[17] Ananthi, S., N. Sirija, R. Shanmugapriya, and M. Rajeswari. "Design and Development of a Solar-Powered UAV Using IoT and Machine Learning." In *E3S Web of Conferences*, vol. 387, p. 05005. EDP Sciences, 2023.

[18] Anaya, Maribel, H. Ceron, Jaime Vitola, D. A. Tibaduiza, and F. Pozo. "Damage classification based on machine learning applications for an un-manned aerial vehicle." *Structural Health Monitoring* (2017).

[19] M. Chen, U. Challita, W. Saad, C. Yin, and M. Debbah, "Artificial Neural Networks-Based Machine Learning for Wireless Networks: A Tutorial," IEEE Communications Surveys & Tutorials, vol. 21, no. 4, pp. 3039–3071, 2019.

[20] I. Macaluso, H. Ahmadi, and L. A. DaSilva, "Fungible Orthogonal Channel Sets for Multi-User Exploitation of Spectrum," IEEE Transactions on Wireless Communications, vol. 14, no. 4, pp. 2281–2293, 2015

[21] H. Ahmadi, Y. H. Chew, N. Reyhani, C. C. Chai, and L. A. DaSilva, "Learning solutions for auction-based dynamic spectrum access in multicarrier systems," Computer Networks, vol. 67, pp. 60–73, 2014

[22] H. Ahmadi, Y. H. Chew, P. K. Tang, and Y. A. Nijsure, "Predictive opportunistic spectrum access using learning based hidden markov models," in 2011 IEEE 22nd International Symposium on Personal, Indoor and Mobile Radio Communications, 2011, pp. 401–405.

[23] M.-A. Lahmeri, M. A. Kishk, and M.-S. Alouini, "Artificial Intelligence for UAV-Enabled Wireless Networks: A Survey," IEEE Open Journal of the Communications Society, vol. 2, pp. 1015–1040, 2021

[24] L. A. L. da Costa, R. Kunst, and E. P. de Freitas, "Q-fanet: Improved q-learning based routing protocol for fanets," Computer Networks, vol. 198, p. 108379, 2021.

[25] D. Gunning, "Explainable artificial intelligence (xai)," Defense advanced research projects agency (DARPA), nd Web, vol. 2, no. 2, p. 1, 2017

[26] W. Guo, "Partially Explainable Big Data Driven Deep Reinforcement Learning for Green 5G UAV," in ICC 2020-2020 IEEE International Conference on Communications (ICC). IEEE, 2020, pp. 1–7.

[27] L. He, A. Nabil, and B. Song, "Explainable Deep Reinforcement Learning for UAV Autonomous Navigation," arXiv preprint arXiv:2009.14551, 2020

[28] J. Konecnˇy, H. B. McMahan, F. X. Yu, P. Richt`arik, A. T. Suresh, ´ and D. Bacon, "Federated Learning: Strategies for Improving Communication Efficiency," arXiv preprint arXiv:1610.05492, 2016

[29] Q.-V. Pham, M. Zeng, T. Huynh-The, Z. Han, and W.-J. Hwang, "Aerial Access Networks for Federated Learning: Applications and Challenges," IEEE Network, vol. 36, no. 3, pp. 159–166, 2022.

[30] H. Zhang and L. Hanzo, "Federated Learning Assisted Multi-UAV Networks," IEEE Transactions on Vehicular Technology, vol. 69, no. 11, pp. 14 104–14 109, 2020

[31] J. Yosinski, J. Clune, Y. Bengio, and H. Lipson, "How transferable are features in deep neural networks?" in Advances in neural information processing systems, 2014, pp. 3320–3328.

[32] G. Fontanesi, A. Zhu, M. Arvaneh, and H. Ahmadi, "A Transfer Learning Approach for UAV Path Design with Connectivity Outage Constraint," IEEE Internet of Things Journal, vol. 10, no. 6, pp. 4998–5012, 2023

[33] X. Zhang, G. Zheng, and S. Lambotharan, "Trajectory Design for UAV-Assisted Emergency Communications: A Transfer Learning Approach," in GLOBECOM 2020-2020 IEEE Global Communications Conference. IEEE, 2020, pp. 1–6

[34] S. Y. Choi and D. Cha, "Unmanned aerial vehicles using machine learning for autonomous flight; state-of-the-art," *Adv. Robot.*, vol. 33, no. 6, pp. 265_277, Mar. 2019.

[35] A. I. Khan and Y. Al-Mulla, "Unmanned Aerial Vehicle in the Machine Learning Environment," *Procedia Comput. Sci.*, vol. 160, pp. 46_53, Jan. 2019.

[36] Cui, Z., & Wang, Y. (2021). UAV Path Planning Based on Multi-Layer Reinforcement Learning Technique. *Ieee Access*, 9, 59486-59497.

[37] Challita U, Ferdowsi A, Chen M, Saad W. Machine Learning for Wireless Connectivity and Security of Cellular-Connected UAVs. *IEEE Wirel Commun.* 2019;26:28-35.

[38] Korki M, Shankar ND, Shah RN, Waseem SM, Hodges S. Automatic Fault Detection of Power Lines Using Unmanned Aerial Vehicle (UAV). Paper presented at: Proceedings of the 1st International Conference on Unmanned Vehicle Systems-Oman UVS, Muscat, Oman; 2019.

[39] Lodeiro-Santiago M, Caballero-Gil P, Aguasca-Colomo R, Caballero-Gil C. Secure UAV- Based System to Detect Small Boats Using Neural Networks. *Complexity.* 2019;2019:1-11.

[40] J. Park, Y. Kim, and J. Seok, "Prediction of information propagation in a drone network by using machine learning," in 2016 International Conference on Information and Communication Technology Convergence (ICTC), 2016, pp. 147–149.

[41] M. Campion, P. Ranganathan, and S. Faruque, "UAV swarm communication and control architectures: a review," Journal of Unmanned Vehicle Systems, vol. 7, no. 2, pp. 93–106, 2018.

[42] Y. Zhou, B. Rao, and W. Wang, "UAV Swarm Intelligence: Recent Advances and Future Trends," Ieee Access, vol. 8, pp. 183 856–183 878, 2020.

[43] F. Fallucchi, M. Gerardi, M. Petito, E. W. De Luca, Blockchain Framework in Digital Government for the Certification of Authenticity, Timestamping and Data Property, in: Proceedings of the 54th Hawaii International Conference on System Sciences, 2021, p. 2307.

[44] A. Simonetta, A. Trenta, M. C. Paoletti, A. Vetrò, Metrics for Identifying Bias in Datasets, SYSTEM (2021).

[45] Al-Haddad, L. A., & Jaber, A. A. (2022). Applications of Machine Learning Techniques for Fault Diagnosis of UAVs. *SYSTEM*, 19-25.

[46] G. Fontanesi, F. Ort´ız, E. Lagunas, V. M. Baeza, M. V´azquez, J. V´asquez-Peralvo, M. Minardi, H. Vu, P. Honnaiah, C. Lacoste et al., "Artificial Intelligence for Satellite Communication and Non-Terrestrial Networks: A Survey," arXiv preprint arXiv:2304.13008, 2023.

CHAPTER 17   ADVANCEMENT IN ML TECHNIQUES AND APPLICATIONS FOR
             UAV SWARM MANAGEMENT

[47]   B. Galkin, J. Kibilda, and L. A. DaSilva, "UAVs as Mobile Infrastructure: Addressing Battery Lifetime," IEEE Communications Magazine, vol. 57, no. 6, pp. 132–137, 2019.

CHAPTER 18

AI Integration in Drone Technology: Revolutionizing Applications in Agriculture, Security, and Beyond

Faisal Rehman[1,2], Asna Rabail[1], Muhammad Hamza Sajjad[1], Shanza Gul[1]

[1]Department of Statistics & Data Science, University of Mianwali, Mianwali, Pakistan

[2]Department of Robotics & Artificial Intelligence, National Univresity of Sciences & Technology, NUST, Islamabad, Pakistan

Correspondence: Faisal Rehman[1,2] (faisalrehman0003@gmail.com)

CHAPTER 18 AI INTEGRATION IN DRONE TECHNOLOGY: REVOLUTIONIZING APPLICATIONS IN AGRICULTURE, SECURITY, AND BEYOND

Introduction

Indeed, thus considered, the combination of artificial intelligence (AI) and drone technology has created a vast range of opportunities that have reallocated and even reinvented different industries and sectors [1]. Refocusing its attention on the relationship between AI and drones, this paper aims to explore how AI has significantly transformed drone usage in various industries.

While these two technological fields intensify their dependence on each other, progress comes to the capabilities, performance, and deployments of unmanned aerial vehicles. This introduction paves the way for a more elaborate discussion by grounding it on the two major technological interfaces, namely, AI and drones, while focusing on the tremendous triple-A role that AI plays in disruptive innovation with drones.

By taking time and examining the role of contemporary artificial intelligence in drone developments and its versatility, the goal of this analysis is to provide insights into future of the AI-driven drones and the numerous chances they bring in every field.

Thus, the combination of technologies such as artificial intelligence and drones is not just a trend of technology innovation moving beyond the automation of processes and the usage of unmanned aerial vehicles [2]. Drones or UAVs which include automatic and pilotless aircraft have also been revolutionized through the incorporation of artificial intelligence/automation to cover a greater level of aeronautics [3].

Artificial intelligence offers a cognitive perspective over draconian technology; these aerial structures feature a cognitive capability to analyze and approach their surroundings at that time [4]. It should also be noted that this mutually beneficial synergy presents an opportunity to expand the use of drones in more than traditional ways, opening a new era of smart and adaptable unmanned aerial systems.

CHAPTER 18   AI INTEGRATION IN DRONE TECHNOLOGY: REVOLUTIONIZING APPLICATIONS IN AGRICULTURE, SECURITY, AND BEYOND

AI's importance in transforming drone applications' use falls in its ability to build and advance UAV functionalities in productivity in numerous sectors [5]. AI enhances drones' perception and decision-making skills, enhancing performance in challenging terrains and new scenarios, particularly when learning to operate in new scenarios [6]. This disruptive fusion makes it possible for drones to play a significant role in other areas of society including the resolution of crises, farming, monitoring of the environment, and so many more transforming how society will engage itself with the aerial systems.

The incorporation of AI scales up the drones to allow them to perform enhanced operations in diverse terrains, undertake more challenging operations, and adapt promptly to prevailing circumstances [7]. For instance, in aerial surveillance, AI enables drones to detect objects, track them, distinguish between patterns of data, and monitor data in real time besides improving its surveillance and scouting functions [8].

In addition, decision-making, path planning, and obstacle avoidance lead to the fact that drones can move through complex geography and keep away from objects independently. This decomposable ability is very useful where integrated robotic systems can be created which can be applied in fields such as farming, bridge inspection, and search operation. These intelligent systems incorporated in drones are capable of providing efficient coverage of large spaces, capturing flaws in infrastructures, and even identifying people in risky regions, making numerous processes safer and more effective [9].

This chapter aims to highlight the existing and estimated roles of artificial intelligence in drone.

This chapter explores the integration of AI into drone technology, focusing on its potential applications in various industries such as farming, medicine, and ecology, from click-to-tingle navigation to new uses.

CHAPTER 18 AI INTEGRATION IN DRONE TECHNOLOGY: REVOLUTIONIZING APPLICATIONS IN AGRICULTURE, SECURITY, AND BEYOND

Specifically, this chapter aims to present the current state of development in the context of the advancing technological application of drones and AI, explore the trends that exist and may further develop, and finally suggest future directions. Knowledge of the trends and issues of AI and drone technology can benefit industry experts, researchers, policymakers, and enthusiasts because it helps to unveil the potential and prospects as well as problems and difficulties of the application of the two significant technologies [11].

In conclusion, the incorporation of artificial intelligence into drone systems represents a marvelous achievement in the realm of unmanned aerial vehicles. Innovatively, the incorporation of AI in enlarged applications of drones is nothing short of transformational throughout various sectors. With this review, we set out on the process of exploring up close this synergistic partnership and delving deeper into the development of this versatile drone technology, the role played by artificial intelligence, as well as different applications that define the unmanned aerial systems.

Here's a summarized version of the sections mentioned:

- Introduction: This chapter introduces the integration of AI in drone technology, aiming to maintain original intent while enhancing readability and coherence.

- Section One: Discusses historical context with clarity and coherence, maintaining precise language.

- Section Two: Focuses on AI enabling drones for autonomous navigation, adapting in real time to enhance effectiveness and autonomy across industries.

- Section Three: Highlights how AI-driven advancements reduce operational costs and enhance performance in surveying and environmental monitoring.

- Section Four: Emphasizes the importance of ensuring rational and responsible AI services for safe and ethical drone operations.

- Section Five: Details how edge computing empowers drones to act quickly without relying on continuous ground-based servers.

- Section Six: Explores how swarm intelligence revolutionizes drone technology, enhancing coordination, scalability, and productivity in various sectors.

- Conclusion: Concludes with the impact of edge computing and swarm intelligence on future unmanned aerial systems, driving innovation and implementation.

- This summary encapsulates the key points and progression of your chapters on AI and drone technology.

Historical Development of AI in Drone Technology

The history of the emergence of phenomena associated with AI and drones, or more accurately the AI in UAVs, can be considered in parallel with the development of the UAVs themselves [12]. The idea of having an unmanned aerial vehicle can be traced back to the early 1900s when the first flights of remotely piloted aircraft were affected. Still, it was within the mid to latter half of the century that the use of drones started gaining some recognition, particularly with the start of the Cold War. This was followed by the improvement in UAV technology, including the availability of reconnaissance drones like the Ryan Model 147 which was operated by the United States during the 1960s.

CHAPTER 18 AI INTEGRATION IN DRONE TECHNOLOGY: REVOLUTIONIZING APPLICATIONS IN AGRICULTURE, SECURITY, AND BEYOND

These early drones were used exclusively for surveillance and were controlled with basic rudimentary controls using radio control [13]. With the progressive development of technology, there was more enhanced development in the functionality of drones. Some of the changes and improvements in the late twentieth century include better navigation systems, increased system endurance and payload, and enhanced capabilities among other factors [14]. These were significant developments that needed to take place to pave the way for the incorporation of AI in drones, a key milestone that could be considered a major event in the history of drones. The integration of AI algorithms further evolved the drones and enhanced their capability to exhibit intelligent unstated aerial systems capable of accomplishing elaborate tasks beyond human operations. It has over time continued to transform different fields and has also extended the areas of application of unmanned aerial vehicles making a new dawn in the history of drones [15]. With the beginnings of drone application, prominent uses were largely associated with military uses with a major emphasis on reconnaissance. They provided an opportunity to gather information without risking human lives and also penetrate enemies' territories for basic information key to military actions. However, there were constraints when these early applications were being pursued. Drones had a lot of problems during the war because their control systems were controlled by human operators, and these systems were not very accurate because they relied on the human eye; also, drones were not very autonomous, though they had some level of autonomy. Crossing from military business, drones flew to the civil fields including farming to address issues of aerial observation of crops and environmental missions to conduct surveys [16].

Nevertheless, the innovative uses of drones were limited and could not be expanded until the incorporation of artificial intelligence which could mark a revolution in the world of drone technology. Basic AI components began to be incorporated into drones during the late parts of the twentieth century through the early twenty-first century. In their infancy, these AI components were concentrated largely on the areas of pathfinding and guidance. Semiautonomous drones powered by basic artificial intelligence enable them to navigate their own flight plan and perform defined tasks [17]. However, some challenges limited the use of drones for nonmilitary purposes; during this time, they include technological limitations, which made it hard to make good drones, and regulatory issues, which restricted the use of drones, and lastly, there was the issue of public perception which was not positive toward drones. In the early stages of the period under analysis, artificial intelligence was mostly used to improve the performance of manual control over drones. Some key ideas in contemporary avionic systems, particularly those regarding self-governing drones able to make decisions on their own, were yet to be developed. However, the improvement of AI in the midfield of the 2000s is noted as a drastic advancement in the implementation of AI with drones [18]. Another key area was intelligent autonomy to make the drones independent and capable of processing data, as well as making decisions on their own in real time. This moved operations from more rigid prescripted maneuvering to newer, malleable, and intelligent control. Real improvements in drones' cognition have been introduced with the help of AI algorithms and, particularly, machine learning [19].

With this enhancement, the drones could undertake analysis of sensor data, object recognition, decision-making, and adaptability to the current environmental conditions. This development provided opportunities for the drones to perform more than surveillance operations and allowed them to perform more complex tasks and missions, which may require the drones to operate independently or think for themselves in some aspects.

The advanced developments in computer vision in the late 2000s and early 2010s gear aided drone technology, a key aspect of artificial intelligence. During this time, drones began to integrate with better quality cameras and sensors as well as computer vision algorithms.

CHAPTER 18 AI INTEGRATION IN DRONE TECHNOLOGY: REVOLUTIONIZING APPLICATIONS IN AGRICULTURE, SECURITY, AND BEYOND

It was a tool that put into their hands all that was necessary to register and even analyze their environment with a precision that could not have been possible before.

With the help of the computer vision technique, the drone was able to perform several operations such as object identification, ground profiling stumbling block identification, and many more. It was less only improving the safety and utilization of drone performances but also the preliminary step for expanding the usage of unique functions in other sectors like rescue missions, identification of structures, ecology, and more [20].

It could be seen that computer vision capabilities can be used in drones to enable them to move along specific areas and objects of interest that it is entitled to capture and post possible obstructions to their movements in real time. This not only increased their functions from surveillance or reconnaissance only but also extended their responsibilities and capabilities of executing far more challenging operations more accurately and efficiently [21].

There was an increased use of machine learning technology along with neural networks in drone technology in the middle of the 2010s. It was more complex, and as data from learning algorithms were inputted into drones, they were capable of adapting to changes in the environment and the drones' performance was characteristic of a progressive enhancement. This paradigm shift is made possible through the use of machine learning frameworks whereby drones could analyze large datasets, identify trends, and make prognoses regarding patterns recognized within the dataset.

Increased practicality of technological innovations such as neural networks, especially deep learning models, resulted in huge advancements in image recognition, meaning drones would be able to identify objects with vast precision [22]. This advancement unlocked new and unique possibilities in a range of fields, and one of them was precision agriculture. Examples in this context: in this field, drones could capture the imagery data and analyze the status of crops to give recommendations for managing farming processes.

CHAPTER 18   AI INTEGRATION IN DRONE TECHNOLOGY: REVOLUTIONIZING APPLICATIONS IN AGRICULTURE, SECURITY, AND BEYOND

The advancement of machine learning along with neural networks helped drones to become more sophisticated and self-sufficient, not only in terms of data gathering and flying but also in different aspects as well. This evolution was marked by the enhanced use of drones in various industries and sectors, thereby increasing their usefulness.

Besides, with the help of the new generation of the artificial intelligence of drones, there are innovations toward support of collaboration. Recent technological advances allow swarms of drones to share information and plan their operations, so that swarms could be useful for large-scale coordinated activities, including surveillance, search and rescue missions, and monitoring the environment.

Drone technology has become common in the present world, and various industries utilize it professionally ranging from delivery services, infrastructure inspection, ecological conservation, and public safety. There is a significant point here; drones are no more mere flying systems but intelligent ones powered by AI to solve problems of various types. This shift is a radical improvement for shaping further development of drone technologies and identifying their further practical applications essential for numerous societal-oriented and industrial problem-solving.

As we are standing at the isle of this crossroad, the process opens with numerous prospects forming a future with the main stroke driven by AI capabilities of drones for the reforming of various sectors as well as the solving of social issues. Such an evolution shows the new prospects of the available drone technology and asserts a new paradigm in the use of drone technology within various sectors.

Applications of AI in Drone Technology

There is a vast number of different areas of artificial intelligence usage in drones, but one of the main aspects is the possibility of achieving autonomous navigation, which means the flight planning will be completed by AI mechanisms. Old drones had fixed paths as they could only be controlled by a predetermined system or script at the developers' discretion. However, with the use of artificial intelligence, drones are capable of self-navigating both within the environment and around obstacles and concurrently controlling the flight path depending on the assigned mission and the environmental constraints and conditions [23].

Using data from onboard GPS, IMUs, and other obstacle detection systems, AI algorithms make or compute flight plans in real time. These algorithms have been developed to consider certain aspects including weather conditions, airspace characteristics, and also geographical terrains to facilitate more efficient and safe flying. When applying AI in elf flight planning, drones become able to maneuver in complex terrains with precision and in this sense expand the potential uses across a range of industries such as surveillance, surveying, and inspecting. This particular development can be regarded as crucial for the enhancement of the general effectiveness of the drones themselves, as well as for the development of the scenarios in which they can be used and their degree of autonomy [18].

A very important use of AI for drone operation is recognition of objects and approach avoidance: piloted drones can only operate autonomously to a limited extent and would need human interference to avoid any barriers, making them highly restricted in the decision-maker's terrain. Traditional drones have their limitations, especially when driven by limited sensors and computer vision systems, as compared to AI drones which are incorporated with enhanced sensors and computer vision and can detect the barriers along their path and avoid them.

These include a building, trees, or other drones, and the AI algorithms utilize the sensors to reveal the obstacles in real time. Once an object is identified, these algorithms compute possible paths for the flight that will help to avoid any encounter while preserving mission goals. This capability is especially useful when, let's say, search and rescue drones are flying through cities or forests to find certain lost or trapped people.

CHAPTER 18   AI INTEGRATION IN DRONE TECHNOLOGY: REVOLUTIONIZING APPLICATIONS IN AGRICULTURE, SECURITY, AND BEYOND

Drone perception is one of the key areas that allow the drones to detect and follow objects in that environment. Self-driving cars use sensors that collect visual data from cameras mounted on the vehicle to analyze objects like cars, pedestrians, or animals. Once the algorithms correctly detect the object, drones are also capable of tracking their progress in real time making situational awareness of distinct uses possible. The combination of AI-based obstacle detection and computer vision in drones makes the operation of drones much safer and improves the efficiency of their functioning in a wide variety of contexts.

Features like identifying and tracking objects are important in several areas such as security, police operations, and monitoring of wildlife among others. With properly installed AI-driven computer vision, police can use drones to follow suspects, and borders, or to census animals in the wild in faster and superior ways than mean utilization of the equipment.

Self-driving drones are highly effective in identifying aspects of an image for multiple functions across sectors. A wide range of applications that include bridge inspection through to air and water pollution detection entails converting visual data collected by drones into a format that is useful for decision-making by the use of AI algorithms.

Automated diction and recognition of possible flaws or deviations in infrastructure objects like bridges, pipelines, or power lines through the use of drones and computer vision systems may help in the inspection process. Several algorithms that have been developed in computer vision are used to assess visual data that shows some signs of damage, corrosion, or excessive usage so that these signs can be attended to before they worsen.

Similarly, for photographic purposes in environmental degradation, for instance, drones can obtain information in a bid to evaluate the status of ecosystems, counting wildlife, or even as an evaluating tool for changes in land usage. Machine learning methods scrutinize this data for signs of environmental threats to monitor the changes in habitats or detect the presence of prohibited practices such as deforestation or hunting protected species.

CHAPTER 18 AI INTEGRATION IN DRONE TECHNOLOGY: REVOLUTIONIZING APPLICATIONS IN AGRICULTURE, SECURITY, AND BEYOND

Figure 18-1 illustrates how each category of drone application branches out from general use cases into specific applications. Each category has its own set of technologies, regulations, and challenges, making drones versatile tools in various industries and sectors.

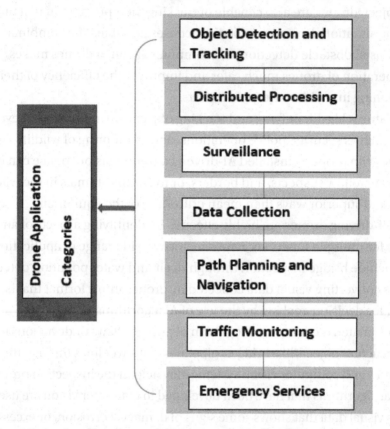

Figure 18-1. *Applications of AI in Drone Technology*

The integration of AI-driven computer vision systems in drones augments the utilities of unmanned aerial systems in responding to a broad spectrum of problems and goals within diverse societal fields.

The use of progressed drones assisted by AI is assisting farmers in gathering crucial data on the status, growth rate, and productivity of crops in their fields. With multispectral or hyperspectral systems onboard,

CHAPTER 18 AI INTEGRATION IN DRONE TECHNOLOGY: REVOLUTIONIZING APPLICATIONS IN AGRICULTURE, SECURITY, AND BEYOND

drones photograph and capture detailed images of agricultural fields, which then can be processed by AI algorithms to determine a crop's health and its condition based on chlorophyll content, moisture, or diseases. The algorithms then utilized the data for developing the crop maps, as well as for flagging stress or nutrient deficiency and suggesting solutions such as irrigation or shifting to fertilization. Drones allow farmers to observe crops from a distance and make decisions based on real-time data, which subsequently helps reduce inputs, increase yield, and be as environmentally friendly as possible.

Also, these drones in particular that are fueled by artificial intelligence allow for the mechanical application of pesticides and controlled measures with regard to farming, making farming more efficient and sustainable [11]. Self-driving robots that are loaded with accurate spraying technologies and artificial intelligence methods can systematically spray the required nutrient or pesticide in a particular region of the field which helps in decreasing the amount of soil pollution. AI algorithms use real-time data gathered from inspectors or onboard sensors to identify correct spraying parameters like the size of the droplets, the angle at which they are applied, and the rate at which they are deposited taking into account the type, growth stage of the crop, and prevailing environmental conditions. This paper illustrates that by accurately focusing on the inputs that farmers require, drones are capable of delivering better yields within a shorter period, thus minimizing the input costs as well as having minimal impacts on the environment.

In conclusion, it claimed that there are various roles of artificial intelligence in drone innovation, including the ones mentioned in this chapter: autonomous navigation, computer vision, and precision farming. Artificial intelligence-powered drones are revolutionizing various fields by optimizing flight trajectories, detecting and avoiding objects, identifying objects of interest, and making informed farm decisions. As AI technology trickles down, drones' enhanced attributes in a business's day-to-day operations will increase, hence the enhancement of possibilities of implementing new technologies and optimization of processes in various fields.

Machine Learning and Drones: Enhancing Capabilities and Applications

In drone technology, predictive maintenance involves using machine learning to decode past data concerning the performance of the drone and analyze the findings with a view of uncovering parts of the system that may soon require elementary work. Flight parameters, as well as any environmental indicators, that aircraft are equipped with sensors and AI algorithms, can point at the need for maintenance or suggest preventive measures. Controlling possible risks makes it possible to reduce time out, enhance the longevity of the drone, and lower maintenance expenses, all of which assure reliability and effectiveness in various uses.

Drone machines use machine learning to optimize maintenance and data analysis through the help of computers [21]. By applying AI to telemetry, sensors' data, and flight logs, feedback on potential failures, which can be corrected early, is given to prevent frequent downtimes and operational costs. Lightning bolt maintenance, material fatigue failure prevention, and timely repairs all enable drones to run with increased reliability and efficiency. The data processing and analysis abilities of the drones are also something machine learning can perform, particularly when it comes to functions like land surveying and mapping and monitoring the environment [6]. These advancements go a long way in enhancing various sectors such as construction and conservation among others since they simplify their operations and help to monitor the surrounding environment. With the growth in machine learning, drones will become better in their numerous uses as they fuel the aerial technology.

CHAPTER 18   AI INTEGRATION IN DRONE TECHNOLOGY: REVOLUTIONIZING APPLICATIONS IN AGRICULTURE, SECURITY, AND BEYOND

AI and drones will thus complement each other in presenting new opportunities. The advancement of unmanned aerial systems has significantly increased productivity and strategic direction in various fields and districts.

Challenges in the Integration of AI and Drones

The integration of AI in drone technology has brought various ethical issues such as infringement of privacy, biased use in law enforcement, laws, and data privacy. It is necessary to establish clear criteria and rules to regulate the operations of drones in risky territories and guarantee rationality, reasonableness, and responsibility concerning AI services. Where drones work across borders, privacy laws must be in synch across the same borders because while safety and security are major issues that govern possibilities of flying drones autonomously and avoiding obstacles, AI algorithms used in navigation and obstacle detection require to be tested and validated across borders.

The use of technologies such as artificial intelligence in driving drone systems raises issues of cybersecurity to a rather high level due to the proliferation of cyberattacks that can cause serious blows to the operation, integrity, and data confidentiality impacts that the systems offer [24]. Tools like locking down communications, information assurance of transmission, and ensuring user identity are an effective way of dealing with cyber threats and other forms of infringements. Thus, effective safety and security can be achieved by the development of thorough risk analysis and protective measures against the possible utilization of AI-based UAVs for obtrusive purposes, as well as adversarial manipulation of the automated decision-making system [3]. The management of possible risks necessitates the continuous cooperation of the stakeholders in the field of drone manufacturing, AI development, and regulation, both at the governmental and nongovernmental levels, as well as cybersecurity specialists to guarantee the safety of operations with the help of AI-enabled drones for the prompt delivery of goods and services across different sectors.

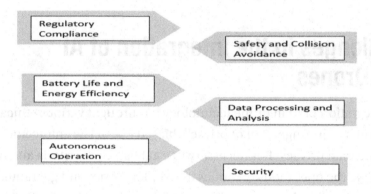

Figure 18-2. Challenges in the Integration of AI and Drones

This diagram visually represents the interconnected challenges of integrating AI with drones. Each challenge area is critical for successful deployment of AI-driven capabilities in drone technology, emphasizing the complexity and interdisciplinary nature of this integration process. Addressing these challenges requires advancements in AI algorithms, hardware capabilities, and regulatory frameworks to unlock the full potential of AI-enhanced drone operations.

Future Prospects of AI in Drone Technology

Another fascinating advancement in AI and drone systems is edge computing; it is the processing of computing and analyzing raw data without sending it to a centralized station or cloud because this form of computing is localized and can work independently or in tandem with drones or other edge devices. Edge computing entails processing data closer to the source and thus cutting on latency when compared to obtaining data from the cloud, making drones' programming more responsive in areas where programs run in real time. By using contemporary drones with integrated high-performance computing, telemetry streams, images, and sensors can be processed immediately. Digitalization enhances drones' decision-making capabilities by eliminating contact-based servers, allowing them to take appropriate actions in changing environments with minimal latency. Dynamic control and decision-making features allow drones to quickly replant and execute tasks that include navigation, avoidance of obstacles, and object recognition based on varying circumstances and adjusted goals. Edge computing improves the effectiveness of drone operations in various sectors to increase application areas created by drones including surveillance, search and rescue, or inspection.

Drones craft environments, requiring information exchange and alignment for efficient task division, resource utilization, and optimal environmental utilization. Swarm intelligence means drones can work on search and rescue tasks, environment surveillance and monitoring, disaster response, and more tasks in a systematic way and easily manageable than handling individual drones. This progressive phenomenon in the sphere of artificial intelligence and drones may potentially disrupt numerous industries and provide broad interoperable and versatile aerial systems able to solve demanding issues within a performed environment.

In off-counter drones and artificial intelligence, swarm intelligence is implemented in such a way that multiple drones can work together and send information on their position velocity and the objectives of the mission [3]. This cooperation improves the speed of activities such as distributed detection, monitoring, localization, and rescue operations. In surveillance, swarm-enabled drones enhance productivity and wastage reduction in farming practices through constant checks on the well-being of crops as well as the correct application of chemical fertilizers, when necessary, through precision spraying. In disaster control, they support reconnaissance missions in the air to search for survivors and deliver effects in unsafe zones. Swarm-based drones work in tandem with robots across various industrial sectors like factory stock checking and building repair and maintenance, to improve system productivity and cut incidences of cost. Trends that are now starting to materialize in unmanned aerial systems include edge computing and swarm intelligence that enhance these to be more adaptive and operate in dynamic environments. These interdependent relationships between, and innovations in, artificial intelligence and drones at work continue to foster progressive developments in unmanned aerial systems.

Future Outlook for AI in Drone Technology

The future of drone technology and their management and organization, primarily based on physical laws and principles, is uncertain. Swarm intelligence enhances distributed sensing, surveillance, task performance, and scalability in various fields, including drone swarms, which can operate in areas like search and rescue. This advancement will have strenuous effects on several fields such as the agricultural sector, environmental monitoring, disaster response, infrastructure inspection, and others. Furthermore, swarm-empowered drones will enable coordinated actions with other industrial robots which will increase flow and overall productivity in a production line. Altogether, the potential developments of the two technologies are expected to continuously build around AI and drone technology and pave novel breakthroughs in the future of unmanned aerial systems and various industries.

CHAPTER 18 AI INTEGRATION IN DRONE TECHNOLOGY: REVOLUTIONIZING APPLICATIONS IN AGRICULTURE, SECURITY, AND BEYOND

The AI forecast predicts significant future changes and impacts on unmanned aerial systems and technology, with key areas of anticipated advancements including:

- Enhanced Autonomy: Future drones will be all smart with some of the real artificial intelligence algorithms to navigate on their own, avoiding obstacles automatically and detecting them. These evolutions will bring drones to fly in high-density environments with low intervention from human beings, thus expanding what is currently possible within delivery services, infrastructure inspections, and even city skies.

- AI-Driven Sensing and Sense-Making: Future drones shall, therefore, be equipped with sound devices and systems to offer computer vision from artificial intelligence algorithms for improved permeability and sensing. These drones will be able to collect and analyze huge amounts of real-time information, opening up fields such as environmental imaging and surveillance, disaster prevention, and agriculture control on a scale yet to be seen.

- Swarm Intelligence: The future of drone technology will be focused on swarm intelligence where many UAVs will be able to act collectively and derive coordinated behavior. A swarm of drones shall complete tasks like cooperative sensing, surveillance, and search and rescue operations through division of labor where cohort drones complete complex operations in parallel with better scalability.

- Edge Computing: Advanced drones in the future will incorporate edge computing to perform real-time data computations and analysis to support instant decision-making and avoid heavy dependence on cloud computing setups. Independent or partially autonomous drones will be able to collect data from sensors and perform a specific level of analysis on board or in nearby edge devices, which can significantly increase the responsiveness and efficiency of management in critical missions.

It acknowledges that the future of both AI and drone technology is highly dependent on advances in technologies and innovation where research collaboration is key involving academic institutions, industries, as well as government institutions. The strategies for cooperation will provide innovations and prompt the completion of new-generation technologies, knowledge of advanced AI algorithms, and methods of sensor technologies and drone platforms.

Public participation, research, and collaboration between industries, academia, and the government play a crucial role in advancing and spreading AI-driven drones. Software companies interact with the leading manufacturers of drones, AI, sensors, and software to improve and enhance the capabilities of their products to meet the needs of target activities. Some are performed in academic institutions, and others involve state-of-the-art research projects that advance the technology.

The state administrations offer financial support and grants, material support, and legal advice and oversee the proper and ethical application. Partnerships contribute to the exchange of best practices, technical experience, and information about trends in laws and rules. This paper acknowledges that regulatory measures touch on safety, security, privacy, and environmental issues concerning the development of future AI and drones. Furthermore, the advancement of AI has revolutionized various sectors, including IoT, the energy sector, quantum computing, and the fields of image and signal analysis [25–27].

AI-driven drone systems require strong regulatory frameworks based on ethical norms and standards to prevent exploitation and ensure transparency. Stakeholders like industry players, civil society, and academics are crucial in building these frameworks. International cooperation and synchronization of regulations are necessary for safety and performance.

Joint work between various authorities will likely foster compatibility, encourage knowledge sharing, and ensure the participants in the AI and drone business function under the same set of rules around the world. Therefore, the expectations of AI on future drone advances provide a foundation for emerging autonomy, sensing, and collaboration of unmanned aerial systems in the future. Joint actions in the fields of research and development will foster the growth and development of innovative AI-driven drone technologies on the global market, whereas the legislation will ensure the proper and legal usage of drone technology to reflect legal and ethical norms. It is now important to note that the advancements of the AI and drones, along with the pressures created by partnerships and governance structures, will define how society will be addressed through unmanned airborne systems in the future.

Conclusion

In further discussions and conclusion, the depicted information and artificial intelligence review in drone technology opens new horizons for unmanned aerial systems and multiple types of usage. From the studies, important insights were identified showing that AI drones exhibit higher levels of autonomy, sensing, and collaboration compared to

CHAPTER 18 AI INTEGRATION IN DRONE TECHNOLOGY: REVOLUTIONIZING APPLICATIONS IN AGRICULTURE, SECURITY, AND BEYOND

previous routes, which makes it possible to perform diverse tasks with high efficiency and accuracy. These applications cover the ministries of agriculture, infrastructural and environment, natural disasters, and safety. Areas such as edge computing, swarm intelligence, and predictive maintenance are now defining the direction and scope of the AI-driven drone application, boosting the innovation and implementation frontiers.

Drone technology powered by artificial intelligence significantly contributes to radical changes in several industries by promoting activities, increasing productivity, as well as supporting the decision-making stage. In agriculture, specialized drones containing AI algorithms for farming can be used to monitor crops, analyze soil conditions, and spray crops in real time enhancing productivity while protecting the environment.

In maintenance, inspection, and 3D mapping of infrastructure, high-performance drones integrated with different sensors and computer vision can evaluate deteriorations, damages, and structural health and make necessary repairs and maintenance planning to guarantee the safety and efficiency of critical assets.

Drones are crucial in disaster response, providing situational awareness, delivering essential commodities, and performing rescue operations in dangerous conditions, ultimately reducing the cost of natural disasters.

In conclusion, it is consisted in this paper that the application of AI to drone technology contributes tremendously to various sectors by using higher technologies that can solve the problems and lead to positive results.

Incorporating AI, which will steadily develop in the future, gauging its usage in drones and its consequent impacts, is incredibly significant. It implies that collaboration with the industry's stakeholders and major institutions and regulation bodies is paramount in sourcing the solutions to problems, advancing innovativeness and entrepreneurship, and ensuring that drones under the new AI technology are reasonable and legal.

CHAPTER 18　AI INTEGRATION IN DRONE TECHNOLOGY: REVOLUTIONIZING APPLICATIONS IN AGRICULTURE, SECURITY, AND BEYOND

If the current understanding of AI involves interdisciplinary fields, openness, and the creation of stringent regulations regarding flying drones, the implementation of artificial intelligence-driven drone capabilities is possible. It will also allow us to respond to societal needs and changes, promote the development of industries, and improve the standard of living.

To sum up, AI in drones presents incredible potential that can be useful for various industries and markets. It can be concluded that by taking action in advance of the progression of these trends, solution seeking, and ethical utilization of AI in the framework of drone operations, we can leverage its potential to the maximum.

AI and UAS integration will significantly impact human development, global challenges, and management, improving manufacturing efficiency, disaster relief, and environmental protection.

A major application of AI-driven UAS can be in the field of precision agriculture which can immensely benefit from the service provided by UAS. Drones with better sensors and AI algorithms can always provide farmers with potential insights about the health of crops, the status of soil, and the required water to be supplied. This leads to applications such as precision farming, where elements such as water and pesticides are used as required in the proper areas to maximize crop yields while using the least number of resources and negative environmental impact possible. Also, looking at the possibilities of using AI-enabled drones for crop health surveillance and potentially even pollination, the prospects of AI in the agricultural industry and sustainable food production for the growing population are imaginable.

Opportunities of AI integration with UAVs: AI-equipped UAS are virtually revolutionary in the field of environmental monitoring and conservation on some fragile biotopes. With the help of avionics, which include sensors and computer vision systems, drones fly high and systematically collect data on changes in vegetation, numbers of animals, as well as other factors in huge territories of land or water, after which they could be processed through artificial intelligence algorithms to discern potential threats in the environment, or trends exhibiting levels of diversity on species, or for eventual conservation. For instance, a drone could be used to track deforestation, mining, or poaching in remote areas and relay important information to authorities immediately, so they can be sent in to prevent further damage to the endangered species and other animals and plant life.

In addition, the use of AI integration in the UAS is expected to greatly transform the practices relating to disasters and emergency management services. After disasters like earthquakes, hurricanes, or wildfires that ravage communities leaving behind scenes of destruction, drones with sensors driven by artificial intelligence algorithms are readily available to survey the devastated areas, detect dangerous areas, and look for signs of life where humans cannot easily gain access. With these drones, one can map disaster areas from the air, create high accurate 3D models, as well as provide rescue teams with immediate visual information helping them coordinate their actions better and faster. Furthermore, AI algorithms and models can play a significant role in analyzing aerial imagery to improve the strategic ability to allocate response resources and ensure a more effective response to future disasters and their impact on the affected community.

In the industrial context, the adoption of advanced technologies such as articulated unmanned aerial systems (UAS) can provide new approaches toward various inspections, monitoring, and maintenance tasks of infrastructures. In utilizing drones with AI algorithms for monitoring various infrastructural structures such as bridges, pipelines, and power lines among others, problems of structural defects, corrosion, or any form of aberration from the normal state can quickly be noted. It is a policy that proactively guards against system failures that are not only time-consuming and costly but also dangerous to the infrastructure and service that the public relies on. AI and UAS can enhance safety, productivity, and environmental sustainability in industrial areas by monitoring processes, detecting spills, and mapping potential risks.

CHAPTER 18 AI INTEGRATION IN DRONE TECHNOLOGY: REVOLUTIONIZING APPLICATIONS IN AGRICULTURE, SECURITY, AND BEYOND

In conclusion, the incorporation of AI into the unmanned aerial system is a revolution to learn from and adapt to when developing solutions to social issues and prospects. Furthermore, as the potential of AI-driven UAS is discovered, it becomes possible to drastically expand the horizons of several fields in terms of increased effectiveness and the use of sustainable and innovative solutions. As these technologies become more refined and standardized, or in other words as they go mainstream, these technologies are poised to be a key to improving human endeavors, solving global challenges, and creating a happier, longer future for humanity.

References

[1] Adel, A. (2023). Unlocking the Future: Fostering Human–Machine Collaboration and Driving Intelligent Automation through Industry 5.0 in Smart Cities. Smart Cities, 6(5), 2742-278.

[2] Alqahtani, H., & Kumar, G. (2024). Machine learning for enhancing transportation security: A comprehensive analysis of electric and flying vehicle systems. Engineering Applications of Artificial Intelligence, 129, 107667.

[3] Johnson, J. (2020). Deterrence in the age of artificial intelligence & autonomy: a paradigm shift in nuclear deterrence theory and practice? Defense & Security Analysis, 36(4), 422-448.

[4] Zhang, T., Hu, X., Xiao, J., & Zhang, G. (2020). A Machine Learning Method for Vision-Based Unmanned Aerial Vehicle Systems to Understand Unknown Environments. Sensors, 20(11), 3245.

[5] Fan, B., Li, Y., Zhang, R., & Fu, Q. (2020). Review on the Technological Development and Application of UAV Systems. Chinese Journal of Electronics, 29(2), 199-207.

[6] Bayomi, N., & Fernandez, J.E. (2023). Eyes in the Sky: Drones Applications in the Built Environment under Climate Change Challenges. Drones, 7(10), 637.

[7] Sharma, S. (2023). Trustworthy Artificial Intelligence: Design of AI Governance Framework. Strategic Analysis, 1-22.

[8] Telli, K., Kraa, O., Himeur, Y., Ouamane, A., Boumehraz, M., Atalla, S., & Mansoor, W. (2023). A Comprehensive Review of Recent Research Trends on Unmanned Aerial Vehicles (UAVS). Systems, 11(8), 400.

[9] Riaz, N., Shah, S. I. A., Rehman, F., & Khan, M. J. (2021). An Intelligent Hybrid Scheme for Identification of Faults in Industrial Ball Screw Linear Motion Systems. IEEE Access, 9, 35136-35150.

[10] Johnson, J. (2020). Artificial Intelligence, Drone Swarming and Escalation Risks in Future Warfare. The RUSI Journal, 165(2), 26-36.

[11] Banafaa, M., Pepeoğlu, Ö., Shayea, I., Alhammadi, A., Shamsan, Z., Razaz, M.A., Alsagabi, M., & Al-Sowayan, S. (2024). A Comprehensive Survey on 5G-and-Beyond Networks With UAVS: Applications, Emerging Technologies, Regulatory Aspects, Research Trends and Challenges. IEEE Access.

[12] Menkhoff, T., Kan, S.N., Tan, E.K., & Foong, S. (2022). Future-proofing students in higher education with unmanned aerial vehicles technology: A knowledge management case study. Knowledge Management & E-Learning: An International Journal, 14(2), 223.

[13] Mohsan, S.A.H., Khan, M.A., Noor, F., Ullah, I., & Alsharif, M.H. (2022). Towards the Unmanned Aerial Vehicles (UAVs): A Comprehensive Review. Drones, 6(6), 147.

[14] Dai, M., Huang, N., Wu, Y., Gao, J., & Su, Z. (2022). Unmanned-Aerial-Vehicle-Assisted Wireless Networks: Advancements, Challenges, and Solutions. IEEE Internet of Things Journal, 10(5), 4117-4147.

[15] Riaz, N., Shah, S. I. A., Rehman, F., & Gilani, S. O. (2020). An Approach to Measure Functional Parameters for Ball-Screw Drives. In Intelligent Technologies and Applications: Second International Conference, INTAP 2019, Bahawalpur, Pakistan, November 6–8, 2019, Revised Selected Papers 2 (pp. 398-408). Springer Singapore.

[16] Emimi, M., Khaleel, M., & Alkrash, A. (2023). The Current Opportunities and Challenges in Drone Technology.

[17] Saeed, R.A., Omri, M., Abdel-Khalek, S., Ali, E.S., & Alotaibi, M.F. (2022). Optimal path planning for drones based on swarm intelligence algorithm. Neural Computing and Applications, 34(12), 10133-10155.

[18] Riaz, N., Shah, S. I. A., Rehman, F., Gilani, S. O., & Udin, E. (2020). A Novel 2-D Current Signal-Based Residual Learning With Optimized Softmax to Identify Faults in Ball Screw Actuators. IEEE Access, 8, 115299-115313.

[19] Ullah, Z., Al-Turjman, F., Mostarda, L., & Gagliardi, R. (2020). Applications of Artificial Intelligence and Machine Learning in smart cities. Computer Communications, 154, 313-323.

[20] Maghazei, O., & Steinmann, M. (2020). Drones in Railways: Exploring Current Applications and Future Scenarios Based on Action Research. European Journal of Transport and Infrastructure Research, 20(3), 87-102.

[21] Humayoun, M., Sharif, H., Rehman, F., Shaukat, S., Ullah, M., Maqsood, H., ... & Chandio, A. H. (2023, March). From Cloud Down to Things: An Overview of Machine Learning in Internet of Things. In 2023 4th International Conference on Computing, Mathematics and Engineering Technologies (iCoMET) (pp. 1-5). IEEE.

[22] Osco, L.P., Junior, J.M., Ramos, A.P.M., de Castro Jorge, L.A., Fatholahi, S.N., de Andrade Silva, J., Matsubara, E.T., Pistori, H., Gonçalves, W.N., & Li, J. (2021). A review on deep learning in UAV remote sensing. International Journal of Applied Earth Observation and Geoinformation, 102, 102456.

[23] Nguyen, D.D., Rohacs, J., & Rohacs, D. (2021). Autonomous Flight Trajectory Control System for Drones in Smart City Traffic Management. ISPRS International Journal of Geo-Information, 10(5), 338.

[24] Riaz, N., Shah, S. I. A., Rehman, F., & Gilani, S. O. (2020). An Intelligent Approach to Detect Actuator Signal Errors Based on Remnant Filter. In Intelligent Technologies and Applications: Second International Conference, INTAP 2019, Bahawalpur, Pakistan, November 6–8, 2019, Revised Selected Papers 2 (pp. 675-683). Springer Singapore.

[25] H. Sharif, F. Rehman and A. Rida, "Deep Learning: Convolutional Neural Networks for Medical Image Analysis - A Quick Review," 2022 2nd International Conference on Digital Futures and Transformative Technologies (ICoDT2), Rawalpindi, Pakistan, 2022, pp. 1-4, doi: 10.1109/ICoDT255437.2022.9787469.

[26] A. Ashfaq, M. Kamran, F. Rehman, N. Sarfaraz, H. U. Ilyas and H. H. Riaz, "Role of Artificial Intelligence in Renewable Energy and its Scope in Future," 2022 5th International Conference on Energy Conservation and Efficiency (ICECE), Lahore, Pakistan, 2022, pp. 1-6, doi: 10.1109/ICECE54634.2022.9758957.

[27] I. Manan, F. Rehman, H. Sharif, N. Riaz, M. Atif and M. Aqeel, "Quantum Computing and Machine Learning Algorithms - A Review," 2022 3rd International Conference on Innovations in Computer Science & Software Engineering (ICONICS), Karachi, Pakistan, 2022, pp. 1-6, doi: 10.1109/ICONICS56716.2022.10100452.

CHAPTER 19

Infrastructure Resilience and Disaster Management

Tayyaba Basri[1]

[1]Department of Electrical Engineering, National University of Computer and Emerging Sciences (NUCES) Peshawar campus, Peshawar, Pakistan,
Email: tayyababasri56@gmail.com

Introduction

Weather is so unpredictable and can seriously interfere with our phones and Internet's functionality. However, advanced algorithms for computers can be used to predict weather patterns. Networks for communication are crucial in today's world. They help us stay connected through things like Internet and the Wi-Fi. Adverse weather can cause internet connection issues and call loss. Traditional weather prediction methods have limitations, but new approaches use AI and advanced computer programs for more accurate forecasts.

CHAPTER 19 INFRASTRUCTURE RESILIENCE AND DISASTER MANAGEMENT

Wireless AI is an intriguing concept that integrates wireless communication and artificial intelligence. It is similar to placing extremely intelligent computer brains at the network's edge, making the network itself smarter. AI uses five main areas through which it uses data to make wireless networks better; these can be categorized into network device AI, access AI, user device AI, and data provenance AI [2]. AI is now a vital component of disaster management, helping to both prevent and respond to natural disasters. By utilizing the AI technologies, such as machine learning (ML), disaster management can anticipate extreme events, create hazard detection designs for real-time monitoring, and make quick decisions. A branch of AI called machine learning makes it possible for software programmers to increase their prediction accuracy without requiring explicit reprogramming. ML uses historic data to predict new output values. Artificial neural networks (ANNs) have shown themselves to be competitive with conventional regression and statistical models [3]. Figure 19-1 shows the network communication and how they help us connected to Internet and Wi-Fi.

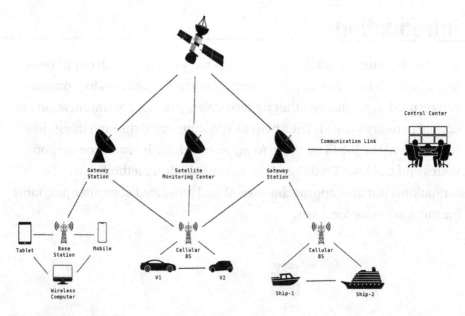

Figure 19-1. Communication network

CHAPTER 19 INFRASTRUCTURE RESILIENCE AND DISASTER MANAGEMENT

Literature Study

Artificial intelligence plays significant role in the vision of next-generation networks (NGNs), which strive to achieve self-optimizing networks zero-touch service management. While AI-enabled neural graph networks (NGNs) have several benefits, they also come with design challenges and critical challenges, especially in dynamic heterogeneous softwarized networks where microservices must be autonomously orchestrated, scaled, and maintained. The development of an improved service-oriented architecture at both the network core and edge levels is being driven by the integration of disruptive technologies with NGNs, such as AI, network softwarization, and hybrid cloud/edge native computing architecture.

To predict real-time edge network flow between a centralized service orchestrator hub and multiple geographically dispersed edge devices, deep learning–based forecast model was used. The results show how accurately the deployed forecaster model predicts edge network flow throughput and latency [4]. There are advantages and disadvantages to combining climate science and artificial intelligence (AI). AI has the potential to greatly enhance the ability to model and comprehend climate systems, but it also introduces new risks, particularly in the form of adversarial machine learning. AI-driven climate models are susceptible to adversarial tactics, which can cause them to predict things like sea levels and temperatures incorrectly. Not only this poses a challenge for modeling, but it also poses a risk to AI systems used in resource management, conversation, and agriculture [5].

References	Advantages	Limitations
[6]	AI can handle massive volume of data, analyze information, and make predictions about things like environmental changes. It is helpful for tasks like merging multiple datasets, eliminating unwanted noise from data, and identifying objects in photos	AI systems require data from satellites, airplanes, or ground instruments. However, this data is under control of government or businesses, or it might not cover long enough time periods; accessing it can be challenging
[7]	Smart cities utilize technology, to keep an eye on the population, welcome people of all backgrounds and disabilities, and act swiftly in the event of an emergency, such as a disaster outbreak	There aren't enough mathematical models for smart cities. This is because the different city boundaries and their interactions with surrounding areas make modeling communities within smart cities complex and difficult to capture the dynamic nature of interactions between smart city components
[8]	AEC's integration of AI, IoT, and big data is critical to achieve Sustainable Development Goals. These technologies contribute to strengthening infrastructure and buildings against external stresses and promote environmentally friendly building techniques	Architects, engineers, data scientists, and computer programmers must collaborate to integrate AI, IoT, and big data in architecture, engineering, and construction (AEC). However, these domains speak different languages; it's not simple to make them function together

(continued)

References	Advantages	Limitations
[9]	AI is a super smart tool that improves our ability to forecast natural disasters. It uses historical data along with current environmental conditions to produce precise forecast regarding events such as earthquakes and hurricanes	AI requires a lot of technology, including computers and Internet connections; using it for environmental conservation can be costly. This could be major issue in smaller communities with less access to technology

Role of AI in Disaster Management

Artificial intelligence (AI) is a critical tool throughout all stages of the disaster response process. AI-powered tools for tracking and mapping, robotics, machine learning, remote sensing, and other application play significant roles in understanding, observing, and managing risks and emergencies. AI, for instance, assists in tracking and mapping disaster-affected areas and analyzes data from multiple sources, including drones and satellites, to determine the extent of damage and plan response efforts effectively. In order to reduce the risks, it also aids in analyzing hotspot area prone to disasters and the planning of smart city infrastructure [10]. Social media platforms are rapidly swamped with massive amount of data following disaster. This flood can overwhelm emergency responders (ERs), especially since a large portion of it may be unnecessary or redundant. The researcher conducted a literature review to understand how AI can analyze large social media data for disaster relief, focusing on text and image classification using convolution neural networks, a common AI algorithm. For this kind of classification task, convolution neural networks (a type of AI algorithm) were commonly used [11]. The application of AI systems in the management of natural disasters, such as hurricanes, flood, fires, and earthquakes, was examined in this study. It classified, compared, and evaluated different datasets related to these disasters according to criteria such as the location and time of their use, AI system created their goals techniques, scope, and ultimate results [12]. Microblogging platforms have become crucial for disseminating critical information during disasters, including both natural and man-made disruptions. People use these platforms to update on the situation, exchange news, check in on loved ones, and plan relief efforts. Following a disaster, individuals use social media to communicate updates in real time. AI system analyzes this flood of information to comprehend better and assist in organizing relief operations. We can better manage calamities and react swiftly to assist those in need by utilizing social media and AI [13].

Cybersecurity in AI-Enhanced Network

The main goal of cybersecurity is to prevent attacks and compromises on our digital resources, which include our networks, systems, and data. Cybercriminals are becoming more skilled at what they do, much as Moore's law predicts that technology will continue to advance quickly. They are constantly improving their attacks tools, making them more affordable and efficient to use [14]. Below table shows the advantages and disadvantages of cybersecurity in AI-enhanced network.

References	Advantages	Disadvantages
[15]	Different cybersecurity issues can be addressed by different deep learning models, such as convolution neural networks. These models are trained via algorithm to make accurate prediction	Neural networks like these can be used in a variety of cybersecurity tasks. But they are sensitive to feature scaling and the need for tuning hyperparameters, which increases computational cost of solving intricate security issues
[16]	About cybersecurity, ChatGPT can assist with threat detection and prevention, user authentication enhancement, security training, and advanced threat intelligence	There might be certain disadvantages to utilizing ChatGPT in cybersecurity, which includes incomplete comprehension of context, vulnerability to social engineering techniques, ability to produce false information, and shallow comprehension of intricate security concepts
[17]	AI has potential to greatly enhance customer and business security in the digital sphere by quickly identifying and addressing threats	Since more personal data is available online, safeguarding against cyberattack is essential to preserving a brand's viability and reputation
[18]	AI is a potent instrument that aids in the defense against constantly changing cyberattacks. By quickly evaluating enormous amount of data and keeping eye on various threats	The fact that biassed data may occasionally be used by AI systems to make decisions is another crucial point, which may give rise to questions about justice and fairness

CHAPTER 19 INFRASTRUCTURE RESILIENCE AND DISASTER MANAGEMENT

Resilient Network Protocol

Being resilient is having the ability to recover swiftly from major setbacks. Because energy is so important to our society, it is imperative that our energy systems be able to quickly recover from unforeseen disruptions. This review examines the language used to describe the resilience of the energy system and how various threats can be evaluated. Main concentration is on two primary threats: cyberattacks, which are increasingly concerning, and extreme weather, which frequently results in power outages. In terms of weather, we consider how events like heat waves and storms impact the availability of energy. With the increasing digitalization of our energy systems, such as smart grids, we can better leverage technology to manage meteorological and technical issues. However, this also exposes to fresh dangers, such as cyberattacks [19]. Data are transferred from a source to a destination in wireless sensor using a number of communication protocol tiers. These protocols manage the physical layer, plan channel slots, restrict access to wireless media, and facilitate application-level communication, among other things. These communication procedures frequently result in high energy consumption at each sensor node, which can significantly reduce overall lifetime of the network [20]. Communication between different real microgrid networks can be impacted by delays and noise. Three consensus protocols are used in their approach to help synchronize the output of various energy resources, such as batteries and solar panels. The maintenance of voltages and frequencies at desired levels even in the event of communication disruptions is ensured by this synchronization. To create these control protocols and guarantee stability in the microgrid system in spite of noise and delays, they make use of mathematical tools and graph theory [21]. Bandwidth management becomes essential in industrial settings where data delivery must be dependable even in the event of network failures. The amount of bandwidth required by each network segment is specified in these contracts. Every contract has a set of monitors who are responsible for monitoring two things: any changes in the requirements while the contract is in operation and any failures of network links. A pathfinding algorithm determines new paths for the data to follow when a monitor notices either of these occurrences. Next, an observer determines whether the revised routes still adhere to the terms outlined in the contracts. If not, it is regarded as a flaw [22].

Integrated Sensor Networks

A new era of technological innovation has been brought about by wireless sensor networks (WSNs), which provide a variety of sensor nodes suited to a wide range of applications. Underground sensors must operate for an extended period of time without frequent maintenance because there are few options for replacing their batteries in difficult-to-reach places like mines or caves. Mobile sensors, capable of traversing different environments, provide real-time data from multiple locations. This allows them to track specific targets or provide insights into dynamic systems [23]. Two main factors have made integrated sensing and communication (ISAC) an essential technology for upcoming wireless systems. First off, high-performance sensing and wireless communication capabilities are required for numerous critical applications in fifth generation (5G) and beyond, including Wi-Fi sensing, autonomous vehicles, and extended reality. Second, the development of massive multiple-input multiple-output (MIMO) and millimeter wave technologies in 5G and beyond leads to communication signals with high resolution in both angular and temporal domains, which makes ISAC possible. As a result, ISAC has attracted a lot of interest from both academia and business [24].

CHAPTER 19 INFRASTRUCTURE RESILIENCE AND DISASTER MANAGEMENT

ISAC offers benefits by integrating communication and radar sensing capabilities into a single system. First off, by combining the two features, it makes effective use of the hardware and limited wireless resources. Better overall performance is achieved by facilitating coordination between sensing and communication tasks [25]. Wireless rechargeable sensor networks (WRSNs) focus on how malware is eliminated from these networks, how infections spread, and how devices are recharged. The study examines the stability of a system, focusing on the relationship between reproductive value and device recharge rate, and considers various malware attack and defense tactics.

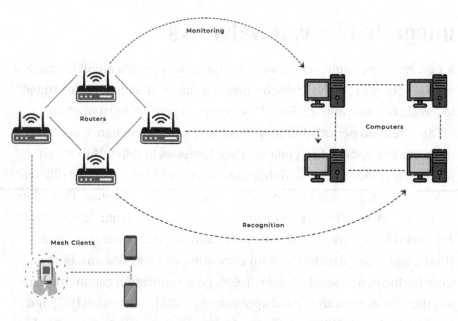

Figure 19-2. Wireless sensor network (WSN)

Future Direction

In envisioning the future direction of AI-enhanced communication networks for disaster resilience, a number of significant trends come to mind. Firstly, the way data is gathered, processed, and disseminated in emergency situations will change as a result of AI algorithms being integrated into communication systems. Drones, satellites, and sensors with AI capabilities can quickly collect a lot of data about the affected areas including population movements, environmental conditions, and damage assessments. Responders are better able to make decisions, thanks to the abundance of data available.

Furthermore, advancement in machine learning algorithms will strengthen communication networks' predictive capacities, enabling the creation of early warning systems that can more accurately predict the frequency and severity of disasters. Artificial intelligence (AI) models can identify regions and communities at risk and enable preemptive measures to lessen the impact of potential disasters by analyzing historical data, weather patterns, and socioeconomic factors. In summary, innovation, collaboration, and adaptability are key components of AI-driven communication networks for preparedness against disasters. These networks could change the way we handle emergencies by utilizing AI to better understand data, predict disasters, coordinate responses, and build resilience. This could save lives and lessen the impact on communities. To fully realize AI's potential in disaster preparedness, however, and to overcome technological, ethical, and legal challenges, researcher must continue to invest in, conduct research, and collaborate on this project.

Conclusion

AI integration into communication networks represents a revolutionary shift in our understanding of disaster resilience. It offers fantastic opportunities to enhance preparedness and response. It's clear that in the future, AI-driven solutions will be even more important in reducing the damaging effects on international communities of both man-made and natural disasters. AI can be used to enhance communication within communities, enabling them to recover more resiliently from adversity. But in order to make this a reality, everyone must be committed to funding, research, and teamwork. It's about making sure AI serves society's interests in hard times.

References

[1] Anaka, R.E., Review of AI-Enhanced Weather Forecasting Application for Communication Networks.

[2] Nguyen, D.C., Cheng, P., Ding, M., Lopez-Perez, D., Pathirana, P.N., Li, J., Seneviratne, A., Li, Y. and Poor, H.V., 2020. Enabling AI in Future Wireless Networks: A Data Life Cycle Perspective. *IEEE Communications Surveys & Tutorials*, 23(1), pp. 553-595.

[3] Kuglitsch, M., Albayrak, A., Aquino, R., Craddock, A., Edward-Gill, J., Kanwar, R., Koul, A., Ma, J., Marti, A., Menon, M. and Pelivan, I., Artificial Intelligence for Disaster Risk Reduction: Opportunities, difficulties, and industry.

[4] Zeb, S., Rathore, M.A., Hassan, S.A., Raza, S., Dev, K. and Fortino, G., 2023. Toward AI-Enabled NextG Networks with Edge Intelligence-Assisted Microservice Orchestration. *IEEE Wireless Communications*, 30(3), pp. 148-156.

[5] Calengor, S., Katragadda, S.P. and Steier, J., 2023. Adversarial Threats in Climate AI: Navigating Challenges and Crafting Resilience. In *Proceedings of the AAAI Symposium Series* (Vol. 2, No. 1, pp. 46-53).

[6] Janga, B., Asamani, G.P., Sun, Z. and Cristea, N., 2023. A Review of Practical AI for Remote Sensing in Earth Sciences. *Remote Sensing, 15*(16), p. 4112.

[7] Bragazzi, N.L., Dai, H., Damiani, G., Behzadifar, M., Martini, M. and Wu, J., 2020. How Big Data and Artificial Intelligence Can Help Better Manage the COVID-19 Pandemic. *International journal of environmental research and public health, 17*(9), p. 3176.

[8] Rane, N., 2023. Integrating Leading-Edge Artificial Intelligence (AI), Internet of Things (IOT), and Big Data Technologies for Smart and Sustainable Architecture, Engineering and Construction (AEC) Industry: Challenges and Future Directions. *Engineering and Construction (AEC) Industry: Challenges and Future Directions (September 24, 2023).*

[9] Chisom, O.N., Biu, P.W., Umoh, A.A. and Obehioye, B., 2024. Reviewing the role of AI in environmental monitoring and conservation: A data-driven revolution for our planet.

[10] Abid, S.K., Sulaiman, N., Chan, S.W., Nazir, U., Abid, M., Han, H., Ariza-Montes, A. and Vega-Muñoz, A., 2021. Toward an Integrated Disaster Management Approach: How Artificial Intelligence Can Boost Disaster Management. *Sustainability, 13*(22), p. 12560.

[11] Nunavath, V. and Goodwin, M., 2018, December. The Role of Artificial Intelligence in Social Media Big Data Analytics for Disaster Management-Initial Results of a Systematic Literature Review. In *2018 5th International Conference on information and communication technologies for disaster management (ICT-DM)* (pp. 1-4). IEEE.

[12] Şimşek, D., Kutlu, İ. and Şık, B., 2023, October. The role and applications of artificial intelligence (AI) in disaster management. In *3rd International Civil Engineering and Architecture Congress*.

[13] Khattar, A. and Quadri, S.M.K., 2020, March. Emerging Role of Artificial Intelligence for Disaster Management Based on Microblogged Communication. In *Proceedings of the International Conference on Innovative Computing & Communications (ICICC)*.

[14] Zeadally, S., Adi, E., Baig, Z. and Khan, I.A., 2020. Harnessing Artificial Intelligence Capabilities to Improve Cybersecurity. *IEEE Access*, 8, pp. 23817-23837.

[15] Ghillani, D., 2022. Deep Learning and Artificial Intelligence Framework to Improve the Cyber Security. *Authorea Preprints*.

[16] Kalla, D. and Kuraku, S., 2023. Advantages, Disadvantages and Risks Associated with ChatGPT and AI on Cybersecurity. *Journal of Emerging Technologies and Innovative Research*, 10(10).

[17] Basnet, A., 2022. Artificial intelligence in cyber security.

CHAPTER 19   INFRASTRUCTURE RESILIENCE AND DISASTER MANAGEMENT

[18] Dai, D. and Boroomand, S., 2022. A Review of Artificial Intelligence to Enhance the Security of Big Data Systems: State-of-Art, Methodologies, Applications, and Challenges. *Archives of Computational Methods in Engineering*, 29(2), pp. 1291-1309.

[19] Jasiūnas, J., Lund, P.D. and Mikkola, J., 2021. Energy system resilience–A review. *Renewable and Sustainable Energy Reviews*, 150, p. 111476.

[20] Dhabliya, D., Soundararajan, R., Selvarasu, P., Balasubramaniam, M.S., Rajawat, A.S., Goyal, S.B., Raboaca, M.S., Mihaltan, T.C., Verma, C. and Suciu, G., 2022. Energy-Efficient Network Protocols and Resilient Data Transmission Schemes for Wireless Sensor Networks—An Experimental Survey. *Energies*, 15(23), p. 8883.

[21] Lai, J., Lu, X., Dong, Z. and Cheng, S., 2021. Resilient Distributed Multiagent Control for AC Microgrid Networks Subject to Disturbances. *IEEE Transactions on Systems, Man, and Cybernetics: Systems*, 52(1), pp. 43-53.

[22] Jhaveri, R.H., Ramani, S.V., Srivastava, G., Gadekallu, T.R. and Aggarwal, V., 2021. Fault-Resilience for Bandwidth Management in Industrial Software-Defined Networks. *IEEE Transactions on Network Science and Engineering*, 8(4), pp. 3129-3139.

[23] Landaluce, H., Arjona, L., Perallos, A., Falcone, F., Angulo, I. and Muralter, F., 2020. A Review of IoT Sensing Applications and Challenges Using RFID and Wireless Sensor Networks. *Sensors*, 20(9), p. 2495.

[24] Liu, A., Huang, Z., Li, M., Wan, Y., Li, W., Han, T.X., Liu, C., Du, R., Tan, D.K.P., Lu, J. and Shen, Y., 2022. A Survey on Fundamental Limits of Integrated Sensing and Communication. *IEEE Communications Surveys & Tutorials*, 24(2), pp. 994-1034.

[25] Cui, Y., Liu, F., Jing, X. and Mu, J., 2021. Integrating Sensing and Communications for Ubiquitous IoT: Applications, Trends, and Challenges. *IEEE Network*, 35(5), pp. 158-167.

[26] Mugunthan, D.S., 2021. Wireless Rechargeable Sensor Network Fault Modeling and Stability Analysis. *Journal of Soft Computing* Paradigm, 3(1), pp. 47-54.

CHAPTER 20

Innovations and Future Directions in UAV Swarm for Protecting Smart Cities

Manish Thakral* (Senior Consultant, Ernst & Young (EY), Mumbai, India, manishthakra@gmail.com)
Trisha Polly (Consultant, Deloitte India, Mumbai, India, trishachakkalamattam@gmail.com)
Nilesh Jadhav (Consultant, KPMG India, Mumbai, India, nillyjadhav209@gmail.com)

Introduction

Smart cities represent the convergence of technology, infrastructure, and data to improve efficiency, sustainability, and quality of life for urban residents. However, the rapid urbanization and digitalization of cities also pose new challenges for security and safety. In response to these challenges, unmanned aerial vehicle (UAV) swarm technology emerges as a transformative solution for protecting smart cities against a diverse array of threats, ranging from security breaches and terrorist attacks to natural disasters and public health emergencies.

CHAPTER 20 INNOVATIONS AND FUTURE DIRECTIONS IN UAV SWARM FOR PROTECTING SMART CITIES

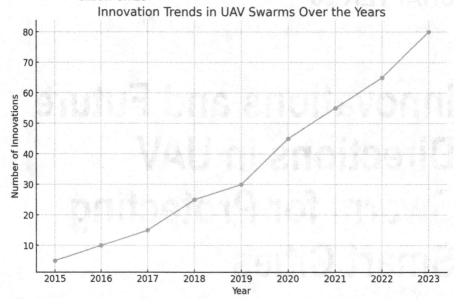

Figure 20-1. *Representation of the Number of Innovations Against the Years 2015-2023*

This chapter explores the innovations and future directions in utilizing UAV swarm technology for safeguarding smart cities. It begins by providing an overview of the concept of smart cities and the security challenges they face in an increasingly interconnected and digitized world. Smart cities leverage technologies such as Internet of Things (IoT), artificial intelligence (AI), and big data analytics to optimize urban operations, enhance sustainability, and improve citizen services. However, these advancements also introduce new vulnerabilities and risks that must be addressed to ensure the safety and security of city residents and infrastructure.

The chapter then introduces the concept of UAV swarms and their potential applications in smart city protection. UAV swarms consist of multiple autonomous drones operating collaboratively to achieve a common goal, such as surveillance, monitoring, or response to emergencies. Unlike traditional single-drone operations, UAV swarms

CHAPTER 20 INNOVATIONS AND FUTURE DIRECTIONS IN UAV SWARM FOR PROTECTING SMART CITIES

offer advantages such as scalability, redundancy, and distributed sensing capabilities, making them well suited for dynamic and complex urban environments.

Next, the chapter explores recent innovations in UAV swarm technology, including advancements in drone design, sensor integration, communication systems, and autonomous navigation algorithms. These innovations enable UAV swarms to operate effectively in urban environments, where they can navigate obstacles, collaborate with each other, and adapt to changing conditions in real time. Furthermore, the integration of AI and machine learning enables UAV swarms to analyze data, identify patterns, and make intelligent decisions autonomously, enhancing their capabilities for smart city protection.

The chapter also discusses the potential applications of UAV swarms in smart city protection, including the following:

> Surveillance and Monitoring: UAV swarms can provide real-time aerial surveillance and monitoring of critical infrastructure, public spaces, and events, enabling early detection of security threats, criminal activities, and safety hazards.
>
> Emergency Response and Disaster Management: In the event of natural disasters, accidents, or public health emergencies, UAV swarms can be deployed for search and rescue operations, damage assessment, and logistics support, facilitating rapid response and recovery efforts.
>
> Infrastructure Inspection and Maintenance: UAV swarms can conduct routine inspections of infrastructure assets such as bridges, buildings, and utilities, detecting defects, monitoring structural integrity, and optimizing maintenance schedules to ensure the reliability and resilience of smart city infrastructure.

CHAPTER 20   INNOVATIONS AND FUTURE DIRECTIONS IN UAV SWARM FOR PROTECTING SMART CITIES

Source of Data

To explore the innovations and future directions in UAV swarm technology for protecting smart cities, a comprehensive range of sources is utilized. These sources provide valuable insights, data, and perspectives on the development, application, and impact of UAV swarm technology in urban security and resilience. The following categories of sources are leveraged:

> Academic Research Papers: Peer-reviewed academic journals and conference proceedings serve as primary sources of scholarly research on UAV swarm technology and its applications in smart city protection. These papers provide in-depth analyses, experimental findings, and theoretical frameworks that contribute to understanding the capabilities, challenges, and opportunities of UAV swarm systems.

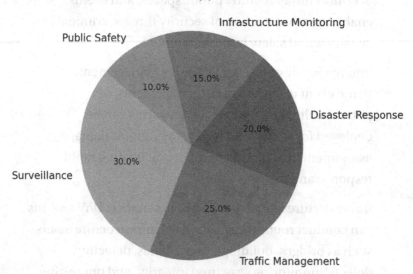

Figure 20-2. *Representation of Distribution of UAV Swarm Applications in Smart Cities*

CHAPTER 20 INNOVATIONS AND FUTURE DIRECTIONS IN UAV SWARM FOR PROTECTING SMART CITIES

Industry Reports and Case Studies: Reports and case studies from industry organizations, consulting firms, and technology providers offer practical insights into the deployment, adoption, and performance of UAV swarm technology in real-world settings. These reports highlight use cases, best practices, and lessons learned from successful implementations of UAV swarm systems for smart city protection.

Government Publications and Policies: Government agencies, regulatory bodies, and policymakers publish reports, guidelines, and policies related to UAV operations, urban security, and smart city initiatives. These publications provide insights into regulatory frameworks, safety standards, and strategic priorities shaping the development and deployment of UAV swarm technology in smart cities.

Organizations and industry consortia create technical standards and specifications for UAV systems, communication protocols, and interoperability requirements, ensuring compatibility, reliability, and safety in smart city protection.

Expert Interviews and Surveys: Interviews with subject matter experts, practitioners, and stakeholders in the fields of UAV technology, urban planning, security, and emergency management offer firsthand perspectives, insights, and experiences regarding the use of UAV swarm technology in smart city protection. Surveys and opinion polls may also be conducted to gather feedback and perceptions from relevant stakeholders.

Open Data Repositories and Datasets: Open data repositories and datasets provide access to real-world datasets, simulations, and testbeds for evaluating UAV swarm algorithms, communication protocols, and mission planning strategies. These datasets enable researchers to validate their models, algorithms, and simulations in diverse urban environments.

Literature Review

The literature on UAV swarm technology for protecting smart cities encompasses a wide range of research studies, technical reports, case studies, and academic publications. This section provides a condensed review of key literature themes, findings, and insights relevant to the innovations and future directions in UAV swarm technology for smart city protection.

Technological Advancements: Numerous studies have highlighted the rapid advancements in UAV swarm technology, including drone design, sensor integration, communication protocols, and autonomous navigation algorithms. Researchers have explored innovative approaches to enhancing swarm coordination, scalability, and resilience in dynamic urban environments.

CHAPTER 20 INNOVATIONS AND FUTURE DIRECTIONS IN UAV SWARM FOR PROTECTING SMART CITIES

Applications in Smart City Protection: A significant body of literature focuses on the diverse applications of UAV swarms in smart city protection, including surveillance, monitoring, emergency response, and infrastructure inspection. Case studies and real-world deployments demonstrate the effectiveness of UAV swarms in enhancing situational awareness, response times, and resource allocation in urban security operations.

Integration with Smart City Infrastructure: Studies have examined the integration of UAV swarm technology with smart city infrastructure, such as IoT networks, sensor networks, and data analytics platforms. Integration with smart city systems enables seamless data sharing, real-time decision support, and automated response actions, enhancing the effectiveness and efficiency of UAV swarm operations.

CHAPTER 20 INNOVATIONS AND FUTURE DIRECTIONS IN UAV SWARM FOR PROTECTING SMART CITIES

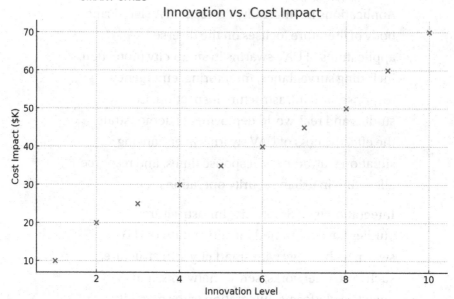

Figure 20-3. *Representation of Innovation Level vs. Cost Impact*

Challenges and Limitations: Despite the promise of UAV swarm technology, literature also highlights various challenges and limitations, including regulatory barriers, safety concerns, privacy implications, and technical complexities. Researchers have identified the need for addressing these challenges through interdisciplinary collaboration, policy frameworks, and technological innovations to realize the full potential of UAV swarms for smart city protection.

Future Directions and Opportunities: The literature points toward exciting future directions and opportunities for UAV swarm technology in smart city protection. These include advancements in AI-driven analytics, edge computing, swarm intelligence, and human-machine interaction, as well as exploring novel applications in areas such as environmental monitoring, crowd management, and public health surveillance.

Experimental Setup

The experimental setup for studying UAV swarm technology in the context of smart city protection plays a crucial role in evaluating the feasibility, effectiveness, and scalability of swarm-based solutions. This chapter delves into the key components, considerations, and methodologies involved in designing and implementing experimental setups for UAV swarm experiments tailored to smart city environments.

Hardware Infrastructure

The hardware infrastructure forms the foundation of the experimental setup, comprising UAV platforms, ground control stations, communication systems, and sensor payloads. Key components of the hardware infrastructure include the following:

> UAV Platforms: Selecting appropriate UAV platforms capable of autonomous flight, collaborative behavior, and payload flexibility is essential for conducting swarm experiments. Fixed-wing, rotary-wing, and hybrid UAVs offer different capabilities suited to various smart city applications, such as aerial surveillance, mapping, and emergency response.

> Ground Control Stations (GCS): GCS serve as the command and control interface for managing and coordinating UAV swarm operations. Advanced

GCS software enables mission planning, real-time monitoring, and telemetry data visualization, facilitating seamless communication and collaboration among swarm members.

Communication Systems: Reliable communication links between UAVs and the GCS are critical for maintaining situational awareness, coordinating swarm behavior, and ensuring safe operation in urban environments. Long-range radio systems, mesh networks, and cellular connectivity provide robust communication channels for UAV swarm experiments.

Sensor Payloads: Equipping UAVs with sensor payloads, such as cameras, LiDAR, thermal imagers, and environmental sensors, enhances their capabilities for surveillance, mapping, and environmental monitoring. Sensor fusion techniques enable multimodal data acquisition and analysis, enriching the information available to swarm members during operations.

Software Platforms

Software platforms play a vital role in designing, simulating, and executing UAV swarm experiments. Key software components include the following:

Simulation Environments: Simulation tools, such as Gazebo, AirSim, and ROS (Robot Operating System), enable virtual modeling and simulation of UAV swarm behavior, environmental dynamics, and mission scenarios. Simulation environments facilitate rapid prototyping, testing, and validation of swarm algorithms and strategies before deployment in real-world settings.

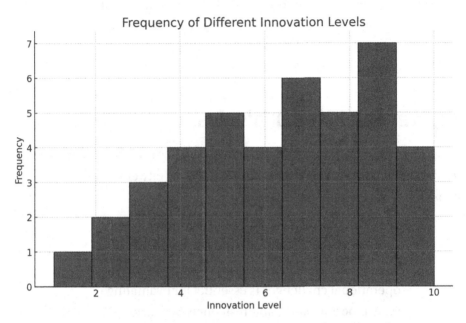

Figure 20-4. *Representation of Innovation Level Against the Frequency*

Mission Planning Software: Mission planning software provides intuitive interfaces for defining mission objectives, waypoints, flight paths, and task assignments for UAV swarm operations. Advanced mission planning algorithms optimize swarm trajectories, resource allocation, and task scheduling to achieve mission objectives efficiently.

CHAPTER 20 INNOVATIONS AND FUTURE DIRECTIONS IN UAV SWARM FOR PROTECTING SMART CITIES

Swarm Control Algorithms: Swarm control algorithms govern the behavior, coordination, and decision-making of individual UAVs within the swarm. These algorithms leverage techniques such as distributed control, consensus algorithms, and artificial intelligence to enable swarm members to collaborate, communicate, and adapt to dynamic environments.

Experimental Methodologies

Experimental methodologies define the procedures, protocols, and metrics for conducting UAV swarm experiments in smart city environments. Key considerations include the following:

Scenario Design: Designing realistic and representative scenarios that simulate smart city security threats, emergency situations, and operational challenges is essential for evaluating swarm performance and resilience. Scenario parameters, such as weather conditions, terrain features, and population density, should be carefully calibrated to reflect urban dynamics accurately.

Data Collection and Analysis: Collecting telemetry data, sensor readings, and performance metrics during UAV swarm experiments enables quantitative analysis of swarm behavior, mission performance, and system reliability. Data analysis techniques, such as statistical analysis, machine learning, and simulation-based modeling, provide insights into swarm dynamics and emergent behavior.

Validation and Verification: Validating and verifying swarm algorithms and strategies through rigorous testing and evaluation is crucial for assessing their effectiveness and robustness in real-world scenarios. Comparative studies, benchmarking tests, and field trials validate swarm performance against predefined benchmarks and requirements.

Legal Considerations

The deployment of UAV swarm technology for protecting smart cities raises a myriad of legal considerations, spanning regulatory compliance, privacy protection, liability management, and ethical governance. This chapter explores the legal landscape surrounding UAV swarm operations in smart city environments and highlights key considerations for researchers, developers, policymakers, and stakeholders.

Regulatory Frameworks

UAV swarm operations are subject to a complex regulatory landscape governed by national aviation authorities, civil aviation regulations, and local ordinances. Regulatory requirements encompass aspects such as airspace management, flight safety, pilot licensing, and operational limitations. Researchers and operators of UAV swarm systems must adhere to applicable regulations and obtain necessary permits and approvals for conducting experiments and deployments in smart cities.

CHAPTER 20   INNOVATIONS AND FUTURE DIRECTIONS IN UAV SWARM FOR PROTECTING SMART CITIES

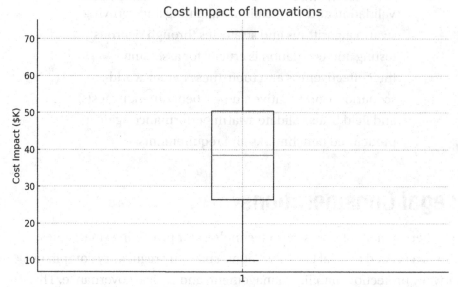

Figure 20-5. Representation of Cost Impact in the UAV Vertical

Privacy Protection

The use of UAV swarm technology for surveillance and monitoring in smart cities raises significant privacy concerns related to data collection, storage, and use. Privacy laws and regulations, such as the General Data Protection Regulation (GDPR) in Europe and the California Consumer Privacy Act (CCPA) in the United States, impose strict requirements on the processing of personal data collected by UAV swarms. Implementing privacy-by-design principles, anonymizing data, and obtaining informed consent from affected individuals are essential for protecting privacy rights and complying with legal requirements.

Liability Management

The deployment of UAV swarm technology introduces liability risks for operators, manufacturers, developers, and users in the event of accidents, property damage, or personal injury. Liability may arise due to negligence, product defects, or failure to adhere to safety standards. Establishing clear liability frameworks, insurance coverage, and risk mitigation strategies is essential for addressing liability concerns and ensuring accountability in UAV swarm operations.

Ethical Governance

Ethical considerations surrounding the use of UAV swarm technology in smart city protection include issues such as autonomy, transparency, accountability, and societal impact. Ethical guidelines, such as the IEEE Ethically Aligned Design framework and the ACM Code of Ethics and Professional Conduct, provide principles and guidelines for responsible and ethical use of UAV swarms. Addressing ethical considerations requires a holistic approach that balances technological innovation with ethical values, human rights, and societal welfare.

Public Engagement and Stakeholder Consultation

Engaging communities, stakeholders, and end users in the development and deployment of UAV swarm systems is essential for building trust, fostering acceptance, and addressing concerns related to privacy, safety, and ethical implications. Public consultation, participatory design workshops, and stakeholder engagement initiatives enable meaningful dialogue, collaboration, and co-creation of UAV swarm solutions that align with societal values and priorities.

CHAPTER 20 INNOVATIONS AND FUTURE DIRECTIONS IN UAV SWARM FOR PROTECTING SMART CITIES

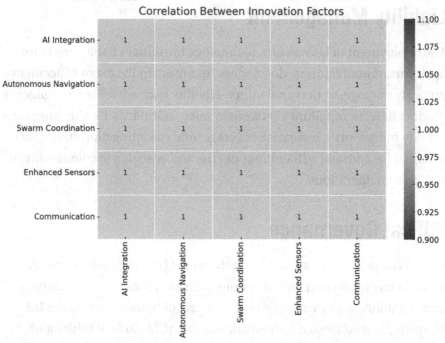

Figure 20-6. Representation of Heat Map for the Different UAV Swarm for Protecting Smart Cities

Limitation and Conclusion

Limitations

While UAV swarm technology holds immense promise for protecting smart cities, several limitations and challenges must be addressed to realize its full potential:

>Regulatory Constraints: Complex regulatory frameworks governing UAV operations pose challenges for researchers and operators seeking to deploy UAV swarm systems in smart

CHAPTER 20 INNOVATIONS AND FUTURE DIRECTIONS IN UAV SWARM FOR PROTECTING SMART CITIES

city environments. Compliance with aviation regulations, airspace restrictions, and privacy laws requires significant time, resources, and expertise.

Technical Complexity: UAV swarm systems involve complex coordination, communication, and control mechanisms, which can be challenging to implement and maintain. Technical issues such as communication latency, sensor integration, and algorithm scalability may limit the effectiveness and reliability of UAV swarm operations.

Safety Concerns: Safety is a paramount concern in UAV swarm operations, particularly in urban environments with dense population centers and critical infrastructure. Mitigating risks of collisions, airspace congestion, and system failures requires robust safety protocols, redundancy measures, and fail-safe mechanisms.

Privacy Implications: The use of UAV swarms for surveillance and monitoring raises privacy concerns related to data collection, storage, and use. Safeguarding privacy rights and complying with data protection laws necessitates careful consideration of privacy-by-design principles and ethical guidelines in UAV swarm deployments.

CHAPTER 20 INNOVATIONS AND FUTURE DIRECTIONS IN UAV SWARM FOR PROTECTING SMART CITIES

Conclusion

In conclusion, UAV swarm technology represents a transformative approach to protecting smart cities against emerging security threats and challenges. Despite the limitations and challenges, the potential benefits of UAV swarm systems for enhancing situational awareness, response capabilities, and resilience in urban environments are substantial.

Looking ahead, addressing regulatory constraints, technical complexities, safety concerns, and privacy implications will be critical for advancing the adoption and deployment of UAV swarm technology in smart city protection. Collaborative efforts between government agencies, industry stakeholders, academic institutions, and civil society organizations are essential for developing regulatory frameworks, technical standards, and best practices that facilitate responsible and ethical use of UAV swarm systems.

Furthermore, continued research and innovation in areas such as autonomy, communication protocols, sensor technologies, and human-machine interaction will drive advancements in UAV swarm technology and unlock new capabilities for smart city protection. Interdisciplinary collaboration, knowledge sharing, and capacity building initiatives will foster a vibrant ecosystem of research, development, and deployment in UAV swarm technology, paving the way for safer, more resilient, and sustainable smart cities of the future. By addressing limitations and embracing opportunities for innovation, stakeholders can harness the full potential of UAV swarm technology to create cities that are safer, more livable, and better prepared to address the complex challenges of the 21st century.

CHAPTER 20   INNOVATIONS AND FUTURE DIRECTIONS IN UAV SWARM FOR PROTECTING SMART CITIES

Future Scope

The future scope of UAV swarm technology for protecting smart cities is vast and encompasses a myriad of opportunities for innovation, research, and application. This chapter explores the emerging trends, technological advancements, and potential applications that will shape the future of UAV swarm systems in safeguarding urban environments.

Advancements in Autonomy and Intelligence

Future UAV swarm systems will exhibit enhanced autonomy, intelligence, and adaptive behavior, enabling them to operate seamlessly in complex urban environments. Advancements in AI, machine learning, and swarm intelligence algorithms will empower UAV swarms to learn from experience, anticipate threats, and optimize mission performance in real time.

Integration with Emerging Technologies

UAV swarm technology will increasingly integrate with other emerging technologies such as 5G networks, edge computing, and IoT devices, enabling seamless connectivity, real-time data processing, and distributed decision-making. Integration with smart city infrastructure and digital twin models will enhance situational awareness, resilience, and response capabilities in urban security operations.

Multidomain Collaboration

Future UAV swarm systems will collaborate with other autonomous systems, including ground robots, autonomous vehicles, and smart sensors, to form multidomain networks for comprehensive smart city protection. Cross-domain collaboration enables synergistic interactions, shared situational awareness, and coordinated response actions across different layers of urban infrastructure.

Environmental Sensing and Monitoring

UAV swarm systems will play a critical role in environmental sensing, monitoring, and disaster management in smart cities. Equipped with advanced sensor payloads, UAV swarms will provide real-time data on air quality, pollution levels, weather patterns, and natural disasters, enabling proactive interventions and informed decision-making to mitigate environmental risks.

Human–Machine Interaction and Augmented Reality

Advancements in human–machine interaction technologies, such as gesture recognition, natural language processing, and augmented reality interfaces, will enable intuitive control and collaboration between operators and UAV swarms. Augmented reality overlays will provide operators with immersive visualizations, real-time analytics, and decision support tools for managing swarm operations in smart cities.

Ethical and Societal Implications

Addressing ethical, legal, and societal implications of UAV swarm technology will remain a priority in future research and development efforts. Responsible innovation, ethical design principles, and stakeholder engagement initiatives will ensure that UAV swarm systems align with societal values, respect privacy rights, and enhance public trust in smart city protection.

Future research in UAV swarm technology will focus on advancing swarm dynamics to achieve greater efficiency, scalability, and robustness in smart city protection. Swarm algorithms will evolve to enable self-organization, adaptive behavior, and fault tolerance, allowing UAV swarms to autonomously adapt to changing environmental conditions, mission requirements, and system constraints. Multiagent coordination mechanisms, such as distributed consensus algorithms and swarm optimization techniques, will enable UAV swarms to collaboratively solve complex tasks, optimize resource allocation, and achieve collective objectives with minimal centralized control.

Resilience and Fault Tolerance

Ensuring the resilience and fault tolerance of UAV swarm systems will be critical for overcoming challenges such as communication failures, sensor malfunctions, and adverse weather conditions in urban environments. Future research will focus on developing resilient swarm architectures, redundancy mechanisms, and decentralized decision-making strategies that enable UAV swarms to continue operating effectively in the face of disruptions and failures. Fault detection, isolation, and recovery techniques will enable UAV swarms to self-diagnose and mitigate faults autonomously, ensuring continuous mission performance and system reliability.

Privacy-Preserving Technologies

Addressing privacy concerns associated with UAV swarm operations will be a key area of focus in future research and development efforts. Privacy-preserving technologies, such as secure multiparty computation, differential privacy, and federated learning, will enable UAV swarms to collect and process data while preserving the privacy and anonymity of individuals. Privacy-enhancing techniques, such as data encryption, anonymization, and access controls, will be integrated into UAV swarm systems to ensure compliance with privacy regulations and protect sensitive information from unauthorized access or disclosure.

Collaborative Governance Models

Developing collaborative governance models and regulatory frameworks for UAV swarm technology will be essential for promoting responsible and ethical use of UAV swarms in smart city protection. Collaborative governance approaches involve engaging stakeholders from government, industry, academia, and civil society in participatory decision-making processes, policy development, and standard-setting initiatives. By fostering dialogue, transparency, and accountability, collaborative governance models enable stakeholders to collectively address regulatory challenges, ethical concerns, and societal implications of UAV swarm technology, ensuring that deployments are aligned with public values and interests.

References

[1] Kaushik, K. (2022). A Novel Approach to Secure Files Using Color Code Authentication. In: Sugumaran, V., Upadhyay, D., Sharma, S. (eds) Advancements in Interdisciplinary Research. AIR 2022. Communications in Computer and Information Science, vol 1738. Springer, Cham. https://doi.org/10.1007/978-3-031-23724-9_4.

[2] Kaushik, K., Singh, V., Manikandan, V.P. (2022). A Novel Approach for an Automated Advanced MITM Attack on IoT Networks. In: Sugumaran, V., Upadhyay, D., Sharma, S. (eds) Advancements in Interdisciplinary Research. AIR 2022. Communications in Computer and Information Science, vol 1738. Springer, Cham. https://doi.org/10.1007/978-3-031-23724-9_6.

[3] K. Kaushik and V. Naik, "Making Ductless-split Cooling Systems Energy Efficient using IoT," 2023 15th International Conference on Communication Systems & Networks (COMSNETS), Bangalore, India, 2023, pp. 471-473, doi: 10.1109/COMSNETS56262.2023.10041408.

[4] Porwal, P., & Devare, M. (2024, August 1). Scientific impact analysis: Unraveling the link between linguistic properties and citations. Journal of Informetrics. https://doi.org/10.1016/j.joi.2024.101526.

[5] Porwal, P., & Devare, M. (2023, September 20). Citation count prediction using weighted latent semantic analysis (wlsa) and three-layer-deep-learning paradigm: a meta-heuristic approach. Multimedia Tools and Applications. https://doi.org/10.1007/s11042-023-16957-8.

[6] P. Porwal and M. Devare, "Citation Classification Prediction Implying Text Features Using Natural Language Processing and Supervised Machine Learning Algorithms," Communications in computer and information science, Jan. 01, 2021. https://doi.org/10.1007/978-981-16-0507-9_46.

[7] K. A. Tatkare and M. Devare, "A Novel Region Duplication Detection Algorithm Based on Hybrid Approach," Social Science Research Network, Jan. 01, 2019. https://doi.org/10.2139/ssrn.3425340.

[8] Fuzzy Probability Model for Quantifying the Effectiveness of the MSW Compost," IEEE Conference Publication I IEEE Xplore, Jan. 01, 2019. https://ieeexplore.ieee.org/document/8945830.

[9] A. Shitole and M. Devare, "Machine Learning Supported Statistical Analysis of IoT Enabled Physical Location Monitoring Data," Springer eBooks, Jan. 01, 2020. https://doi.org/10.1007/978-3-030-41862-5_13.

[10] M.. Devare and M.. Thakral, "Enhancing Automatic Speech Recognition System Performance for Punjabi Language through Feature Extraction and Model Optimization", Int J Intell Syst Appl Eng, vol. 12, no. 8s, pp. 307–313, Dec. 2023.

[11] "Designing an Automatic Speech Recognition System for the Minor Age Group," IEEE Conference Publication I IEEE Xplore, Dec. 15, 2023. https://ieeexplore.ieee.org/abstract/document/10431311.

[12] M. Thakral, R. K. Singh, and K. Kaushik, "Integration of Blockchain Technology and Intelligent System for Potential Technologies," Chapman and Hall/CRC eBooks, Aug. 23, 2022. https://www.taylorfrancis.com/chapters/edit/10.1201/9781003193425-6/integration-blockchain-technology-intelligent-system-potential-technologies-manish-thakral-rishi-raj-singh-keshav-kaushi.

[13] M. Thakral and S. Singh, "A Secure Bank Transaction Using Blockchain Computing and Forest Oddity," Blockchain technologies, Jan. 01, 2022. https://link.springer.com/chapter/10.1007/978-981-19-1960-2_6.

[14] Thakral, M., Singh, R.R., Singh, S.P. (2022). An Extensive Framework Focused on Smart Agriculture Based Out of IoT. In: Choudhury, A., Singh, T.P., Biswas, A., Anand, M. (eds) Evolution of Digitized Societies Through Advanced Technologies. Advanced Technologies and Societal Change. Springer, Singapore. https://doi.org/10.1007/978-981-19-2984-7_12.

[15] M. Thakral, R. K. Singh, and B. V. Kalghatgi, "Cybersecurity and Ethics for IoT System: A Massive Analysis," Transactions on computer systems and networks, Jan. 01, 2022. https://link.springer.com/chapter/10.1007/978-981-19-1585-7_10.

[16] Singh, R.R., Thakral, M., Kaushik, S., Jain, A., Chhabra, G. (2022). A Blockchain-Based Expectation Solution for the Internet of Bogus Media. In: Hemanth, D.J., Pelusi, D., Vuppalapati, C. (eds) Intelligent Data Communication Technologies and Internet of Things. Lecture Notes on Data Engineering and Communications Technologies, vol 101. Springer, Singapore. https://doi.org/10.1007/978-981-16-7610-9_28.

[17] M. Thakral, A. Jain, V. Kadyan and A. Jain, "An Innovative Intelligent Solution Incorporating Artificial Neural Networks for Medical Diagnostic Application," 2021 Sixth International Conference on Image Information Processing (ICIIP), Shimla, India, 2021, pp. 529-532, doi: 10.1109/ICIIP53038.2021.9702631.

[18] M. Thakral, R. R. Singh, A. Jain and G. Chhabra, "Rigid Wrap ATM Debit Card Fraud Detection Using Multistage Detection," 2021 6th International Conference on Signal Processing, Computing and Control (ISPCC), Solan, India, 2021, pp. 774-778, doi:10.1109/ISPCC53510.2021.9609521.

[19] I. Girish, A. Kumar, A. Kumar and A. M, "Driver Fatigue Detection," 2020 IEEE 17th India Council International Conference (INDICON), New Delhi, India, 2020, pp. 1-6, doi: 10.1109/INDICON49873.2020.9342456.

[20] Prajeesha and A. M, "EDGE Computing Application in SMART GRID-A Review," 2021 Second International Conference on Electronics and Sustainable Communication Systems (ICESC), 2021, pp. 1-6, doi: 10.1109/ICESC51422.2021.9532792.

[21] D. M. Udeshi, S. G. L. Divakarla, N. C. Rajdev and A. M, "Wind Speed Forecasting using Hybrid Model," 2022 IEEE 7th International conference for Convergence in Technology (I2CT), Mumbai, India, 2022, pp. 1-5, doi: 10.1109/I2CT54291.2022.9823995.

[22] Tammana, A., Amogh, M.P., Gagan, B., Anuradha, M., Vanamala, H.R. (2021). Thermal Image Processing and Analysis for Surveillance UAVs. In: Kaiser, M.S., Xie, J., Rathore, V.S. (eds) Information and Communication Technology for Competitive Strategies (ICTCS 2020). Lecture Notes in Networks and Systems, vol 190. Springer, Singapore. https://doi.org/10.1007/978-981-16-0882-7_50.

[23] S. Balasubramanian, R. Kashyap, S. T. CVN and M. Anuradha, "Hybrid Prediction Model For Type-2 Diabetes With Class Imbalance," 2020 IEEE International Conference on Machine Learning and Applied Network Technologies (ICMLANT), Hyderabad, India, 2020, pp. 1-6, doi: 10.1109/ICMLANT50963.2020.9355975.

[24] M. Yakasiri, A. M and K. B. K, "Comparative Analysis of Markov Chain and Polynomial Regression for the Prognostic Evaluation of Wind Power," 2020 IEEE International Conference for Innovation in Technology (INOCON), Bangluru, India, 2020, pp. 1-5, doi: 10.1109/INOCON50539.2020.9298374.

[25] P. J, V. M, V. G. Pai and A. M, "Comparative Analysis of Marker and Marker-less Augmented Reality in Education," 2020 IEEE International Conference for Innovation in Technology (INOCON), Bangluru, India, 2020, pp. 1-4, doi: 10.1109/INOCON50539.2020.9298303.

[26] M. Yakasiri, J. Avrel, S. Sharma, M. Anuradha and B. K. Keshavan, "A Stochastic Approach for the State-Wise Forecast of Wind Speed Using Discrete-Time Markov Chain," TENCON 2019 - 2019 IEEE Region 10 Conference (TENCON), Kochi, India, 2019, pp. 575-580, doi: 10.1109/TENCON.2019.8929529.

[27] Gunjan Chhabra, Ajay Prasad & Venkatadri Marrabenta (2022) Comparison and performance evaluation of human bio-field visualization algorithm, Archives of Physiology and Biochemistry, 128:2, 321-332, DOI: 10.1080/13813455.2019.1680699.

[28] D. M. Udeshi, S. G. L. Divakarla, N. C. Rajdev and A. M, "Wind Speed Forecasting using Hybrid Model," 2022 IEEE 7th International conference for Convergence in Technology (I2CT), Mumbai, India, 2022, pp. 1-5, doi: 10.1109/I2CT54291.2022.9823995.

[29] Tammana, A., Amogh, M.P., Gagan, B., Anuradha, M., Vanamala, H.R. (2021). Thermal Image Processing and Analysis for Surveillance UAVs. In: Kaiser, M.S., Xie, J., Rathore, V.S. (eds) Information and Communication Technology for Competitive Strategies (ICTCS 2020). Lecture Notes in Networks and Systems, vol 190. Springer, Singapore. https://doi.org/10.1007/978-981-16-0882-7_50.

Index

A

AANET, *see* Aeronautical ad hoc network (AANET)
Accountability, 143, 387, 498, 504
Acoustic sensors, 565–567
Address Resolution Protocol (ARP), 379
Ad hoc networks
    AI technologies
        flying ad hoc network, 301–303
        IoT network connectivity, 307
        MANET, 297–300
        resource management, 317
        robot ad hoc networks, 303–305
        ship ad hoc networks (SANET), 309, 310
        vehicular ad hoc network, 300–301
        vehicular networks, 297
    civil/military domains, 286
    configurations, 285
    contribution points, 287
    cyberattack detection system, 314–315
    design infrastructure/resources, 287
    device communication, 317
    federated techniques, 317
    Internet of Things (IoT), 294–296
    literature review, 315, 316
        attacks, 290
        design infrastructure, 290
        disadvantages, 293
        drawbacks/limitations, 290–294
        FANET, 290
        limitations, 289
        nodes (vehicles), 288
        packet dropping attacks, 289
        routing connection, 289
        routing protocols, 289
        techniques, 288
        vehicular CCN, 289
        Wi-Fi-connection, 290
    mobile nodes, 286
    node positioning, 286
    security, 317
    variations, 287

INDEX

Ad hoc networks (*cont.*)
    wireless communication, 310–313
        data transmission, 312
        logistic regression analysis, 311
        protocols/techniques, 311
        roadside units, 311
        security models, 311
        underwater sensor networks, 313
Ad hoc on-demand distance vector (AODV), 297, 299
Aeronautical ad hoc network (AANET), 286, 290
AG-IoT intelligence
    cyber threats
        differential privacy (DP), 378
        inference attack, 378
        strategies, 378
        training phase, 377
    digital threats
        control system, 376
        navigational systems, 375, 376
        perception-targeted attacks, 374
        planning layer, 376
        planning/navigation/control layers, 374
        repercussions, 374
        software aspects, 373
        traffic signals, 374, 375
    electronics advancement, 362
    ethical considerations, 382
        assistive UAVs, 384
        autonomous, 383, 384
        construction site, 385, 386
        healthcare industry, 382, 383
        manufacturing sector/issues, 386
    IoT/smart cities
        DoS and DDoS attacks, 379, 380
        network server, 380
        selective forwarding attack, 380
        traffic sniffing, 378
        wormhole attack, 381
    key functions, 362
    legal issues, 387
    literature review, 364–367
    modern routing protocols, 365
    physical threats
        attacks, 369
        jamming attacks, 369, 370
        LiDAR sensors, 372
        manipulation attacks, 372, 373
        OpenCV framework, 373
        spoofing, 370–372
        vulnerabilities, 373
        wireless communication technologies, 369
    privacy concerns/security issues, 364, 366, 381, 382
    security issues, 367, 368
    significant risks, 389
    smart homes, 363
    social analysis, 365

INDEX

STAR supervisor, 362
threat landscape, 368
AI, *see* Artificial intelligence (AI)
AI-based anomalous detection systems, 406
Airframe, 546, 547
Ambient sensors, 657
Analysis-ready dataset, 461–462
ANNs, *see* Artificial neural networks (ANNs)
Ant colony optimization (ACO), 14, 256, 300, 567–569
Antennas, 620
Ant pheromone trails, 567
AODV, *see* Ad hoc on-demand distance vector (AODV)
Application programming interfaces (APIs), 218
Application-specific innovations, 571–572
Artificial bee colony (ABC), 14, 297
Artificial fish swarm algorithm (AFSA), 15
Artificial general (AG) intelligence
See also AG-IoT intelligence
Artificial intelligence (AI), 12–13, 128, 287, 293, 297–310, 430, 503, 511, 550, 570, 593, 619, 639, 642, 645, 670, 698, 751, 754, 768
 ad hoc networks (*see* Ad hoc networks)
 applications, 729–734
 capabilities, 198
 challenges, integration, 735, 736
 cognitive perspective, 722
 combination, 722
 communication technologies, 27
 computers/wireless networks, 306
 data processing, 600
 decision-making/coordination, 13, 723
 in disaster management, 755–756
 drone applications, 723
 drone technology, 723
 explainability/transparency, 27
 future outlook, 738–741
 future prospects, 736–738
 historical development, 725–729
 horizons/technological innovations, 114
 human–swarm interaction, 28
 incorporation, 723, 724
 innovative technologies, 12
 integration, 10–11
 IoT networks, 293–296
 ML (*see* Machine learning (ML))
 ML and drones, 734, 735
 models, 761
 multimodal learning, 27
 multiple types, 741
 neural networks, 729
 optical/acoustic communication, 27
 quantum communication, 27

797

INDEX

Artificial intelligence (AI) (*cont.*)
   real-world deployments, 94–95
   regulatory frameworks, 741
   in resource management, 753
   robot ad hoc networks, 304
   robotics, 28
   sensor fusion, 230
   surveillance/monitoring system, 248
   swarm intelligence, 14, 15
   technological application, 724
   traditional approaches, 293
   trends/directions, 28
Artificial neural networks (ANNs), 139, 297, 752
Augmented reality (AR)/virtual reality (VR)
   case study analysis, 210
   coordination complexity, 209
   data overload, 209
   data source, 211
      communication networks, 214
      control responsiveness, 213
      environmental information, 212
      real-time video feeds, 213
      telemetry/sensor data, 212
   directions/opportunities, 229
   ethical/societal considerations, 229
   experimental setup, 218
      cloud-based demonstration, 223
      comparative studies, 224
      data acquisition/analysis tools, 221
      data collection/analysis, 222
      development tools, 220
      experimental methodologies, 221–223
      hardware infrastructure, 219, 220
      input devices, 219
      performance evaluation, 224
      simulation engines, 221
      software platforms, 220, 221
      tasks/scenarios, 222
      usability testing, 223
      user recruitment/training, 221
      validation/evaluation, 223, 224
   future scope, 230
      advancements, 230
      community engagement/stakeholders, 232
      education/workforce development, 232
      emerging application, 231
      ethical/societal implications, 231
      research/innovation, 232
      telepresence/remote collaboration, 231
   hardware/software, 229
   human–machine interaction, 210

immersive training/
    simulation, 211
interdisciplinary
    collaboration, 229
legal considerations
    ethical considerations, 226
    integration, 224
    intellectual property
        rights, 226
    liability risks, 225
    privacy laws/regulations, 225
    regulatory landscape, 225
limitations/challenges, 227
    cost/accessibility, 228
    human factors, 227
    regulatory compliance, 228
    technological
        constraints, 227
literature review, 214
    advancements/future
        directions, 217, 218
    challenges/limitations,
        216, 217
    disaster response/
        emergency
        management, 215
    diverse applications,
        214, 215
    hardware innovations, 217
    human factors, 217
    interdisciplinary
        collaboration, 218
    interoperability/
        integration, 217

software developments, 218
technological
    constraints, 216
real-time visualizations, 208
representation, 208
technology comparison, 209
visualization, 209, 210
Authentication, 71, 348, 422, 496,
    498, 555
Authentication attacks,
    421–422, 424
Authorization, 71, 118, 421,
    496, 498
Automated forest restoration
    (AFR), 613
Autonomous decision-making, 69,
    433, 506
Autonomous navigation, 50,
    65, 66, 570
Autonomous systems, 426,
    550–551, 572
Autonomy, 7, 9, 785
Autopilot, 412, 550, 595

B

Base stations (BS), 694
Battery models, 479
Battery performances, 477
Battery technology, 22
Behavioral models, 409
Beyond the line of sight
    (BLOS), 552
Big data analytics, 29, 130, 674

INDEX

Biosensors, 658
Bitcoin, 491
Black Hole Attack (BHA), 297
Blockchain application
   authentication, 348
   benefits
      data sharing/
         coordination, 344
      scalability, 345
      transparency, 345
   conceptual model, 337
   consensus approaches, 337
   consensus mechanisms, 348
   data storage/access control
      decentralized storage, 347
      smart contracts, 347
   definition, 337, 338
   energy efficiency, 350
   identity registration/
      management, 347
   interactions, 338
   interoperability, 351
   network security threats, 334
   operational protocols, 349
   primary objectives, 335, 336
   research directions, 352
   revolutionary approach, 334
   scalability/performance,
      349, 350
   secure communication,
      338–340, 345, 346
   security, 338
      chain of blocks, 342
      decentralization, 343
      dynamic network
         topology, 341
      immutability, 342, 343
      integration, 340
      resource constraints, 341
      scalability, 340
      transparency, 344
   swarm operation, 334
   UAV swarms architecture,
      337, 338
Blockchain technology, 407, 430,
   431, 490, 619, 640
   authentication and
      authorization, 496
   data integrity, 495
   decentralized
      coordination, 495
   environmental impact, 509
   ethical considerations, 508
   public perception and
      acceptance, 508
   regulatory and ethical
      issues, 507–510
   reliability and safety, 509
   secured communication,
      495–496
   security challenges
      cybersecurity threats, 506
      data privacy, 507
      physical security, 507
   technical challenges
      autonomous decision-
         making, 506
      communication, 506

INDEX

energy efficiency, 506
scalability, 505
UAV swarm security,
491–494, 510–515
Byzantine fault tolerance
(BFT), 496

C

California Consumer Privacy Act
(CCPA), 780
Camera sensors, 556–558
Carrier sense multiple access
(CSMA), 257
Cascaded short-term memory
(C-LSTM), 140
Cellular communication, 552, 553
Centralized vs. decentralized
approaches, 25
Central processing unit (CPU), 548
Chemical sensors, 657
Closed-circuit television (CCTV)
cameras, 17
Cloud computing
algorithmic and physical
structures, 446
challenges, 445
data collection and
questions, 458–462
IoT, 654, 655
multidisciplinary approach, 445
preliminary development and
validation, 450, 451
regional analysis, 464–469

research trends
publications, 456–458
simulation and real-world
implementation, 469–481
simulation trends, 462–464
UAV operations, 444
UAV swarms, 443
Cluster routing scheme
(CRSF), 302
CNN, *see* Convolutional neural
networks (CNNs)
Cognitive radio systems, 52
Collaborative governance
models, 788
Commercial UAVs, 538, 544
Communication, 457, 497, 506,
661, 662, 664, 672, 693, 694,
698, 712, 735, 759
channels, 513
modules, 602
networks, 761
protocols, 603, 604
systems, 599
Communication protocols
TCP/IP, 553
UDP, 554
Communication systems, 550,
551, 776
cellular, 552, 553
radio frequency (RF), 551, 552
SATCOM, 552
security, 554, 555
Communication technology, 526
Connectivity, 183, 506

801

INDEX

Controlled environment
   risk mitigation, 448
   safe exploration, 448
Convolutional neural networks
      (CNNs), 64, 70, 72, 139, 140, 294, 700
Cost-effectiveness, 642, 644, 661
   financial savings, 448
   resource allocation, 449
Cryptography, 512–513
Cutting-edge technologies, 636
Cyberattack detection system
   attack patterns, 314
   K-median clustering, 315
   SA-IDPS, 314
   supervised/deep learning, 314
Cyberattacks, 404, 410, 493, 496, 735, 758
Cyber-physical systems (CPS), 407, 416, 431, 643
Cybersecurity, 410, 445, 493, 513, 617, 735, 736, 756–757
   at-risk profile, 409
   attacks, 408–409
   autonomy and decision-making, 426
   communication security, 425
   data protection and privacy, 425
   e-commerce, 408
   fly-off behavior, 409
   guidelines, 431
   human factors, 428, 429
   integration and interoperability, 427
   issues, 617
   machine learning, 410
   regulatory and ethical considerations, 427
   resilience, 427
   swarm operations, 407
   UAVs, 404
   UAV swarm operations, 408–411
   vulnerability management, 426
Cyber threats, 404, 735
   authentication attacks, 421, 422
   data transmission, 418–420
   GPS spoofing, 419, 420
   malware and software vulnerabilities, 422, 423
   operational capacity, 418
   physical attacks, 423, 424

D

DAS, *see* Distributed antenna systems (DAS)
Data collection, 604–606
Data integrity, 495
Data privacy, 507
Data processing, 600
Data protection and privacy, 425
Dataset refining steps, 459–461
Data transmission, 418–420, 552, 708
Data transmission link attacks, 424
DDoS, *see* Distributed Denial of Service (DDoS)

INDEX

Decision-making, 506
Decision-making algorithms, 407, 416, 417, 426, 430, 493, 553, 636, 639, 641, 655, 735
Deep convolutional neural network (DCNN), 134
Deep embedded median clustering (DEMC), 315
Deep learning (DL), 48–49, 134, 139, 140, 314, 692
  ad hoc IoT networks, 293–294
  challenges, 65
  definition, 64
  detection/classification, 139
  integration, 139
  neural networks, 64, 139
  structured data, 140
  techniques, 64
Deep Q-network (DQN), 376
Deep reinforcement learning (DRL), 312, 313, 375, 376
Defense-based incorporating drones, 408
Denial-of-service (DoS) attacks, 286, 308, 369, 379, 380, 406, 408, 411, 493, 615
Disaster management, 608, 609, 755–756
Distributed antenna systems (DAS), 177
Distributed Denial of Service (DDoS), 286, 308, 379, 380
DL, *see* Deep learning (DL)
DRL, *see* Deep reinforcement learning (DRL)
Drones, 408, 409, 490, 524, 526, 555, 556, 649, 711, 725, 726, 729
  aerial robots, 92
  analysis/usage, 113–116
  application, 4, 93, 94
  communication tool, 3
  coordinated missions, 49
  definition, 2
  goods and services, 736
  ML, 149, 734, 737 (*see also* Machine learning (ML))
  multispectral/hyperspectral systems, 732
  perception, 731
  security/surveillance, 108, 109
  self-driving, 731
  surveillance, 257
  technology, 729
  See also Unmanned aerial vehicles (UAVs)

E

Eavesdropping attacks, 616
Edge computing, 407, 431, 601, 620, 655, 725, 736, 740, 742
Electric sensors, 659
Electromagnetic battle management (EMBM), 408
Electronic health records (EHRs), 670

803

INDEX

Emergency medical services (EMS), 17
Emergency responders (ERs), 755
Emerging technologies, 785
Enhanced Mobile Broadband (eMBB), 175
Environmental sensors, 602
Explanatory artificial intelligence (XAI), 143, 144, 704

F

Failure mode and effect analysis (FMEA), 77
FANET, *see* Flying ad hoc network (FANET)
Fault tolerance, 787
Federal Aviation Administration (FAA), 225
Federated learning (FL), 142, 144, 147, 704, 705
    ad hoc IoT networks, 294
Feedforward Networks, 64
Fish swarm optimization algorithm (FSOA), 305
5G technologies
    agriculture, 182
    applications, 193
        delivery/logistics, 195
        disaster management, 195
        emergency response, 194
        environmental monitoring, 194
        infrastructure inspection, 194
        precision agriculture, 194, 195
        public safety, 194
        search/rescue operations, 193
        security/surveillance, 193
        urban planning/infrastructure development, 195
        wildlife conservation, 194
    autonomy, 185
    bandwidth, 172
    capabilities, 198
    collaboration, 185
    connectivity, 186
        data transmission, 186
        device connectivity, 187
        edge computing, 187
        features, 186
        lower latency, 186
        network slicing, 187
        reliability, 186
    coordination algorithm, 179
        ad hoc networks, 180
        autonomous capabilities, 180
        communication systems, 180
        consensus, 179
        leader–follower models, 179
        flocking algorithms, 179
        mesh networks, 180
    cybersecurity, 197
    data transmission
        capabilities, 188
        decision-making process, 188

INDEX

delegation, 189
IoT communication, 189–192
operational context, 188
search/rescue, 188
deployment/infrastructure
cloud/edge computing, 177
DAS nodes, 177
data processing, 178
key components, 176
small cell networks, 176
edge computing, 172
energy consumption, 198
evaluation, 173
FANET, 302
features, 173, 175, 176
field testing, 185
historical progression, 173–175
impressive perspectives, 200
integration, 115–117, 178
high bandwidth/low
latency, 178
operations, 178
real-time data processing, 178
reliability, 179
seamless connectivity, 178
miniaturization, 199
network interference/
congestion, 197
network slicing, 175, 176
principles, 179
research/development, 185
search patterns
communication, 181
coordination, 181
data capturing/
processing, 181
forest management, 182
pollution detection, 182
time-consuming
problem, 180
wildlife monitoring, 181
simulation/modeling, 185
swarming system, 171
swarm intelligence, 199
technical challenges, 183
connectivity/
communication, 183
consensus-based
methods, 184
coordination/control, 184
redundancy, 184
solutions/approaches, 183
technological issues, 195
connectivity/
communication, 195
consideration, 196
intelligence, 197
operations/
movements, 196
optimization, 196
Fixed-wing UAVs, 541, 542
Flight control system, 549, 550
Flying ad hoc network (FANET),
286, 290, 301–303, 365, 710
cluster head (CH), 302
connection, 301
constant connection, 301
data transmission, 303

805

INDEX

Flying ad hoc network
(FANET) (cont.)
efficacy/reliability, 302
5G technology, 302
literature review, 316
peer-to-peer wireless, 303
Fog computing model, 654–655

G

Gazebo ROS package, 451
General Data Protection
Regulation (GDPR), 780
Generic Middleware for Smart City
Applications
(GMSCA), 635–636
Genetic algorithms (GAs),
300, 567–569
Global Positioning System (GPS),
371, 413, 501, 602, 644, 730
Google Maps, 644
GPS spoofing, 419, 420, 424
Ground control station (GCS), 550,
616, 775
Guidance navigation and control
(GNC), 707

H

Hardware attacks, 618
Hardware-based
experiments, 462–464
Hardware infrastructure, 775, 776
Hardware testing, 472, 473

Head-mounted displays (HMDs),
219, 230
Hierarchical nested personalized
federated learning
(HN-PFL), 129
Hierarchical personalized
federated learning
(HN-PFL), 142
Human-centric security, 432
Human factors, 428, 429
Human–machine interaction
technologies, 786
Human–machine interfaces
(HMI), 432
Hybrid approaches, 456
Hybrid UAVs, 543–546
Hydraulic sensors, 658
Hyperspectral cameras, 558

I, J

Identifier sensors, 658
Inertial measurement units
(IMUs), 212, 372, 549, 730
Information and communication
technologies (ICT),
638, 644
Information-centric networking
(ICN), 289
Innovation, 432–434, 448, 462,
618–620, 670, 673, 722, 728,
733, 740, 768, 774, 781, 785
Integrated sensing and
communication (ISAC), 760

Integrated sensor networks, 759–760
Integration, 427, 481, 504, 505, 635, 636, 638, 659, 674, 693, 706, 711, 726, 732
    See also Artificial intelligence (AI)
Integration of IoT
    decision-making, 592
    future trends, 618–620
    and GCS, 593
    regulatory and ethical issues, 617
    smart cities, 591
    solutions, 617
    technical challenges, 614
    with UAV swarms, 591, 598–600
Intelligence, surveillance and reconnaissance (ISR), 412, 414, 533, 537
Intelligence, surveillance, target acquisition and reconnaissance (ISTAR), 503
Intelligent transportation systems (ITS), 300
Interdisciplinary collaboration, 432–434
Internet of Things (IoTs), 12, 32, 95, 115–117, 189–192, 768
    ad hoc networks, 307, 308
    blockchain technology, 297
    data streams, 294
    deep learning approaches, 294
    dynamic resource management, 296
    features, 294
    federated learning, 295
    peer-to-peer exchanges, 296
    reinforcement learning, 296
    wireless communication, 308
    AG (*see* Artificial general (AG) intelligence)
    applications/capabilities, 192
    capabilities, 189
    centralized management/control, 192
    connectivity, 190
    data exchange, 191
    massive integration, 190
    scalability/flexibility, 191
Internet of vehicles (IoV), 382
Interoperability, 427, 571, 672
Intrusion detection systems (IDSs), 71, 75, 379, 695
IoTs, *see* Internet of Things (IoTs)
IoT technology, 634
    in healthcare services, 664–670
    related works, 635, 636
    sensing technologies, 656–659
    smart cities, 636–646, 650–652
    in UAV swarms, 601, 602
    in urban infrastructure, 659–664

K

Kettering Bug, 412, 499

L

Liability, 225–226, 781
Light detection and ranging (LiDAR) sensors, 560–562
Line of sight (LOS), 267
Lithium-polymer (LiPo), 548
Logistical planning, 450
Logistic regression (LR), 709
Long short-term memory (LSTMs), 294
LoRa, 603, 605
Low-altitude long endurance (LALE) UAVs, 536, 537
Low-altitude short endurance (LASE) UAVs, 535, 536

M

Machine learning (ML), 430, 504, 511, 570, 572, 619, 692, 693, 752, 761
  adaptive behavior, 51
  adaptive mission planning, 67
  ad hoc networks, 286
  advantages, 49, 79
  aggregation approaches, 148
  anomaly identification, 132
  anomaly/intrusion detection algorithms, 74
  applications, 706, 707
    fault diagnosis, 710
    resource and network optimization, 709
    security and surveillance, 708
    swarm UAV system, 709, 710
    UAV path planning, 707, 708
  autonomous navigation/control, 50, 66
  battery recharging/scheduling, 76
  black boxes, 146
  black box model, 143
  blockchain principles, 144
  capabilities, 50, 142, 145
  challenges, 80
  characteristics, 79
  classification
    RL, 702, 703
    semi-supervised learning, 703
    supervised learning, 699, 700
    unsupervised learning, 701
  cloud-based approaches, 143
  clusters, 132
  collaborative decision, 51
  collision avoidance method, 50
  communication/networking
    components, 68
    distributed computing, 69
    operations, 69
    optimization, 69
  computational limitations, 143
  computational resources, 147
  concepts/techniques, 61
  control interfaces, 77
  cooperative learning, 149
  data collection/analysis, 66
  data protection/privacy, 147

DL, 48, 139, 140 (*see also* Deep learning (DL))
and drones, 728, 734, 735
fault detection, 77
fault-tolerant/self-adapting control, 128
federated learning, 144, 147, 148
future research, 711, 712
geo-distributed device clusters, 132
history, 79
image identification/object recognition, 129
information systems/electronic markets, 145
integration/response
    collaborative threat management, 74
    real-time threat response, 73
IoT, ad hoc networks, 309
literature review, UAV, 694–698
MAV (*see* Micro aerial/air vehicles (MAVs))
modern techniques, 703–706
monitoring system, 274
motonomy, 146
on-board/battery management, 148
operations, 143
predictive analytics/optimization algorithms, 68
real-time response/security threats, 75
real-world deployments, 103
reinforcement learning, 63, 131, 137–139
resource optimization, 51–52
safety/security/reliability, 128
security architectures, 142
security measures, 144
security protocols, 146
security/threat detection, 70
    anomaly detection, 71
    authentication/authorization, 71
    behavioral analysis, 73
    electronic threat detection, 72
    encryption/secure communication, 70
    physical threat detection, 72
sensitive information, 142
smart city protection, 140
    monitoring systems, 140
    SAR operations, 141
    surveillance models, 140
    traffic management, 141, 142
supervised learning, 61, 62, 129–131, 133–135
surveillance/monitoring system, 248
suspicious applications, 145
swarm intelligence, 14, 15, 67
techniques, 80
technological advancements/innovations, 26
traditional methods, 145

INDEX

Machine learning (ML) (*cont.*)
   training/transmission
      quality, 133
   UAVs, 693
   unsupervised learning, 62,
      63, 135–137
   utilization/algorithms, 52
   vehicle data analysts, 145
   vehicular ad hoc network
      (VANET), 300
   XAI techniques, 144
Malfunction, 428, 556
Malware
   and software exploits, 422, 423
   UAV, 424
Marine integrated communication
   network system (MICN), 305
Maritime wireless mesh network
   (MWMN), 305
Markov decision process
   (MDP), 704
Massive machine-type
   communication
   (mMTC), 175
Massive multiple-input multiple-
   output (MIMO), 759
Material stress testing, 476
MATLAB UAV toolkit, 451
MAVs, *see* Micro aerial/air
   vehicles (MAVs)
Medium-altitude long endurance
   (MALE) UAVs, 537
Micro aerial/air vehicles (MAVs)
   amalgamation, 57

   battery recharging/
      scheduling, 76
   collisions, 59
   comprehensive overview, 57
   conceptual structure, 59
   control interfaces, 77
   control laws, 58
   design protocol, 58
   environment, 57
   local sensing/control, 56
   neural networks, 59
   progressive level, 60
   requirements/
      constraints, 60
Microservices, 753
ML, *see* Machine learning (ML)
Mobile ad hoc networks
   (MANET), 286, 289,
   297–300, 317, 365
   Ant-OBS technique, 299
   attacking node, 297
   black hole attack, 297
   Black Hole Attack (BHA), 297
   communication, 299
   data transmission, 298
   literature review, 316
   nodes, 299
   techniques, 300
   wireless communications, 299
Modern machine learning
   techniques
   FL, 704, 705
   TL, 705
   XAI, 704

Monitoring system, *see* Surveillance/monitoring system
Motion sensors, 659
Motion tracking systems, 219
Multiagent deep deterministic policy gradient (MADDPG), 70
Multidomain networks, 785
Multifactor authentication (MFA), 71
Multilayer perceptron (MLP), 700
Multirole UAVs, 502–503

N, O

Named data networking (NDN), 289
Navigation, 550, 551, 723, 724, 726, 733, 735, 737
Neural graph networks (NGNs), 753
Neural network (NN), 708
Next-generation networks (NGNs), 753

P

Parameter exploration, 449
Particle swarm optimization (PSO), 256, 300, 567–569
Passive optical networks (PONs), 12
Parental Change Control RPL (PCC-RPL), 381

Performance discrepancy, 452
Physical attacks, 423, 424
Physical security, 507
Ping of Death (PoD), 286
Postman Moving Voronoi Coverage (PMVC), 136
Power line inspection, 612
Power management, 600
Power source, 548
Presence sensors, 658
Privacy-preserving technologies, 787
Privacy protection, 780, 781
Proof of stake (PoS), 496
Propulsion system, 547
Public engagement, 781, 782
Public healthcare surveillance, 670
Public participation, 740
Public perception, 508

Q

Quality of Service (QoS), 297, 365
Quantum Key Distribution (QKD), 513

R

Radar sensors, 562, 563
Radio frequency (RF) communication, 551, 552
Radioplane OQ-2, 499
Reactive handover coordination system (RHCRB), 144

INDEX

Real-world deployments
  across diverse sectors, 111
  agriculture/infrastructure
    inspection, 113
  aims/contributions, 96
  auto/decision-making, 115
  beneficial directions, 95
  case studies, 107
  central processing tasks, 92
  challenges/
    implications, 109–111
  classification, 92
  cost analysis, 113, 114
  drones/aerial robots, 92
  economic/environmental/
    social considerations,
    111–114
  environmental surveillance, 93
  ethical dimensions/public
    safety, 110
  global positioning
    technology, 109
  horizons/technological
    innovations, 114–117
  methodological approach, 100
    case study research, 101
    data sources/collection
      techniques, 101
    drone building, 101, 102
    error analysis tools/
      metrics, 103
    research design
      rationale, 100–101
    survey approach, 101
  military/defense systems, 93
  predeployment planning, 118
  search/rescue, 93, 107
  security/surveillance, 108, 109
  strategic recommendations, 117
  surveillance purposes, 112
  technical aspects
    centralized architecture, 105
    communications networks/
      protocols, 104, 105
    control systems, 103
    coordination, 107
    key values, 103
    navigation/sensing
      capabilities, 106
    routing protocols, 105
    swarm intelligence
      algorithms, 106
  technological system, 94, 95
  theoretical background/
    literature review, 97
    advantages, 98
    decentralization, 98–99
    effective communication, 98
    fundamental concepts, 97
    historical milestones, 99, 100
    network theory
      principles, 100
    potential capabilities, 99
    principles/characteristics, 97
    scalability, 97
    swarm intelligence, 97–98
    theoretical frameworks, 100
  ultrafast data speeds, 116

INDEX

urban system/infrastructure management
  planning side/development, 108
  traffic flow, 108
  transforming cities, 108
Received signal strength (RSS), 695
Recurrent neural networks (RNNs), 64, 75, 294, 700
Regional analysis, 464–469
Regular neural networks (RNNs), 140
Regulatory compliance, 507
Regulatory constraints, 449
Regulatory frameworks, 779, 780
Reinforcement learning (RL), 63, 131, 137–139, 145, 304, 702, 703, 707
  advantage, 138
  algorithms, 131, 138
  contributions, 138
  dynamic environments, 137
  obstacles, 137
  safety issues, 138
  search/rescue operations, 138
Reliability concerns, 453
Remote-controlled bomb, 412
Remote patient monitoring (RPM)
  applications, 667–669
  IoT, 666
  overview, 665
  patterns and irregularities, 665
  treatment strategy, 664
Repeatability, 449

Resilience, 427, 634, 636, 640, 655, 673, 758, 759, 787
Return on investment (ROI), 113, 114
Return-to-home (RTH) program, 409
RL, *see* Reinforcement learning (RL)
Roadside units (RSUs), 288
Robot ad hoc networks (RANETs), 286
Robust communication systems, 571
Role-based access control (RBAC), 71
Rotary-wing UAVs, 542
RPM, *see* Remote patient monitoring (RPM)
Ryan Model 147, 725

S

Safety concerns, 112, 450, 783
SANETs, *see* Ship ad hoc networks (SANETs)
Satellite communication (SATCOM), 501, 502, 552
Scalability, 505, 512, 528–529, 709, 738, 739
Search and rescue (SAR), 141, 609
Secure communication
  data encryption/identity verification, 346
  security, 345, 346

INDEX

Secure communication (*cont.*)
   smart contracts/message exchanges, 346
   uninterrupted communication, 339
   vulnerabilities, 339, 340
Self-driving robots, 733
Semi-supervised learning, 703
Sensor fusion techniques, 230
Sensor technologies, UAV-based, 526, 740
   acoustic sensors, 565–567
   camera sensors, 556–558
   radar sensors, 562, 563
   thermal imaging, 558–560
   ultrasonic sensors, 563–565
Ship ad hoc networks (SANETs), 286, 309, 310
   cognitive radio (CR), 309, 310
   frequencies, 309
   ITU-R M.184-1, 309, 310
   node density, 309
   send/receive signals, 309
Simulation, 709
   agents employed, 480
   cost-effective exploration, 473
   environments, 447
   factors, 447–450
   fidelity, 469
   future researches, 474, 475
   hardware devices, 471
   and hardware validation, 455–457
   performance evaluation, 468, 469
   and real-world conditions, 477, 478
   real-world disruptions, 475, 476
   tools, 473
   trends, 462–464
   in UAV swarm research, 452–454
   validation, 473, 474
Simultaneous localization and mapping (SLAM), 376
Single Shot Detector (SSD), 61
Smart approach for intrusion detection and prevention system (SA-IDPS), 314
Smart cities, 636
   challenges
      networking, 673
      security and privacy, 672
      smart sensors, 672, 673
   components, 641, 642
      agriculture, 644
      energy, 645
      health, 643
      homes, 646
      industry, 643, 644
      services, 642
   evolution, 637–640
   IoT architectures, 651–653
   protection, 769, 773
   UAVs in
      applications, 649
      characteristics, 649
      electronics, 647
      infrastructure, 647
      worldwide market, 648

INDEX

Smart hospitals and healthcare facilities, 669, 670
Smart sensors, 672, 673
Smart system (cities)
   innovative technologies, 12
   interconnection, 5
   ITS applications, 4
   scenarios, 2
Smart Tissue Autonomous Robot (STAR), 362
Smart transportation systems, 644
Software-defined radio (SDR), 409
Software platforms, 776–778
Spoofing attacks, 420, 421
Stakeholder consultation, 781, 782
Supervised learning, 61, 62, 145, 699, 700
   anomaly detection, 135
   categories, 133–135
   classification process, 133
   clustering/principal component, 130
   crime data/sensor perception, 130
   distributed learning, 131
   energy consumption costs, 131
   integration, 133
   object detection/classification, 129–131
   object detection model, 133
   predictive capabilities, 130
   traffic management, 133, 134
Support vector machines (SVMs), 62, 297, 300, 709

Surveillance/monitoring system applications, 258
   agriculture management/environments, 262
   commercial opportunities, 257
   communication network, 257
   delivery systems, 259
   disaster response/recovery, 263
   entertainment, 258
   feasible/cost-effective solutions, 260
   healthcare services, 261
   remote sensing, 259–260
   safety/security, 262
   search/rescue missions, 262
   surveillance, 260
   traffic monitoring/management, 261
aspects, 242
autonomous system, 248
CCTV/sensor networks, 247
current technologies, 247–249
end-to-end security, 244
further research, 278
privacy perspective, 245
real-life scenarios
   air quality, 277
   carnivals, 278
   concepts/patterns, 277
   implementation, 276
   natural disasters, 277
   reduction, 277

INDEX

Surveillance/monitoring system (*cont.*)
    reconstruction phase, 263
    regulation and compliance, 271–273
    smart city projects
        access control mechanisms, 243
        advantages, 242
        agriculture, 246
        framework, 244
        fundamental object, 245
        medical devices/telehealth services, 242–243
        privacy/ethical issues, 244
        problems/issues, 243
        role of, 245–247
        traffic patterns, 246
        urban construction/planning, 242
        utilization, 243
    swarm technology, 254
        advantages, 254
        algorithms/protocols, 256–257
        centralized/decentralized coordination, 255
        communication strategies, 257
        consensus algorithms, 256
        definition/concepts, 254
        direct/indirect communication, 255, 256
    technical issues
        batteries, 268–271
        challenges/solution, 263, 269
        communication channel, 264
        communication/information, 265
        cost-benefit structure, 263
        data collection, 264
        edge computing, 268
        end user data, 267
        interservice operation, 266
        inventory management, 265
        line of sight (LOS), 267
        low interoperability, 266
        network system, 267
        self-controller systems, 266
        speed/security issues, 264
        storage/processing power, 266
    trends, 273
        acceptance/adoption, 272
        ethical issues, 272
        functionalities, 274
    UAV methods
        autonomous, 253
        decision-making, 253
        fixed-wing drones, 251
        hybrid VTOL drones, 251
        multirotor drones, 250
        operations, 254
        quadcopters/hexacopters, 250, 251
        technical features, 249

INDEX

technological
    advancement, 252–254
    types of, 250
Swarm application, 2
    AI (*see* Artificial intelligence (AI))
    algorithms/control techniques, 6
    autonomous/
        semiautonomous, 46
    autonomous urban mobility
        solutions, 29
    balloons, 47
    big data analytics, 29
    bridges, roads, and
        buildings, 18–19
    buildings/infrastructure, 29
    CCTV cameras, 17
    characteristics, 53
        adaptability, 54
        autonomy, 53
        benefits, 55
        collective intelligence, 54
        key capabilities/benefits, 55
        robustness, 54
        scalability, 53
    city development, 16
    classification, 46
    communication network, 6
    communication systems, 11
    conventional technologies, 56
        MAV swarm, 56–60
    coordination protocols, 10
    cybersecurity, 30
    emergency response
        systems, 17, 18
    fundamentals, 6–7, 53
    hardware components, 9
    healthcare/biomedicine, 30
    history of, 47, 48
    inspection/maintenance, 18
    integration, 29
    likelihood approach, 2
    ML (*see* Machine learning (ML))
    monitoring public spaces, 17
    parking system, 29
    real-world (*see* Real-world
        deployments)
    robotics, 30
    RoF system-level solutions, 12
    software, 10–11
    surveillance systems/
        security, 16
    technologies, 2
    urban mobility, 19
    urban planning/management, 31
        global cooperation, 31
        long-term benefits, 31
        standardization/
            scalability, 32
Swarm communication, 555, 556
Swarming systems, 404, 444, 482
Swarm intelligence, 739, 742
    concepts and methods, 567, 568
    real-time control and
        adaptation, 569
Swarm robotics, 467
Swarms of UAV, 709, 710
    automation systems, 525
    autonomy, 528, 529

817

INDEX

Swarms of UAV (*cont.*)
   classification, 533–539
   components, 526
   cooperation, 528, 529
   intelligence, 526
   natural phenomenon, 527
   scalability, 528
Swarms of unmanned aerial vehicles, *see* Unmanned aerial vehicles (UAV)
Synthetic aperture radar (SAR) technology, 563

T

Technological integration
   IoT devices, 601, 602
Telehealth, 670
Telemetry, 606
Testbeds, 453, 454
Thermal imaging sensors, 558–560
Threat detection, 430
Transfer learning (TL), 705
Transmission control protocol/internet protocol (TCP/IP), 553
Transparency, 427, 741

U

UAS, *see* Unmanned aircraft system (UAS)
UAVs, *see* Unmanned aerial vehicles (UAVs)
UAV swarm, 596, 597

UAV swarm security
   AI-and ML-based techniques, 511
   communication channels, 513
   construction monitoring and infrastructure inspection, 610
   and distributed ledger technologies, 511–512
   emergency response, 608, 609
   environment impact, 514
   ethical and regulatory frameworks, 514
   ethical and societal implications, 786
   evironmental monitoring, 609
   military intelligence, 608
   quantum computing, 512–513
   remote sensing, 609
   rescue and search, 609
   security and monitoring, 607–608
   traffic monitoring and management, 611, 612
Ultrasonic sensors, 563–565
Underwater wireless sensor network (UWSN), 305–307
Unmanned aerial vehicles (UAVs), 404, 446, 524–526, 594–596
   ad hoc networks, 286
   advanced sensor integration, 570
   AG (*see* Artificial general (AG) intelligence)

INDEX

AI-powered onboard systems, 415
application
   civil, 540, 541
   commercial, 538
   military, 538, 539
AR/VR (*see* Augmented reality (AR)/virtual reality (VR))
bandwidth/tolerance, 4
biocomplexity, 404
blockchain (*see* Blockchain application)
characteristics
   autonomy, 7–9
   cooperation, 7–9
   scalability, 7–9
classification
   large, 534, 535
   medium, 533
   mini-, 532
   nano-and micro-, 532
   size and weight, 529
cold war period, 412, 413
comparative analysis, 25, 26
in coordination, 404
damage assessments, 405
decision-making, 407
disaster response, 23, 24
drones, 1–4
early developments, 411–412
emergence, 415–417
ethical concerns, 22
ethical governance, 781
ethical guidelines/privacy protections, 23
evolution
   development, swarms, 502, 503
   early developments, 499–500
   integration, 505–506
   military applications, 499–500
   multirole UAVs, 502–503
   technological advancements, 501–502
experimental methodologies, 778, 779
experimental setup, 775–779
factors, 447–450
ITS applications, 4
key factor, 415
legal considerations, 779–782
limitations, 782–784
literature review, 772–775
ML (*see* Machine learning (ML))
modern UAVs, 414, 415
multiple types, 543–546
proposed solutions, 22, 23
protection, 406
publications, 456–458
real-time insights, 405
real world, 91
regulatory frameworks, 21, 571
scenarios, 404
security, 405
self-organized networks, 20
self-sufficient analytics, 411
simulation and hardware validation, 455–457

INDEX

Unmanned aerial vehicles (UAVs) (cont.)
   simulation experiences, 452–454
   smart cities, 646–650
   source of data, 770–773
   surveillance system, 249–254
   swarm operation, 333
   swarms (see Swarm application)
   swarms architecture, 336, 337
   swarm security, 491–494
   swarm technology, 570, 767
   technical challenges, 20, 21
   technological advancements, 413, 414
   technologies, 20
Unmanned aircraft system (UAS), 411, 743, 744
Unsupervised learning, 62, 63, 129, 130, 135–137, 145
   advantages, 136
   characteristics, 135
   cooperative, 135
   integration, 136
   security, 137
   utilization, 136
Unsupervised learning, ML, 701
Urban air mobility (UAM), 231
Urban infrastructure, 659, 660
   smart environment monitoring, 663
   smart grid, 664
   smart infrastructure, 661
   smart parking, 661, 662
   waste management system, 662

Urban planning, 613
User datagram protocol (UDP), 554
User equipment (UEs), 694
Unsupervised learning, 129

V

VANETs, see Vehicular ad hoc networks (VANETs)
Variational autoencoder (VAE), 137
Vehicle to infrastructure (V2I), 644
Vehicle to pedestrian (V2P), 644
Vehicular ad hoc networks (VANETs), 286, 288, 300–301
   AI integration, 301
   cognitive radio (CR), 301
   cybersecurity threats, 300
   literature review, 316
   reinforcement learning, 312
   roadside unit (RSU), 300
   wireless communication, 310
Vehicular CCN (VCCN), 289
Vertical takeoff and landing (VTOL), 46, 250, 542
Very low-altitude (VLA) UAVs, 534, 535
Video data surveillance (VDS), 274
Video surveillance systems (VSS), 244

INDEX

Virtual reality (VR), *see* Augmented reality (AR)/virtual reality (VR)
Vulnerability management, 426

W, X, Y

Waze, 644
Weather, 751
Web of Science (WoS), 464
Webots simulator system, 451
Wi-Fi, 604
Wireless ad hoc networks (WANETs), 285–286
Wireless communication systems, 286
Wireless power transfer (WPT), 619
Wireless rechargeable sensor networks (WRSNs), 760
Wireless sensor networks (WSNs), 759, 760

Z

ZigBee, 603, 605

Printed in the United States
by Baker & Taylor Publisher Services